Materials Horizons: From Nature to Nanomaterials

Series Editor

Vijay Kumar Thakur, School of Aerospace, Transport and Manufacturing, Cranfield University, Cranfield, UK

Materials are an indispensable part of human civilization since the inception of life on earth. With the passage of time, innumerable new materials have been explored as well as developed and the search for new innovative materials continues briskly. Keeping in mind the immense perspectives of various classes of materials, this series aims at providing a comprehensive collection of works across the breadth of materials research at cutting-edge interface of materials science with physics, chemistry, biology and engineering.

This series covers a galaxy of materials ranging from natural materials to nanomaterials. Some of the topics include but not limited to: biological materials, biomimetic materials, ceramics, composites, coatings, functional materials, glasses, inorganic materials, inorganic-organic hybrids, metals, membranes, magnetic materials, manufacturing of materials, nanomaterials, organic materials and pigments to name a few. The series provides most timely and comprehensive information on advanced synthesis, processing, characterization, manufacturing and applications in a broad range of interdisciplinary fields in science, engineering and technology.

This series accepts both authored and edited works, including textbooks, monographs, reference works, and professional books. The books in this series will provide a deep insight into the state-of-art of Materials Horizons and serve students, academic, government and industrial scientists involved in all aspects of materials research.

More information about this series at http://www.springer.com/series/16122

Agbaje Lateef · Evariste Bosco Gueguim-Kana ·
Nandita Dasgupta · Shivendu Ranjan
Editors

Microbial Nanobiotechnology

Principles and Applications

Springer

Editors
Agbaje Lateef
Department of Pure and Applied Biology
Ladoke Akintola University of Technology
Ogbomoso, Nigeria

Evariste Bosco Gueguim-Kana
Department of Microbiology
University of KwaZulu-Natal
Pietermaritzburg, South Africa

Nandita Dasgupta
Department of Biotechnology
Institute of Engineering and Technology
Lucknow, Uttar Pradesh, India

Shivendu Ranjan
Faculty of Engineering
and the Built Environment,
Institute for Intelligent Systems
University of Johannesburg
Johannesburg, South Africa

ISSN 2524-5384 ISSN 2524-5392 (electronic)
Materials Horizons: From Nature to Nanomaterials
ISBN 978-981-33-4776-2 ISBN 978-981-33-4777-9 (eBook)
https://doi.org/10.1007/978-981-33-4777-9

This Springer imprint is published by the registered company Springer Nature Singapore Pte Ltd.
The registered company address is: 152 Beach Road, #21-01/04 Gateway East, Singapore 189721, Singapore

Preface

Milestones and developments in microbiology have been defined by constant changes in the outlook of the discipline, from being a basic science course in the nineteenth century to the applied exploitation of microbes that gave birth to microbial technology as a component of biotechnology in the twentieth century and its entry into the field of nanoscience in the twenty-first century to herald the sub-discipline of microbial nanobiotechnology that exists at the interface of basic microbiology, applied microbiology/microbial technology and nanoscience. The unique attributes of microbes, high multiplication rates, diversity and tolerance to challenging environmental factors, efficient and versatile metabolic machineries, controllable growth and amenable genetic manipulation have been regularly exploited to advance the applications of microbes in nanobiotechnology. Being a new emerging area, there is dearth of reading texts on microbial nanobiotechnology.

The current endeavour is therefore to produce a textbook that would serve as a vital resource material on the contributions of micro-organisms to advances in nanotechnology, establishing their applications in diverse areas of biomedicine, environment, biocatalysis, food and nutrition, and renewable energy. It attempts to bridge the gap between life sciences and nanotechnology towards stimulating the interests of applied microbiologists and biochemists in nanotechnology. It would serve as a good compendium documenting the impacts of micro-organisms in nanotechnology, leading to further developments in the sub-discipline of "microbial nanobiotechnology". The text would be of immense use to graduate students and researchers of microbiology, biochemistry and nanotechnology.

In a lucid manner, authors have discussed several topical issues to X-ray the potentials, challenges and advances of microbial nanobiotechnology. In the chapter "Microbial Nanobiotechnology: The Melting Pot of Microbiology, Microbial Technology and Nanotechnology", Lateef et al. introduced microbial nanobiotechnology as a melting pot of microbiology, microbial technology and nanotechnology bringing to the fore the interrelatedness and addressed challenging issues in the development of microbial nanobiotechnology. Asafa et al. gave an expose of characterization techniques in nanotechnology in the chapter "Characterization Techniques in Nanotechnology: The State of the Art", while in the chapter

"Precision Microbial Nanobiosynthesis: Knowledge, Issues, and Potentiality for the In Vivo Tuning of Microbial Nanomaterials" Grasso et al. provided an insightful discussion on precision microbial nanobiosynthesis to produce tuned nanomaterials from microbes deploying optimization and genetic engineering. Chapter "Current Advances in Fungal Nanobiotechnology: Mycofabrication and Applications" by Shaheen et al. discussed advances in mycosynthesis of nanoparticles, the chapter "Environmental Nanobiotechnology: Microbial-Mediated Nanoparticles for Sustainable Environment" by Darwesh et al. dealt with environmental remediation using microbe-mediated nanoparticles, and in the chapter "Nanotechnology in Bioprocess Development: Applications of Nanoparticles in the Generation of Biofuels", Sanusi et al. dwelt on applications of nanoparticles in bioprocess development with the focus of biofuel technology.

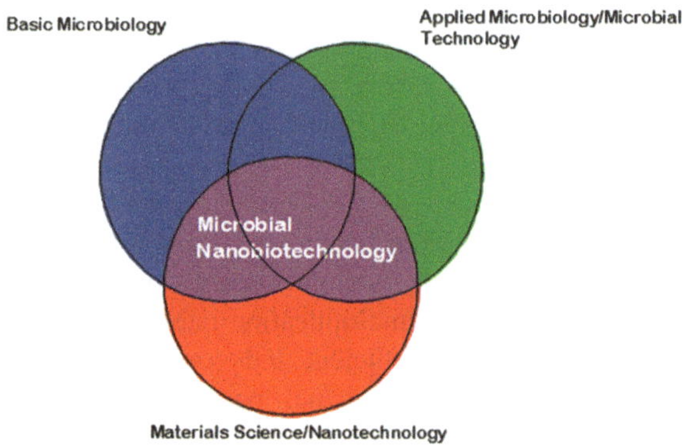

Microbial nanobiotechnology: the melting pot of microbiology, microbial technology and nanotechnology

In the chapter "Microbial Enzymes in Nanotechnology and Fabrication of Nanozymes: A Perspective", the interplay of exploitation of enzymes in microbial nanobiosynthesis and nanomaterials as nanozymes for catalytic applications was documented by Elegbede and Lateef, while in the chapter "Microalgal Nanobiotechnology and Its Applications—A Brief Overview", Adelere and Lateef provided an overview of microalgae in nanobiotechnology. In the chapter "Mushroom Nanobiotechnology: Concepts, Developments and Potentials", Adebayo et al. discussed mushroom nanobiotechnology: concepts, developments and potentials, while Jaffri and Ahmad in the chapter "Microbial-Mediated Nanoparticles for Sustainable Environment: Antimicrobial and Photocatalytic Applications" described antimicrobial and photocatalytic applications of microbe-mediated nanoparticles. Adebayo et al. made submission on beneficial microbes to produce nanoparticles for nanomedical applications in the chapter "Beneficial Microbes as Novel Microbial Cell Factories in Nanobiotechnology: Potentials in Nanomedicine"; Raj et al. addressed the applications of microbe-mediated nanoparticles in agriculture in the chapter "Applications of Microbe-Based Nanoparticles in Agriculture: Present

State and Future Challenges", while Donia et al. in the chapter "Microbial Nanobiotechnology in Nanocatalysis: Degradation of Pollutants and Sensing Applications" discussed applications of microbe-mediated nanoparticles for degradation of pollutants and sensing. Lastly, in the chapter "Application of Microbial-Synthesized Nanoparticles in Food Industries", Ingle et al. reviewed the applications of microbial-synthesized nanoparticles in the food industries.

We welcome the readers to the exciting contributions by various authors from Nigeria, South Africa, Egypt, India, Pakistan, Brazil and Italy.

Thanks for reading.

Ogbomoso, Nigeria — Agbaje Lateef, Editor-in-Chief

Pietermaritzburg, South Africa — Evariste Bosco Gueguim-Kana, Co-editor

Lucknow, India — Nandita Dasgupta, Co-editor

Johannesburg, South Africa — Shivendu Ranjan, Co-editor

Contents

About the Editors

Prof. Agbaje Lateef is the Head of LAUTECH Nanotechnology Research Group (NANO+) at Ladoke Akintola University of Technology, Nigeria. He obtained his B.Tech. in Pure and Applied Biology, M.Tech. in Biotechnology and Ph.D. in Microbiology in 1997, 2001 and 2005, respectively, from Ladoke Akintola University, Nigeria. He received advanced training in fermentation technology, enzyme technology, biocatalysis, functional foods and molecular biology at Central Food Technological Research Institute, Mysore, India. He has more than twenty years of research and teaching with interest in industrial microbiology and biotechnology, especially fermentation processes, enzyme technology and nanobiotechnology. He has more than one hundred publications in reputable journals and books to his credits, more than fifty of which are in nanobiotechnology.

Prof. Evariste Bosco Gueguim-Kana research focuses on microbial process technology with interest in biorefinery of agricultural and industrial wastes for biofuels & bioproducts, application of artificial intelligence in bioprocess development, bioreactor design and nanobiotechnology. He has attracted and trained various national and international postgraduate students in the field of bioprocess technology. Prof. Gueguim-Kana has authored approximately 90 publications in national and international peer-review journals and book contributions. His current Google Scholar h-factor is 30, and he has been cited more than 2200 times. He has received the award of a top published researcher of the University of KwaZulu-Natal.

Dr. Nandita Dasgupta is currently Assistant Professor at the Department of Biotechnology, Institute of Engineering and Technology (IET), India. She is also serving as Senior Research Associate (Visiting) at Faculty of Engineering & Built Environment, University of Johannesburg, South Africa. Her areas of interest include micro/nanomaterial fabrication and its applications in various fields – medicine, food, environment, agriculture and biomedical. She has published six edited books and one authored book. She has also authored many chapters and also published many scientific articles in international peer-reviewed journals.

Dr. Shivendu Ranjan completed his B.Tech. and Ph.D. in Biotechnology from VIT University, Vellore, India. He is currently Director (Honorary), Centre for Technological Innovations & Industrial Research at SAIARD, India. He is also serving as Senior Research Associate (Visiting) at Faculty of Engineering & Built Environment, University of Johannesburg, South Africa. His research interests include micro/nanobiotechnology applied in pharmaceuticals, nutraceuticals and environment. He is Associate Editor of Environmental Chemistry Letters – a Springer journal. He has written two authored books and also serves as an editorial board member for journals of national and international repute.

Microbial Nanobiotechnology: The Melting Pot of Microbiology, Microbial Technology and Nanotechnology

A. Lateef, O. M. Darwesh, and I. A. Matter

1 Introduction

One of the ultimate goals of the life sciences, whether in the past or in the present, is to solve the problems that face humans to improve the quality of life. From thousands of years ago, sciences such as chemistry, astronomy, engineering and biology were used to create great civilizations such as those of ancient Egypt and China, where humans had many facilities and luxuries that were not available before. Continuing this noble approach, scientists, either individually or in integrated groups, are working in a constant challenge to obtain the best results that benefit humanity through expanding our knowledge of existing sciences as well as the development of new aspects of sciences. For example, despite the primitive uses of microbial fermentation in ancient times (in some industries of food and alcoholic beverages), a tangible development in microbiological science occurred after the use of the optical microscope to see the microbes in 1676. Later, with the invention of the electron microscope (and many other modern analytical devices), the study of microbes (that are invisible to naked human eyes) became more accessible to many scholars all over the world (Ball et al. 2019). Through the overlap and integration between microbiology and other sciences, the exploitation of microbes in many technological applications has become the main feature of the so-called microbial biotechnology. Microbial biotechnology has allowed microbes to be exploited in the production of many environmentally friendly products in the medical, industrial and agricultural sectors, as well as the removal of various pollutants that threaten the sustainability of the environment and

A. Lateef (✉)
Laboratory of Industrial Microbiology and Nanobiotechnology, Department of Pure and Applied Biology, Ladoke Akintola University of Technology, Ogbomoso, Nigeria
e-mail: alateef@lautech.edu.ng; agbaje72@yahoo.com

O. M. Darwesh · I. A. Matter
Agricultural Microbiology Department, National Research Centre, Elbohouth St, Dokki, Cairo 12622, Egypt

A. Lateef et al. (eds.), *Microbial Nanobiotechnology*, Materials Horizons: From Nature to Nanomaterials, https://doi.org/10.1007/978-981-33-4777-9_1

human life (Lateef et al. 2008a, b, 2010a, b, 2012, 2015a, b; Abou-Shanab et al. 2012; Ganaie et al. 2014; Hoballah et al. 2014; Saber et al. 2014; Adeoye et al. 2015, 2016; El-Baz et al. 2016; Matter et al. 2016; Ojoawo et al. 2017; Darwesh et al. 2018a; Adelere and Lateef 2019; Bamigboye et al. 2019; Elegbede and Lateef 2018).

Recently, with the advent of the concept of nanotechnology, microbiologists have begun to explore the possibilities of microbial nanotechnology applications to maximize the multiple uses of microbes. In addition to the possibility of producing new by-products, and maximizing the productivity of traditional microbial products, the application of microbes themselves (fungi, bacteria and microalgae) in the biosynthesis of nanomaterials for various applications has been studied on a large scale (Lateef and Adeeyo 2015; Lateef et al. 2015c, d, 2016a; Ojo et al. 2016; Oladipo et al. 2017a, b; Darwesh et al. 2019a, b, 2020a; Matter et al. 2019; Elegbede et al. 2018, 2019, 2020; Elegbede and Lateef 2019a; Kim et al. 2020).

Despite the diversity and multiplicity of the existing and potential fields that combine microbiology and material nanoscience, there are many obstacles that face scholars and researchers in this promising field. These obstacles for microbiology specialists and students are mainly represented in the lack of experience, weak training and insufficient undergraduate and pre-university curricula in the field of material nanosciences (Elegbede and Lateef 2019b). Here, in this chapter, from the point of view of microbiologists, the expected challenges, obstacles and potential solutions related to the field of microbial nanotechnology are briefly discussed.

2 Microbiology as a Discipline: Basic and Applied Concepts in Microbiology

Microbiology is the study of microscopic organisms (microbes), as any living organism that is single cell, cell gathering or no cells (non-cellular). This includes eukaryotes like fungi and protists, and prokaryotes. Microbiology typically includes the study of immune system. Generally, immune systems interact with pathogenic microbes. It is a broad term that includes bacteriology, mycology, virology, immunology, parasitology and other branches (Fig. 1). A microbiologist is a specialist in microbiology and its related topics. Microbiological procedures usually must be aseptic and use a variety of tools like light microscopes with a combination of stains.

The information gained by microbiologists in basic studies can be applied to many medicinal, agricultural, environmental and commercial endeavours. Microbiology is the study of microbes, which affect almost every aspect of life on the earth. In addition, there are huge commercial and medicinal benefits in understanding microbes. The application of this understanding is known as applied microbiology. There are many different types of applied microbiology which can be briefly defined as follows:

Medical microbiology is the study of the pathogenic microbes and the role of microbes in human illness. This includes the study of microbial pathogenesis and

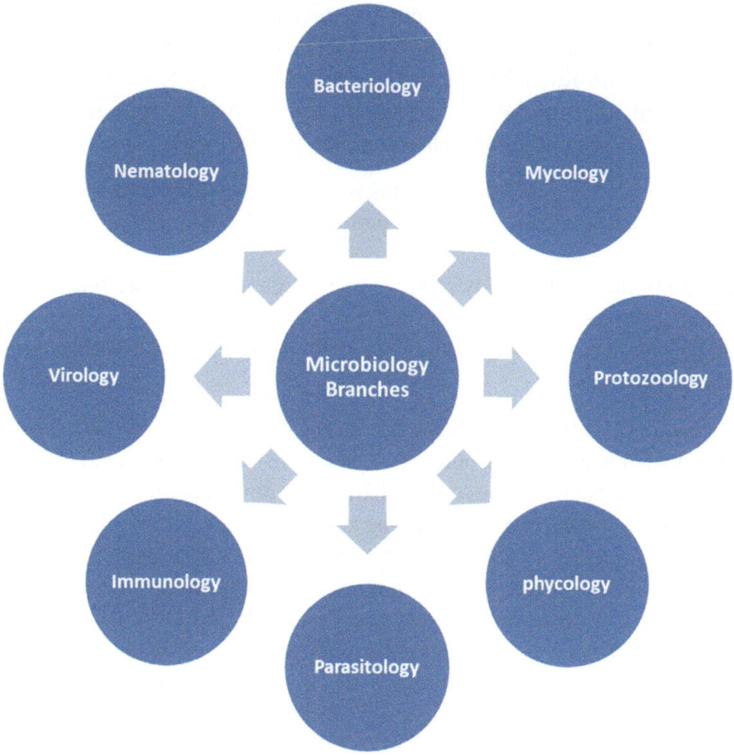

Fig. 1 Basic branches of microbiology

epidemiology and is related to the study of disease pathology and immunology (Mohamed et al. 2015; Sultan et al. 2016).

Pharmaceutical microbiology is the study of microorganisms that are related to the production of antibiotics, enzymes, vitamins, vaccines and other pharmaceutical products. Pharmaceutical microbiology also studies the causes of pharmaceutical contamination and spoilage (Barakat et al. 2015; Darwesh et al. 2019c).

Industrial microbiology is the exploitation of microbes for use in industrial processes. Examples include industrial fermentation and wastewater treatment. This is closely linked to the biotechnology industry or white biotechnology. This field also includes brewing which is an important application of microbiology (Emam et al. 2020).

Microbial biotechnology is the manipulation of microorganisms at the genetic and molecular level to generate useful products, and generally relates to the exploitation of microbes to render goods and services for mankind.

Food and dairy microbiology is the study of microorganisms causing food spoilage and food-borne illnesses and their control. It also entails studying microorganisms that can produce foods, for example, by fermentation (Darwesh et al. 2018b;

Sadek et al. 2018) and deployment scientific techniques to attain food safety in the production of foods and feeds (Adewoye and Lateef 2004; Lateef et al. 2010a, b; Lateef and Gueguim-Kana 2014; Lateef and Ojo 2016).

Agricultural microbiology is the study of agriculturally relevant microorganisms, and this field can be further classified into the following sub-fields:

- Plant microbiology and plant pathology—the study of the interactions between microorganisms and plants and plant pathogens (Elshahawy et al. 2018; Darwesh et al. 2019d).
- Soil microbiology—the study of those microorganisms that are found in the soil and their activities (Darwesh et al. 2020b).
- Veterinary microbiology—the study of the role of microbes in veterinary medicine or animal production.

Environmental microbiology is the study of the function and diversity of microbes in their natural environments. This involves the characterization of key bacterial habitats such as the rhizosphere and phyllosphere, soil and groundwater ecosystems, open oceans or extreme environments (extremophiles). This field includes other branches of microbiology such as: microbial ecology (microbially mediated nutrient cycling), geomicrobiology (microbial diversity), water microbiology (the study of those microorganisms that are found in water), aeromicrobiology (the study of airborne microorganisms), petroleum microbiology (the study of microbes involved in the formation of crude oil, exploration and degradation), and epidemiology and public health (the study of the incidence, spread and control of disease) (Lateef 2004; Darwesh et al. 2014, 2015).

3 Microbial Technology: An Overview

Microorganisms have been exploited for their specific biochemical and physiological properties from the earliest times for baking, brewing and food preservation and more recently for producing antibiotics, solvents, amino acids, feed supplements and chemical feedstuffs. Over time, there has been continuous selection by scientists of special microbial strains, based on their efficiency to perform a desired function. Recent developments in molecular biology and genetic engineering could provide novel solutions to long-standing problems. Over the past decade, scientists have developed the techniques to move a gene from one organism to another, based on discoveries of how cells store, duplicate and transfer genetic information. Microbial and biochemical technology journals are the major sources of knowledge for young and aspiring generations who are keen in pursuing their careers in science. This system provides easy access to networks of scientific journals. Authors who contribute their scholarly works to microbial and biochemical technology journals gain remarkable reputation as the research scholars explore these works extensively. This process assures considerable impact factor for the journal and good reputation to the authors that add value to their Academic Performance Index (API) Score.

4 Nanotechnology: An Overview

Nanotechnology is an interdisciplinary field and integrates science and technology using materials at nanoform. It spots on the synthesis of materials at 1–100 nm scale with application in agriculture, medical, pharmaceutical, environmental and other various fields (Lateef et al. 2018; Elegbede and Lateef 2020; Lateef 2020). This technology is growing day by day, and the researchers as scientists work hard daily to introduce new concepts in this technology through either synthesis or applications. The emerging research materials are in the research stage and would be used in nanometre scale devices. Development of characterization technology is needed to understand the composition, structure, morphology and other requirements at nanodimension. At the nanometre scale, the physical, chemical and biological properties of nanomaterials are fundamentally different from those of individual atoms, molecules and bulk materials. They differ significantly from other materials due to the increased surface area and quantum effects. A larger surface area usually results in more reactive chemical properties and also affects the mechanical or electrical properties of the materials (Thassu et al. 2007).

Molecular nanotechnology, also called molecular manufacturing, describes engineered nanosystems or machines based on molecular scale. It is associated with molecular assembler: a machine that can produce a wanted structure or device atom by atom using the principles of mechanosynthesis. The premise was that molecular scale biological analogies of traditional machine components demonstrated molecular machines were possible. By the countless examples found in biology, it is known that sophisticated, stochastically optimized biological machines can be produced.

It is hoped that developments in nanotechnology will make possible their construction by some other means, maybe using biomimetic principles. However, Drexler and other researchers (Phoenix 2005) have proposed that advanced nanotechnology, although initially implemented by biomimetic incomes, eventually could be based on mechanical engineering principles, namely a manufacturing technology based on the mechanical functionality of these components (like gears, motors, bearings and structural members) that could enable programmable, positional assembly to atomic specification. In general, it is very difficult to assemble devices on the atomic scale, as one has to position atoms on other atoms of comparable size and stickiness. Another view, put forth by Carlo Montemagno, is that future nanosystems will be hybrids of silicon technology and biological molecular machines. Richard Smalley argued that mechanosynthesis is impossible due to the difficulties in mechanically manipulating individual molecules (Belkin et al. 2015).

5 The Confluence: Microbial Nanobiotechnology

Nanomaterials are increasingly being used in new products and devices with a great impact on different fields from sensoristics to biomedicine. Biosynthesis of nanomaterials by microorganisms is recently attracting interest as a new, exciting approach towards the development of 'greener' nanomanufacturing compared to traditional chemical and physical approaches. During the last decade, metal nanoparticles have gained immense popularity due to their characteristic physicochemical properties, as well as containing antimicrobial, anti-cancer, catalytic, optical, electronic and magnetic properties. Nanoparticles are biosynthesized when the microorganisms grab target ions from their environment and then turn the metal ions into the element metal through enzymes generated by the cell activities. It can be classified into intracellular and extracellular synthesis according to the location where nanoparticles are formed (Simkiss and Wilbur 1989; Mann 2001). The interrelatedness of microbiology, microbial technology and nanotechnology for the creation of microbial nanobiotechnology is illustrated in Fig. 2.

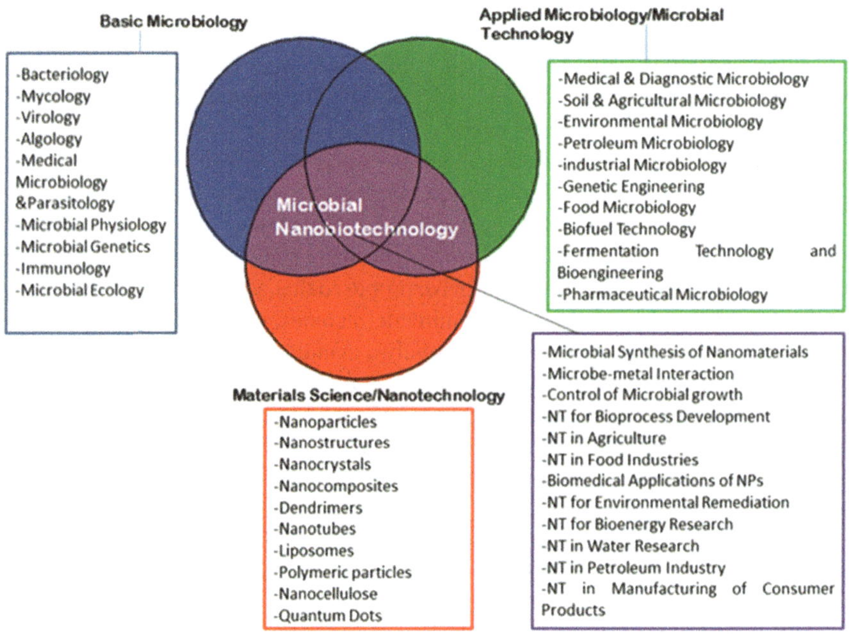

Fig. 2 Interrelatedness of microbiology, microbial technology and nanotechnology for the creation of microbial nanobiotechnology

6 What Are on Offer? Microbial Synthesis of Nanoparticles, Control of Microbes and Applications in Different Areas of Sub-disciplines of Microbiology

A lot of microorganisms have shown the ability to synthesize exclusive nanostructured materials, like bio-mineralized nanostructures as silicified frustules (Kröger and Poulsen 2008), calcified coccoliths, magnetosomes (Yan et al. 2017) and organic nanomaterials like microbial nanocellulose (Hasanin et al. 2018), exopolysaccharide nanoparticles and bacterial nanowires (Malvankar and Lovley 2012). The microbial-mediated biosynthesis of nanomaterials has been extensively explored showing many advantages and features including: (i) synthesized nanomaterials have defined chemical composition, size and morphology, (ii) biosynthesis is performed at mild physicochemical conditions, (iii) easy handling and cultivation of microbial cells and possibility of cell culture scale-up, (iv) possibility of in vivo tuning nanomaterial characteristics by changing key parameters of cell culture operational set-up or through genetically engineering tools (Prasad et al. 2016). In order to enable a broad applicability of microbial-mediated biosynthesis of nanomaterials as a real alternative to 'traditional' synthetic approaches to nanomanufacturing, many hurdles still need to be overcome: a reduction of polydispersity of nanoparticles, a more complete characterization of biocapping layer agents, effectiveness of removal procedures of biocapping layer and nanomaterial purifications, standardization of microbial cell culture protocols for reproducibility of nanosynthesis processes, as well as production costs and yields.

Overreaching the challenge for the development of reliable eco-friendly nanotechnologies for nanomaterial synthesis is of utmost importance for future exploitations of broad-impact nanostructured-based technologies and applications, like innovative optical and electrochemical (bio)sensoristic devices and therapeutic and diagnostic applications of nanostructured materials, e.g. for drug delivery, in vivo and in vitro imaging and development of antimicrobial and anti-tumoral drugs (Kiessling et al. 2014). The biological synthesis techniques have emerged as the biological methods of NPs preparation involving the application of different microorganisms and their enzymes, plant products and extracts derived from animals as shown in Fig. 3 (Adelere and Lateef 2016; Lateef et al. 2016b; Akintayo et al. 2020). The natural biogenic metallic nanoparticle synthesis is divided into two categories—(1) Bioreduction: carried out by the reduction of metal ions into the biologically stable form using microorganisms and their enzymes. The formed metallic nanostructures are stable and inert in nature that can be safely separated from contaminated sample; (2) biosorption carried out using a metal cation in aqueous media allows binding with organism cell wall that further leads to the formation of stable NPs because of cell wall or peptide interaction. It has many advantages like cost-effectiveness, eco-friendly and easy scalability for large-scale production, and does not involve the use of high pressure, energy, temperature and/or toxic chemicals.

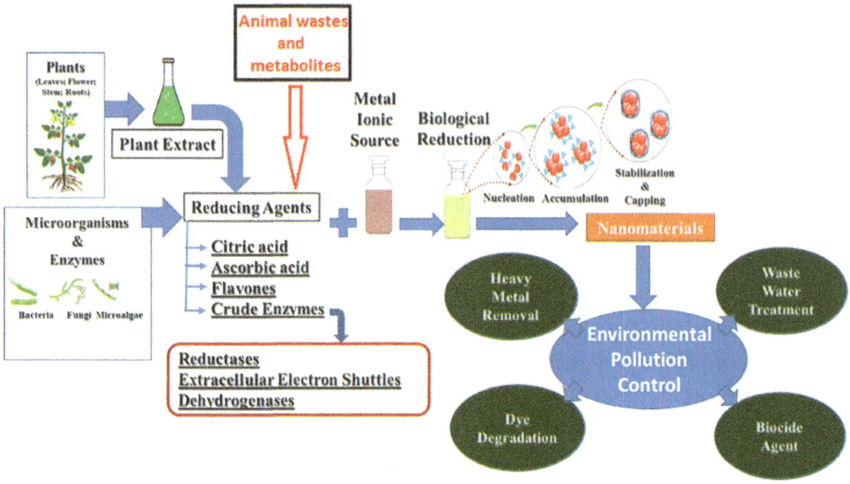

Fig. 3 Biological synthesis of metal oxide nanoparticles

7 Challenges in Microbial Nanobiotechnology

The challenges of development in microbial nanobiotechnology are multifactorial: firstly from the emerging nature of nanotechnology, existence of the sub-discipline at the interface of microbiology and materials science that requires a new understanding/knowledge, low level of expertise in microbial nanobiotechnology, deficit in curriculum of microbiology in relation to studies and contents in materials science and reduced access to analytical instruments that are needed for nanotechnology research, especially in several developing countries of the world. These challenges have contributed to low level of exploitation of microbes in nano-biosynthesis beyond microbe-mediated synthesis and antimicrobial profiling of nanomaterials despite the huge metabolic machineries of microbes, their diversity and amenable control potentials. Thus, a new knowledge is desirable to advance the potentials of microbes in nanobiotechnology; these include advanced physiology, genetic engineering, optimization studies and system and computational biology which are to be deployed within the context of materials science for the paradigm shift in microbiology research. These engagements would herald microbes as new 'nanofactories' to develop novel nano-based products and processes for several medical, industrial, agricultural and environmental applications.

8 Ways Out

To address the challenges as opined by Elegbede and Lateef (2019b), concerted efforts must be made by scholars, institutions, academic societies, funding agencies and governments to advance the studies of nanotechnology in the developing nations. Specifically, advances in microbial nanobiotechnology would require greater efforts, knowing fully that sub-field exists at the interface of nanotechnology and microbiology. Some of the precise ways to address the challenges are hereby discussed.

8.1 Curriculum Development

To popularize nanotechnology research among learners of microbiology, there is the need to re-engineer the curriculum at various levels to incorporate topics on materials science and nanotechnology to expose the learners to nanotechnology and expand their horizons. At undergraduate level, nanobiotechnology can be infused into relevant courses such as biotechnology or an introductory course on nanobiotechnology can be created at higher level. At postgraduate level, it would be imperative to have a course on nanobiotechnology. Among other things, the courses could cover topics such as:

- Introduction to nanotechnology and historical perspectives.
- Overview of nanomaterials (natural and man-made), types, sources and properties (physical, chemical and biological).
- Overview of fabrication of nanomaterials by physical, chemical and biological methods.
- Characterization techniques in nanotechnology.
- Biosynthesis of nanomaterials using plant, animal and microbial resources—advantages of bio-, green, one-pot, facile and benign synthesis.
- Strategies to produce tuned microbe-mediated nanomaterials: optimization studies and genetic engineering.
- A survey of applications of nanomaterials in different areas of human endeavours.
- Nanotoxicity and nanotoxicology.
- Fate of nanomaterials in the environment and safety concerns.
- Nanotechnology and sustainable development.
- Practical sessions on green synthesis of nanomaterials and applications.

Suffice to state that curriculum development and curriculum re-engineering approaches have been deployed to address the gap in the curriculum of science subjects at secondary school and engineering courses at tertiary institutions in different countries to promote learning of nanotechnology across several disciplines (Alford et al. 2009; Zheng et al. 2009; Yawson 2010; Mohammad et al. 2012; Murcia 2013; Quirola et al. 2018).

8.2 Manpower Development and Training via Workshops and Short Training

The curriculum of studies in microbiology at both undergraduate and postgraduate levels does not address sufficiently the needed exposure to materials science and nanotechnology. However, various investigations in microbiology employ the use of series of techniques that are basic tools in nanotechnology research; these include spectroscopy (UV–visible spectrophotometry and Fourier transform infrared spectroscopy) and electron microscopy (scanning and transmission) which are well covered in the curriculum. The handling of other techniques such as X-ray diffraction, dynamic light scattering, energy-dispersive X-ray spectroscopy and thermogravimetric analysis would require some level of competence in these techniques whose coverage is beyond the scope of microbiology discipline. Thus, acquisition of knowledge about them can be accessed by advanced scholars of microbiology through short training and workshops on nanotechnology that are handled by core specialists and researchers in nanotechnology. These workshops and training are available at international, regional and national levels in different countries. In addition, the training can also be organized at institutional level by cognate departments: typically chemistry, physics, materials science and engineering, metallurgical engineering and mechanical engineering. Similarly, through interdisciplinary research, specialized workshops can be organized for the participating members to acquaint them of these techniques and import to microbial nanobiotechnology.

The aforementioned methods have been found to work with encouraging results. At the institution of the lead author, a multidisciplinary research group that consisted of microbiologists, botanist, zoologist, physicists, biochemist and mechanical engineers has worked out basic workshop modules with theoretical and practical exposition on the synthesis, characterization and applications of nanoparticles. The group, LAUTECH Nanotechnology Research Group (*NANO*$^{+}$) has organized workshops on nanotechnology in 2017, 2018 and 2019 with good attendance (www.lautechnanotech.com). The success story is that many of the attendees at these workshops have become proficient in nanotechnology and initiated researches in this area in not less than twenty universities within Nigeria. The group also instituted an annual conference on nanotechnology with broad themes and audiences to provide discourse platform for nanotechnology engagement in a broad-based manner. Thus, opportunities can be easily created to provide a learning environment to acquire the needed knowledge to conduct research in nanotechnology by non-specialists in materials science. Such trainings have also been advanced by scholars (Wang and Hsu 2004; Fonash et al. 2006; Furlan et al. 2013; Malsch 2014). Moreover, stimulating lectures by nano-biotechnologists can be organized for students in life sciences to introduce them to the concepts of nanobiotechnology and its relevance to their disciplines and research endeavours (Lateef 2020). With the advances in ICT, such talk can be delivered via webinar.

Beyond these, national and international associations and societies in microbiology must rise to the occasion to recognize microbial nanobiotechnology as a subfield of endeavour, and articulate issues relating to its practice, growth and development. A section should be created for the sub-discipline to thrive, reputable international platforms should be established for publication of research outcomes in the area, and established researchers in the field of microbial nanobiotechnology must also rise to the occasion to write and publish textbooks on topical issues in the field. In addition, opportunities should be made available for young scholars and postdoctoral fellows to pursue career in the bourgeoning field.

8.3 Interdisciplinary Research and International Collaboration

Obviously, engagement in nanotechnology research by life scientists generally and microbiologists in particular entails interdisciplinary pursuit. Primarily, understanding of the concepts of materials science, nanoscience, nanotechnology and instrumentation is requisite for meaningful research in nanotechnology. Therefore, microbiologists must endeavour to integrate their research activities within the enclave of interdisciplinary research which can be exploited among disciplines of natural sciences, food sciences and nutrition, agriculture, environmental sciences, medical and allied sciences. In some specific instances, wider coverage that integrates physical and chemical sciences as well as engineering may be necessary to conduct some investigations. It is interesting to note that due to the ubiquitous nature of microbes and their negative and positive impacts in all facets of the ecosystem, most disciplines in engineering, natural, agricultural, environmental and medical sciences have been impacted by studies on microbes. Thus, it becomes easier for a microbiology research to be integrated within these disciplines. Table 1 illustrates some of these potential interdisciplinary activities.

Aside institutional interdisciplinary research, advances in ICT have made it easier for collaboration beyond borders. Both national and international collaborations can be exploited for research activities to address the dearth of sophisticated equipment that are needed for nanotechnology investigations on the one hand, and to seek for advanced training in established institutions through short training and fellowships. Similarly, bilateral and multilateral relationships among developed and developing nations can provide opportunities for research collaboration in nanotechnology through the engagement of science diplomacy (Ezekiel 2020), to address the wide gap in nanotechnology research between the north and the south. Such engagements have been reported in the literature (Appelbaum and Parker 2008; Katz et al. 2009; Yazdi and Zarj 2013; Ezema et al. 2014; Sgouros and Stavrou 2019; Friedersdorf 2020; Yu and Jen 2020).

Table 1 Some interdisciplinary research activities involving microbial nanobiotechnology

Disciplines	Applications/interdisciplinary research activities
Food science and technology	Antimicrobial packaging materials, nano-sensors, nano-nutrients, nutraceuticals and nano-nutrition
Chemistry and chemical engineering	Nanodegradation, nanoremediation, nanocatalysis, nano-enhanced oil recovery, antimicrobials, water research, nanocomposites, films and foam research, nanoceramics, nano-sensors, polymers, bioprocess development, separation technology research, functionalized materials and biofuel research
Civil engineering and architecture	Water research, environmental engineering, green structures, pollution studies, nanocomposites, nano-paints, functionalized materials, nano-cements, nano-concrete, nano-sensors and nanoremediation
Biochemistry	Antioxidant research, metabolic engineering, cancer research, phytochemicals, nutraceuticals, nanotoxicology and diabetes research
Medicine, diagnostics, virology and pharmaceuticals	Antimicrobials, drug resistance research, nanosanitizers, nano-delivery system, nano-drugs, imaging, diagnosis, biomedical and tissue engineering
Agriculture	Nanosanitizers for tissue culture, nanofertilizers, nanopesticides, nanofeeds, phytochemicals, plant and animal nutrition, antimicrobials, antifouling, antiodour and smart farming
Mechanical and agricultural engineering	Nanocomposites, antimicrobials, polymers, renewable energy, antifouling and anticorrosion research, biodegradable films, biofuel research, water research
Fine and applied arts	Antimicrobials, protection against deterioration and degradation, nanoceramics, nanocomposites, functionalized clothing, nano-paints
Physics and electronics	Radiation and health physics, renewable energy research, photoluminescence, nanocatalysis, quantum dots, energy storage devices and nano-sensors
Biological sciences	Antimicrobials, phytonutrients, animal feeds, toxicology, cancer research, taxonomy, fisheries, nanoremediation, phytochemicals, nanopesticides, water research and pollution
Product-based investigations	Assessment of several nano-based products for their microbial quality and antimicrobial properties. These include paints, fabrics, cosmetics, packaging materials, films, foams, agricultural products, soaps and detergents, toothpastes, drugs, filters, ceramics, sanitizers, coated instruments, etc.

9 Future Trends

Though a good number of studies in nanotechnology in relation to microbiology have largely focussed on the use of nanomaterials to control microbial growth, newer areas of investigations have emerged. These include microbial synthesis of nanoparticles and the exploitation of microbe-mediated nanoparticles for myriad of applications, immobilization of microbial enzymes using nanoparticles for improved biocatalysis and utilization of nanoparticles for bioprocess development to improve the yields of microbial metabolites and products such as the generation of biofuels (Sanusi et al. 2020). It is envisaged that nanotechnology would play more prominent roles in future in enhancing the capabilities of microbes in several bioprocesses and transformation that would include but not limited to microbial remediation, biodegradation, biofuel generation, biocatalysis, microbial enhanced oil recovery and plant growth promotion with positive impacts on microbial technology and processes derived therefrom. Biosynthesis of nanomaterials by microbes would be impacted by genetic engineering to elucidate the mechanisms involved in their synthesis, to manufacture tuned nanomaterials of specific attributes and functionality and to position microbes as efficient nanofactories to produce nanomaterials at industrial scale. These areas of investigations can become the hotbed of research activities in the nearest future with lots of potentials to create new sub-specialties and widen the scope of applications of microbes through the production of microbe-inspired nano-based products and processes.

In the continued efforts to defeat microbial pathogens of man, animals and plants, nanomaterials have emerged as bug killers with profound effects due to the multifaceted assaults on pathogens which have proven to be more advantageous to the narrow range of activities exhibited by antimicrobial drugs. Thus, either by acting singly or in synergistic activities with established drugs, nanomaterials can be efficiently deployed to fight the scourge of drug resistance among microbes and even in cancer cells. Investigations in this area shall promote the development of nano-based antimicrobial agents for applications in health care, pharmaceuticals, environment, agriculture and food processing. Another budding area of development is in the applications of nanomaterials for the sensing of chemicals with particular relevance in the detection of microbial cells and toxins through nano-sensors. The emergence of these smart systems would enhance the development of novel packaging materials and promote real-time detection of microbes and their metabolites with varying degrees of applications in food industries, pharmaceuticals, clinical practice and environment.

In basic microbiology, the incorporation of nutrients in culture media in nanoformulated forms would spur new investigations in microbial physiology, proteomics, gene expression and metabolomics to fully understand the impacts of nanomaterials on metabolic fluxes of microbial cells and their activities. On this basis, microbes can be utilized in the evaluation of toxicity of nanomaterials and their environmental impacts. In addition, the influences of nanomaterials on the performances of varying microbial metabolites and products would be another enterprising area of engagement. For instance, it would be interesting to know the impacts of nanomaterials on the activities of metabolites such as nisin, bacteriocin, surfactin, extracellular polysaccharides, siderophores and enzymes to mention a few, when these metabolites are biogenically employed to synthesize nanoparticles. Studies have shown that with the nanoparticles acting as carriers of bioactives and with improved surface area-to-volume ratio, the activities of the metabolites could be greatly enhanced even with infinitesimal amounts of the metabolites. Certainly, the future portends a great opportunity for microbiology and microbial technology in nanotechnology research and development which must be taken up by microbiologists.

10 Conclusion

In this chapter, we have discussed the connections among microbiology, microbial technology and nanotechnology in a lucid manner to herald the sub-discipline of microbial nanobiotechnology: a new concept in the fields of microbiology and nanotechnology. The imperatives of extending the frontiers of knowledge to be acquired by microbiologists in order to play rewarding roles in the developments of microbial nanobiotechnology have been highlighted. It is a considered opinion that such enterprise would not only advance the scientific study and exploitation of microbes, but would also position microbiologists with greater opportunities in the field of nanotechnology and allied disciplines. Thus, budding microbiologists are encouraged to take shot at nanotechnology with the view of establishing microbiology as a key player in the field.

References

Abou-Shanab RAI, Khalafallah MA, Emam NF, Aly MA, Abou-Sdera SA, Matter IA (2012) Characterisation and identification of carbofuran-utilising bacteria isolated from agricultural soil. Chem Ecol 28:193–203. https://doi.org/10.1080/02757540.2011.628317

Adeeyo AO, Lateef A, Gueguim-Kana EB (2016) Optimization of the production of extracellular polysaccharide from the Shiitake medicinal mushroom *Lentinus edodes* (Agaricomycetes) using mutation and a genetic algorithm-coupled artificial neural network (GA-ANN). Int J Med Mushrooms 18(7):571–581. https://doi.org/10.1615/IntJMedMushrooms.v18.i7.20

Adelere IA, Lateef A (2016) A novel approach to the green synthesis of metallic nanoparticles: the use of agro-wastes, enzymes, and pigments. Nanotechnol Rev 5(6):567–587. https://doi.org/10.1515/ntrev-2016-0024

Adelere IA, Lateef A (2019) Degradation of keratin biomass by different microorganisms. In: Sharma S, Kumar A (eds) Keratin as a protein biopolymer. Springer series on polymer and composite materials. Springer International Publishing AG, Cham, Switzerland, pp 123–162. ISBN 978-3-030-02900-5, https://doi.org/10.1007/978-3-030-02901-2_5

Adeoye AO, Lateef A, Gueguim-Kana EB (2015) Optimization of citric acid production using a mutant strain of *Aspergillus niger* on cassava peel substrate. Biocatal Agric Biotechnol 4(4):568–574. https://doi.org/10.1016/j.bcab.2015.08.004

Adewoye SO, Lateef A (2004) Assessment of the microbiological quality of *Clarias gariepinus* exposed to an industrial effluent in Nigeria. Environmentalist 24:249–254. https://doi.org/10.1007/s10669-005-1000-7

Akintayo GO, Lateef A, Azeez MA, Asafa TB, Oladipo IC, Badmus JA, Ojo SA, Elegbede JA, Gueguim-Kana EB, Beukes LS, Yekeen TA (2020) Synthesis, bioactivities and cytogenotoxicity of animal fur-mediated silver nanoparticles. IOP Conf Ser: Mater Sci Eng 805:012041. https://doi.org/10.1088/1757-899X/805/1/012041

Alford KJ, Calati F, Clarke A, Binks PN (2009) Creating a spark for Australian science through integrated nanotechnology studies at St. Helena secondary college. J Nano Educ 1(1):68–74. https://doi.org/10.1166/jne.2009.005

Appelbaum RP, Parker RA (2008) China's bid to become a global nanotech leader: advancing nanotechnology through state-led programs and international collaborations. Sci Public Policy 35(5):319–334. https://doi.org/10.3152/030234208X319366

Ball AS, Patil S, Soni S (2019) Introduction into nanotechnology and microbiology. In: Methods in microbiology, vol 46. Academic Press, pp 1–18. https://doi.org/10.1016/bs.mim.2019.04.003

Bamigboye CO, Oloke JK, Dames JF, Burton M, Lateef A (2019) Optimization of the process for producing biomass and exopolysaccharide from the king tuber oyster mushroom, *Pleurotus tuber-regium* (Agaricomycetes), for biotechnological applications. Int J Med Mushrooms 21(4):311–322. https://doi.org/10.1615/IntJMedMushrooms.2019030357

Barakat KM, Mattar MZ, Sabae SZ, Darwesh OM, Hassan SH (2015) Production and characterization of bioactive pyocyanin pigment by marine *Pseudomonas aeruginosa* OSh1. Res J Pharm Biol Chem Sci 6(5):933–943

Belkin A, Hubler A, Bezryadin A (2015) Self-assembled wiggling nano-structures and the principle of maximum entropy production. Sci Rep 5:8323. https://doi.org/10.1038/srep08323

Darwesh OM, Moawad H, Abd El-Rahim WM, Barakat OS, Sedik MZ (2014) Bioremediation of textile reactive blue (RB) azo dye residues in wastewater using experimental prototype bioreactor. Res J Pharm Biol Chem Sci 5(4):1203–1219

Darwesh OM, Moawad H, Barakat OS, Abd El-Rahim WM (2015) Bioremediation of textile reactive blue azo dye residues using nanobiotechnology approaches. Res J Pharm Biol Chem Sci 6(1):1202–1211

Darwesh OM, Eida MF, Matter IA (2018a) Isolation, screening and optimization of L-asparaginase producing bacterial strains inhabiting agricultural soils. Biosci Res 15:2802–2812

Darwesh OM, Sultan YY, Seif MM, Marrez DA (2018b) Bio-evaluation of crustacean and fungal nano-chitosan for applying as food ingredient. Toxicol Rep 5:348–356. https://doi.org/10.1016/j.toxrep.2018.03.002

Darwesh OM, Matter IA, Eida MF (2019a) Development of peroxidase enzyme immobilized magnetic nanoparticles for bioremediation of textile wastewater dye. J Environ Chem Eng 7:102805. https://doi.org/10.1016/j.jece.2018.11.049

Darwesh OM, Matter IA, Eida MF, Moawad H, Oh YK (2019b) Influence of nitrogen source and growth phase on extracellular biosynthesis of silver nanoparticles using cultural filtrates of *Scenedesmus obliquus*. Appl Sci 9:1465. https://doi.org/10.3390/app9071465

Darwesh OM, Barakat KM, Mattar MZ, Sabae SZ, Hassan SH (2019c) Production of antimicrobial blue green pigment Pyocyanin by marine *Pseudomonas aeruginosa*. Biointerf Res Appl Chem 9(5):4334–4339. https://doi.org/10.33263/BRIAC95.334339

Darwesh OM, El-Hawary AS, El Kelany US, El-Sherbiny GM (2019d) Nematicidal activity of thermostable alkaline protease produced by *Saccharomonospora viridis* strain Hw G550. Biotechnol Rep 24:e00386. https://doi.org/10.1016/j.btre.2019.e00386

Darwesh OM, Ali SS, Matter IA, Elsamahy T, Mahmoud YA (2020a) Enzymes immobilization onto magnetic nanoparticles to improve industrial and environmental applications. In: Methods in enzymology, vol 630, pp 481–502. Academic Press. https://doi.org/10.1016/bs.mie.2019.11.006

Darwesh OM, El-Maraghy SH, Abdel-Rahman HM, Zaghloul RA (2020b) Improvement of paper wastes conversion to bioethanol using novel cellulose degrading fungal isolate. Fuel 262:116518. https://doi.org/10.1016/j.fuel.2019.116518

El-Baz FK, Gad MS, Abdo SM, Abed KA, Matter IA (2016) Performance and exhaust emissions of a diesel engine burning algal biodiesel blends. Int J Mechan Mechatr Eng 16:150–157

Elegbede JA, Lateef A (2018) Valorization of corn-cob by fungal isolates for production of xylanase in submerged and solid state fermentation media and potential biotechnological applications. Waste Biomass Valor 9(8):1273–1287. https://doi.org/10.1007/s12649-017-9932-y

Elegbede JA, Lateef A (2019a) Green synthesis of silver (Ag), gold (Au) and silver-gold (Ag-Au) alloy nanoparticles: a review on recent advances, trends and biomedical applications. In: Verma DK, Goyal MR, Suleria HAR (eds) Nanotechnology and nanomaterial applications in food, health and biomedical sciences. Apple Academic Press Inc./CRC Press, Taylor and Francis Group, Oakville, Ontario, Canada, pp 3–89. ISBN 978-1-77188-764-9, https://doi.org/10.1201/9780429425660-1

Elegbede JA, Lateef A (2019b) Green nanotechnology in Nigeria: The research landscape, challenges and prospects. Ann Sci Technol 4(2):6–38. https://doi.org/10.2478/ast-2019-0008

Elegbede JA, Lateef A (2020) Nanotechnology in the built environment for sustainable development. IOP Conf Ser: Mater Sci Eng 805:012044. https://doi.org/10.1088/1757-899X/805/1/012044

Elegbede JA, Lateef A, Azeez MA, Asafa TB, Yekeen TA, Oladipo IC, Adebayo EA, Beukes LS, Gueguim-Kana EB (2018) Fungal xylanases-mediated synthesis of silver nanoparticles for catalytic and biomedical applications. IET Nanobiotechnol 12(6):857–863. https://doi.org/10.1049/iet-nbt.2017.0299

Elegbede JA, Lateef A, Azeez MA, Asafa TB, Yekeen TA, Oladipo IC, Hakeem AS, Beukes LS, Gueguim-Kana EB (2019) Silver-gold alloy nanoparticles biofabricated by fungal xylanases exhibited potent biomedical and catalytic activities. Biotechnol Prog 35(5):e2829. https://doi.org/10.1002/btpr.2829

Elegbede JA, Lateef A, Azeez MA, Asafa TB, Yekeen TA, Oladipo IC, Aina DA, Beukes LS, Gueguim-Kana EB (2020) Biofabrication of gold nanoparticles using xylanases through valorization of corncob by *Aspergillus niger* and *Trichoderma longibrachiatum*: antimicrobial, antioxidant, anticoagulant and thrombolytic activities. Waste Biomass Valor 11(3):781–791. https://doi.org/10.1007/s12649-018-0540-2

Elshahawy I, Abouelnasr HM, Lashin SM, Darwesh OM (2018) First report of *Pythium aphanidermatum* infecting tomato in Egypt and its control using biogenic silver nanoparticles. J Plant Protec Res 15(2):137–151. https://doi.org/10.24425/122929

Emam HE, Darwesh OM, Abdelhameed RM (2020) Protective cotton textiles via amalgamation of cross-linked zeolitic imidazole frameworks. Ind Eng Chem Res 59(23):10931–10944. https://doi.org/10.1021/acs.iecr.0c01384

Ezekiel IP (2020) Engaging science diplomacy for nanotechnology development in Africa. IOP Conf Ser: Mater Sci Eng 805:012039. https://doi.org/10.1088/1757-899X/805/1/012039

Ezema IC, Ogbobe PO, Omah AD (2014) Initiatives and strategies for development of nanotechnology in nations: a lesson for Africa and other least developed countries. Nanoscale Res Lett 9(1):133. https://doi.org/10.1186/1556-276X-9-133

Fonash SJ, Fenwick D, Hallacher P, Kuzma T, Nam WJ (2006) Education and training approach for the future nanotechnology workforce. In: IEEE conference on emerging technologies—nanoelectronics, pp 235–236. https://doi.org/10.1109/NANOEL.2006.1609719

Friedersdorf LE (2020) Developing the workforce of the future: how the national nanotechnology initiative has supported nanoscale science and engineering education in the United States. IEEE Nanotechnol Mag 14(4):13–20. https://doi.org/10.1109/MNANO.2020.2994799

Furlan R, Rosa LG, Vedrine-Pauléus J (2013) Implementing a training program for the nanotechnology workforce at the university of Puerto Rico. J Nano Educ 4(1–2):63–66. https://doi.org/10.1166/jne.2012.1026

Ganaie MA, Lateef A, Gupta US (2014) Enzymatic trends of fructooligosaccharides production by microorganisms. Appl Biochem Biotechnol 172(4):2143–2159. https://doi.org/10.1007/s12010-013-0661-9

Hasanin MS, Mostafa AM, Mwafy EA, Darwesh OM (2018) Eco-friendly cellulose nanofibres via first reported Egyptian *Humicola fuscoatra* Egyptia X4: isolation and characterization. Environ Nanotechnol Monit Manag 10:409–418. https://doi.org/10.1016/j.enmm.2018.10.004

Hoballah E, Saber M, Matter I, Zaghloul A (2014) Bioremediation of polychlorinated biphenyl (PCBs) in a sewaged soil by certain remediative amendments followed by phytoremediation. Res J Pharm Biol Chem Sci 5:91–103

Katz E, Solomon F, Mee W, Lovel R (2009) Evolving scientific research governance in Australia: a case study of engaging interested publics in nanotechnology research. Pub Underst. Sci 18(5):531–545. https://doi.org/10.1177/0963662507082016

Kiessling F, Mertens ME, Grimm J, Lammers T (2014) Nanoparticles for imaging: top or flop? Radiology 273:10–28. https://doi.org/10.1148/radiol.14131520

Kim YE, Matter IA, Lee N, Jung M, Lee YC, Choi SA, Lee SY, Kim JR, Oh YK (2020) Enhancement of astaxanthin production by *Haematococcus pluvialis* using magnesium aminoclay nanoparticles. Bioresour Technol 307:123270. https://doi.org/10.1016/j.biortech.2020.123270

Kröger N, Poulsen N (2008) Diatoms—from cell wall biogenesis to nanotechnology. Annu Rev Genet 42:83–107. https://doi.org/10.1146/annurev.genet.41.110306.130109

Lateef A (2004) The microbiology of a pharmaceutical effluent and its public health implications. World J Microbiol Biotechnol 20(2):167–171. https://doi.org/10.1023/B:WIBI.0000021752.29468.4e

Lateef A (2020) Applications of nanoparticles: perspectives from nanobiotechnology. Being an international webinar organized by the post graduate department of chemistry, and IQAC, TATA College, Chaibasa, India on 27 Aug 2020, pp 1–77. https://doi.org/10.13140/RG.2.2.21106.32966

Lateef A, Gueguim-Kana EB (2014) Quality assessment and hazard analysis in the small-scale production of poultry feeds in Ogbomoso, Southwest. Nigeria Qual Assur Saf Crops Foods 6(1):105–113. https://doi.org/10.3920/QAS2012.0209

Lateef A, Adeeyo AO (2015) Green synthesis and antibacterial activities of silver nanoparticles using extracellular laccase of *Lentinus edodes*. Not Sci Biol 7(4):405–411. https://doi.org/10.15835/nsb749643

Lateef A, Ojo MO (2016) Public health issues in the processing of cassava (*Manihot esculenta*) for the production of '*lafun*' and the application of hazard analysis control measures. Qual Assur Saf Crops Foods 8(1):165–177. https://doi.org/10.3920/QAS2014.0476

Lateef A, Oloke JK, Gueguim-Kana EB, Oyeniyi SO, Onifade OR, Oyeleye AO, Oladosu OC, Oyelami AO (2008a) Improving the quality of agro-wastes by solid state fermentation: enhanced antioxidant activities and nutritional qualities. World J Microbiol Biotechnol 24(10):2369–2374. https://doi.org/10.1007/s11274-008-9749-8

Lateef A, Oloke JK, Gueguim-Kana EB, Oyeniyi SO, Onifade OR, Oyeleye AO, Oladosu OC (2008b) *Rhizopus stolonifer* LAU 07: a novel source of fructosyltransferase. Chem Pap 62(6):635–638. https://doi.org/10.2478/s11696-008-0074-3

Lateef A, Davies TE, Adelekan A, Adelere IA, Adedeji AA, Fadahunsi AH (2010a) *Akara* Ogbomoso: microbiological examination and identification of hazards and critical control points. Food Sci Technol Int 16(5):389–400. https://doi.org/10.1177/1082013210366894

Lateef A, Oloke JK, Gueguim-Kana EB, Sobowale BO, Ajao SO, Bello BY (2010b) Keratinolytic activities of a new feather-degrading isolate of *Bacillus cereus* LAU 08 isolated from Nigerian soil. Int Bioremed Biodegr 64(2):162–165. https://doi.org/10.1016/j.ibiod.2009.12.007

Lateef A, Oloke JK, Gueguim-Kana EB, Raimi OR (2012) Production of fructosyltransferase by a local isolate of *Aspergillus niger* in both submerged and solid substrate media. Acta Aliment 41(1):100–117. https://doi.org/10.1556/AAlim.41.2012.1.12

Lateef A, Adelere IA, Gueguim-Kana EB (2015a) *Bacillus safensis* LAU 13: a new source of keratinase and its multi-functional biocatalytic applications. Biotechnol Biotechnol Equip 29(1):54–63. https://doi.org/10.1080/13102818.2014.986360

Lateef A, Adelere IA, Gueguim-Kana EB (2015b) The biology and potential biotechnological applications of *Bacillus safensis*. Biologia 70(4):411–419. https://doi.org/10.1515/biolog-2015-0062

Lateef A, Adelere IA, Gueguim-Kana EB, Asafa TB, Beukes LS (2015c) Green synthesis of silver nanoparticles using keratinase obtained from a strain of *Bacillus safensis* LAU 13. Int Nano Lett 5(1):29–35. https://doi.org/10.1007/s40089-014-0133-4

Lateef A, Ojo SA, Akinwale AS, Azeez L, Gueguim-Kana EB, Beukes LS (2015d) Biogenic synthesis of silver nanoparticles using cell-free extract of *Bacillus safensis* LAU 13: antimicrobial, free radical scavenging and larvicidal activities. Biologia 70(10):1295–1306. https://doi.org/10.1515/biolog-2015-0164

Lateef A, Ojo SA, Oladejo SM (2016a) Anti-candida, anti-coagulant and thrombolytic activities of biosynthesized silver nanoparticles using cell-free extract of *Bacillus safensis* LAU 13. Process Biochem 51(10):1406–1412. https://doi.org/10.1016/j.procbio.2016.06.027

Lateef A, Ojo SA, Elegbede JA (2016b) The emerging roles of arthropods and their metabolites in the green synthesis of metallic nanoparticles. Nanotechnol Rev 5(6):601–622. https://doi.org/10.1515/ntrev-2016-0049

Lateef A, Ojo SA, Elegbede JA, Akinola PO, Akanni EO (2018) Nanomedical applications of nanoparticles for blood coagulation disorders. In: Dasgupta N, Ranjan S, Lichtfouse E (eds) Environmental nanotechnology. Environmental chemistry for a sustainable world, vol 14. Springer International Publishing AG, Cham, Switzerland, pp 243–277. ISBN 978-3-319-76089-6, https://doi.org/10.1007/978-3-319-76090-2_8

Malsch I (2014) Nano-education from a European perspective: nano-training for non-R&D jobs. Nanotechnol Rev 3(2):211–221. https://doi.org/10.1515/ntrev-2013-0039

Malvankar NS, Lovley DR (2012) Microbial nanowires: a new paradigm for biological electron transfer and bioelectronics. Chemsuschem 5:1039–1046. https://doi.org/10.1002/cssc.201100733

Mann S (2001) Biomineralization: principles and concepts in bioinorganic materials chemistry. Oxford University Press, Oxford, UK

Matter IA, Darwesh OM, El-baz FK (2016) Using the natural polymer chitosan in harvesting *Scenedesmus* species under different concentrations and cultural pH values. Int J Pharm Biol Sci 7:254–260

Matter IA, Bui VK, Jung M, Seo JY, Kim YE, Lee YC, Oh YK (2019) Flocculation harvesting techniques for microalgae: a review. Appl Sci 9:3069. https://doi.org/10.3390/app9153069

Mohamed AA, Ali SI, Darwesh OM, El-Hallouty SM, Sameeh MY (2015) Chemical compositions, potential cytotoxic and antimicrobial activities of *Nitraria retusa* methanolic extract sub-fractions. Int J Toxicol Pharmacol Res 7(4):204–212

Mohammad AW, Lau CH, Zaharim A, Omar MZ (2012) Elements of nanotechnology education in engineering curriculum worldwide. Procedia-Soc Behav Sci 60:405–412. https://doi.org/10.1016/j.sbspro.2012.09.398

Murcia K (2013) Secondary school students' attitudes to nanotechnology: what are the implications for science curriculum development? Teach Sci 59(3):15–21

Ojo SA, Lateef A, Azeez MA, Oladejo SM, Akinwale AS, Asafa TB, Yekeen TA, Akinboro A, Oladipo IC, Gueguim-Kana EB, Beukes LS (2016) Biomedical and catalytic applications of gold and silver-gold alloy nanoparticles biosynthesized using cell-free extract of *Bacillus safensis* LAU 13: antifungal, dye degradation, anti-coagulant and thrombolytic activities. IEEE Trans Nanobiosci 15(5):433–442. https://doi.org/10.1109/TNB.2016.2559161

Ojoawo SO, Lateef A, Oyeniran FA, Kupoluyi OT, Opatola OS, Daramola JO (2017) Bioaccumulation of heavy metals in steel processing industrial effluents using *Bacillus safensis* LAU 13. J Environ Biotechnol Res 6(1):58–63

Oladipo IC, Lateef A, Elegbede JA, Azeez MA, Asafa TB, Yekeen TA, Akinboro A, Gueguim-Kana EB, Beukes LS, Oluyide TO, Atanda OR (2017a) Enterococcus species for the one-pot biofabrication of gold nanoparticles: characterization and nanobiotechnological applications. J Photochem Photobiol B 173:250–257. http://dx.doi.org/10.1016/j.jphotobiol.2017.06.003

Oladipo IC, Lateef A, Azeez MA, Asafa TB, Yekeen TA, Akinboro A, Akinwale AS, Gueguim-Kana EB, Beukes LS (2017b) Green synthesis and antimicrobial activities of silver nanoparticles using cell-free extracts of *Enterococcus* species. Not. Sci. Biol. 9(2): 196-203. http://dx.doi.org/10.15835/nsb929938

Phoenix C (2005) Nanotechnology: developing molecular manufacturing archived 2005-09-01 at the wayback machine. crnano.org

Prasad R, Pandey R, Barman I (2016) Engineering tailored nanoparticles with microbes: Quo vadis? Wiley Interdiscip. Rev Nanomed Nanobiotechnol 8:316–330. https://doi.org/10.1002/wnan.1363

Quirola N, Marquez V, Tecpan S, Baltazar SE (2018) Didactic proposal to include nanoscience and nanotechnology at high school curriculum linking physics, chemistry and biology. J Phys: Conf Ser 1043:012050. https://doi.org/10.1088/1742-6596/1043/1/012050

Saber M, Hoballah E, Matter I, Zaghloul A (2014) Bioremediation of chlorinated hydrocarbons in a sewaged soil by certain remediative amendments followed by phytoremediation. Res J Pharm Biol Chem Sci 5:83–90

Sadek ZI, Abdel-Rahman MA, Azab MS, Darwesh OM, Hassan MS (2018) Microbiological evaluation of infant foods quality and molecular detection of *Bacillus cereus* toxins relating genes. Toxicol Rep 5:871–877. https://doi.org/10.1016/j.toxrep.2018.08.013

Sanusi IA, Suinyuy TN, Lateef A, Gueguim-Kana EB (2020) Effect of nickel oxide nanoparticles on bioethanol production: process optimization, kinetic and metabolic studies. Process Biochem 92:386–400. https://doi.org/10.1016/j.procbio.2020.01.029

Sgouros G, Stavrou D (2019) Teachers' training in developing nanoscience and nanotechnology teaching modules in the context of a community of learners. In: Bridging research and practice in science education. Springer, Cham, pp 339–356. https://doi.org/10.1007/978-3-030-17219-0_21

Simkiss K, Wilbur KM (1989) Biomineralization. Academic, New York, NY, USA

Sultan YY, Ali MA, Darwesh OM, Embaby MA, Marrez DA (2016) Influence of nitrogen source in culture media on antimicrobial activity of *Microcoleus lacustris* and *Oscillatoria rubescens*. Res J Pharmaceut Biol Chem Sci 7(2):1444–1452

Thassu D, Deleers M, Pathak Y (eds) (2007) Nanoparticulate drug delivery systems, vol 166, 1st ed. Drugs and The Pharmaceutical Sciences Informa Healthcare, New York

Wang GJ, Hsu CC (2004) Toward a comprehensively interdisciplinary nanoscience and nanotechnology personnel training program. In: International conference on engineering education and research, Olomouc, Czech Republic pp 27–30

Yan L, Da H, Zhang S, Lopez VM, Wang W (2017) Bacterial magnetosome and its potential application. Microbiol Res 203:19–28. https://doi.org/10.1016/j.micres.2017.06.005

Yawson RM (2010) Skill needs and human resources development in the emerging field of nanotechnology. J Vocat Educ Training 62(3):285–296. https://doi.org/10.1080/13636820.2010.499474

Yazdi FS, Zarj MA (2013) Nanotechnology: new horizons, approaches, and challenges. In: Science and innovations in Iran. Palgrave Macmillan, New York, USA, pp 115–138. https://doi.org/10.1057/9781137030108_6

Yu HP, Jen E (2020) Integrating nanotechnology in the science curriculum for elementary high-ability students in Taiwan: evidenced-based lessons. Roeper Rev 42(1):38–48. https://doi.org/10.1080/02783193.2019.1690078

Zheng W, Shih HR, Lozano K, Pei JS, Kiefer K, Ma X (2009) A practical approach to integrating nanotechnology education and research into civil engineering undergraduate curriculum. J Nano Educ 1(1):22–33. https://doi.org/10.1166/jne.2009.008

Characterization Techniques in Nanotechnology: The State of the Art

T. B. Asafa, O. Adedokun, and T. T. Dele-Afolabi

1 Spectroscopy

Spectroscopy is the branch of science that deals with the examination and quantification of spectra generated when matters interact with or release electromagnetic radiation. Different types of interaction such as absorption, scattering, and reflection occur due to the different wavelengths of electromagnetic wave used. Hence, salient information can be deduced from different spectra. Here, we present briefly the spectroscopy techniques that are most pertinent in the study of nanomaterials and nanoparticles characteristics.

1.1 UV-vis Absorption Spectroscopy

Ultraviolet visible *(*UV-vis*)* spectroscopy or absorption spectroscopy or reflectance spectroscopy is a quantitative analysis technique commonly employed to characterize both nanomaterials. It involves measurement of attenuation of UV-vis light beam after moving through a sample or reflecting from a sample surface (Slater 1994). The instrument employed in UV-vis spectroscopy is referred to as a UV-vis

T. B. Asafa (✉) · T. T. Dele-Afolabi
Department of Mechanical Engineering, Ladoke Akintola University of Technology, PMB 4000, Ogbomoso, Nigeria
e-mail: tbasafa@lautech.edu.ng

O. Adedokun
Department of Physics, Ladoke Akintola University of Technology, PMB 4000, Ogbomoso, Nigeria

T. B. Asafa · O. Adedokun
Nanotechnology Research Group, Ladoke Akintola University of Technology, PMB 4000, Ogbomoso, Nigeria

A. Lateef et al. (eds.), *Microbial Nanobiotechnology*, Materials Horizons: From Nature to Nanomaterials, https://doi.org/10.1007/978-981-33-4777-9_2

spectrophotometer, and it measures and compares the initial light intensity before interaction with a sample (I_0) and after passing through it (I). The ratio of $\frac{I}{I_0}$ is known as the transmittance (usually expressed as a percentage, $\%T$). The absorbance (A) is expressed as function of transmittance as shown in Eq. 1:

$$A = -\log\left(\frac{\%T}{100}\right) \tag{1}$$

The spectroscopic technique operates on the principle of the Beer–Lambert law given as:

$$A = \log_{10}\left(\frac{I_0}{I}\right) = \varepsilon \cdot c \cdot L \tag{2}$$

where 'A' is the measured absorbance, I_0 is the initial light intensity before interaction with a sample at a given wavelength, I is the transmitted intensity, L is the path length through the sample, c is the concentration of the absorbing samples, and ε is a constant referred to as the "molar absorptivity" or "extinction coefficient." The electron in an atom excites and moves from lower energy to higher energy state when light of sufficient energy falls on the material. The energy state of an electron can be categorized into the lowest unoccupied molecular orbit (LUMO) and the highest unoccupied molecular orbit (HOMO). The recognition of the evolution of bonds being formed is connected with the specific behavior of material sample which is under investigation. Various bonds formation during electronic transition for materials is depicted in Fig. 1. The approximate wavelength ranges for electronic transitions, in addition to their possible list of bonds, functional groups, or molecules responsible for these transitions are given in Table 1. The strongest bond is sigma (σ) while the weakest bond is pi (π), and all rely on the nature of bond formed in the material.

UV-vis spectroscopy analysis provides ample information for scientists and practitioners, which includes: identification of chromophore functional group, determination of unknown compound, purity of the material, and determination of energy band gap. Chromophores are materials with chemical bonds and functional groups

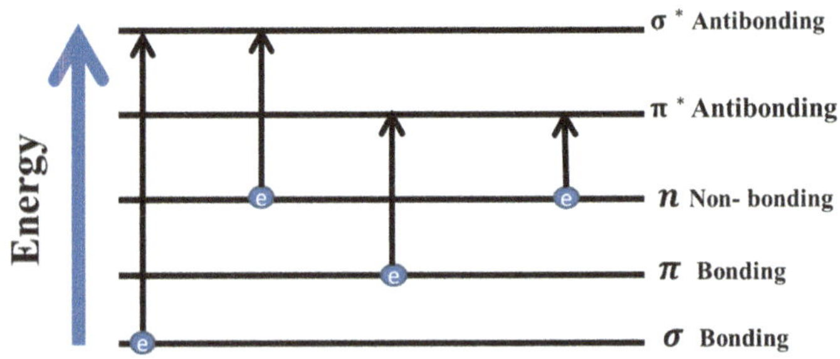

Fig. 1 Electronic transition involving different molecular orbits

Table 1 Molecular orbital *n*, σ, and π in electronic transitions

Transition	Range of wavelength (nm)	Examples
$\sigma \rightarrow \sigma^*$	<200	C–H, C–C,
$n \rightarrow \sigma^*$	160–260	H_2O, CH_3OH, CH_3Cl
$\pi \rightarrow \pi^*$	200–500	C=C, C=O, C=N, C≡C
$n \rightarrow \pi^*$	250–600	C=O, C=N, N=N, N=O

that absorb ultraviolet and visible radiation. The band gap (E_g) can be estimated by assuming direct transition between conduction band and valence band. Theory of optical absorption gives the correlation between the absorption coefficients α and the photon energy $h\upsilon$ for direct allowed transition (Pankove and Kiewit 1972) as:

$$(\alpha h\upsilon)^2 = A\left(h\upsilon - E_g\right) \tag{3}$$

The band gap (direct) is determined when straight line of the '$(\alpha h\upsilon)^2$' against 'hυ' plot based on Eq. 3 is extrapolated to intercept the energy axis at $\alpha = 0$.

A typical UV-vis absorption experiment for a material in liquid is described as depicted in Fig. 2. A monochromatic beam of light from a source is split into two beams, whereby one of them is moved through the sample and the other moves through the reference. The two beams are directed back to the detectors, after transmission across the sample and reference, through which they are compared to give an output (Kalantar-zadeh and Fry 2008). The basis for this measurement is the distinction between the two signals. The samples in liquid are usually analyzed in a cell known as cuvette (made of flat and fused quartz faces). The quartz type of cuvette is commonly employed because it is transparent to both UV and visible light.

UV-vis absorption spectra for AgNPs biosynthesized using cocoa bean extract and *Bacillus safensis* cell-free extract (Fig. 3a, b) were reported elsewhere (Azeez et al. 2017; Lateef et al. 2015). The absorption spectra for AgNPs showed maximum

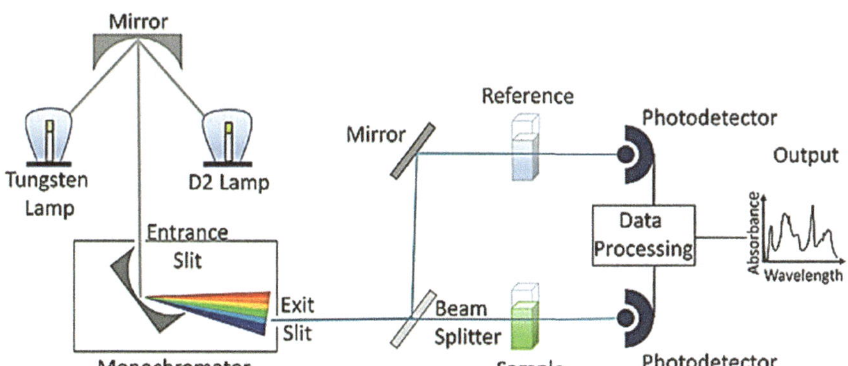

Fig. 2 Schematic diagram for a double beam UV-vis spectrophotometer (Rocha et al. 2018)

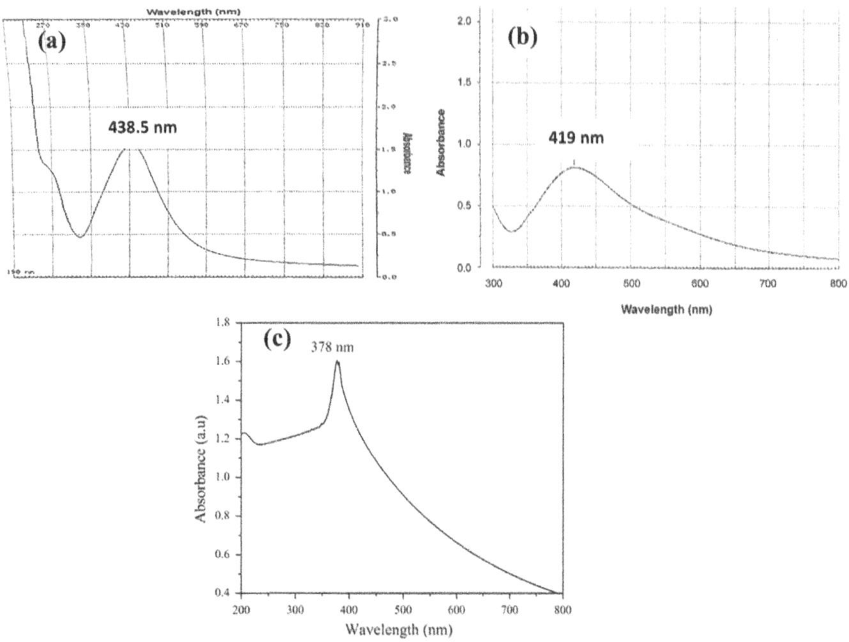

Fig. 3 UV-vis absorption spectra for **a** AgNPs biosynthesized using cocoa bean extract (Azeez et al. 2017), **b** AgNPs biosynthesized using *Bacillus safensis* cell-free extract (Lateef et al. 2015), **c** ZnO nanoparticles (Pudukudy and Yaakob 2015)

peaks at wavelength 438.5 and 419 nm for AgNPs biosynthesized using cocoa bean extract and *B. safensis* extract, respectively, which is a signature of surface plasmonic resonance properties of AgNPs. For ZnO nanoparticles powder dispersed in distilled water, the UV-vis spectrum (Fig. 3c) showed peak characteristics of hexagonal wurtzite ZnO nanoparticles at 378 nm indicating infrared shift compared to that of bulk ZnO with its calculated energy band of 3.38 eV (Pudukudy and Yaakob 2015). The absorption spectral of different mono-metallic and bimetallic nanoparticles varies widely, and these are determined through UV-vis measurement. The transformation in the color formation during the reduction metal ions to zerovalent forms causes the absorption of UV-vis light, which signifies the formation of nanoparticles.

1.2 Fourier Transform Infrared Spectroscopy

Fourier transform infrared spectroscopy (FTIR) is an analytical technique that depends on the principle of infrared spectroscopy, and it has been employed to study various nanomaterials. It is a technique of data measurement whereby spectra are collated on the basis of measurements of the coherence of a radiative source, using time-domain or space-domain measurements of the electromagnetic radiation. Thus, an FTIR spectrometer concurrently collects spectral data in a broad spectral range. These raw data are then transformed into the actual spectrum via Fourier transformation based on Wiener–Khinchin theorem (Iniewski 2009). The FTIR spectrometer depends on interferences of different frequencies of light to generate a spectrum. It comprises of a light source, interferometer, sample compartment, and detector among others (Fig. 4). The electromagnetic radiation is produced from light source which passes the sample through the interferometer and reaches the detector. The signal is eventually transferred to a computer whereby mathematical function known as Fourier transform is carried out to convert the intensity-versus-time spectrum into intensity-versus-frequency spectrum. FTIR is mostly useful for identifying chemicals with their chemical bonds (functional groups). The wavelength of the light absorbed is a characteristic of the chemical bond. Invariably, FTIR can be employed to quantify constituents of an unknown mix either solids, liquids, or gases. The solid samples can be ground and mixed with potassium bromide (about 5% of the weight of the sample) and pressed to form a pellet. Then, the sample is mounted in the sample compartment for readings. FTIR spectra of pure compounds are so unique that they are like a molecular 'fingerprint' (Atkins and De Paula 2009).

The functionalization of nanoparticles surface is a distinctive approach to examine the rare properties of nanoparticles and to improve its application (Roy and Fendler 2004; Abad et al. 2005; Armelao et al. 2007). However, the functionalized nanoparticles characterization by FTIR analysis may be tedious to show the chemical functional groups between the functionalized molecules and the nanoparticles present on a catalog of identified vibrational frequencies. This is dependent on the novelty of the chemical bonding that is in existence between the nanomaterial and functionalized molecules, and hence, an approach must be put in place to show evidence about the vibrational resonant information between the nanosized particles and grafted molecules (Baudot et al. 2010). The functional group corresponding to the as-prepared carbon nanoparticles (CPs) is depicted in Fig. 5a. The FTIR spectrum shows a broad characteristic OH peak at 3340 cm^{-1}, the existence of C–H stretching

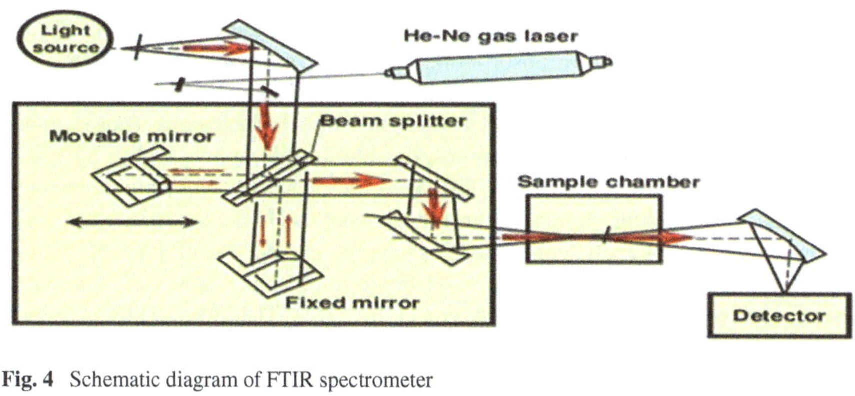

Fig. 4 Schematic diagram of FTIR spectrometer

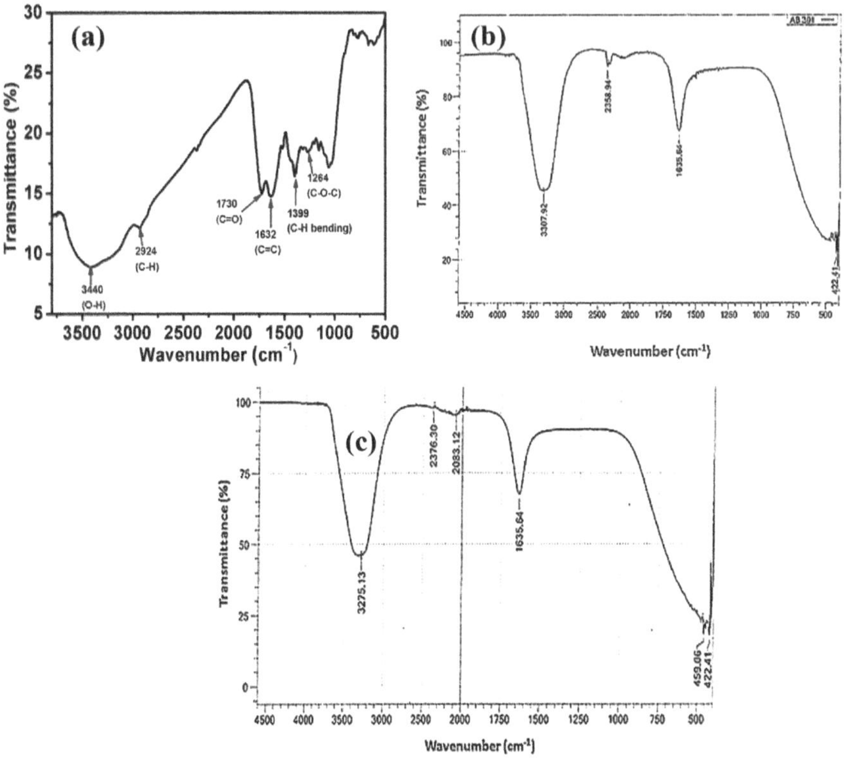

Fig. 5 FTIR spectra of **a** carbon nanoparticles (Adedokun et al. 2017a, b), **b** biosynthesized silver nanoparticles (Lateef et al. 2015), **c** biosynthesized AgNPs using cocoa bean extract (Azeez et al. 2017)

and bending vibration at 2924 cm^{-1} and 1399 cm^{-1}, respectively. The presence of carboxyl group is associated to C=C stretching modes at 1730 cm^{-1}. The asymmetric and symmetric stretching vibrations of C–O–C band at 1264 cm^{-1} and C=C stretching mode at 1632 cm^{-1} are confirming the existence of hydrophilic oxygen-containing functional groups and hydrophobic moieties (Adedokun et al. 2017a, b). Figure 5b shows the FTIR absorption spectrum for biosynthesized AgNPs. The distinct band at 3308 and 1636 cm^{-1} are the bonding vibration associated with N–H group (amines) of protein and the deformation atoms, respectively. These amines identified in the cell-free extract were responsible for the capping and stabilization of the AgNPs (Lateef et al. 2015). FTIR spectrum of AgNPs biosynthesized using cocoa bean extract (Fig. 5c) shows an intense peak at 3275 and 1635 cm^{-1} indicating protein and phenolic compounds as capping and stabilization agents. These two prominent peaks are typical of N–H bonding of amines and C=C stretch of alkenes or C=O stretch of amides (Azeez et al. 2017). Therefore, FTIR is a good analytical technique to elucidate the biomolecules responsible for the formation of nanoparticles, as these molecules are borne on the surfaces of nanoparticles.

1.3 Energy Dispersive X-ray Spectroscopy

Energy dispersive X-ray spectroscopy (EDS or EDX) is a qualitative, quantitative and non-destructive chemical microanalytical technique which provides information on the chemical composition of a nanomaterial sample. X-ray emission, the energy fingerprint of each element, is stimulated by the irradiation of the solid material surface with a high-energy focused X-ray beam (Bergström 2015; Shindo and Oikawa 2002). A typical EDX system has the following components: X-ray source and detector, sample holder, the pulse processor, and the analyzer (Fig. 6). An electron beam is focused on the sample from either a scanning electron microscope (SEM) or a transmission electron microscope (TEM). The electrons from the primary beam penetrate the sample and interact with the atoms from which it is made. Whenever an EDX system is employed in conjunction with SEM, the backscattered electrons produce compositional contrast, emanating from elements of various atomic numbers, and their distribution enhances the recognition of all elements and their proportions (Goldstein et al. 1981). As in SEM imaging, the electron beam can be rastered within the area of the sample to produce X-ray element distribution mapping. The reliability and efficiency of the analysis rely on the precision and response of the equipment, the spatial resolution, and the method employed in sample preparation (DeBeer and Neese 2013).

The X-rays are detected by an energy-dispersive detector which displays the signal as a spectrum of intensity (X-ray count rate) versus X-ray energy. The energies of the characteristic X-rays allow the elements adding up in the sample to be identified, while the intensities of the peaks of characteristic X-ray allow the concentrations of the elements to be quantified. The EDX analysis detection limit in the SEM relies on the composition of the sample under analysis, but it generally falls in the range 0.1–0.5 wt%. The TEM detection limits of ~0.01–0.1 wt% can also be achieved. Examples of EDX spectra are presented in Fig. 7. The EDX spectrum and elemental distribution for TiO_2 nanoparticles (Fig. 7a) indicate a corresponding peak of titanium (Ti) and oxygen (O) which were quantitatively identified as 71.72% and 28.28%, respectively (Adedokun et al. 2017a, b). The EDX spectrum and mapping of the

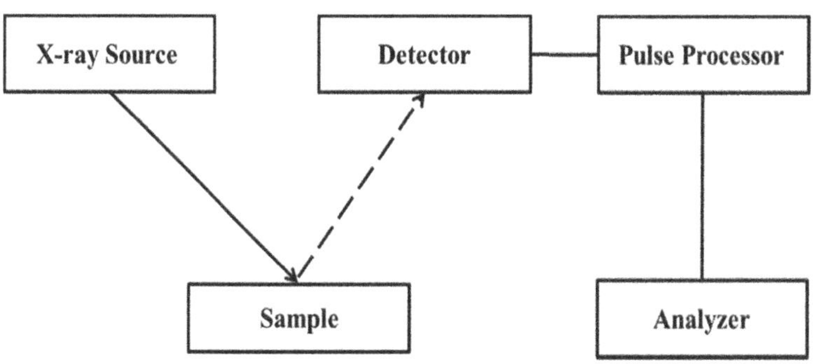

Fig. 6 Schematic diagram of basic EDX system

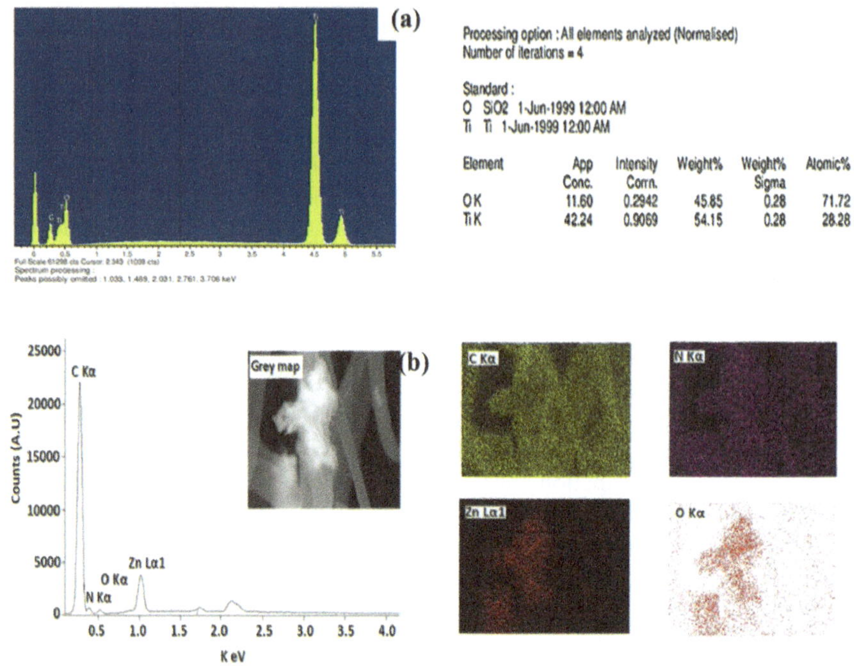

Fig. 7 EDX spectrum of **a**, TiO_2 nanoparticles (Adedokun et al. 2017a, b), **b** PAN/ZnO nanofiber composite with elemental mapping (Tissera et al. 2018)

PAN/ZnO nanofiber composite are also shown in Fig. 7b (Tissera et al. 2018). Peaks correspond to zinc (Zn), oxygen (O), carbon (C), and nitrogen (N) were present at higher counts. The selected elemental mapping shows that ZnO nanoparticle is assumed to be encapsulated with the structure of the nanofiber. Zn and O signals are due to the existence of ZnO nanoparticles, while signals of C and N confirm their presence in polymeric PAN nanofiber. Thus, EDX is an important tool to detect and quantify the elemental components of nanomaterials.

1.4 Photoluminescence Spectroscopy

Photoluminescence (PL) depicts the process by which a chemical compound contained in a material takes in a photon energy enable it transit from a lower to a higher energy state and then releases the photon thereby returning to a lower energy state. Quantum mechanics describes an excitation to a higher electronic energy state and a return to a lower energy state accompanied by the release of photons (McNaught and Wilkinson 1997). PL spectroscopy has emerged as an important tool for studying the optical and electronic properties such as band gap, impurity, and defects in solid-state materials appropriate for optoelectronics application (Gilliland

1997). Two forms of photoluminescence are available; fluorescence and phosphorescence. Fluorescence involves de-excitation of a chemical substance by emitting photons while transiting to its ground state. Some of the absorbed energy is dissipated, so that the emitted photons of light are of lower energy than those absorbed. However, phosphorescence is a chemical transition in which the absorbed energy undergoes intersystem crossing into a state with a different multiplicity of spin. The lifetime of phosphorescence (10^{-4} to 10^{-2} s) is usually much longer than that of fluorescence (10^{-8} to 10^{-4} s). The principle of photoluminescence system has been described elsewhere (Harris 2010). A monochromator is used for selection of excitation wavelength, and luminescence is observed through the second monochromator positioned at 90° to the incident light to minimize the intensity of scattered light reaching the detector. If the excitation wavelength is fixed and the emitted radiation is scanned, an emission spectrum is generated. The application of photoluminescence for semiconductor material is depicted in Fig. 8.

The PL spectra of anatase TiO_2 nanoparticles, obtained with an excitation wavelength of 310 nm, are displayed in Fig. 9a. TiO_2 is an indirect band gap semiconductor

Fig. 8 Schematic diagram of process in semiconductor material (McNaught and Wilkinson 1997)

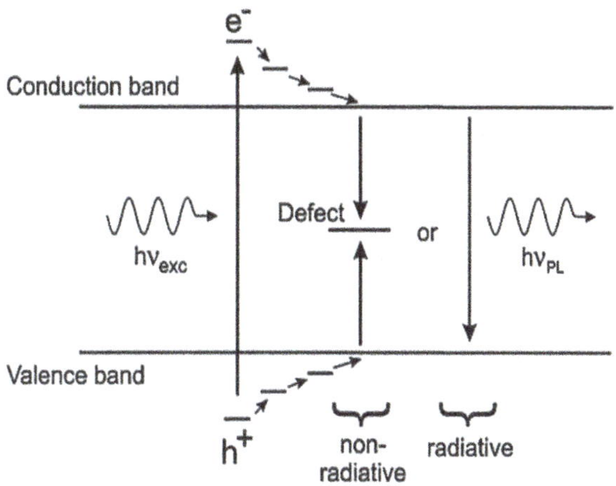

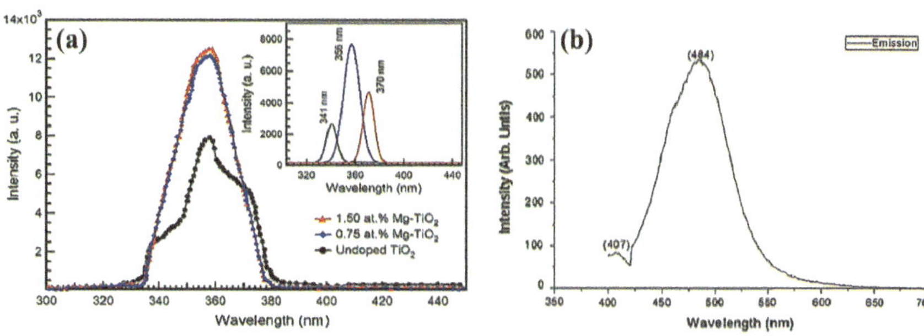

Fig. 9 Photoluminescence spectra for **a** undoped and Mg-doped TiO_2 nanoparticles (Giahi et al. 2019), **b** ZnO nanopowder (Taunk et al. 2015)

associated with radiative recombination of charge carriers. This means that the electrons in the valence band (VB) receive sufficient energy to transfer to the conduction band (CB) whenever light of sufficient energy (longer or equal wavelength to its band gap) irradiated on it (Giahi et al. 2019). Therefore, emission of energy as PL radiation occurs when electrons go back from CB to VB. As evident from Fig. 9a, a strong peak is identified around 356 nm that can be attributed to band–to–band transition (Giahi et al. 2019). The inset indicates the deconvoluted spectrum for pure TiO_2 nanoparticles. The small emission noticed at 341 nm could be attributed to the recombination of electrons–holes in surface defects or surface oxygen vacancies (Shariatzadeh and Moradi 2014). The emission of intermediate band transition observed in titania could be seen as a shoulder peak at 370 nm. It is observed that intensity of the peak becomes higher with the increase of the Mg dopants amount.

The room temperature emission PL spectrum of ZnO nanopowder excited at 235 nm is shown in Fig. 9b. It has emission peaks in blue region around 407 nm and broad blue region around 484 nm. The emission peak at 407 nm emanates from the recombination of free excitons through an exciton–exciton collision process or due to intrinsic defects such as oxygen and zinc interstitials (Taunk et al. 2015). The broad blue emission band at 484 nm is usually attributed to the radiative recombination of a photo-generated hole with an electron occupying the oxygen vacancy (Shariatzadeh and Moradi 2014). The possible cause of the visible emission in the ZnO has been attributed to surface state of the nanomaterials (Taunk et al. 2015). However, it is accepted that the PL emission is caused by the recombination of excited electrons and holes, and the lower PL intensity may imply the lower recombination rate of electrons and holes under light irradiation.

1.5 X-ray Diffraction (XRD) Technique

X-ray diffraction (XRD) is one of the powerful techniques extensively used for determination of crystal structure, grain size, nature of crystalline phase, and lattice parameters for nanoparticles based on X-ray radiation (Mourdikoudis et al. 2018). These high-energy (100 eV–100 keV) radiations are generated when a focused electron beam accelerated across a high voltage field bombards a stationary or rotating solid target. They penetrate deep into a given material and give information about the structural arrangement of atoms and molecules (Cullity and Stock 2001). The lattice parameters of nanomaterials are evaluated by employing the Scherrer equation using the broadening of the most intense peak from the XRD data. The diffraction angle and intensity of the peaks of the XRD pattern are then compared to the available data from Joint Committee on Powder Diffraction Standards (JCPDS, presently known as the International Center for Diffraction Data, ICDD) database. Common targets used in X-ray tubes include Cu and Mo, which emits 8 and 14 keV X-rays with corresponding wavelengths of 1.54 Å and 0.8 Å, respectively.

X-rays primarily interact with electrons in atoms, and during the course of interaction, some photons from the incident beam will be reflected away or diffracted

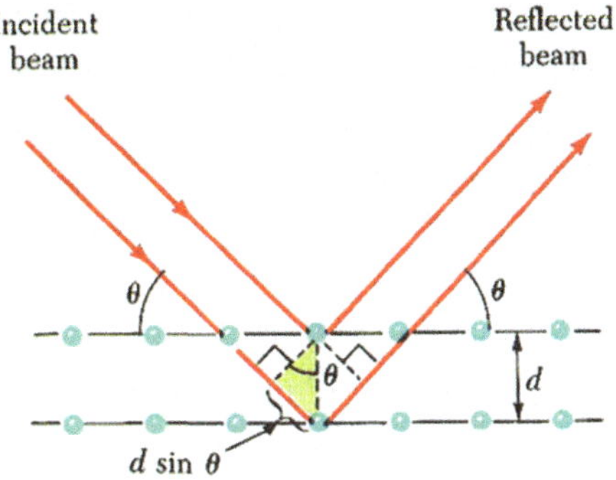

Fig. 10 X-ray reflections from a crystal with the atoms in the two or more different planes

from the direction where they originally travel (Fig. 10). The scattered X-rays carry information about the electron distribution in materials. If the atoms are arranged in a periodic fashion, as in crystals, the diffracted waves will consist of sharp interference maxima (peaks) with the same symmetry as in the distribution of atoms. Measuring the diffraction pattern therefore allows one to deduce the distribution of atoms in a material. The peaks in an X-ray diffraction pattern are directly related to the atomic distances.

For a given set of lattice planes with an inter-plane spacing of 'd,' Bragg's provide the condition for a diffraction (peak) to occur (Eq. 4):

$$2d \sin\theta = n\lambda \tag{4}$$

where 'λ' is the wavelength of the X-ray, 'θ' the scattering angle, and 'n' an integer representing the order of the diffraction peak. The relation between inter-plane spacing (*d*-value) and lattice parameters (*a*, *b* and *c*) for cubic and hexagonal system is:

$$\frac{1}{d^2} = \frac{h^2}{a^2} + \frac{k^2}{b^2} + \frac{l^2}{c^2} \tag{5}$$

$$\frac{1}{d^2} = \frac{4}{3}\left(\frac{h^2 + hk + k^2}{a^2}\right) + \frac{l^2}{c^2} \tag{6}$$

The crystallite size *D* can be determined from Scherer equation given (Scherrer 1918) as:

$$D = \frac{K\lambda}{\beta_{hkl} \cos\theta} \tag{7}$$

where *K* is a constant taken to be equal to 0.90, λ is the wavelength of the incident X-rays ($\lambda = 1.540$ Å), β_{hkl} is the sample full-width-at-half-maximum (FWHM, in radian) of the most intense peak at diffracting angle θ, and θ is the Bragg's angle position in degree. Further, the dislocation density S, a measure of number of defects and vacancies in the crystal, and micro-strain (ε) are determined from Eqs. 8 and 9.

$$S = \frac{1}{D^2} \tag{8}$$

$$\varepsilon = \frac{\beta cos\theta}{4} \tag{9}$$

For more analysis, the volume of the nanoparticles V, volume of hexagonal unit cell υ, and number of unit cells present in a particle (ratio V/υ) are calculated from the relations:

$$V = \frac{4}{3}\pi\left(\frac{D}{2}\right)^3 \tag{10}$$

$$\upsilon = \frac{3a^2c}{2} \tag{11}$$

$$n = 1.621\left(\frac{D^3}{a^2c}\right) \tag{12}$$

The XRD pattern for ZnO nanopowder shown in Fig. 11a agreed well with the reported reference patterns of wurtzite ZnO (hexagonal phase, JCPDS file 75-0576). The mean crystallite size was evaluated for the highest diffraction peak along the <101> plane to be 34.2 nm (Adedokun et al. 2018). For the TiO_2 powder, the XRD pattern (Fig. 11b) matched with TiO_2 anatase tetragonal phase (JCPDS file 71-1168). The mean crystallite size was 39.10 nm (Adedokun et al. 2017a, b).

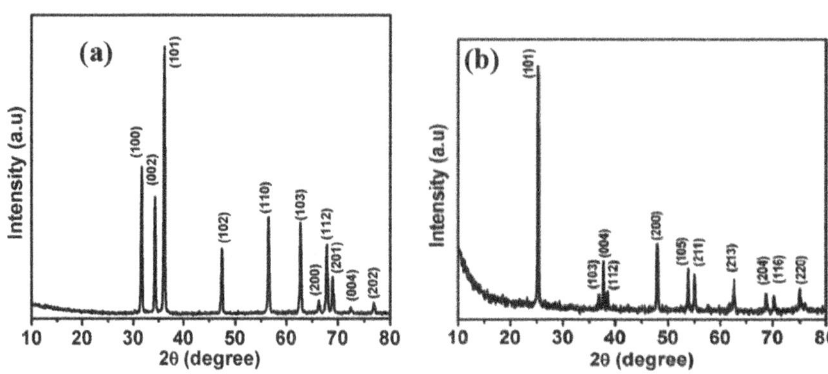

Fig. 11 XRD pattern of **a** ZnO nanopowder (Adedokun et al. 2018), **b** TiO_2 powder (Adedokun et al. 2017a, b)

1.6 Raman Spectroscopy

Raman spectroscopy is a method used to study vibrational, rotational, and other low-frequency modes in materials thereby providing information about molecular signature needed to identify and quantify samples (Gardiner 1989). In this technique, a monochromatic light source (usually laser) is shined on a sample, and the scattered light is detected. In most cases, the frequency of the scattered light and that of excitation source is the same for large fraction of the scattered light (Rayleigh or elastic scattering). About 5–10% of the light intensity incident on the sample which constituted a small fraction of the scattered light has a shift in energy from the laser frequency because of the interactions between the incident electromagnetic wave and the vibrational energy levels of the molecules in the sample (Raman scattering). Plotting the intensity of this "shifted" light against frequency results in a Raman spectrum of the sample (Nakamoto 2008). Stokes lines are noticed in the Raman spectrum whenever the frequency of incident radiation is greater than frequency of scattered radiation. Otherwise, anti-Stokes lines become visible in the Raman spectrum. Scattered radiation is usually estimated at right angle to incident radiation (Skoog et al. 2007; Smith and Dent 2005; Willard et al. 1952). The schematic diagram of Raman spectrometer is shown in Fig. 12.

In the past, the experimental and theoretical efforts in Raman spectroscopy have been focused on the fundamentals of inelastic scattering, and it is employed for studying molecular structure. However, several developments of chemical measurements raise the interest in Raman spectroscopy especially for analysis of a wide range of forensic samples (Laserna 2001). Raman spectroscopy can be used for both qualitative as well as quantitative purpose. While qualitative analysis involves measurement of frequency of scattered radiations, quantitative analysis has to do with the measurement of intensity of scattered radiations (Laserna 2001; Skoog et al. 2007;

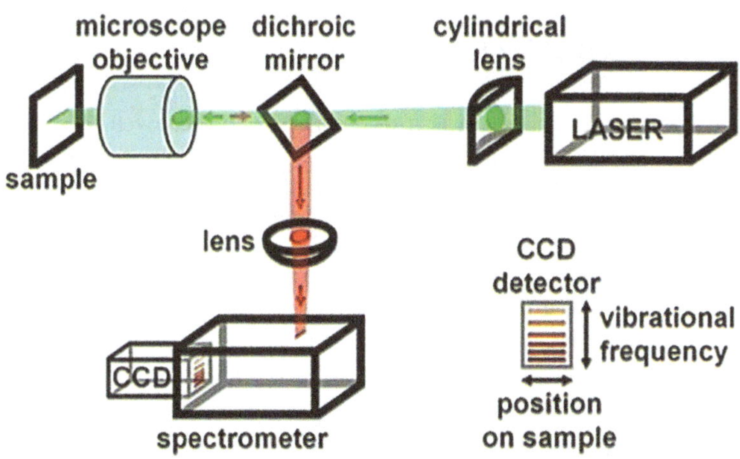

Fig. 12 Schematic diagram of Raman spectrometer (Downes and Elfick 2010)

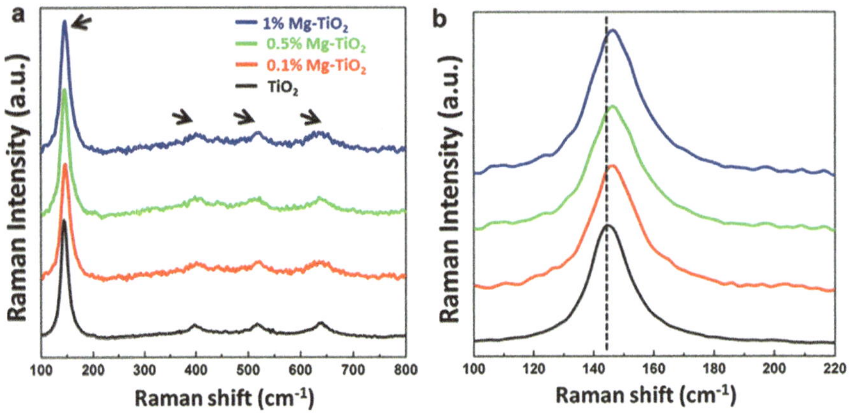

Fig. 13 Raman spectra of the pure and Mg-TiO_2 hollow spheres at lower Raman shift between **a** 100 and 800 cm^{-1}, **b** 100 and 220 cm^{-1} (Gao et al. 2017)

Smith and Dent 2005). Raman analysis is a powerful technique that is sensitive to the second phases and clusters of heterogeneous materials in matrix.

Figure 13 shows Raman spectra of pure anatase TiO_2 and Mg-doped TiO_2 samples. The characteristic bands common to anatase TiO_2 are observed at 144 (E_g), 399 (B_{1g}), 516 ($A_{1g} + B_{1g}$), and 639 cm^{-1} (E_g) bands (Gao et al. 2017). It is observed that the Raman bands for all the Mg–TiO_2 are almost at the same position as observed in pure anatase TiO_2, and there are no others bands existed in the Raman spectra of Mg–TiO_2 hollow sphere. But the broadening and small shift of Raman bands observed in Fig. 13(b) indicated that the Mg ions in the Mg–TiO_2 samples are homogenously substituted Ti in TiO_2 lattice.

1.7 Dynamic Light Scattering

Dynamic light scattering (DLS) is a technique widely deployed to characterize nanoparticles-in-dispersion that enable monitoring of dispersion effects such as effect of salt or pH on colloidal stability, impact of thermal or chemical stresses on protein denaturation, or size and distribution of nanoparticles (Bauer et al. 2016; Kaszuba et al. 2008; Moore et al. 2015). DLS is a non-invasive technique that requires comparatively low quantity of samples and provides reliable estimates of the quality of preparation rapidly. DLS has been applied to study protein RNA, homogeneity of proteins, and their complexes (Stetefeld et al. 2016). DLS measures diffusion of particles moving under Brownian motion and converts this to size and a size distribution using Stokes–Einstein relationship. In principle, when laser light from a DLS equipment passes through macromolecules, the incident light scatters in all directions, and scattering intensity is recorded by a detector. The monochromatic incident light undergoes Doppler broadening as the macromolecules are in continuous motion in solution (Hermodson 1996). The scattered light will either result in mutually destructive phases or cancel each other out or in mutually constructive phases to produce a detectable signal (Berne and Pecora 1977). The particle size and distribution are evaluated from the signal. The schematic diagram of the basic principle of DLS is shown in Fig. 14.

DLS measurement was employed to compare the particles size distribution of ZnO and TiO_2 nanoparticles under different sonication times (Fig. 15). Accordingly, particle size decreases with increased sonication time for both ZnO and TiO_2

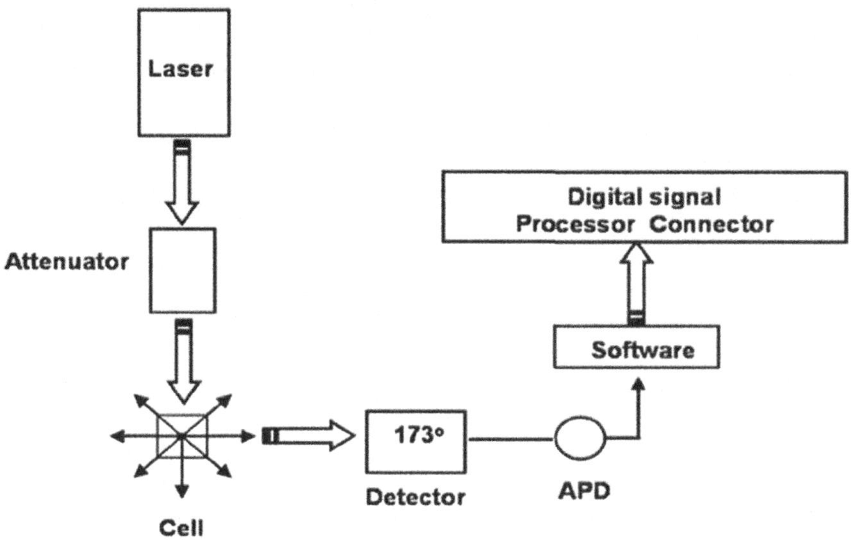

Fig. 14 Schematic diagram of basic principle of DLS

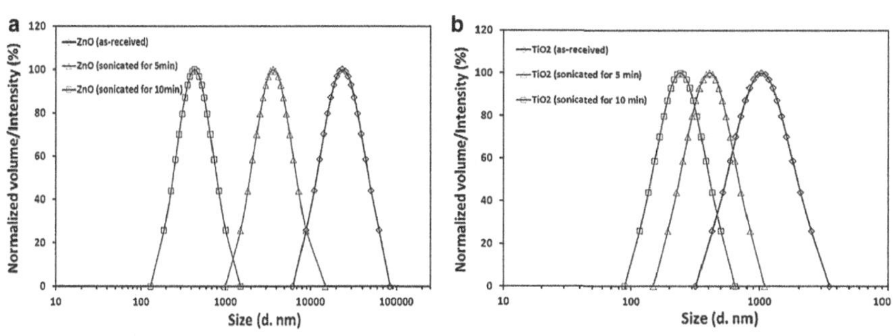

Fig. 15 Dynamic light scattering spectra of particle size and distribution for: **a** ZnO and, **b** TiO_2 (González et al. 2017)

nanoparticles. Within the sonication time range of 0–10 min, the smallest particle sizes of 242 ± 41 nm and 436 ± 80 nm for TiO_2 and ZnO, respectively, were obtained with 10 min of sonication indicating reduction of particle agglomeration (González et al. 2017).

1.8 X-ray Photoelectron Spectroscopy

X-ray photoelectron spectroscopy (XPS) is a surface analytical technique that measures the elemental composition, empirical formula, chemical state, and electronic state of elements that exist within the surface of material (Welker 2012). XPS is the most commonly used technique in areas of materials science, chemistry, nanomaterial, and chemical engineering to assess surface chemistry, bonding structure, and composition of surfaces and interfaces (Greczynski and Hultman 2020). XPS spectra are obtained by irradiating a material with a beam of X-rays while simultaneously measuring the kinetic energy and number of electrons escaping within the 10 nm of the top surfaces. The strength of this technique relies on the fact that the chemical environment of an atom has a pronounced effect on the assessed binding energies of core-level electrons (termed chemical shift) (Sokolowski et al. 1958). XPS can be deployed to analyze semiconductors, glasses, polymers, ceramics, catalysts, metal alloys, and bones among others.

XPS is based on photoelectric effect in which the binding energy of a core-level electron is overcome by the energy of an impinging soft X-ray photon, and the core-level electron is excited and ejected from the surface of the material. The kinetic energies of the ejected photoelectrons are measured by an electron spectrometer. By invoking conservation of energy, the biding energy E_B can be related to the kinetic energy E_k via the following equation (Lee and Flynn 2006; Einstein 1905; Rutherford and Andrade 1914):

$$E_B = hv - E_k - \varphi \tag{13}$$

where h is the planks constant, φ is the spectrometer work function, and υ is the light frequency.

In Fig. 16, an introduction of the light source over the sample generates the X-rays, resulting into photoelectrons ejection and photons annihilation. Electron optics is used as the collector of the photoelectrons. Energy in the electron energy analyzer will then be employed to sort the collected photoelectrons. The number of electrons per energy interval is then transduced to a current by an electron detector. Subsequently, the spectrum of suitable electronics converts and processes the photocurrent. The experiment is usually conducted under ultra high vacuum (UHV) conditions (Lee and Flynn 2006).

Figure 17 shows the results of XPS characterization for surface compositions and chemical states of pure ZnO and Ag doped ZnO films. It is observed that Zn 2p spectra

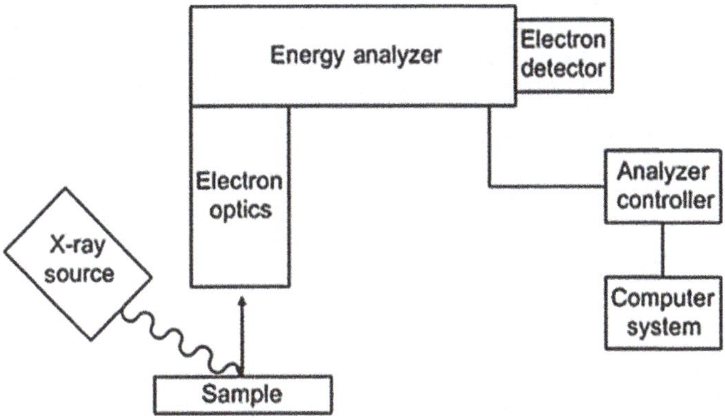

Fig. 16 Schematic diagram of an X-ray photoelectron spectrometer (Lee and Flynn 2006)

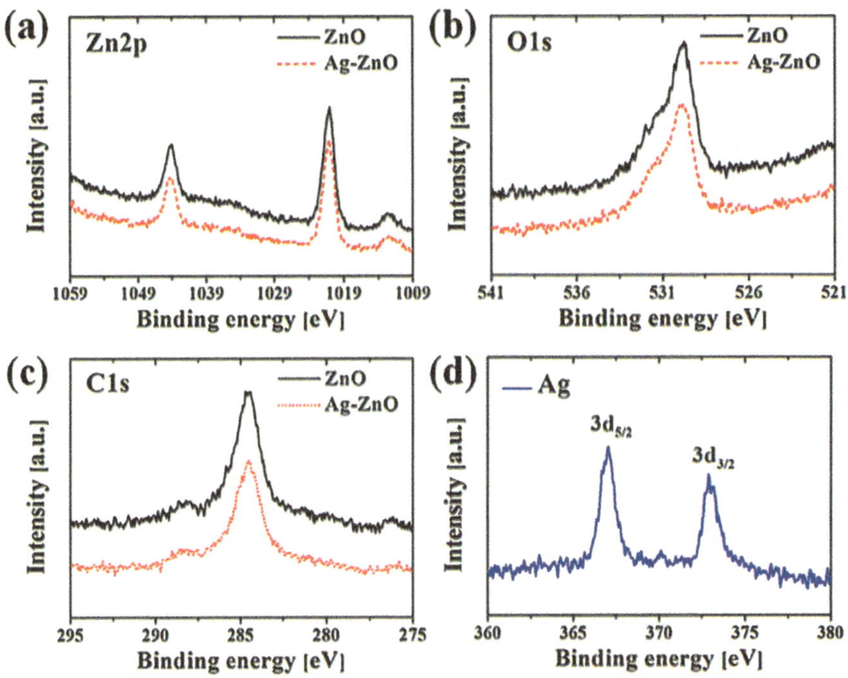

Fig. 17 XPS spectra of pure ZnO and Ag doped ZnO nanostructures (Patil et al. 2015)

for ZnO and Ag-ZnO samples display doublets peaks at 1044.3 and 1021.1 eV which is ascribed to the Zn 2p5/2 and Zn 2p3/2 lines, respectively (Fig. 17a). The difference between two bonding energies implies that Zn is in a +2 oxidation state. The O 1s spectra of ZnO and Ag-ZnO samples with a peak at a binding energy of 529.75 eV are shown in Fig. 17b. Figure 17c displays a peak at 285 eV for the C 1s spectra of the ZnO and Ag–ZnO samples which is ascribed to the presence of carbon on the surface of the films (Patil et al. 2015). The difference between the peaks at 372.9 (Ag 3d5/2) and 367 eV (3d3/2) implies the presence of metallic Ag and Ag–Zn–O ternary compounds, respectively, as shown in Fig. 17d.

2 Microscopy

Microscopy plays an important role in the field of nanotechnology as it is employed to generate images for the morphological studies of nanomaterials including surface roughness, topology, exact particles size, distribution, and thickness. Microscopy techniques can be categorized into three branches namely optical, electron, and scanning probe microscopy. Each of these can be further divided as presented in Table 2. For nanotechnology, the conventional optical microscope is not useful since the size ranges of nanomaterials are below the diffraction limit of visible light. This section describes various types of microscopy techniques, their modes of operation, and their applications in the field of nanotechnology.

2.1 Scanning Electron Microscopy

Since the advent of electron microscopy in 1932 following the works of Ernst Ruska which earned him a shared Nobel prize in Physics in 1986 (Nobel Prize 1986), advances in visualization of microscopic and submicroscopic details have been on the rise. In life sciences, sub-cellular structures and viruses are being studied with

Table 2 Techniques in microscopy

Techniques	Acronym	Applications
Scanning electron microscopy	SEM	Surface imaging and morphology
Field emission scanning electron microscopy	FESEM	Surface imaging and morphology; films thickness
Transmission electron microscopy	TEM	Particle size and distribution, morphology, crystal structure
High-resolution transmission electron microscopy	HRTEM	Atomic and crystal structure, lattice defects
Atomic force microscopy	AFM	Surface structure and imaging, topology and roughness
Scanning transmission electron microscopy	STEM	Surface structure and imaging
Magnetic force microscopy	MFM	Magnetic materials analysis
Scanning tunneling microscopy	STM	Surface structure and imaging

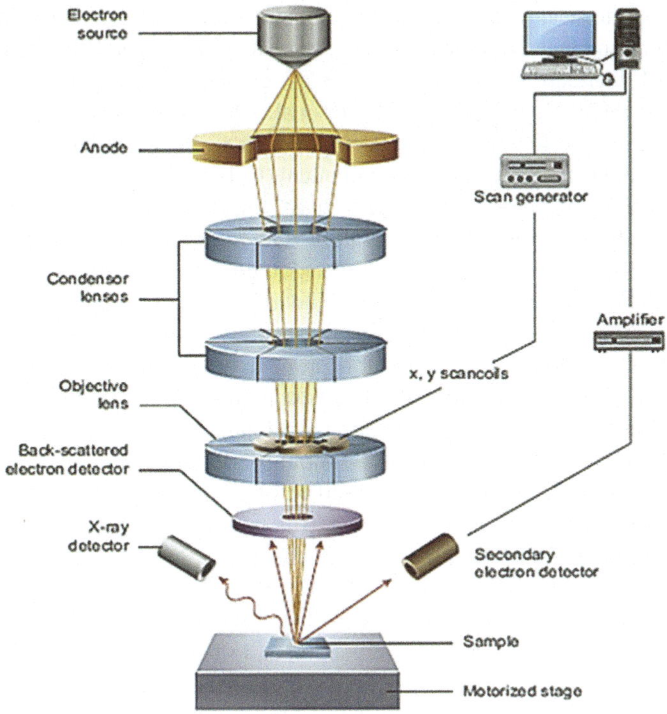

Fig. 18 Schematic diagram of the core components of an SEM microscope (Inkson 2016)

great accuracy. The scanning electron microscope (SEM) is one of the widely used microscopic methods in the field of nanoscience and nanotechnology. Figure 18 shows the core components of SEM consisting of electron source, electron lenses, sample stage, detectors for all signals of interest, and data output devices as well as infrastructural requirements such as power supply, vacuum system, cooling system, vibration-free floor, and room free of ambient magnetic and electric fields (Egerton 2005). The capacity of SEM is dependent on which detectors that it accommodate with most SEMs having at least a secondary electron detector.

The principle of SEM relies on a pseudo three-dimensional image which is developed point-by-point and line-by-line from secondary electrons (Bancroft and Gamble 2008). The emission of these electrons takes place as the specimen reacts with the incident electron beam that scans the specimen in a square raster pattern. Interactions occur between electrons and sample, and electrons and photons emitted from the sample. Signals in the form of secondary electrons (that produce SEM images showing texture, morphology, and topography of samples), backscattered electrons (that are used to determine crystal structures and orientations of minerals and for rapid-phase discrimination), and photons/characteristic X-rays (that are used for elemental analysis and continuum X-rays) are generated from the surface of the sample. SEM analysis is a non-destructive testing technique in which the X-rays generated by electron interactions with samples do not lead to volume loss of the sample indicating that the same materials can be analyzed repeatedly (Beane 2004).

For SEM, there are two basic types of electron sources that can be used for the formation of the electron beam (Goldstein et al. 1992). Conventional SEMs utilize thermionic emission sources that are equipped with tungsten, lanthanum hexaboride (LaB_6), or more recently with a cerium filament (CeBix). With thermionic emission, electrons become adequately energetic to leave the metal cathode after the wire or rod is heated beyond a specific temperature. The tungsten filament is the simplest in development and operation, less costly, and highly reliable, and its properties are comprehensible. However, the tungsten cathode has relatively low brightness, and the operating time is constrained. Meanwhile, the LaB_6 cathode has higher brightness and the operating lifetime is longer, but is prone to contamination, more expensive, and the processing is more difficult. The best is the CeBix filament which offers equal performance with better stabilization and longer life than the LaB_6 filament.

Careful preparation of specimen is essential for good electron microscopy analysis, as well as some additional time used to optimize specimen preparation and cleanliness can prevent days of wasted effort in conducting microscopy and image analysis (Echlin 1999). In other words, contaminated or dirty specimens result to the production of SEM images of the surface dirt. Best practices for the avoidance of contamination include specimen storage in clean environments (sealed boxes, desiccators, and vacuum packs), humidity reduction in desiccating media, and soaking of oil-contaminated specimen in solvents and ultrasonic agitation (Inkson 2016). After cleaning, non-conducting specimens often require ultrathin coating using carbon, gold, and platinum, which creates a path to conduct away accumulated electrostatic charge. Sample size is often limited to 10 cm in horizontal dimension and 40 mm in vertical dimensions. Samples must not be oxidized when placed in a vacuum of about 10^{-5} to 10^{-6} torr and should not outgas. Conventional SEM scans surfaces within approximately 1 cm to 5 microns while magnification ranges between 20X and 30,000X with a spatial resolution of 50–100 nm (Beane 2004; Egerton 2005). Asides surface scanning, SEM can also perform analyses of a chosen point across the surface of the sample for the purpose of qualitative or semi-quantitative determination of chemical compositions (using EDS), crystalline structure, and crystal orientations (using EBSD).

Most of the studies in nanotechnology research and related fields require extensive utilization of scanning electron microscope to quantify and qualify the surface structures of developed materials. Figure 19 presents the images of carbon nanotubes and CoPt multilayer nanowires (Kohl and Reimer 2008), silicon germanium (SiGe), nanocantilevers (Asafa et al. 2013a, b), and nanosized silicon (Si) (Durowoju et al. 2019). In these images, the structure of the materials, orientation, size, and distribution are well quantified.

Fig. 19 Morphology characterization using secondary electron contrast **a** carbon nanotubes on a holey carbon film and, **b** CoPt multilayer nanowires (Kohl and Reimer 2008), **c** nanocantilevers made from SiGe and, **d** SiC particles (Durowoju et al. 2019; Fang et al. 2016)

2.2 *Field Emission Scanning Electron Microscopy*

Field emission scanning electron microscope (FESEM) is an instrument which extracts various information from the sample surface but with greater resolution and a much higher energy range compared to SEM. The emitter type (i.e., thermionic emitter and field emitter) differentiates FESEM from SEM. The thermionic sources used by SEM are marred by drawbacks such as low brightness, thermal drift during operation, and evaporation of the cathode material. A field emission gun (FEG) also known as a cold cathode emitter has exhibited capacity to address the drawbacks of the thermionic emitter. With field emission (FE) sources of electrons, the electrons can leave the very sharp cathode tip due to an electron tunneling effect in a high electric field created at the tip of the cathode (Meerbeek et al. 2000).This FE source offers a significantly enhanced resolution due to its extremely focused high- and low-energy electron beams. Its brighter electron source and narrower electron beam enable scanning at low accelerating voltages (0.02–5 kV) thereby minimizes the charging effect on non-conductive specimens helping avoid damage to electron-beam-sensitive samples. Consequently, images can be acquired without any or with

Fig. 20 FESEM micrographs for **a** Al_2O_3 nanoparticles, **b** TiO_2 nanoparticles (Ali et al. 2016), **c**, **d** as-prepared Cu_2O product at lower and higher magnification (Sun et al. 2014)

limited conductive coating on sample surface. In addition, FESEM uses in-lens detectors which are optimized to work at high resolution and very low acceleration potential which is fundamental for getting the maximum performance from the equipment. FESEM has a long operating lifetime, but requires a much higher vacuum and is more easily contaminated. Therefore, it needs to be "flashed" very frequently.

Good examples of FE-SEM images are shown in Fig. 20. The morphology of Al_2O_3 and TiO_2 nanoparticles (Fig. 20 a, b) shows that the nanoparticles were fairly spherical in shape providing a very good rolling performance in the medium where they are utilized (Ali et al. 2016). The FESEM image of Cu_2O acquired at lower magnification (Fig. 20c) indicates uniform and monodisperse highly symmetric polyhedral microcrystals while a single particle observed at and higher magnification shows the cuboctahedral configuration of each particle (Sun et al. 2014).

2.3 Transmission Electron Microscopy

The transmission electron microscope (TEM) is an instrument used for analyzing internal microstructure of specimen and evaluation of nanostructures such as particles, fibers, thin films, and imaging of atoms (Inkson 2016; Tsang et al. 2005; Carter and Williams 2016; Spence et al. 2006; Goodhew et al. 2001). Figure 21 presents the core components of the TEM microscope which includes the electron gun, electrostatic lenses, and a transmitted electron detection system. The electron gun in a TEM has the capacity to accelerate electrons using 80–300 kV accelerating voltage to give

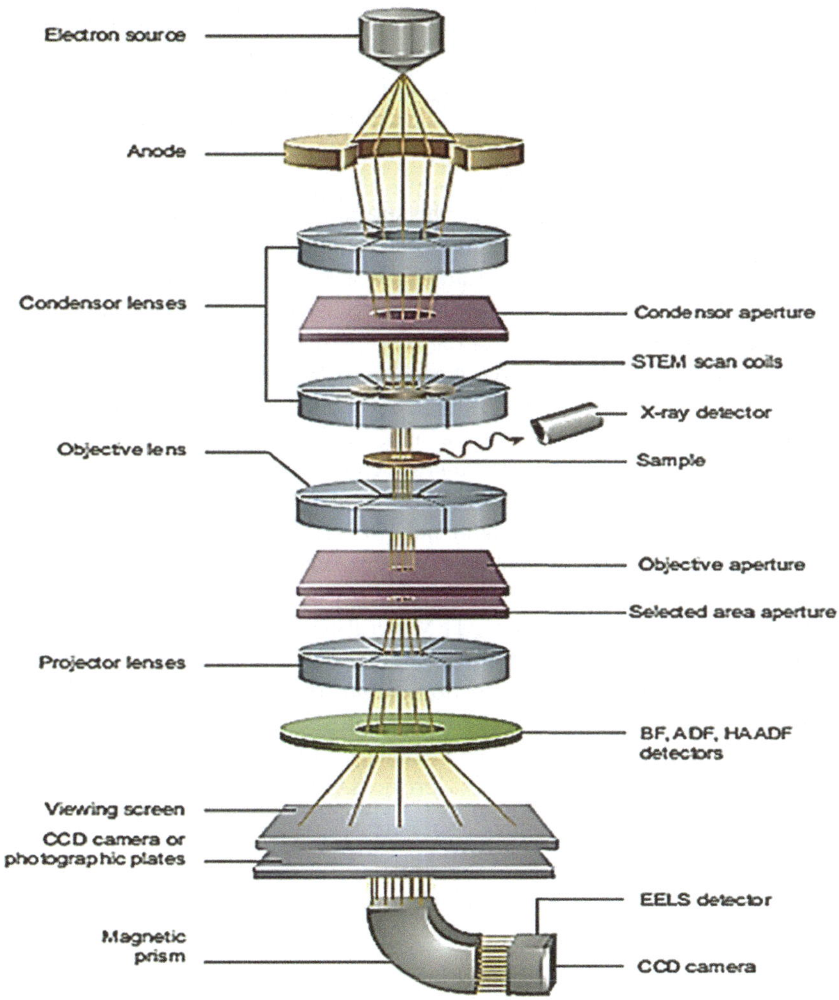

Fig. 21 Schematic diagram of the core components of TEM microscope (Inkson 2016)

them sufficient energy capable of passing through up to 1 mm of material (Kohl and Reimer 2008). The operation of TEM requires the irradiation of the whole specimen, which is thin enough to transmit at least 50% of the incident electrons (Bancroft and Gamble 2008). The transmitted electrons are then directed by a series of lenses resulting into a magnified, two-dimensional image of the sample. Meanwhile, every TEM specimens must have areas of electron transparent material for transmission of electrons. Preparation of TEM specimen is a diverse and highly complex task. Several investigations have generated a series of methods for creating thin material from a thick one (Ayache et al. 2010). Materials having heavy elements composition (high Z-number) may require $\leq$100 nm thickness for transmission electrons. Hence, non-destructive characterization using TEM is only feasible for specimens including nanoparticles, nanofibers, and light materials such as molecules and nanocarbons that are transparent to electron in their natural state. If some degree of destructive specimen preparation is acceptable, then as it is in the case of the TEM in which a wide range of thinning techniques such as electrochemical dissolution, chemical etching, ion sputtering, mechanical abrasion, and controlled fracture can be applied (Ayache et al. 2010; Echlin 1999).

TEM has been used to investigate the size and shape of nanomaterials which are beyond the capability of SEM. For instance, the shape and size of nanoparticles such as gold, silver, gold–silver, and many more as well as those of nanotubes, nanofibers, and several thin films have been investigated. Figure 22a, b presents the TEM images of silver nanoparticles (AgNPs) produced from pod extract of *Cola nitida* (Lateef et al. 2016) and silver–gold (Ag–AuNPs) alloy nanoparticles biosynthesized using cell extracted from *B. safensis* (Ojo et al. 2016). The particles are considered spherical in shape with sizes range within 12–80 nm for AgNPs and 13–80 nm for Ag–AuNPs. The hallow tubular structure of multiwalled carbon nanotubes (MWCNTs) with an average diameter of 30–50 nm is shown in Fig. 22c (Palanisamy et al. 2014). Figure 22d shows ultrathin poly-$Si_{11}Ge_{89}$ film of columnar grain structure. The direction of the grain boundary (GB) explains the growth orientation (Asafa et al. 2014).

In addition to its general function, TEM can also utilize the selected area electron diffraction (SAED) mode to determine crystalline nature of the samples. In this case, amorphous phase appears as diffused rings while crystalline phase is seen as bright spots. Poly-nanocrystallines are viewed as small spots making up rings. The spot arises from Bragg reflection from an individual crystallite. The crystalline phase of each sample is obtained from its characteristic SAED pattern.

2.4 High-Resolution Transmission Electron Microscopy

The high-resolution transmission electron microscope (HRTEM), which utilizes the high-resolution imaging mode of TEM, is an instrument used to directly image atomic structure of a specimen (Spence et al. 2006). HRTEM has capacity for atomic scale

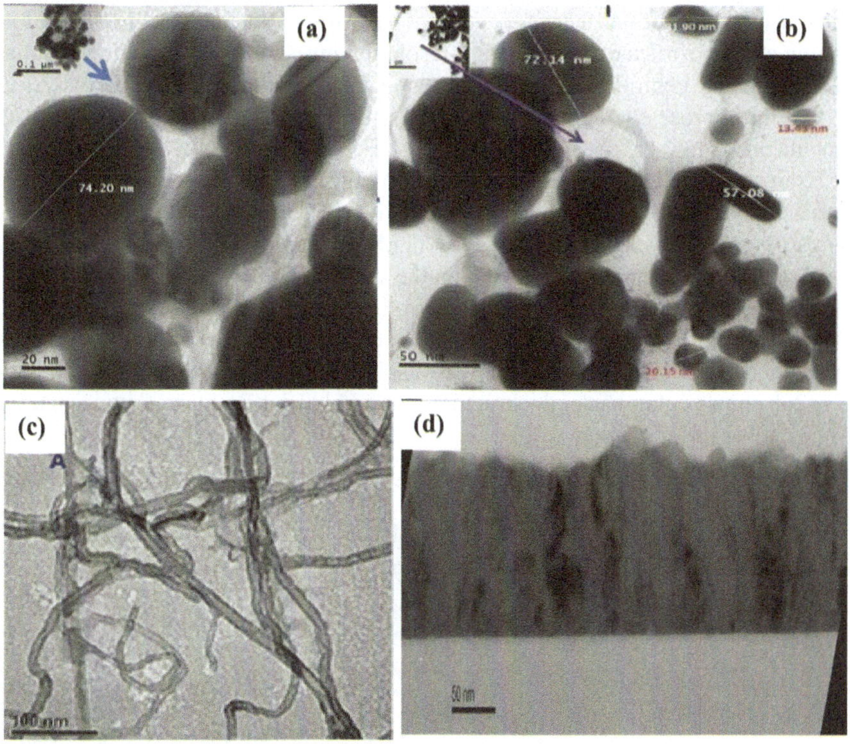

Fig. 22 TEM images of **a** silver nanoparticles produced from pod extract of *Cola nitida* (Lateef et al. 2016), **b** silver–gold alloy nanoparticles biosynthesized using cell-free extract of *Bacillus safensis* (Ojo et al. 2016), **c** multiwalled carbon nanotubes (MWCNTs) (Palanisamy et al. 2014), **d** poly-$Si_{11}Ge_{89}$ film showing grain size, growth orientation, and grain boundary (GB) (Asafa et al. 2014)

evaluation of materials including semiconductors, metals, nanoparticles, and sp^2-bonded carbon like graphene and carbon nanotubes. There are two basic stages for the process of image formation in HRTEM. The incident electrons experience interactions with atoms of the specimen, which involves both the elastic and inelastic scattering process. The wavelength of the electron exiting the surface of the sample is then transmitted via the objective lens and later the magnifying lenses of the electron microscope to form the final enlarged image. For ultrathin specimens having thickness of $\leq$100 nm, the elastic scattering dominates over inelastic scattering. The electron waves interact with the crystal lattice diffract and form complex interference patterns which can be observed at magnification $\geq$400 k (Buseck et al. 1989; Spence et al. 2006). With the use of the right operating conditions and sample preparation, high-resolution images are easily interpretable with respect to projections of individual atomic-column positions. Since irregularities at the nanoscale level have a significant effect on bulk behavior, the HRTEM has shown to be very powerful and essential tool for the characterization of nanostructured materials. Further information can be extracted from the high-resolution investigations, including peculiar insights into the controlling effect of structural discontinuities on a range of physical and chemical processes such as phase transformations, oxidation reactions, epitaxial growth, and catalysis.

Examples of HRTEM images are shown in Fig. 23, where atomic spacing, interplanar spacing, and grain orientation are obvious. Figure 23a is a sample of time-series transmission electron microscopy (TEM) image depicting the formation of agglomerated Sn nanoparticles (NPs) on the surface of SnO_2 nanotubes upon lithiation (Cheong et al. 2017). As time passes, individual as-formed SnNPs combine with each other to form into SnNPs agglomerates. The lattice spacing and grain orientation of both Sn and SnO_2 are discernable. The ultrathin poly-$Si_{11}Ge_{89}$ film of Fig. 23b shows columnar grain structure along {111} lattice plane (lattice spacing

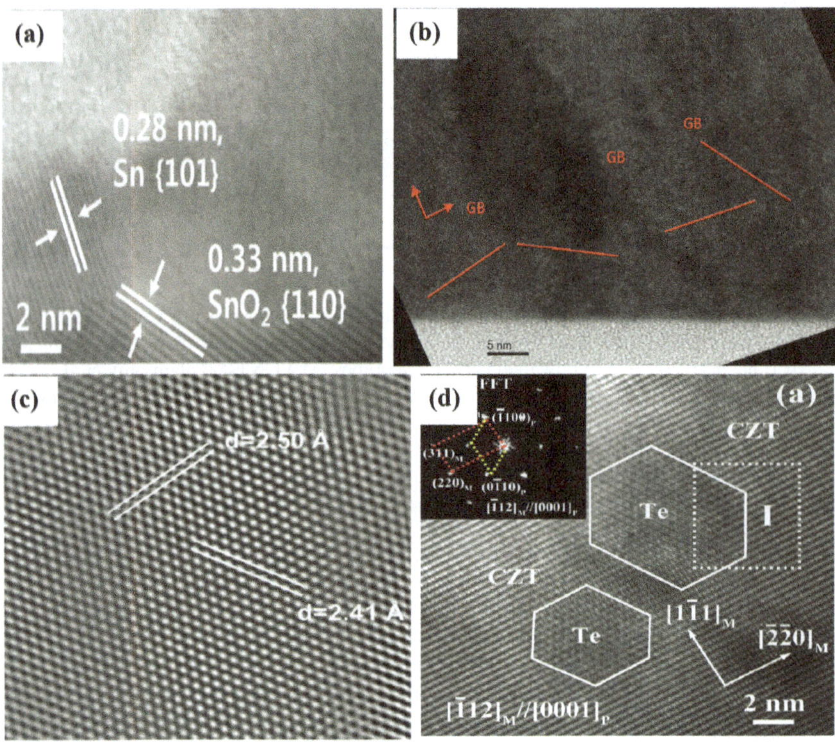

Fig. 23 High-resolution TEM image showing **a** the formation of agglomerated Sn nanoparticles (NPs) on the surface of SnO_2 nanotubes upon lithiation (Cheong et al. 2017), **b** ultrathin poly-$Si_{11}Ge_{89}$ film of columnar grain structure along (Asafa et al. 2014), **c** silver nanoparticle that show interplanar distances consistent with a crystalline structure of type 4H hexagonal after reconstruction with digital micrograph (Virgen-Ortiz et al. 2015), **d** the precipitates along 112 M zone axis (Zhou et al. 2018)

of 0.323 nm) and grain growth in <110> direction (Asafa et al. 2014). The direction of the grain boundary (GB) further explains the growth direction. Figure 23c is the HRTEM image of silver nanoparticle obtained via reconstruction of the crystalline structure using digital micrograph (Virgen-Ortiz et al. 2015). The interplanar distances are in line with a crystalline structure of type 4H hexagonal (index Miller of the electron diffraction is shown in the inset). High-resolution TEM (HRTEM) images of Te precipitates along 112 M zone axis are shown in Fig. 23d (Zhou et al. 2018).

2.5 *Scanning Transmission Electron Microscopy*

Scanning transmission electron microscopy (STEM) uses the combination of the principles for transmission electron microscopy and scanning electron microscopy and can be conducted on either type of instrument. Some TEM equipment have installed scan coils for scanning electron beam across the specimen (Keyso et al. 1998; Nellist 2011). Similar to the TEM, STEM can be performed on very thin specimens and concentrates primarily at beam electrons transmitted by the specimen. One of the primary advantages of STEM over TEM is that it utilizes other signals that cannot be effectively utilized in TEM which include secondary electrons, scattered beam electrons, characteristic X-rays, and electron energy loss. Unlike TEM, the electron beam in STEM is focused on a small spot (having a spot size range of 0.05–0.2 nm) which is then scanned over the specimen in a raster illumination system constructed in such a way that the specimen is illuminated at each point with the beam parallel to the optical axis. The raster pattern of the beam across the specimen makes STEM suitable for the imaging of biological specimen in the high contrast of annular dark-field images, which can allow imaging of biological specimens without the need for staining. Advanced STEM equipment particularly those fitted with extra lenses to correct spherical aberration in the electron beam (STEM probe correction) coupled with ultra-stable scan electronics means that the electron beam spot can be sequentially positioned with accuracy down to different atom columns (Nellist 2011).

An example of STEM image is that of nanostructured copper-supported mesoporous carbon FDU-15 shown in Fig. 24 (Sahin et al. 2018). The STEM images (Fig. 24a, b) together with the bright-field STEM images (Fig. 24b, d) showed how Cu nanoparticles were distributed in the FDU-15 support. Further analysis also showed that some nanoparticles are crossed by holes (Sahin et al. 2018).

2.6 *Atomic Force Microscopy*

Atomic force microscopy (AFM) is a kind of scanning probe microscopy having resolution on the order of fractions of a nanometer (Cappella and Dietler 1999).

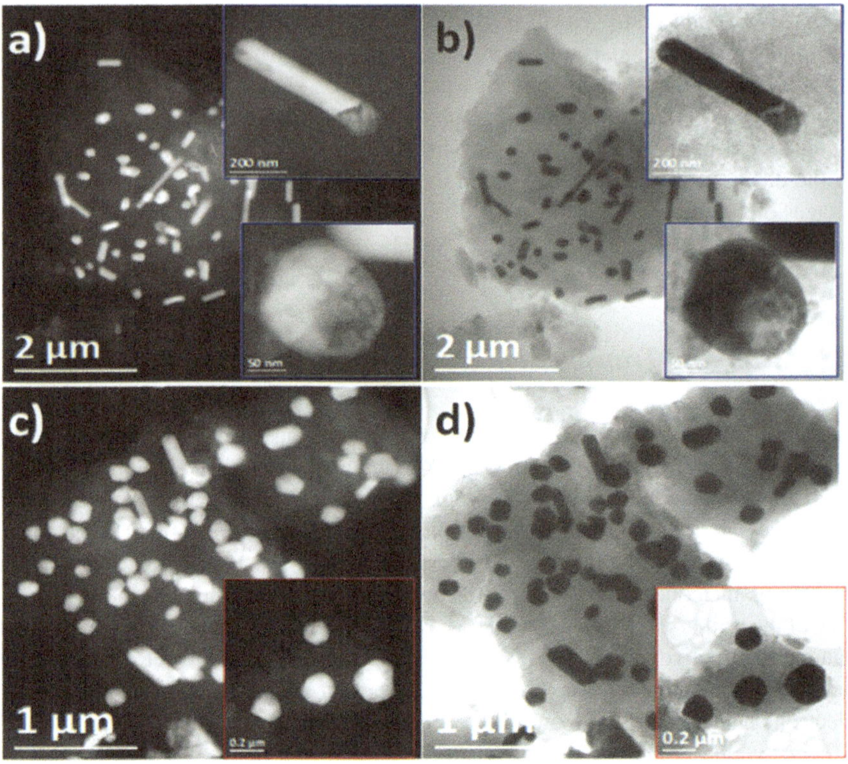

Fig. 24 Scanning transmission electron microscopy (STEM) images in high-angle annular dark and bright fields for **a**, **b** Cu/FDU-15-350, **c**, **d** Cu/FDU-15-500 samples (Sahin et al. 2018)

Ever since its invention, the applications of the AFM have broadened and cover a wide range of materials and biomedical research areas. Figure 25a shows the core components of the AFM instrument. The probe has a sharp tip with a radius within the range of 2–20 nm integrated with a cantilever (Fig. 25b) (Ishida et al. 2016). The probe is scanned over the sample surface utilizing a piezoelectric scanner with the capacity to expand or shrink in response to the voltage applied. The reflected laser beam behind the cantilever is detected using a four-segment photodiode detector. By utilizing the position of the incident beam at the detector, evaluation can then be made using the vertical deflections and lateral distortions of the cantilever. The operator evaluates the signals produced by the photodetector for the adjustment of the z-displacement of the piezo scanner in order to maintain a constant feedback control parameter. Hence, different images of the sample surface are provided by the mapping of the resulting z-piezo movements in the x and y dimensions.

AFM has three main capacities: (i) force measurement, (ii) topographic imaging, and (iii) manipulation. It can be deployed to measure the forces between the probe and the sample on the basis of their separating distance. This can further be applied to conduct force spectroscopy for measuring stiffness (Young's modulus) of samples.

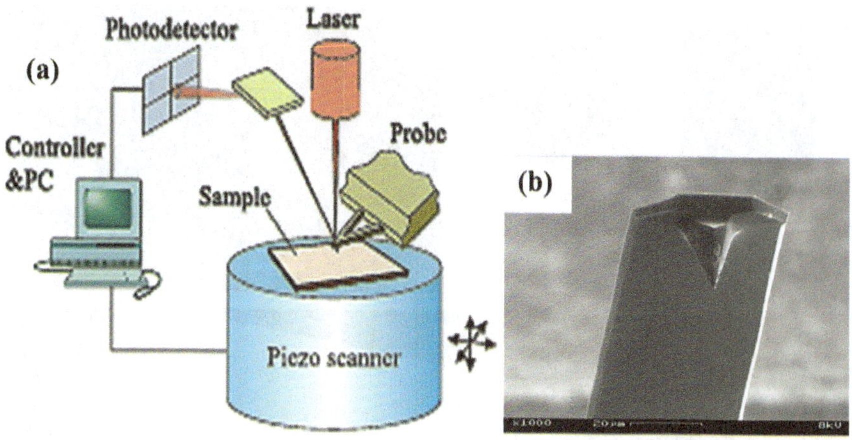

Fig. 25 **a** Schematic diagram of the core components of AFM microscope, **b** AFM cantilever with magnified tip (Ishida et al. 2016)

For imaging, the probe's reaction to the force imposition by the sample can be utilized for 3D image formation (topography) of a sample surface at a high resolution. This is obtained through raster scanning the position of the sample in line with the tip and recording the height of the probe corresponding to a constant probe-sample interaction. For manipulation, the forces between tip and sample can as well be employed for a controlled change in the properties of the sample such as atomic manipulation, scanning probe lithography, and local stimulation of cells.

Meanwhile, the primary advantage of AFM is its non-reliance on complicated and special specimen preparation including chemical fixation, dehydration and drying, or conductive metal coating, as is applicable for TEM and SEM analysis. AFM measurements can be conducted without using a conductive coating and under normal atmospheric pressure conditions or under water in fluid cells. This advantage gives room for easy application of AFM for analyzing biological materials (Ishida and Craig 2018; Marinello et al. 2019; Stylianou et al. 2019) and has in recent times resulted in its application in dental materials science among other (Fang et al. 2016; Fujii and Okajima 2019; Meredith et al. 2016). Figure 26a presents the topography of silver nanoparticles deposited on glass in both 2D and 3D (Mocanu et al. 2013). The profile of the cross-section along the arrow in panel which is also shown indicates the height and the diameter of the nanoparticles. Images of peak force error and height of *Escherichia coli* is shown in Fig. 26b (Huang et al. 2015). The shape of the bacteria can be deduced from the images. Figure 26c shows the cauliflower morphology of poly-SiGe ultrathin film deposited on silicon waver (Asafa et al. 2014) while the nanocantilevers fabricated from ~98 nm thick SiGe film is shown in Fig. 26d (Asafa et al. 2014). In all these images, the shape and size of the nanomaterials can be seen clearly.

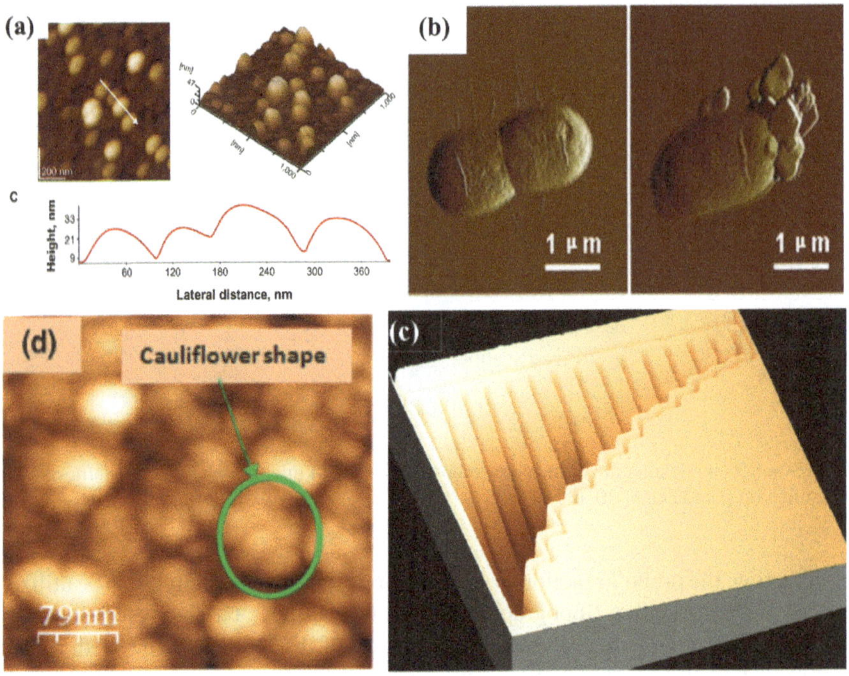

Fig. 26 AFM images of **a** 2D, 3D topography of silver nanoparticles on glass, profile of the cross section along the arrow in panel is shown (Mocanu et al. 2013), **b** peak force error and height images of *E. coli* (Huang et al. 2015), **c** poly-SiGe ultrathin films shown cauliflower morphology (Asafa et al. 2014), **d** 98 nm thick SiGe nanocantilevers (Asafa et al. 2014)

2.7 *Scanning Tunneling Microscopy*

Similar to the AFM, the scanning tunneling microscopy (STM) belongs to the family of scanning probe microscopy techniques that is used for imaging surfaces at the atomic level. Unlike AFM that uses conductive cantilever, STM uses sharpened and conducting wire. This implies that both the sample and the tip must be conductors or semiconductors, and so cannot work on insulating surface. The operation principle of STM is built on the concept of quantum tunneling. A sharp metal tip is fitted to the upper part of a piezo-drive for controlling the height of the tip above a surface (Jia et al. 1986). As soon as the tip gets to the sample surface by about 10 Å, electrons then tunnel through the vacuum barrier or simply the space placed between tip and sample. With the application of bias voltage on the sample, a tunneling current is measured through the tip that is highly sensitive to the distance between the tip and the surface. Meanwhile, two other pizeodrives are employed in scanning the tip in two lateral dimensions. A feedback controller is used for adjusting the height of the tip to keep constant the tunneling current. In the course of the tip scanning on the sample surface, the height of the tip (the supplied voltage to the pizeodrive) is recorded as an STM image representing the surface topography. The STM has capacity for three functions which are surface imaging, tunneling spectroscopy, and tip manipulation.

STM has been deployed in numerous research studies such as semiconductor surfaces where the atomic resolution and unit cells become visible. The STM images of Si(111)7 × 7 (Fig. 27a, b) shown 12 protrusions in each unit cell while the image acquired at negative biased (−0.5 V) indicated non-equivalence as regards the adatoms in the two halves of a unit cell (Jia et al. 1986). The image of $\mathrm{Te}/\mathrm{Si}(111)-\left(2\sqrt{3}\times 2\sqrt{3}\right)R30^{o}$ STM reconstruction (Fig. 27c) (Lupke et al. 2018) shows atomically flat terraces with vacancy defects and occasional adatom on top. The black diamond dots indicate the threefold-symmetric unit cell with a lattice constant of 1.33 nm. In Fig. 27d, the atomic resolution STM image clearly resolves individual defects (dark triangles) present in a high native concentration in mechanically exfoliated MoS_2 single layer of on Au(111) substrate(Vancsó et al. 2016).

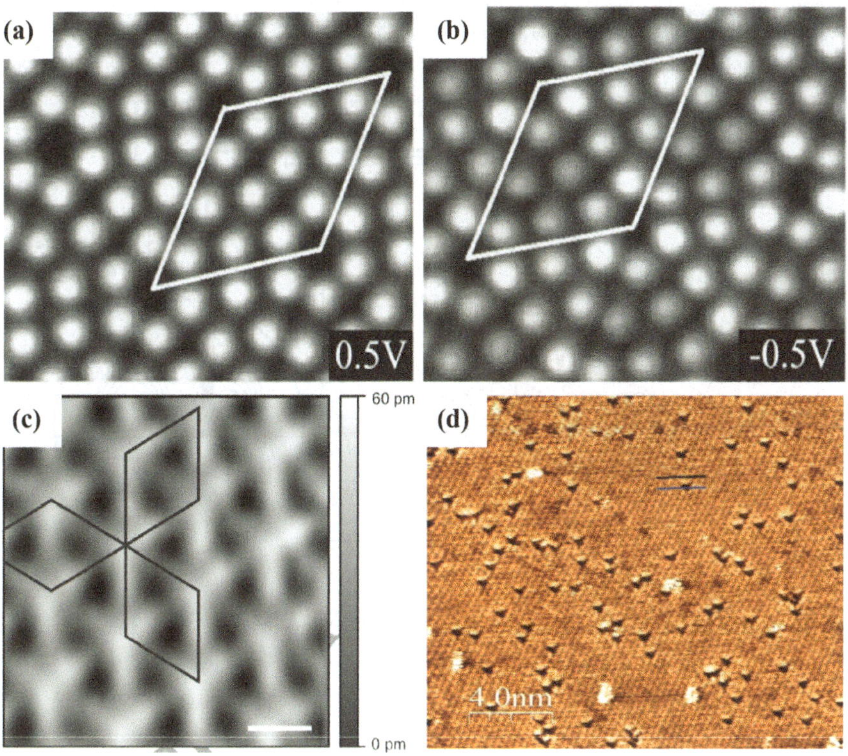

Fig. 27 Atomically resolved STM images of **a**, **b** Si(111)7 × 7 surface at bias voltage of + 0.5 V and −0.5 V with a unit cell of size 2.7 nm × 2.7 nm outline (Jia et al. 1986), **c** $\frac{\mathrm{Te}}{\mathrm{Si}(111)}-\left(2\sqrt{3}\times 2\sqrt{3}\right)R30^{\circ}$ reconstruction (Lupke et al. 2018), **d** native point defects in MoS_2 (Vancsó et al. 2016)

2.8 Magnetic Force Microscopy

The principle of magnetic force microscopy (MFM) is quite similar to that of the AFM. Figure 28 shows the core components of the MFM instrument. The technique is utilized for imaging magnetization patterns with high resolution and minimal sample preparation (Rugar et al. 1990). MFM employs a sharp magnetic tip attached to a flexible cantilever which is positioned near the sample surface (usually within 10–500 nm) and interacts with the stray field that emanates from the sample. The image is produced through scanning of the tip laterally with respect to the sample and carrying out force measurement (or force gradient) as a function of position. Major scan areas are within the range of 1–200 μm, with imaging times on the order of 5–30 min (Abelmann et al. 2005). The interaction strength is measured by monitoring the cantilever motion using a high-sensitivity sensor, such as an optical interferometer or tunneling sensor. MFM has successfully been utilized in nanotechnology application areas including analysis of nanowires for utility as promising magnetic composites and evaluation of magnetotactic bacterium with cell-mineralized magnetic nanoparticles covered by lipid-protein membrane (Hendrych et al. 2007).

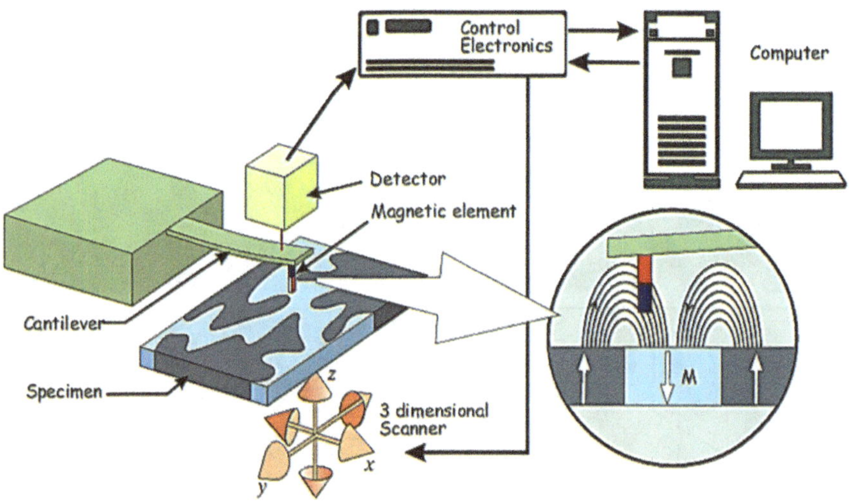

Fig. 28 Schematic diagram of the core components of MFM microscope (Abelmann et al. 2005)

3 Other Characterization Techniques

3.1 Mechanical Characterization

Mechanical characterization refers to testing performed to precisely measure the mechanical properties of materials. For nanotechnology applications, it is essential to measure the exact mechanical properties as a means of evaluating the reliability of products developed with nanomaterials, as these properties are closely related with their reliability metrics, such as the service life of the product. There are several tests in existence, many of which are standardized and vary for the material to be tested and the intended area of application. In this section, the testing procedures for four different mechanical characterization techniques including Young's modulus, abrasive wear measurement, nanoindentation, and microhardness are presented.

3.1.1 Young's Modulus or Modulus of Elasticity

Young's modulus is a measure of material stiffness using the rate of change of strain as a function of stress. Generally, Young's modulus can be obtained from the slope of the tensile stress and tensile strain or the compressive stress and compressive strain retrieved during tension/compression tests conducted using a universal testing machine (Asafa et al. 2015). However, at nanoscale, this approach is not feasible because sample of interest is either a thin film deposited or a substrate or nanoparticles that cannot be mounted on tensile machine. Several studies have shown that the properties of nanomaterials differ from those of the bulk due to enhanced surface-to-volume ratio (Asafa et al. 2014; Namba 2008; Reiss et al. 1991). Also, these properties are influenced when the structural dimensions are scaled down to a length scale that is comparable to the electron mean free paths (Asafa et al. 2014; Namba 2008; Reiss et al. 1991). In this case, predictions based on classical theories such as Euler–Bernoulli theory and Hooke's law become invalid as the results differ from ones.

The use of surface acoustic wave (SAW) (Schneider et al. 2000) and nanoindentation techniques (Oliver and Pharr 1992) are considered accurate and reliable for characterizing Young's moduli of ultrathin films. In SAW experiment, a nitrogen laser with pulse duration of 3 ns and a wavelength of 337 nm is used to generate wide band surface wave pulses. Example of such a step is shown in Fig. 29a (Schneider et al. 1998). These surface acoustic waveforms are then detected at different distance across the sample. The Fourier transform of the waveforms results into the frequency-dependent phase from where the phase velocity and modulus are obtained via dispersion equation (Asafa et al. 2014; Schneider et al. 1998). The values of the Young's modulus obtained by SAW have been established to be accurate and close to those obtained by membrane deflection method (Fig. 29b) for metallographically polished and annealed steel 42CrMo4 (Schneider et al. 1998). Another good example of SAW technique is shown in Fig. 29c where evolutionary study was conducted on influenced

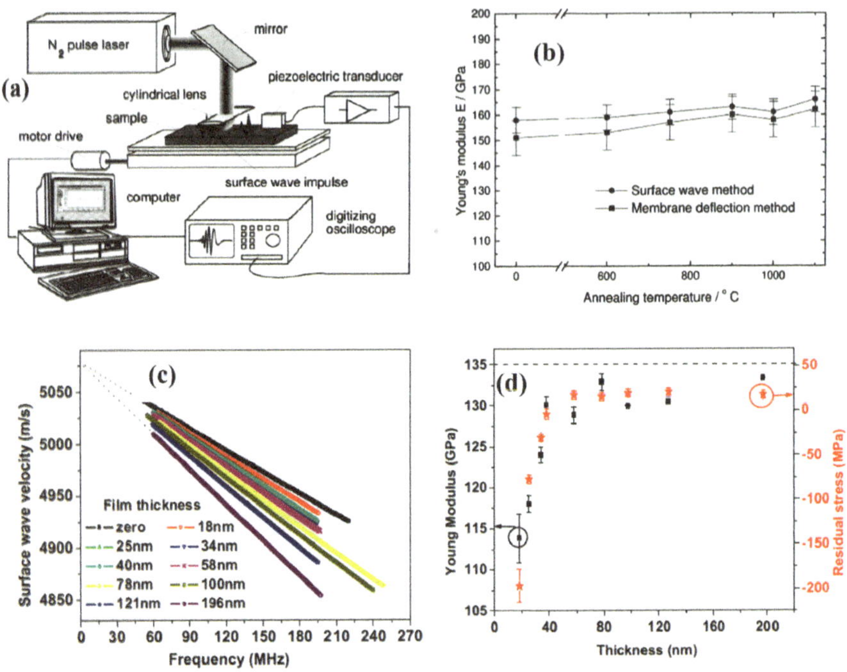

Fig. 29 **a** Schematic of experimental setup for laser acoustic wave (LAW) technique, **b** correlation of the measurement of the Young's modulus measured with LAWave and membrane deflection method (Schneider et al. 1998), **c** the dispersion curves and, **d** Young's modulus and residual stress versus film thickness for poly-SiGe films. The dash line is the theoretical Young's modulus for poly-$Si_{11}Ge_{89}$ films (Asafa et al. 2014)

by size effect on the Young's modulus of poly-SiGe film. According to Fig. 29c, the changes in the slopes of the dispersion curves indicate slight changes in the Young's moduli as film thickens. The Young's modulus is 113 GPa when the film is 18 nm thick, which increases gradually to 130 GPa at thickness of 40 nm (Fig. 29d). For thickness greater than 40 nm, the moduli are fairly constant but slightly lower than the theoretical bulk value of 135 GPa.

3.1.2 Microhardness

Microhardness testing of a material is the measurement of its resistance to permanent indentation of microscopic areas under static and dynamic load. Microhardness testing is conducted using applied load of 1 N or less on the indenter resulting in indentation depth of <70–100 μm. The Knoop and Vickers microhardness testing are the common techniques employed for measuring the microhardness value of materials. For the Vickers microhardness testing, a well-polished, pointed, square-based pyramidal diamond is utilized as the indenter, and HV is calculated by using Equation:

$$\mathrm{HV} = \frac{KP}{l^2} \tag{14}$$

where K is 1.854, P is the applied load (Kg), and l is the average length of the diagonal (mm).

In contrast, the Knoop microhardness testing employs a well-polished, rhombic-based pyramidal diamond as the indenter. This indenter produces a diamond-shaped indentation where the ratio of long to short diagonals is 7:1. Contrary to the Vickers microhardness test, the Knoop hardness number (HK) can be evaluated by dividing the applied load with the projected area of the indentation. Hence, HK is measured using the same equation above, where L is the length of the longer diagonal and the constant K is equal to 14.229.

3.1.3 Nanoindentation

The hardness property of materials at nanoscales has been of great interest in recent times due to the rapidly evolving field of nanotechnology. For this reason, the emergence of nanoindentation technique, which is an extension of microhardness test, has taken the place of conventional hardness measurements to meet the requirements for the investigation of new materials. Nanoindentation technique, otherwise referred to as depth-sensing indentation technique, was first proposed and developed by Oliver and Pharr (1992). This technology relies on the theory of contactelasto-mechanics. Based on the load–displacement data retrieved from the tests, both the Young's modulus and hardness can be obtained from the slope of the initial portion of the unloading curve as well as the ratio of the peak load to the projected contact area of the indent, respectively. Meanwhile, not only hardness and elastic modulus can be evaluated from the loading–unloading curve, but also information such as viscoelasticity, creep, fracture toughness, strain-hardening effect, residual stress, phase transition, and dislocation movement can be obtained (Qian and Zhao 2018).

A typical MTS Nanoindenter XP and its schematic diagram are shown in Fig. 30a and b, respectively. Details of the operating principle is available elsewhere (Qian and Zhao 2018). Figure 30c is an example of Young's moduli and hardness obtained from nanoindentation experiments. Based on the load-indentation (Fig. 30c) and Oliver and Pharr (1992) approach, the modulus-indentation (Fig. 30d) and the hardness-indentation (Fig. 30e) curves for ten different positions on each sample indicate that the Young's modulus differs slightly across the films (Asafa et al. 2013a, b). The elastic modulus as a function of indentation depth for nanoporous zeolite thin films is also in Fig. 30f (Chow et al. 2015).

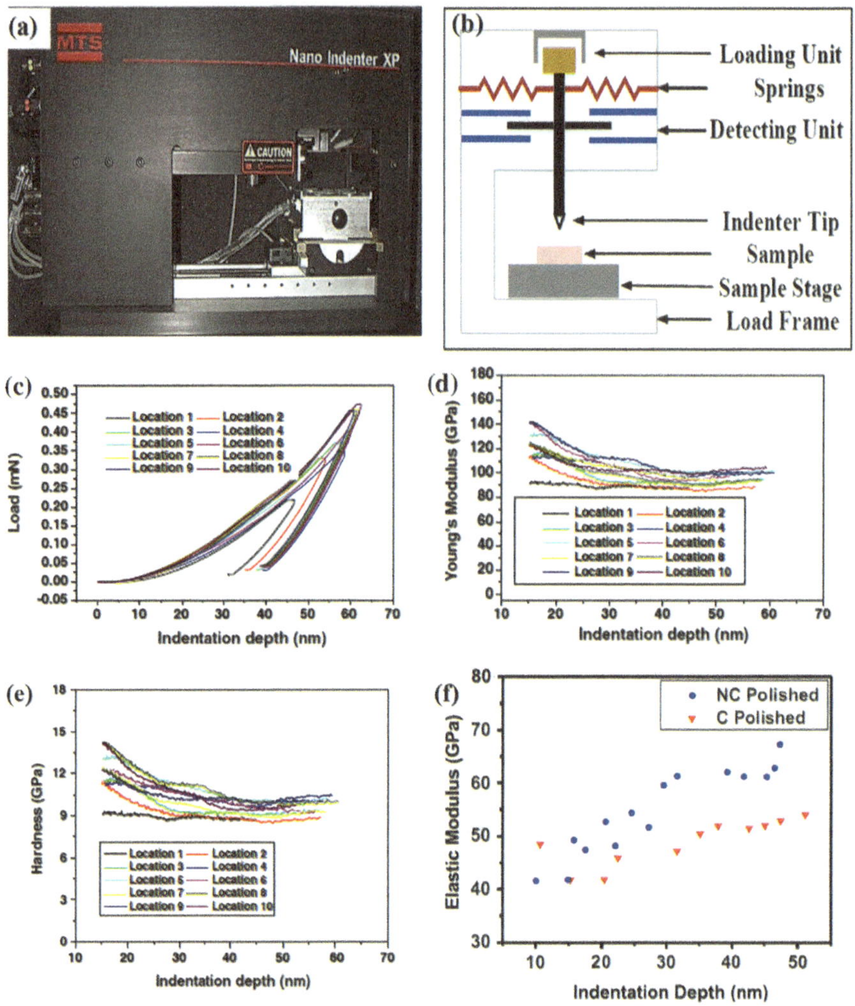

Fig. 30 **a** Nanoindenter MTS-XP, **b** schematic diagram of a nanoindenter (Qian and Zhao 2018), **c** load-indention curves for ~100 nm thick poly-SiGe film, **d** Young's modulus against indentation depth, **e** hardness versus indentation depth (Asafa et al. 2013a, b) and, **f** elastic modulus of nanoporous zeolite thin films (Chow et al. 2015)

3.1.4 Abrasive Wear Measurement

Abrasion wear is caused by the displacement of materials due to the presence of hard particles between or attached to one or both of two surfaces in relative motion. The degree of abrasive wear can be detected by the way which the particles traverse the worn surfaces, i.e., the particles may roll and/or slide over the surface. Hence, two major modes are popular which include two-body and three-body abrasion wear. The former involves material removal by abrasive particles that are held fixed while being moved across a surface. In the case of the latter, abrasive particles are embedded between two surfaces that may rotate and slide as they get in touch with the wearing surface. Some of the existing tests for abrasive wear measurement include but not limited to, pin-on-disk wear test, pin-on-drum abrasive wear test, repeated impact wear test, and taber test (Gobind and Parshad 2015). Abrasive wear measurement can be grouped into four categories namely (a) mass loss measurement which includes weight difference between original specimen and specimen after wear exposure, (b) linear measurement which includes one-dimensional changes in the geometry of interacting tribo-elements perpendicular to their common contact area, (c) area measurement which includes two-dimensional changes of cross sections of interacting tribo-elements perpendicular to their common contact area, and (d) volume measurement of wear which includes three-dimensional changes of geometric regions of interacting tribo-elements adjacent to their common contact area (Ruff1992). Other common wear parameters in use include, wear-time ratio (i.e., wear velocity), wear rate (i.e., wear volume per unit of sliding distance), and wear coefficient, which is expressed in the following equation:

$$K = \frac{W}{F \times S} \tag{15}$$

where W wear volume (in m^3), F is the applied load (in N), and S is the sliding distance (in m).

Abrasive wear measurement has been used to quantify surface resistance to plastic deformation especially when such a surface is modified with nanomaterials (Ali et al. 2016; Gualco et al. 2016; Asafa et al. 2020a, b). Figure 31 shows that the wear surface of nanostructured iron-based alloy deposited by flux cored arc welding process with (a) and without (b, c) Ar-20%CO_2 shielding did not present plastic deformation on the surface. The direction of the abrasive force and the subsequent cracks and brittle fractures are observed. Samples deposited without gas shielding (b, c) show large surface deformation. In Fig. 31d, the influence of AgNPs on the surface resistance of modified paint is obvious and that within the range of concentration of AgNPs considered, 0.175 wt% gave minimum weight loss and maximum number of cycles to scratch 9.31 g of painted surface.

3.2 Electrical Characterization

One of the major application areas of nanotechnology and a driving force for nanoscience is the semiconductor industry which is an important industry when building integrated electronic systems utilized in applications ranging from computers, cell phones, digital cameras, electronic instrumentation for medical diagnostics among others. With the ever-increasing miniaturization of semiconductor

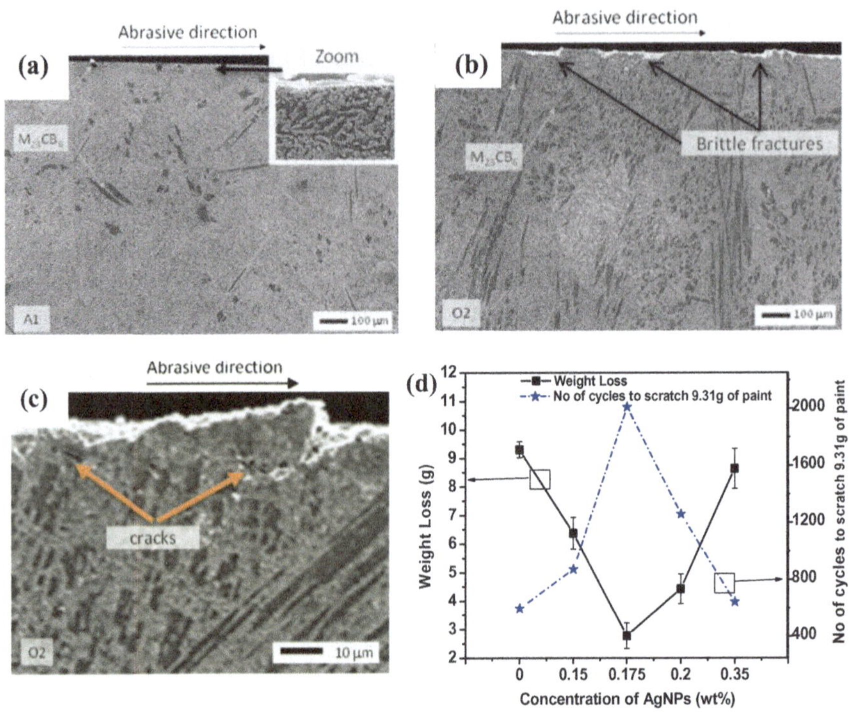

Fig. 31 Abrasive wear on nanostructured iron-based alloy deposited by flux cored arc welding process with **a** and without, **b**, **c** Ar-20%CO_2 shielding (Gualco et al. 2016), **d** influence of concentration of AgNPs solution on weight loss and scratch resistance of embedded emulsion paint (Asafa et al. 2020a, b)

materials, scientists have been investigating the potential of nanomaterials, especially carbon nanotubes as a viable means of enhancing their performance for several applications. Hence, for the assessment of material quality and device reliability, it is important to perform certain electrical characterization techniques so as to determine all the necessary parameters. This section presents two major techniques for electrical characterization of semiconductors including the four-point probe technique for resistivity measurement and the Hall effect principle for the carrier concentration and Hall mobility measurements.

3.2.1 Four-Point Probe

The four-point probe is a technique used to measure the resistivity of a layer of materials that are often used in electronic devices. It is so called because the four points touches the sample surface. The four-point probes are made lined in a straight line with the distance between probes arranged such that each of the probes has the same distance. A constant electric current is streamed along the sample through two outermost probes. If the sample has resistance, there will be a voltage drop when the current flows along the sample. The voltage change is measured through two inner probes. Figure 32a shows KLA-Tencor OmniMap RS75 four-point probe often used to measure the average film resistance. The general mechanism of measurement based on four-point probe is shown in Fig. 32b. For a sample of semi-infinite volume, a very thin layer (thickness t « s), and equal interprobe spacing *s*, the sheet resistivity ρ is given by Eq. 16:

$$\rho = \frac{\pi t}{ln2}\left(\frac{V}{I}\right) \tag{16}$$

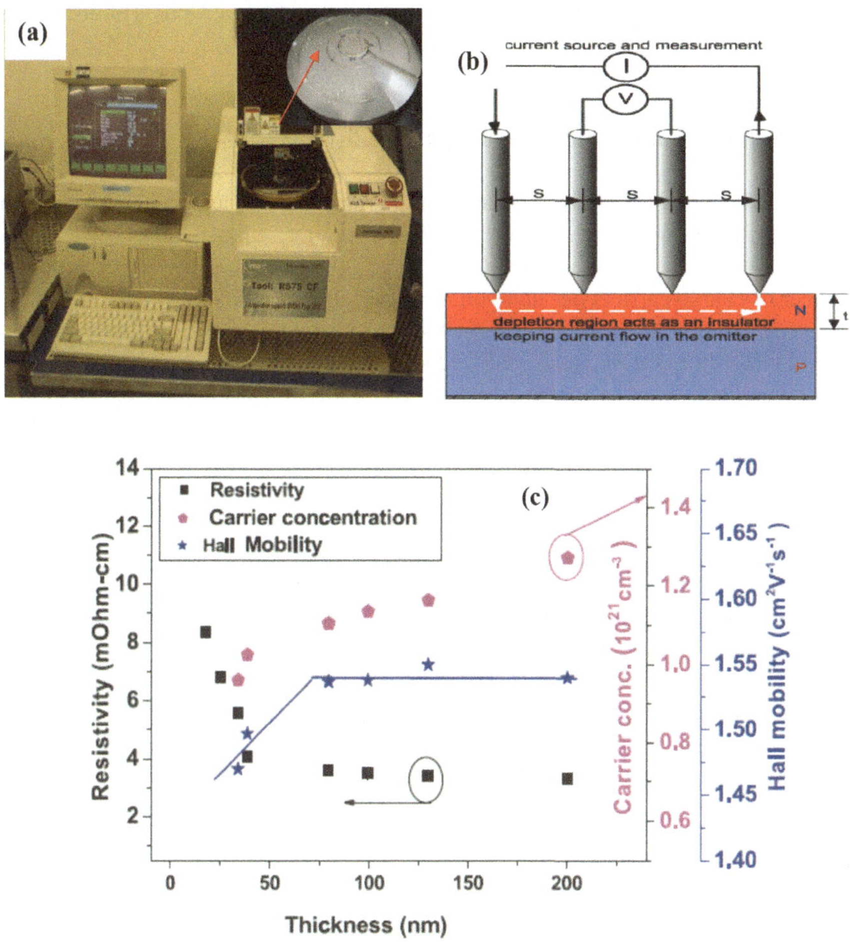

Fig. 32 **a** KLA-Tencor OmniMap RS75 four-point probe (the inset is the measuring table), **b** mechanism of measurement showing the arrangement of the four probes on sample, **c** resistivity, hall mobility, and carrier concentration against thickness of ultrathin poly-SiGe film (Asafa et al. 2014)

The resistivity of a thin film is independent of the interprobe spacing s, so the sheet resistance R_s is given by the following equation:

$$R_s = \frac{\rho}{t} = k\left(\frac{V}{I}\right) \tag{17}$$

where the geometric factor $k = \frac{\pi}{ln2}$, which is 4.53 for a semi-infinite thin sheet which may differ for non-ideal samples. The resistivity of poly-SiGe film measured with four-point probe, showing in Fig. 32c, decreases asymptotically with the film thickness and fairly stabilizes at a thickness of 40 nm. The trend observed can also be used to describe the stages of resistivity evolution which are: initial stage (regime I), transient stage (regime II), and stagnation stage (regime III). Details of the results are available elsewhere (Asafa et al. 2014).

3.2.2 Carrier Concentration and Hall Mobility

Hall mobility and carrier concentration are parameters that are obtained based on the Hall effect principle which explains the behavior of charge carriers when they are exposed to electricity and magnetic fields. Carrier concentration refers to the number of charge carriers per unit volume while Hall mobility is a measure of how fast electrons or holes can travel in a metal or in a semiconductor. Hall mobility is a critical factor in electronic and semiconductor devices/materials as it helps to determine switching frequency in transistors, photoconductive gain in photodetectors, and transport properties in solar cells and emitting devices (Ponce et al. 2020). The hall mobility can be estimated by Eq. 18:

$$\mu_n \text{ or } \mu_p = \sigma n R_H \tag{18}$$

where μ_n is hall mobility for electrons, μ_p is hall mobility for holes, σ is conductivity of the material in the conductor, n is number of charge carriers per unit volume, and R_H is the hall coefficient.

In theory, the Hall effect is generated when a magnetic field of magnitude B is applied perpendicular to a moving carrier (van der Pauw 1958). Because of the magnetic field, the moving carrier is deflected perpendicularly to both the magnetic field and the plane the carrier was traveling in (Fig. 33a). The Lorenz force, $F = -qv \times B$ where q is the carrier charge (1.602×10^{-19} C), introduces a potential difference or Hall voltage, V_H, across the sample resulting in an electric field (Lin et al. 1981). Measurement of the Hall voltage is done following the van der Pauw technique (Fig. 33b) where a constant current is forced through contacts positioned opposite to each other while the Hall voltage is measured across the other two. The magnitude of the Hall voltage, V_H, is related to the current I traveling through the sample, the applied field B, and the sheet density of carriers N_s by the following equation.

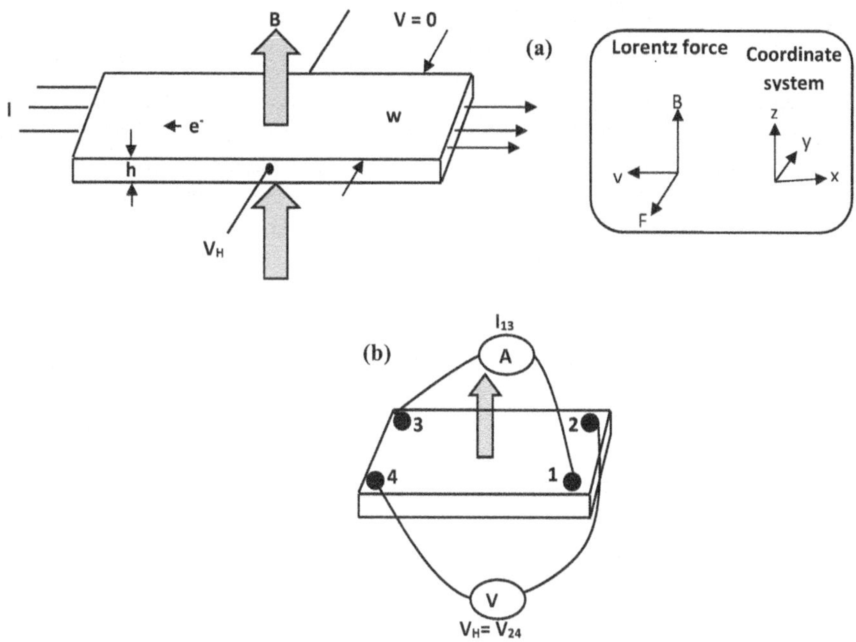

Fig. 33 **a** Hall voltage V_H is generated due to the Lorenz force F when a magnetic field B is applied across a sample of thickness h and width w, the Lorentz force and the coordinate system are shown, **b** illustration of the van der Pauw Hall effect measurement geometry (Asafa 2013)

$$|V_H| = \frac{IB}{qN_s} \tag{19}$$

The carrier mobility is related to the resistivity ρ and carrier concentration N $(= N_s{*}h)$ as follows:

$$\mu = \frac{1}{qN\rho} \tag{20}$$

The Hall mobility and carrier concentration shown in Fig. 32c were measured using a Lakeshore 7507 Hall effect system at a room temperature of 20 °C. The samples are ~1 cm × 1 cm in dimension, and four contacts are made with indium oxide powder and then backed in an oven at 380 °C for 1 min. Each of the samples is fixed to the Hall measurement probe and then placed in-between two magnets. A constant current of 1.0 mA is applied across the sample while the Hall voltage is measured for the magnetic fields between −1 and +1 T. All errors of consistence are less than 0.2%. From the Hall voltage, the carrier concentration is calculated. According to Fig. 32c, the Hall mobility for films thicker than 50 nm is significantly the same while the carrier concentration increases marginally with thickness (Asafa et al. 2014).

3.3 Thermal Characterization

Thermal characterization studies the changes of the properties of materials with temperature. The changes include both the physical and chemical changes which take place over a wide range of temperature. Generally, materials particularly nanomaterials are designed to be used under various service conditions such as arctic cold to tropical heat, in corrosive environments, variable humidity, and under load (stress), which require different operating temperatures. Therefore, it is important to characterize these nanomaterials and their behavior over a range of temperatures to determine their suitability for specific purposes and to determine the appropriate operating temperature range. The obtained results can then be used to forecast the safe operating conditions and the expected lifetime for the products. This section presents four major techniques for thermal characterization of nanomaterials including the thermal conductivity, thermal gravimetric analysis, differential scanning calorimetry, and differential thermal analysis.

3.3.1 Thermal Conductivity and Expansion

Thermal conductivity is a temperature-dependent property which measures the capacity of a material to conduct heat while thermal expansion denotes ability of materials to change shape upon heating. Therefore, assessing material response to temperature change is important especially in thermal devices and electronics where heat conduction, removal, and expansion are important (Asafa et al. 2020a, b; Durowoju et al. 2019). In electronics, materials with high thermal conductivity and low coefficient of thermal expansion are needed as heat sinks to dissipate heat with minimal thermal expansion (Schubert et al. 2007). Generally, the rate of heat transfer is lower for materials of low thermal conductivity than those of high thermal conductivity (Huang 2017).

Thermal conductivity of materials can be measured using different approaches with each method depending on the thermal properties and the medium temperature. In general, there are two basic techniques of measurement. First is the steady-state technique where measurement is performed when material that is analyzed is in complete equilibrium. In as much as there is no temperature change during the measurement at the various points as the sample is tested, the heat flow through the section of the sample is the same. Furthermore, the thermal conductivity of the sample can be estimated from the temperature gradient and heat flux. Second is the dynamic measurement where the temperature change rate overtime is measured to obtain the thermal diffusivity (Liu and Chen 2014). Thermal conductivity is derived based on the specific heat capacity of the material. There are four main types of instruments available to measure thermal conductivity: guarded hot plate based on Lee Disk setup, hot wire, modified hot wire, and laser flash diffusivity. They differ in technique, sample size, testing time, capability, and methodologies of measurement. Coefficient of thermal expansion (CTE) can be calculated from the linear shrinkage

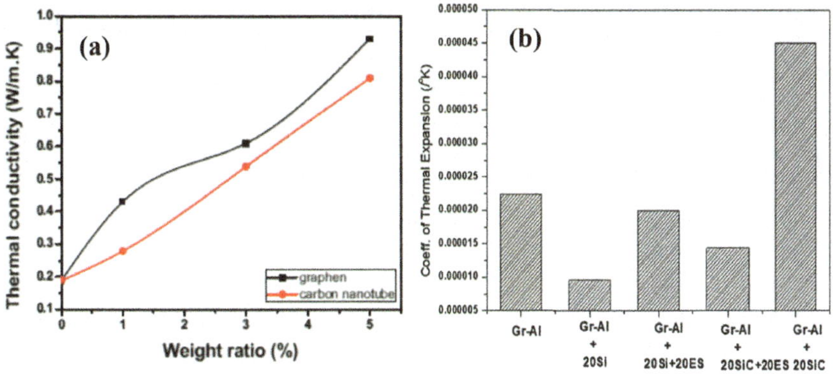

Fig. 34 **a** Thermal conductivity modified epoxy resin as a function of weight ratio of graphene and carbon nanotubes (Hussein et al. 2018), **b** coefficient of thermal expansion of sintered graphite–aluminum (Gr–Al) modified with silicon (Si) and silicon carbide (SiC) (Durowoju et al. 2019)

during sintering processing of composite powders using $\mathrm{CTE} = \left(\frac{\epsilon_{T_{\mathrm{room}}} - \epsilon_{T_{\mathrm{max}}}}{T_{\mathrm{room}} - T_{\mathrm{max}}} \times \frac{1}{100}\right)$ (Largiller et al. 2011; Maca et al. 2008; Olivier et al. 2014) where $\epsilon_{T_{\mathrm{room}}}$ and $\epsilon_{T_{\mathrm{max}}}$ are linear shrinkages at the end of the cooling and dwell process, respectively, while T_{room} and T_{max} are the temperature at the end of the cooling dwell process, respectively.

Figure 34a shows the thermal conductivity coefficients of nanocomposites obtained via the Lee's Disk experiment (Hussein et al. 2018). The thermal conductivity increases linearly with an increased weight fraction of the fillers for both graphene and CNT nanofibers added to epoxy resin to make nanocomposites. The variation in coefficient of thermal expansion (CTE) of the sintered graphite aluminum samples is presented in Fig. 34b (Durowoju et al. 2019). CTE decreased by about 50% when 20 wt% Si was added to Gr–Al which is attributed to the low CTE of Si (2.56×10^{-6}/K) (Serway 1998). However, upon addition of ES to the composite, the CTE increased close to that of Gr–Al composite. This is because the CTE of eggshell is higher (5.4×10^{-6}/K) (Dhanoa et al. 1996) than that of Si (Serway 1998).

3.3.2 Thermal Gravimetric Analysis

Thermal gravimetric analysis (TGA) measures the amount and the rate of weight change (mass loss/gained) of a material with respect to temperature or time in controlled environments. For this system, sample preparation is very significant in obtaining satisfactory result. It is suggested that maximizing the surface area of the sample in a TGA pan improves resolution and reproducibility of weight loss temperatures as the sample weight affects the accuracy of weight loss measurements. Generally, 10–20 mg of sample is preferred in most applications. However, if the sample has volatile elements or compounds, 50–100 mg of sample is considered

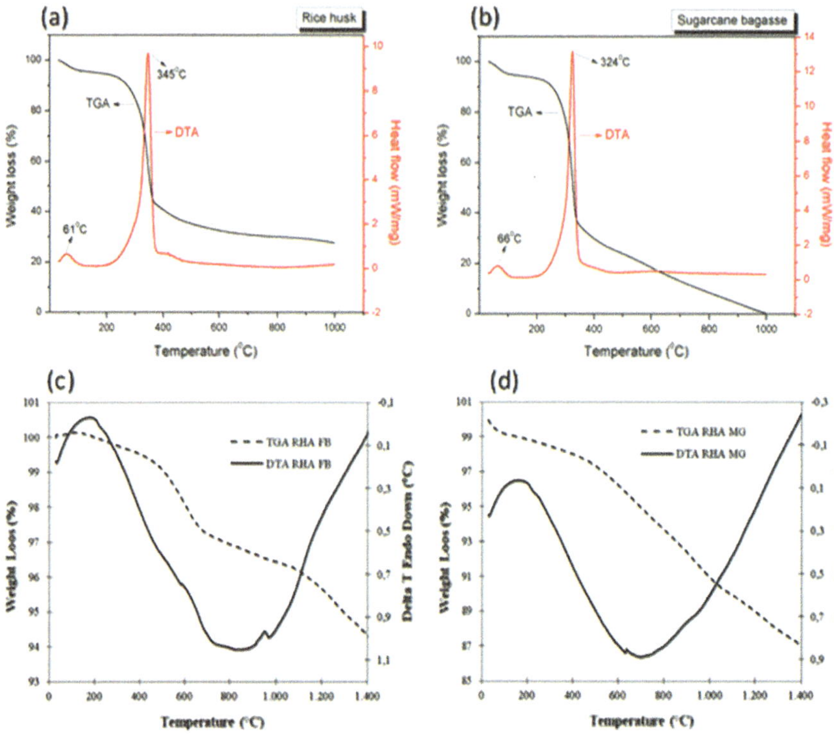

Fig. 35 TGA/DTA curves of **a** rice husk, **b** sugarcane bagasse (Dele-Afolabi et al. 2018a, b), **c**, **d** for two rice husks ashes, one obtained by combustion in a slanted moving grate reactor (RHA-MG) and the other in a fluidized bed reactor (RHA-FB) (Fernandes et al. 2017)

adequate. Moreover, most TGA instruments have baseline drift of ±0.025 mg which is ±0.25% of a 10 mg sample.

In thermogravimetric analysis, the sample is heated in a given environment (air, N_2, CO_2, He, Ar, etc.) at controlled rate. The temperature is increased at a constant rate for a known initial weight of the substance, and the changes in weights are determined as a function of temperature at different time interval. This plot of weight change against temperature is termed thermogravimetric curve. Figure 35 presents the combined thermogravimetric analysis curves (TGA) for rice husk and sugarcane bagasse (Dele-Afolabi et al. 2018a, b).

3.3.3 Differential Thermal Analysis

Differential thermal analysis is a technique where the temperature of the substance under investigation is compared with the temperature of a thermally inert material as a function of temperature/time. Therefore, both the sample and reference material are heated up under carefully controlled conditions. If there is zero temperature difference between the sample and reference material, then the sample does not undergo any chemical or physical change. On the other hand, if any reaction (physical or chemical change) takes place, temperature difference will occur between sample and reference material. Some changes involved absorption of heat by the sample, and are referred to as endothermic. Examples of endothermic changes are phase changes including melting (fusion), vaporization, sublimation, and some transitions between two different crystal structures for a material. Chemical reactions such as dehydration, decomposition, oxidation–reduction, and solid–state reactions can also result in endothermic changes. Other changes result in heat that is released by the sample and are termed exothermic. Examples of exothermic changes are phase changes including freezing (crystallization) and some transitions between different crystal structures. Chemical reactions such as decomposition, oxidation–reduction, and chemisorption can also result in exothermic changes.

Typical examples of TGA/DTA curves are presented in Fig. 35. According to Fig. 35a, b for rice husk and sugarcane bagasse (Dele-afolabi et al. 2018a, b), TGA curves indicate that the decomposition of both rice husk and sugarcane bagasse is gradual below 200 °C, and becomes rapid within the ranges of 316–345 °C and 300–324 °C, respectively. DTA curves indicate exothermic peaks at 61 and 345 °C for rice husk and 66 and 324 °C for sugarcane bagasse at higher heat flow. For two rice husks, one obtained by combustion in a slanted moving grate reactor (RHA-MG), one in a fluidized bed reactor (RHA-FB), and gradual weight loss is observed according to the TGA curves (Fernandes et al. 2017). DTA for RHA-FB shows an exothermic peak before 200 °C and endothermic peak near 800 °C. For the RHA-MG, the endothermic reaction due to decomposition occurs before 800 °C.

3.3.4 Differential Scanning Calorimetry

Differential scanning calorimetry (DSC) measures the temperatures and heat flows involving transitions in materials due to time and temperature change in a controlled atmosphere. The analysis offers quantitative and qualitative information about physical and chemical changes that involve endothermic or exothermic processes, or changes in heat capacity. DSC measures glass transitions, melting and boiling points, crystallization time and temperature, percent crystallinity heats of fusion and reactions, specific heat capacity, oxidative/thermal stability, rate and degree of cure, reaction kinetics, and purity. Samples to be tested usually within a weight range of 3–5 mg can be in the bulk solid, powder (pressed), and liquid forms. Figure 36 presents the DSC scans for carbon nanotubes-reinforced Sn-5Sb composite lead-free solder showing a slight reduction in melting temperature of composite solders compared to the unreinforced solder system (Fernandes et al. 2017).

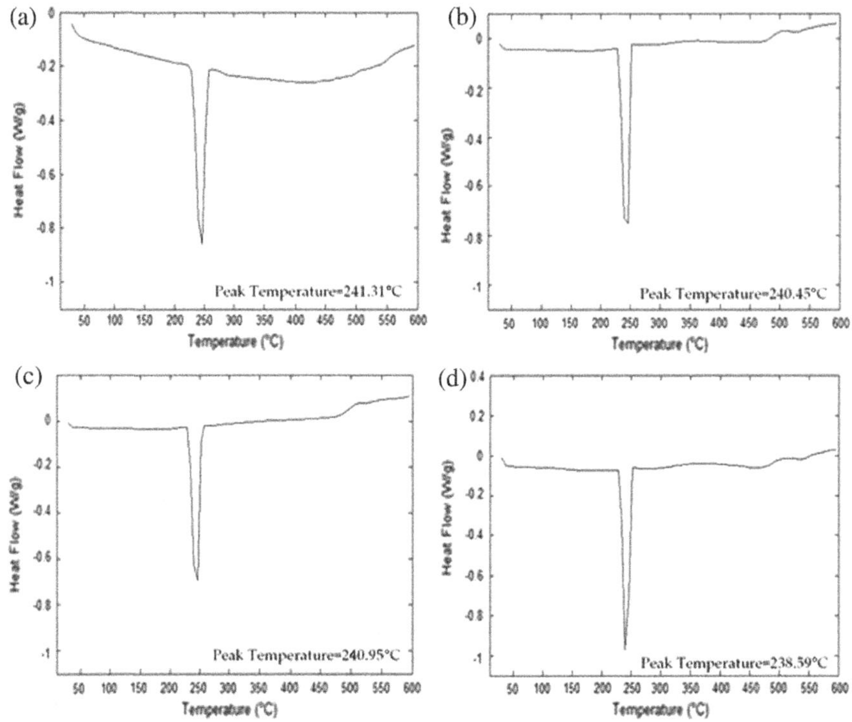

Fig. 36 DSC scans of **a** Sn-5Sb, **b** Sn-5Sb-0.01CNT, **c** Sn-5Sb-0.05CNT, and **d** Sn-5Sb-0.1CNT composite lead-free solders (Fernandes et al. 2017)

4 Conclusion

This chapter contains various techniques that are available and can be utilized for impactful nanotechnology research. Researchers can confidently choose from the quantum of available characterization techniques relevant for their studies. In most studies, a few numbers of techniques can be combined to ensure a meaningful research. Characterization involving spectroscopy (such as UV-visible, FTIR, XRD, EDS), microscopy (SEM, TEM, AFM), and thermal (such as TGA and DTA) are mostly encountered in studies involving nanomaterials. Large number of references provided can also serve as sources of information, especially when more details about a specific characterization technique is required.

References

Abad JM, Mertens SF, Pita M, Fernández VM, Schiffrin DJ (2005) Functionalization of thioctic acid-capped gold nanoparticles for specific immobilization of histidine-tagged proteins. J Am Chem Soc. https://doi.org/10.1021/ja042717i

Abelmann L, Van Den Bos A, Lodder C (2005) Magnetic force microscopy—towards higher resolution. In: Magnetic microscopy of nanostructures. Springer, Berlin, Heidelberg, pp 253–283

Adedokun O, Roy A, Awodugba AO, Devi PS (2017a) Fluorescent carbon nanoparticles from Citrus sinensis as efficient sorbents for pollutant dyes. Luminescence 32:62–70. https://doi.org/10.1002/bio.3149

Adedokun O, Sanusi YK, Awodugba AO (2017b) Pigment extracts of citrus peels as light sensitizers for dye-sensitized solar cells. J Mater Sci Appl 3:1–7. https://doi.org/10.1088/1755-1315/173/1/012040

Adedokun O, Awodele MK, Sanusi YK, Awodugba AO (2018) Natural dye extracts from fruit peels as sensitizer in ZnO-based dye-sensitized solar cells. In: IOP Conference Series: Earth and Environmental Science. https://doi.org/10.1088/1755-1315/173/1/012040

Ali MKA, Xianjun H, Mai L, Qingping C, Turkson RF, Bicheng C (2016) Improving the tribological characteristics of piston ring assembly in automotive engines using Al_2O_3 and TiO_2 nanomaterials as nano-lubricant additives. Tribol Int 103:540–554. https://doi.org/10.1016/j.triboint.2016.08.011

Armelao L, Bertagnolli H, Bleiner D, Groenewolt M, Gross S, Krishnan V, Sada C, Schubert U, Tondello E, Zattin A (2007) Highly dispersed mixed zirconia and hafnia nanoparticles in a silica matrix: first example of a ZrO_2–HfO_2–SiO_2 ternary oxide system. Adv Funct Mater. https://doi.org/10.1002/adfm.200600458

Asafa TB (2013) Development of ultrathin polycrystalline silicon germanium films as structural layers for applications in nanoelectromechanical systems. Unpublished PhD Thesis, King Fahd University of Petroleum and Minerals, Dharaan

Asafa TB, Bryce G, Severi S, Said, SAM, Witvrouw A (2013a) Multi-response optimization of ultrathin poly-SiGe films characteristics for nano-electromechanical systems (NEMS) using the grey-Taguchi technique. Microelectron Eng 111:229–233. https://doi.org/10.1016/j.mee.2013.03.171

Asafa TB, Witvrouw A, Morcos BS, Vanstreels K, Said SAM (2013b) Influence of germanium incorporation on the structural and electrical properties of boron-doped ultrathin poly-Si_{1-x} Gex films deposited by chemical vapour deposition. Appl Phys A 1–7. https://doi.org/10.1007/s00339-013-8141-1

Asafa TB, Witvrouw A, Schneider D, Moussa A, Tabet N, Said SAM (2014) Thickness effect on the structural and electrical properties of poly-SiGe films. Mater Res Bull 49:102–107. http://dx.doi.org/10.1016/j.materresbull.2013.08.054

Asafa TB, Durowoju MO, Oyewole AA, Solomon SO, Adegoke RM, Aremu OJ (2015) Potentials of snail shell as a reinforcement for discarded aluminum based materials. Int J Adv Sci Technol 84:1–8. http://dx.doi.org/10.14257/ijast.2015.84.01

Asafa TB, Durowoju MO, Madingwaneng KP, Diouf S, Sadiku ER, Shongwe MB, Olubambi PA, Ismail OS, Ajala MT, Oladosu KO (2020a). Gr–Al composite reinforced with Si_3N_4 and SiC particles for enhanced microhardness and reduced thermal expansion. SN Appl Sci 1026. https://doi.org/10.1007/s42452-020-2838-5

Asafa TB, Odediji RA, Salaudeen TO, Lateef A, Durowoju MO, Azeez MA, Yekeen TA, Oladipo IC, Irshad HM, Abbas SH (2020b) Physico-mechanical properties of emulsion paint embedded with silver nanoparticles. Bull Mater Sci (In Press)

Atkins P, De Paula J (2009) Physical chemistry 8th edn. https://doi.org/10.1021/ed056pa260.1

Ayache J, Beaunier L, Boumendil J, Ehret G, Laub D (2010) Sample preparation handbook for transmission electron microscopy: techniques. Springer Science & Business Media

Azeez MA, Lateef A, Asafa TB, Yekeen TA, Akinboro A, Oladipo IC, Gueguim-Kana EB, Beukes LS (2017) Biomedical applications of cocoa bean extract-mediated silver nanoparticles as antimicrobial, larvicidal and anticoagulant agents. J Clust Sci 28(1):149–164. https://doi.org/10.1007/s10876-016-1055-2

Bancroft J D, Gamble M (2008). Theory and practice of histological techniques. Elsevier health sciences

Baudot C, Tan CM, Kong JC (2010) FTIR spectroscopy as a tool for nano-material characterization. Infrared Phys Technol. https://doi.org/10.1016/j.infrared.2010.09.002

Bauer KC, Göbel M, Schwab ML, Schermeyer MT, Hubbuch J (2016) Concentration-dependent changes in apparent diffusion coefficients as indicator for colloidal stability of protein solutions. Int J Pharm. https://doi.org/10.1016/j.ijpharm.2016.07.007

Beane R (2004) Using the scanning electron microscope for discovery based learning in undergraduate courses. J Geosci Educ 52:250–253. https://doi.org/10.5408/1089-9995-52.3.250

Berne BJ, Pecora R (1977) Dynamic light scattering. John Wiley, New York, pp 376. Polym J. https://doi.org/10.1002/pi.4980090216

Bergström J (2015) Mechanics of solid polymers: theory and computational modeling. https://doi.org/10.1016/c2013-0-15493-1

Buseck P, Cowley J, Eyring L (1989) High-resolution transmission electron microscopy and associated techniques. Oxford University Press

Cappella B, Dietler G (1999) Force-distance curves by atomic force microscopy. Surf Sci Rep 34:1–104. https://doi.org/10.1016/S0167-5729(99)00003-5

Carter CB, Williams DB (2016) Transmission electron microscopy: diffraction, imaging, and spectrometry. Springer International Publishing

Cheong JY, Chang H, Kim J, Kim C, Seo HK, Shin JW, Yuk JM, Lee JY, Kim I (2017) In situ high-resolution transmission electron microscopy (TEM) observation of Sn nanoparticles. Microsc Microanal 23:1107–1115. https://doi.org/10.1017/S1431927617012739

Chow G, Miller P, Wang J (2015) Correlation between laser-induced surface acoustic waves and nanoindentation on elastic modulus measurement of a nanoporous zeolite thin film. Exp Mech 55:647–650. https://doi.org/10.1007/s11340-014-9969-9

Cullity BD, Stock SR (2001) Elements of X-ray diffraction, 3rd edn. Prentice Hall. https://doi.org/citeulike-article-id:3998040

DeBeer S, Neese F (2013) X-ray spectroscopy. In: Comprehensive inorganic chemistry II (second edition): from elements to applications. https://doi.org/10.1016/b978-0-08-097774-4.00918-9

Dele-Afolabi TT, Hanim MAA, Norkhairunnisa M, Sobri S, Calin R, Ismarrubie ZN (2018a) Significant effect of rice husk and sugarcane bagasse pore formers on the microstructure and mechanical properties of porous Al_2O_3/Ni composites. J Alloys Compd 743:323–331. https://doi.org/10.1016/j.jallcom.2018.01.230

Dele-Afolabi TT, Sobri S, Calin R, Zahari I (2018b) Effect of agro-waste pore formers on the microstructure, hardness, and tensile properties of porous alumina ceramics. Int J Appl Ceram Technol 15:1060–1071. https://doi.org/10.1111/ijac.12874

Dhanoa PS, Puri VM, Anantheswaran RC (1996) Thermal and mechanical properties of eggshell under different treatment. Trans ASAE 39:999–1004. https://doi.org/10.13031/2013.27588

Downes A, Elfick A (2010) Raman spectroscopy and related techniques in biomedicine. Sensors 10(3):1871–1889. https://doi.org/10.3390/s100301871

Durowoju MO, Asafa TB, Sadiku E, Diouf S, Shongwe M, Olubambi P, Oladosu K, Ogbemudia A, Babalola M (2019) Improving mechanical and thermal properties of graphite—aluminium composite using Si, SiC and eggshell particles. J Compos Mater 1–12. https://doi.org/10.1177/0021998319892058

Echlin P (1999) Handbook of sample preparation for scanning electron microscopy and X-ray microanalysis. Springer, Berlin, Heidelberg

Egerton RF (2005) Physical principles of electron microscopy : an introduction to TEM, SEM, and AEM. Springer

Einstein A (1905) On a heuristic point of view about the creation and conversion of light, translated from German by Haar. T Ann Phys. https://doi.org/10.1002/andp.19053220607

Fang J, Wang C, Li Y, Zhao Z, Mei L (2016) Comparison of bacterial adhesion to dental materials of polyethylene terephthalate (PET) and polymethyl methacrylate (PMMA) using atomic force microscopy and scanning electron microscopy. Scanning 38(6):665–670. https://doi.org/10.1002/sca.21314

Fernandes IJ, Sánchez FA, Jurado JR, Kieling AG, Rocha TLAC (2017) Physical, chemical and electric characterization of thermally treated rice husk ash and its potential application as ceramic raw material. Adv Powder Technol 28:1228–1236. https://doi.org/10.1016/j.apt.2017.02.009

Fujii Y, Okajima T (2019) Calibrating the Young's modulus of soft materials with surface tilt angle measured by atomic force microscopy. AIP Adv 9:015028. https://doi.org/10.1063/1.5046372

Gao L, Li Y, Ren J, Wang S, Wang R, Fu G, Hu Y (2017) Passivation of defect states in anatase TiO_2 hollow spheres with Mg doping: realizing efficient photocatalytic overall water splitting. Appl Catal B Environ 202:127–133. https://doi.org/10.1016/j.apcatb.2016.09.018

Gardiner DJ (1989) Introduction to raman scattering. In: Practical raman spectroscopy. https://doi.org/10.1007/978-3-642-74040-4_1

Giahi M, Pathania D, Agarwal S, Ali GAM, Chonge KF, Gupta VK (2019) Preparation of Mg-doped TiO_2 nanoparticles for photocatalytic degradation of some organic pollutants. Stud Univ Babes-Bolyai Chem 64(1):7–18. https://doi.org/10.24193/subbchem.2019.1.01

Gilliland GD (1997) Photoluminescence spectroscopy of crystalline semiconductors. Sci Eng R Rep, Mater. https://doi.org/10.1016/s0927-796x(97)80003-4

Gobind GN, Parshad J (2015) Techniques of measuring wear for bulk materials and advanced surface coatings. IOSR J Mech Civ Eng (IOSR-JMCE) 12(2):101–106. https://doi.org/10.9790/1684-1226101106

Goldstein JI, Newbury DE, Echlin P, Joy DC, Fiori C, Lifshin E, Goldstein JI, Newbury DE, Echlin P, Joy DC, Fiori C, Lifshin E (1981) Materials specimen preparation for SEM and X-ray microanalysis. In: Scanning electron microscopy and X-ray microanalysis. https://doi.org/10.1007/978-1-4613-3273-2_9

Goldstein JI, Newbury DE, Echlin P, Joy DC, Romig AD, Lyman CE, Fiori C, Lifshin E (1992) Sample preparation for biological, organic, polymeric, and hydrated materials. In: Scanning electron microscopy and x-ray microanalysis, Springer, Boston, MA, pp 571–670

González N, Del Àngels Custal M, Rodríguez D, Riba JR, Armelin E (2017) Influence of ZnO and TiO2 particle sizes in the mechanical and dielectric properties of vulcanized rubber. Mater Res 20(4):1082–1091. https://doi.org/10.1590/1980-5373-MR-2017-0178

Goodhew PJ, Humphreys J, Beanland R (2001). Electron microscopy and analysis, 3rd edn. CRC Press

Greczynski G, Hultman L (2020) X-ray photoelectron spectroscopy: towards reliable binding energy referencing. Prog Mater Sci 107 (July 2019):100591. https://doi.org/10.1016/j.pmatsci.2019.100591

Gualco A, Svoboda HG, Surian ES (2016) Study of abrasive wear resistance of Fe-based nanostructured hardfacing. Wear 360–361:14–20. https://doi.org/10.1016/j.wear.2016.04.011

Harris DC (2010) Quantitative chemical analysis, 8th edition. W. H. Freeman and Company. https://doi.org/10.1021/jf103070m

Hendrych A, Kubínek R, Zhukov AV (2007) The magnetic force microscopy and its capability for nano-magnetic studies—the short compendium. Mod Res Educ Top Microsc 805–811

Hermodson M (1996) Current protocols in protein science. In: Proteins: structure, function and genetics. https://doi.org/10.1002/prot.340240303

Huang Y (2017) Electrical and thermal properties of activated carbon fibers. In: Activated carbon fiber and textiles, vol 7. Elsevier Ltd. https://doi.org/10.1016/B978-0-08-100660-3.00007-9

Huang Q, Wu H, Cai P, Fein JB, Chen W (2015) Atomic force microscopy measurements of bacterial adhesion and biofilm formation onto clay-sized particles. Sci. Rep. 5:1–12. https://doi.org/10.1038/srep16857

Hussein SI, Abd-Elnaiem AM, Asafa TB, Jaafar HI (2018) Effect of incorporation of conductive fillers on mechanical properties and thermal conductivity of epoxy resin composite. Appl Phys A Mater Sci Process 124(7). https://doi.org/10.1007/s00339-018-1890-0

Iniewski K (2009) Medical imaging: principles, detectors, and electronics. In: Medical imaging: principles, detectors, and electronics. https://doi.org/10.1002/9780470451816

Inkson BJ (2016) Scanning electron microscopy and transmission electron microscopy for materials characterization. In: Materials characterization using nondestructive evaluation (NDE) methods. Elsevier Ltd. https://doi.org/10.1016/B978-0-08-100040-3.00002-X

Ishida N, Craig VSJ (2018) Direct measurement of interaction forces between surfaces in liquids using atomic force microscopy. Kona Powder Part J, July, 1–14. https://doi.org/10.14356/kona.2019013

Ishida N, Kusaka Y, Fukasawa T, Shinto H (2016) Atomic force microscopy for measuring interaction forces in biological materials. In: Ohshima H (ed) Encyclopedia of biocolloid and biointerface science, vol 1, pp 59–67

Laserna J (2001) An introduction to raman spectroscopy: introduction and basic principles. http://www.Spectroscopynow.Com/Spy/Basehtml/SpyH/1,1181,6-1-12-0-0-News_detail-0-321,00.Html

Jia J, Yang W, Xue Q (1986) Scanning tunneling microscopy. In: Jia J, Yang W, Xue Q (eds) Optical microscopy, scanning probe microscopy, ion microscopy and nanofabrication, pp 55–112

Kalantar-zadeh K, Fry B (2008) Characterization techniques for nanomaterials. In: Kalantar-zadeh K, Fry B (eds) Nanotechnology-enabled sensors. https://doi.org/10.1007/978-0-387-68023-1_5

Kaszuba M, McKnight D, Connah MT, McNeil-Watson FK, Nobbmann U (2008) Measuring sub nanometre sizes using dynamic light scattering. J Nanoparticle Res. https://doi.org/10.1007/s11051-007-9317-4

Keyso RJ, Garratt-Reed AJ, Goodhew PJ, Lorimer GW (1998) Introduction to scanning transmission electron microscopy. CRC Press

Kohl H, Reimer L (2008) Transmission electron microscopy: physics of image formation. Springer, Berlin Heidelberg

Largiller G, Dong L, Bouvard D, Carry CP, Gabriel A (2011) Constitutive modeling of the behavior of cermet compacts during reaction sintering. Powder Technol 208:496. https://doi.org/10.1016/j.powtec.2010.08.049

Lateef A, Ojo SA, Akinwale AS, Azeez L, Gueguim-Kana EB, Beukes LS (2015) Biogenic synthesis of silver nanoparticles using cell-free extract of *Bacillus safensis* LAU 13: antimicrobial, free radical scavenging and larvicidal activities. Biol 70(10):1295–1306. https://doi.org/10.1515/biolog-2015-0164

Lateef A, Azeez MA, Asafa TB, Yekeen TA, Akinboro A, Oladipo IC, Azeez L, Ajibade SE, Ojo SA, Gueguim-kana EB, Beukes LS (2016) Biogenic synthesis of silver nanoparticles using a pod extract of *Cola nitida* : antibacterial and antioxidant activities and application as a paint additive. J Taibah Univ Sci 10(4):551–562. https://doi.org/10.1016/j.jtusci.2015.10.010

Lee H-L, Flynn NT (2006) X-ray photoelectron spectroscopy BT. Appl Solid State Spectrosc, Handb. https://doi.org/10.1007/0-387-37590-2_11

Lin JF, Li SS, Linares LC, Teng KW (1981) Theoretical analysis of Hall factor and Hall mobility in p-type silicon. Solid State Electron 24:827–833. https://doi.org/10.1016/0038-1101(81)90098-8

Liu PS, Chen GF (2014) Porous materials: processing and applications. Butterworth-Heinemann

Lupke F, Dolezal J, Cherepanov V, Ostadal I, Tautz FS, Voigtlander B (2018) Surface structures of tellurium on Si(111)–(7 $\times$ 7) studied by low-energy electron diffraction and scanning tunneling microscopy. Surf Sci 681:130–133. https://doi.org/10.1016/j.susc.2018.11.016

Maca K, Pouchly V, Boccaccini AR (2008) Sintering densification curve-a practical approach for its construction from dilatometric shrinkage data. Sci Sinter 40:117. https://doi.org/10.2298/SOS0802117M

Marinello F, Storia AL, Mauriello G, Passeri D (2019) Atomic force microscopy techniques to investigate activated food packaging materials. Trends Food Sci Technol 87:84–93. https://doi.org/10.1016/j.tifs.2018.05.028

McNaught AD, Wilkinson A (1997) Compendium of chemical terminology: IUPAC recommendations. International Union of Pure and Applied Chemistry. https://doi.org/10.1351/goldbook.i03352

Meerbeek B, Van Inoue S, Yoshida Y, Perdigao J (2000) Microscopy investigations, techniques, results, limitations. Am J Dent 13:4–17

Meredith L, Farella M, Lowrey S, Cannon RD, Mei L (2016) Atomic force microscopy analysis of enamel nanotopography after interproximal reduction. Am J Orthod Dentofac Orthop 151(4):750–757. https://doi.org/10.1016/j.ajodo.2016.09.021

Mocanu A, Pasca RD, Tomoaia G, Garbo C, Frangopol PT, Horovitz O, Tomoaia-cotisel M (2013) New procedure to synthesize silver nanoparticles and their interaction with local anesthetics. Int J Nanomed 8:3867–3874. https://doi.org/10.2147/IJN.S51063

Moore CJ, Montón H, O'Kennedy R, Williams DE, Nogués C, Crean C, Gubala V (2015) Controlling colloidal stability of silica nanoparticles during bioconjugation reactions with proteins and improving their longer-term stability, handling and storage. J Mater Chem B. https://doi.org/10.1039/c4tb01915f

Mourdikoudis S, Pallares RM, Thanh NTK (2018) Characterization techniques for nanoparticles: comparison and complementarity upon studying nanoparticle properties. Nanoscale. https://doi.org/10.1039/c8nr02278j

Nakamoto K (2008) Infrared and raman spectra of inorganic and coordination compounds. In: Infrared and raman spectra of inorganic and coordination compounds. https://doi.org/10.1002/9780470405888

Namba Y (2008) Electrical conduction of thin metallic films with rough surface. J Appl Phys 39:6117–6118. https://doi.org/10.1063/1.1656136

Nellist PD (2011) The principles of STEM imaging. Scanning transmission electron microscopy: imaging and analysis. Springer, New York, NY, pp 91–115

Nobel Prize in Physics (1986) https://www.nobelprize.org/prizes/physics/1986/summary/

Ojo SA, Lateef A, Azeez MA, Oladejo SM, Akinwale AS, Asafa TB, Yekeen TA, Akinboro A, Oladipo IC, Gueguim-kana EB, Beukes LS (2016) Biomedical and catalytic applications of gold and silver-gold alloy nanoparticles biosynthesized using cell-free extract of *Bacillus safensis* anti-coagulant and thrombolytic activities. IEEE Trans Nanobiosci 15(5):433–442. https://doi.org/10.1109/TNB.2016.2559161

Oliver WC, Pharr GM (1992) An improved technique for determining hardness and elastic modulus using load and displacement sensing indentation experiments. J Mater Res 7(6):1564–1583. https://doi.org/10.1557/JMR.1992.1564

Olivier G, Jesus GJ, Benjamin D, Tobias K, Gabi S, Jan R, Nathias H (2014) Field-assisted sintering technology. Adv Eng Mater 95:425

Palanisamy S, Cheemalapati S, Chen S (2014) Amperometric glucose biosensor based on glucose oxidase dispersed in multiwalled carbon nanotubes/graphene oxide hybrid biocomposite. Mater Sci Eng C 34:207–213. https://doi.org/10.1016/j.msec.2013.09.011

Pankove JI, Kiewit DA (1972) Optical processes in semiconductors. J Electrochem Soc. https://doi.org/10.1149/1.2404256

Patil SS, Tamboli MS, Patil DR (2015) Green approach for hierarchical nanostructured Ag-ZnO and their photocatalytic performance under sunlight green approach for hierarchical nanostructured Ag–ZnO and their photocatalytic performance under sunlight. Catal Today 260:126–134. https://doi.org/10.1016/j.cattod.2015.06.004

Ponce S, Li W, Reichardt S, Giustino F (2020) First-principles calculations of charge carrier mobility and conductivity in bulk semiconductors and two-dimensional materials. Rep Prog Phys 83:036501. https://doi.org/10.1088/1361-6633/ab6a43

Pudukudy M, Yaakob Z (2015) Facile synthesis of quasi spherical ZnO nanoparticles with excellent photocatalytic activity. J Clust Sci 26(4):1187–1201. https://doi.org/10.1007/s10876-014-0806-1

Qian L, Zhao H (2018) Nanoindentation of soft biological materials. Micromachines 9(654):1–24. https://doi.org/10.3390/mi9120654

Reiss G, Hastreiter E, Bruckl H, Vancea J (1991) Thickness-dependent thin-film resistivity: application of quantitative scanning-tunneling-microscopy imaging. Phys Rev B 43:5176–5179. https://doi.org/10.1103/PhysRevB.43.5176

Rocha FS, Gomes AJ, Lunardi CN, Kaliaguine S, Patience GS (2018) Experimental methods in chemical engineering: ultraviolet visible spectroscopy—UV-vis. Can J Chem Eng 96(12):2512–2517. https://doi.org/10.1002/cjce.23344

Roy D, Fendler J (2004) Reflection and absorption techniques for optical characterization of chemically assembled nanomaterials. Adv Mater 16(6):479–508. https://doi.org/10.1002/adma.200306195

Ruff AW (1992) Wear measurement. In: ASM handbook. ASM International

Rugar D, Mamin HJ, Guethner P, Lambert SE, Stern JE, Rugar D, Mamin HJ, Guethner P, Lambert SE, Stern JE, Mcfadyen C (1990) Magnetic force microscopy: general principles and application to longitudinal recording media. J Appl Phys 68:1169. https://doi.org/10.1063/1.346713

Rutherford E, Andrade EN da C (1914) The spectrum of the penetrating γ rays from radium B and radium C. London, Edinburgh, Dublin Philos. Mag. J. Sci. 28(164):163. https://doi.org/10.1080/14786440808635207

Sahin NE, Comminges C, Le Valant A, Kiener J, Parmentier J, Napporn TW, Melinte G, Ersen O, Kokoh KB (2018) One-pot soft-template synthesis of nanostructured copper-supported mesoporous carbon FDU-15 electrocatalysts for efficient CO_2 reduction. ChemPhysChem 19:1371–1381. https://doi.org/10.1002/cphc.201701352

Scherrer P (1918) Bestimmung der Größe und der inneren Struktur von Kolloidteilchen mittels Röntgenstrahlen. Nachrichten von der Gesellschaft der Wissenschaften zu Göttingen. *Mathematisch-Physikalische Klasse*

Schneider D, Schultrich B, Scheibe H-J, Ziegele H, Griepentrog M (1998) A laser-acoustic method for testing and classifying hard surface layers. Thin Solid Films 332(1–2):157–163. https://doi.org/10.1016/S0040-6090(98)00988-2

Schneider D, Witke Th, Schwarz Th, Schoneich B, Schultrich B (2000) Testing ultra-thin films by laser-acoustics. Surf Coatings Technol 126(2–3):136–141

Schubert T, Ciupiński MJ, Weidmüller H, Weissgärber T, Kieback B (2007) Advanced composite materials for heat sink applications. In: Proceedings of the euro powder metallurgy congress and exhibition, Euro PM 2007, vol 1, pp 319–324

Serway RA (1998) Principles of physics, 2nd ed. Saunders College Pub, Fort Worth, Texas, London, p 602

Shariatzadeh B, Moradi O (2014) Surface functionalization of multiwalled carbon nanotubes with chitosan and magnesium oxide nanoparticles by microwave-assisted synthesis. Polym Compos. https://doi.org/10.1002/pc.22866

Shindo D, Oikawa T (2002) Analytical electron microscopy for materials science. https://doi.org/10.1007/978-4-431-66988-3

Skoog DA, Holler FJ, Crouch SR (2007) Principles of instrumental analysis, 6th ed. https://doi.org/10.1017/cbo9781107415324.004

Slater B (1994) UV spectroscopy: techniques, instrumentation, data handling. Spectrochim Acta Part A Mol Spectrosc. https://doi.org/10.1016/0584-8539(94)80101-0

Smith E, Dent G (2005) Modern raman spectroscopy—a practical approach. https://doi.org/10.1002/0470011831

Sokolowski E, Nordling C, Siegbahn K (1958) Chemical shift effect in inner electronic levels of Cu due to oxidation. Phys Rev. https://doi.org/10.1103/PhysRev.110.776

Spence JCH, Kolar HR, Hembree G (2006) Imaging dislocation cores—the way forward. Philos Mag 86(29–31):4781–4796. https://doi.org/10.1080/14786430600776322

Stetefeld J, McKenna SA, Patel TR (2016) Dynamic light scattering: a practical guide and applications in biomedical sciences. Biophys Rev. https://doi.org/10.1007/s12551-016-0218-6

Stylianou A, Kontomaris S, Grant C, Alexandratou E (2019). Review article atomic force microscopy on biological materials related to pathological conditions. Scanning 1–25. https://doi.org/10.1155/2019/8452851

Sun S, Zhang H, Tang L, Zhang X, Yang Z (2014) One-pot fabrication of novel cuboctahedral Cu_2O crystals enclosed by anisotropic surfaces with enhancing catalytic performance †. Phys Chem Chem Phys 16:20424–20428. https://doi.org/10.1039/C4CP03381G

Taunk PB, Das R, Bisen DP, Tamrakar RK (2015) Structural characterization and photoluminescence properties of zinc oxide nano particles synthesized by chemical route method. J Radiat Res Appl Sci 8(3):433–438. https://doi.org/10.1016/j.jrras.2015.03.006

Tissera ND, Wijesena RN, Sandaruwan CS, de Silva RM, de Alwis A, de Silva KMN (2018) Photocatalytic activity of ZnO nanoparticle encapsulated poly(acrylonitrile) nanofibers. Mater Chem Phys 204:195–206. https://doi.org/10.1016/j.matchemphys.2017.10.035

Tsang DKL, Marsden BJ, Fok SL, Hall G (2005) Graphite thermal expansion relationship for different temperature ranges. Carbon 43(14):2902–2906. https://doi.org/10.1016/j.carbon.2005.06.009

Vancsó P, Magda GZ, Pető J, Noh J, Kim Y (2016). The intrinsic defect structure of exfoliated MoS 2 single layers revealed by scanning tunneling microscopy. Sci Rep 6 (29726):1–7. https://doi.org/10.1038/srep29726

van der Pauw LJ (1958) A method of measuring the resistivity and Hall coefficient on lamellae of arbitrary shape. Philips Tech Rev 20:220–224

Virgen-Ortiz A, Limon-Miranda S, Soto-Covarrubias MA, Apolinar-Iribe A, Rrodriguez E, Iniguez-Palomarez R (2015) Biocompatible silver nanoparticles synthesized using Rumex hymenosepalus extract decreases fasting glucose levels in diabetic rats. Dig J Nanomater Biostruct 10(3):927–933

Welker RW (2012) Size analysis and identification of particles. In: Developments in surface contamination and cleaning: detection, characterization, and analysis of contaminants. https://doi.org/10.1016/b978-1-4377-7883-0.00004-3

Willard HH, Merritt LL, Dean JA (1952) Instrumental methods of analysis. Bunseki Kagaku. https://doi.org/10.2116/bunsekikagaku.1.202

Zhou B, Jie W, Wang T, Kou Z, Zhao D, Yin L, Yang F, Xi S, Zha G, Yin Z (2018) Observations on nanoscale Te precipitates in CdZnTe crystals grown by the traveling heater method using high resolution transmission electron microscopy. Crystals 8:1–6. https://doi.org/10.3390/cryst8010026

Precision Microbial Nanobiosynthesis: Knowledge, Issues, and Potentiality for the In Vivo Tuning of Microbial Nanomaterials

G. Grasso, D. Zane, and R. Dragone

1 Introduction

The large-scale production and application of nanomaterials are currently mainly restrained by the low environmental sustainability of nanomanufacturing processes based on traditional top-down or bottom-up chemical and physical approaches. The microbial-mediated biosynthesis of nanomaterials is a promising biotechnological-based nanomanufacturing process for the transition toward the development of more sustainable and innovative industrial nanomanufacturing. The achievement of a suitable applicative development and the scaling-up of microbial nanobiosynthesis include at least three different current research key drivers, namely green chemistry, white biotechnology, and synthetic biology. 'Green chemistry' can be defined as the invention, design, and application of chemical products and processes to reduce or to eliminate the use and generation of hazardous substances. 'White biotechnology' is a branch of biotechnology that exploit living cells, including microorganisms, to synthesize products at industrial level using less energy, producing less waste and/or using the by-products, residues, or wastes from other production processes, e.g., agro-wastes industrial by-products. 'Synthetic biology' is a multidisciplinary science that aims to engineer organisms to make them acquire new abilities and functions. Therefore, a full development of microbial nanobiosynthesis is a fascinating emerging prospect for future sustainable nanomaterial production at industrial scale. The ability to biosynthesize metallic, metal oxides nanoparticles (hereinafter abbreviated as 'NPs'), and quantum dots (QDs) has been reported in the literature for a wide variety of microorganisms such as bacteria, yeasts, molds, and microalgae (Li

G. Grasso (✉) · D. Zane · R. Dragone
Consiglio Nazionale delle Ricerche—Istituto per lo Studio dei Materiali Nanostrutturati c/o Dipartimento di Chimica, 'Sapienza', Università di Roma, P. le Aldo Moro 5, 00185 Rome, Italy
e-mail: gerardo.grasso@cnr.it

A. Lateef et al. (eds.), *Microbial Nanobiotechnology*, Materials Horizons: From Nature to Nanomaterials, https://doi.org/10.1007/978-981-33-4777-9_3

et al. 2011; Khan et al. 2018; Hulkoti and Taranath 2014). The biosynthesis of nanomaterials with defined dimension, morphology and composition, the mild physicochemical conditions required for biosynthetic process as well as the easily handling and cultivation of microbial cells and possibility of cell culture scale-up are some of the main advantages and features described in the literature for microbial-mediated nanobiosynthesis.

In addition, some microorganisms have shown the capability to biosynthesize unique nanostructured materials, namely frustules, magnetosomes, nanocellulose, and nanowires. Frustules are periodic and hierarchical 3D-porous micro-nanostructures naturally biosynthesized by unicellular microalgae called 'diatoms.' Today, the 'diatom nanotechnology' is a rapidly evolving research field whose aims to fully exploit the unique optical and optoelectronic properties of frustules and it is also oriented to the development of new frustules-based functionalized nanostructures (Panwar and Dutta 2019). Bacterial magnetosomes are intracellular nanocrystals of Fe_3O_4 and/or Fe_3S_4 surrounded by an organic coating layer derived from the bacterial phospholipid bilayer membrane. In particular, the bacterial Fe_3O_4 magnetosomes possess several interesting features: They are permanently magnetic at ambient temperature with a narrow size range and consistent crystal morphology (Yan et al. 2017).Bacterial nanocellulose is a 3D network of cellulose nanofibrils produced by aerobic acetic bacteria like those belonging to the genus *Gluconacetobacter*. Bacterial nanocellulose is another unique nanomaterial of microbial origin that has recently attracted great attention for biomedical applications (Sharma and Bhardwaj 2019).

Bacterial nanowires are conductive proteinaceous pilus-like nanostructures involved in extracellular electron transport processes of anaerobic dissimilatory metal-reducing bacteria belonging to *Geobacter* and *Shewanella* genera and present in some aerobic bacteria like *Pseudomonas aeruginosa* and aerobic photosynthetic cyanobacteria. Bacterial nanowires exhibit some interesting bioelectronic properties, with high exploitation potential in the near future (Ilshadsabah and Suchithra 2019). Both (bio)sensoristics and biomedical sectors would undoubtedly take advantages for the development of broad-impact nanostructured-based technologies and applications (Dragone et al. 2017; Grasso et al. 2020; Prosa et al. 2020). However, the reproducibility of nanosynthesis processes, the low cost-effectiveness ratio, limited production yields and the inhomogeneity in the size and the shape of microbial-synthesized nanoparticles are just some of the hurdles that still need to be overcome in order to enable a broad applicability of microbial-mediated nanobiosynthesis. Interestingly, by changing some key parameters of cell culture operational set up or through genetically engineering of selected microbial strains, a possible in vivo tuning of size, morphology and composition of nanomaterial can be achieved, as well as the enhancement in nanomaterials production yields. Thus, such microbiological and genetic strategies could represent concrete solutions for the problems currently encountered in microbial nanobiosynthesis as well as the opportunity for future developments and promising uses of these fascinating materials. In this chapter, we reported an overview of scientific literature about current main acquired knowledge, challenges, and potentiality of microbial-mediated biosynthesis, with particular reference to the isolation and screening of new microorganisms with potential for nanobiosynthesis and possible microbiological, biochemical and genetic optimizations of microbial nanobiosynthesis.

2 Microbiological Aspects in Nanobiosynthesis: Isolation, Screening, and Culture Conditions of Useful Microorganisms

The isolation and screening of microbial strains for nanomaterials biosynthesis have recently emerged as an interesting and growing research area for the exploring possible future improvements of nanobiosynthesis. Brayner et al. (2007) have compared the synthetic ability of AuNPs, AgNPs, PdNPs, and PtNPs by three different cyanobacteria *Calothrix pulvinata,* strain ALCP 745A, *Anabaena flosaquae* strain ALCPB24, and *Leptolyngbya foveolarum,* strain ALCP 671B isolated from a sample of a black soiling developed on a wall. The selection of cyanobacterial genus has played a key role in determining the NPs production yield as well as the NPs size (Brayner et al. 2007). The screening of 20 cyanobacterial strains for the SeNPs synthesis has been performed by Afzal et al. (2019). All tested cyanobacterial extracts have been capable to synthesize SeNPs but *Arthrospira indica* SOSA-4 cell extract has shown the ability to synthesize 11.8 nm-sized SeNPs in 48 h.

2.1 Nanobiosynthesis by Extremophiles

The use of extremophilic microorganisms has shown promising results for the biosynthesis of a wide variety of nanoparticles. Extremophiles microorganisms are highly metabolically and biochemically adapted microorganisms for broad range extreme conditions of temperature, salinity, pH, pressure, metal concentrations, etc. Many extremophiles microorganisms have been shown the ability to synthesize nanoparticles starting both from precursor solutions prepared in laboratory and from precursors presents in real environmental samples. Concerning the latter, the biosynthesis of nanoparticles mediated by extremophiles microorganisms might also has interesting prospective for possible application in bioremediation and waste management through the synthesis of value-added nanomaterials, e.g., the recovery of metals from mine wastes (Wong-Pinto et al. 2020), or the synthesis of magnesium oxide nanoparticles from olive mill wastewaters (Hamimed et al. 2020).

Extremophiles like thermophiles and psychrophiles microorganisms have shown a great potential for nanobiosynthetic applications (Tiquia-Arashiro and Rodrigues 2016). In particular, Antarctic environments have recently emerged as an extraordinary source of psychrophilic microorganisms for biotechnological applications, including the synthesis of nanomaterials (Yarzàbal 2016). A number of 12 psychrotolerant bacteria belonging to *Pseudomonas*, *Psychrobacter,* and *Shewanella* genera have been isolated from Antarctic samples by Plaza et al. (2016). All these bacterial strains have shown high heavy metal resistance and the ability to biosynthesized CdTeQDs. The synthesis of AgNPs using culture supernatants of five psychrophilic bacteria *Pseudomonas antarctica, Pseudomonas proteolytica, Pseudomonas meridiana, Arthrobacter kerguelensis,* and *Arthrobacter gangotriensis* have been reported by Shivaji et al. (2011). The thermophilic *Geobacillus* strain ID17 isolated from Deception Island, a volcanic island in Antarctica, has been exploited for the synthesis of Au NPs by Correa-Llantén et al. (2013).

Other extremophilic microorganisms useful for nanobiosynthesis have been isolated from many other natural environments but also from industrial wastewaters, mining sites, and contaminated sites. Some recent examples described in the literature include: (i) the synthesis of AgNPs and TeNPs has been reported for the haloarchaeaon *Halococcus salifodinae* BK3 isolated from the Ribandar salt pans of Goa (Srivastava et al. 2013, 2015), (ii) the thermophilic *Ureibacillus thermosphaericus* isolated from Ramsar geothermal hot springs with high potential for AgNPs biosynthesis (Juibari et al. 2011), (iii) a strain *Brevibacterium casei* isolated from dairy industrial waste has shown the ability to synthesize Au and Ag NPs (Kalishwaralal et al. 2010), (iv) the AgNPs biosynthesis from heavy metal-resistant bacterium *Alcaligenes faecalis* isolated from sewage wastewater (Abo-Amer et al. 2015), and (v) from *Bacillus flexus* S-27 isolated from Ag-rich soil samples from silver mining areas (Priyadarshini et al. 2013), (vi) the AuNPs synthesis by *Stenotrophomonas maltophilia* (Acc No GQ220749), a novel bacterial strain isolated from soil samples from the Singhbhum gold mines (Nangia et al. 2009), (vii) the selenite-reducing bacterium *Bacillus cereus* strain AJK3 isolated from a contaminated lake that receives industrial effluents and domestic sewage has shown the ability to synthesize SeNPs (Kora 2018) and biosynthesis Ag and Ag-AuNPs by metal-tolerant *Bacillus safensis* LAU 13 isolated from feather waste dump site (Lateef et al. 2015a, b, 2016; Ojo et al. 2016).

Furthermore, it should be mentioned the need of future investigation about the nanobiosynthetic potential of the halotolerant fungi species, i.e., fungal species able to tolerate ionic stress and able to grow in high salt concentrations environments (Tiquia-Arashiro 2019). Microorganisms that are useful for nanosynthesis can also be isolated from daily environments and even from unexpected sources. For instance, one of the most studied extremophiles, *Deinococcus radiodurans,* was discovered in 1956 in a meat can. *D. radiodurans* is particularly attractive for AgNPs, Au–Ag bimetallic NPs, and magnetic iron NPs synthesis (Kulkarni et al. 2015; Li et al. 2016a, b; Kim et al. 2019). For the synthesis of many different nanomaterials, some interesting examples of microorganisms used in food and beverage manufacturing and/or isolated from traditional foods have been reported in the literature. Semjonovs et al. (2017) have described a high efficient synthesis of bacterial nanocellulose by *Komagataeibacter rhaeticus* P 1463 isolated from kombucha, a traditional beverage prepared by fermenting sugared tea with a symbiotic culture of acetic bacteria and fungi. The efficiency in bacterial nanocellulose synthesis by *K. rhaeticus* P 1463 has been compared with two reference strains for nanocellulose production, *Komagataeibacter xylinus* DSM 46604 and *Komagataeibacter hansenii* DSM 5602T.

Fernández et al. (2019) have obtained similar results with a strain of *Komagataeibacter intermedius* JF2 isolated from commercial wine vinegar. The synthesis of biomineralized nanostructure of MnOx by *Neurospora crassa,* a type of red bread mold, has been reported by Li et al. (2016b). The ability in nanobiosynthesis of hierarchical porous spore carbon nanomaterials with the potential application in lithium–sulfur batteries has been described by Zhong et al. (2018) for the filamentous fungus *Aspergillus oryzae*, frequently used in the production of Japanese traditional foods like tofu, soy sauce, and sake. Interestingly, the synthesis of AgNPs by human gut microbiota has been recently reported by Yin et al. (2019) for the first time. Similarly, Deshpande et al. (2015) have previously reported the results for AuNPs synthesis by human skin microbiota.

2.2 *The Influence of Cultural Conditions on Microbial Nanobiosynthesis*

The modification of key culture conditions and the standardization of microbial cell culture growth protocols are potentially crucial for the control, implementation, and tuning over microbial-biosynthesized nanomaterials. The main results of research concerning aspects related to changes and optimization of culture conditions for microbial nanobiosynthesis are summarized in Table 1. Alam et al. (2017) have screened the effects on the synthesis of AgNPs by *Aspergillus niger* using three different liquid culture media: Czapex Dox, Yeast-Malt Extract, and Potato Dextrose Broth. The results obtained from UV-vis absorption spectrum measurements have suggested that Czapex Dox broth was the most suitable culture medium to promote *A. niger* cell growth and AgNPs. The effect of Nutrient Broth and Mueller Hinton Broth on the size and morphology of AgNPs synthesized by *Klebsiella pneumoniae* KCTC 2242 and *Escherichia coli* DH5α has been reported by Luo et al. (2018). The nanoparticles production yield has been higher for *K. pneumoniae* KCTC 2242 in Mueller Hinton Broth, the diameter of AgNPs synthesized from Nutrient Broth medium has been significantly smaller and nanoparticles crystallinity has been higher in those biosynthesized in Mueller Hinton Broth.

Abdul et al. (2014) have evaluated the effect of different culture media (MRS broth, JKH_1 broth and Whey solution) on TiO_2NPs synthesis using the bacterium *Lactobacillus crispatus*. The results have shown that the TiO_2NPs synthesized in MRS broth medium had a smaller average size of 70.98 nm, compared to TiO_2NPs biosynthesized in the other two culture media. Furthermore, the addition of 10% glucose in JKH1 medium has resulted in a 20% decreasing of the average size of TiO_2NPs compared to the normal JKH1 medium. The influence of biotic and abiotic factors of cell culture operational setup on nanomaterials biosynthesis has been extensively reported in literature. These factors can include (i) microbial cell

Table 1 Cell culture conditions for microbial biosynthesis of different nanomaterials

Microorganisms	Nanomaterials	Changes in cell culture conditions	Results	References
Aspergillus niger	AgNPs	Type of culture medium	Promotion of cell growth and NPs synthesis	Alam et al. (2017)
Klebsiella pneumoniae KCTC 2242 *Escherichia coli* DH5α	AgNPs	Type of culture medium	Effects on the size and morphology of NPs	Luo et al. (2018)
Lactobacillus crispatus	TiO_2NPs	Type of culture medium	Effects on the size of NPs	Abdul et al. (2014)
Trichoderma WL-Go	SeNPs	Time, precursor concentration, pH	Optimization in NPs synthesis	Diko et al. (2020)
Pseudomonas aeruginosa strain RB	CdSe NPs	pH, T, NaCl concentration	Improvements in purity and production rate of NPs	Ayano et al. (2015)
Saccharomyces cerevisiae BY4742	CdSe QDs	Growth phase, time, precursor concentration	Increase in QDs production	Brooks and Lefebvre (2017)
Shewanella oneidensis MR-1	Ag_2S NPs	Precursor concentrations, no cell washing after growth	Increase in NPs production yield	Voeikova et al. (2018)
Escherichia coli DH5α	AgNPs	Precursor concentration, T and pH	Optimization in NPs synthesis	Gurunathan et al. (2009)
Deinococcus radiodurans	AgNPs	Precursor concentration and pH	Effects on NPs morphology, composition, size and production yield	Chen et al. (2017)
Shewanella algae ATCC 51181	AuNPs	pH	Effects on the NPs morphology and the cellular site of Au deposition	Konishi et al. (2007)
Verticillium luteoalbum DSM 63545	AuNPs	pH	Effects on the NPs size, shape and number	Gericke and Pinches (2006)

(continued)

Table 1 (continued)

Microorganisms	Nanomaterials	Changes in cell culture conditions	Results	References
Morganella psychrotolerans	AgNPs	T	Effects on NPs morphology	Ramanathan et al. (2011)
Trichoderma viride	AgNPs	T	Effect on size and monodispersity of NPs	Mohammed Fayaz et al. (2009)
Yarrowia lipolytica NCIM 3589	AuNPs, Au nanoplates	Cell concentration, salt concentration and T	Effect on the type and size of NPs; effect on the release of Au nanostructures from the yeast cell wall into the aqueous phase	Pimprikar et al. (2009)
Desmodesmus abundans and *Spirulina platensis*	AgNPs	CO_2 mitigation systems	Enhancement in NPs production	Mora-Godínez et al. (2020)
Various Cyanobacteria and Chlorophyta strains	AgNPs	Light/dark condition	Effects on the ability to form NPs	Patel et al. (2015)
Coscinodiscus wailesii	Frustules	Ni concentration	Alteration in frustules pore sizes	Townley et al. (2007)
Pinnularia sp Thalassiosira weissflogii	Frustules	TiO_2 addition	Doping of frustules	Jeffryes et al. (2008a), Lang et al. (2013)
Nitzschia sp. *Pinnularia* sp.	Frustules	$Ge(OH)_4$ or GeO_2 addition	Doping of frustules	Rorrer et al. (2005); Qin et al. (2008); Jeffryes et al. (2008b, c)
Coscinodiscus granii	Frustules	Intensity and spectral characteristics of light	Effects on frustule valves thickness, Si concentration, frustule diameter, foramen characteristics; changes in photonic properties of frustules	Su et al. (2015, 2017, 2018a, b)
Gluconobacter xylinus	Bacterial nanocellulose	Type of culture medium	Enhanced conversion of the carbon source to bacterial nanocellulose	Chandrasekaran et al. (2017)

(continued)

Table 1 (continued)

Microorganisms	Nanomaterials	Changes in cell culture conditions	Results	References
Komagataeibacter xylinus MSKU 12 strain	Bacterial nanocellulose	Use of alternative low-cost culture medium	Biosynthesis of bacterial nanocellulose	Naloka et al. (2020)
Komagataeibacter xylinus PTCC 1734 strain	Bacterial nanocellulose	Use of alternative low-cost culture medium	Biosynthesis of bacterial nanocellulose	Barshan et al. (2019)
Komagataeibacter rhaeticus	Bacterial nanocellulose	Use of alternative low-cost culture medium	Biosynthesis of bacterial nanocellulose	Machado et al. (2018)
Magnetospirillum genus	Bacterial magnetosomes	O_2 concentration in culture medium	Enhancement in magnetosomes production yield	Heyen and Schüler (2003), Liu et al. (2010), Zhang et al. (2011), Sun et al. (2008)
Magnetospirillum magneticum AMB-1	Bacterial magnetosomes	Addition of Iron-chelating agents	Enhancement in magnetosomes production; modification of size and chain length; improved heating properties	Alphandéry et al. (2012)

concentration; (ii) precursor(s) concentration(s); (iii) pH; and (iv) temperature. For instance, it is widely known that environmental factors including carbon source, pH, and temperature can strongly influence the composition of the fungal secretome (McCotter et al. 2016).

Fungal secretome include all the proteins and enzymes secrete by fungi, for either anchored to the cell surface or freed in the extracellular environment. Fungal secretome have been already investigated for promising biotechnological applications, including fungi-mediated biosynthesis of nanoparticles (see Sect. 3). The optimal conditions for the extracellular reduction of SeO_2 to Se^0 for SeNPs synthesis in the culture broth of the fungus *Trichoderma* WL-Go have been recently investigated by Diko et al. (2020). As reported in many works for different microorganisms, optimal conditions for nanoparticles synthesis not always match with optimal conditions for microbial growth. One example has been provided by the results obtained from a study aimed to the identification of the optimal conditions of pH, temperature, and NaCl concentration the bacterium *P. aeruginosa* RB to achieve high purity and high synthesis rate for CdSe NPs synthesis (Ayano et al. 2015). Brooks and Lefebvre (2017) have defined the optimal culture conditions for CdSe quantum dots (CdSe QDs) synthesis by the yeast *Saccharomyces cerevisiae* BY4742. The CdSe QDs have been produced most efficiently when *S. cerevisiae* cells enter into the stationary phase, incubated for 6 h in 1 mM selenite and then incubated in fresh growth medium containing 3 mM Cd (II). The maximum of CdSe QDs synthesis was reached at 84 h, resulting in an increased CdSe QDs production of 70% respect to the previous works (Brooks and Lefebvre 2017).

The work reported by Voeikova et al. (2018) has highlighted how the composition of the nutrient medium could affect the efficiency of Ag_2S NPs biosynthesis using the bacterium *Shewanella oneidensis* MR-1. Through the use of 10 mM concentrations for both the precursors $AgNO_3$ and $Na_2S_2O_3$ and the elimination of cell washing step after the cells have grown in rich nutrient media, the production yield value has increased up to 50–60% compared to 15–20% of a previous described protocol (Voeikova et al. 2018). The effects of precursor concentration, temperature, and pH of the culture supernatant on AgNPs synthesis and particle sizes have been also described by Gurunathan et al. (2009) for *E. coli* strain DH5α. The optimal condition of 5 mM concentration of Ag^+ ions, temperature of 60 °C, pH 10 have shown the maximum synthesis of AgNPs through reduction of Ag^+ ions by the culture supernatant of *E. coli* DH5α.

Chen et al. (2017) have highlighted the relationships between changes in precursor concentration and pH and the AgNPs biosynthesis in *D. radiodurans*, highlighting how the tuning of such specific abiotic factors could provide control over nanoparticles morphology, composition, size, and production yield. In particular, the stationary phase culture of *D. radiodurans* exposed for 2 h to 0.1 and 0.5 mM $AgNO_3$ has shown about 2-times increase in AgNPs production yield compared to the control. When the pH of the supernatant was increased from 7 to 9, the AgNPs production is also increased by about 4 times. These results have shown that tuning of specific environmental stressors could provide control over morphology, composition, and size of nanoparticles synthesized by *D. radiodurans* (Chen et al. 2017).

Konishi et al. (2007) have also described the importance of pH as controlling factor over the morphology of AuNPs and the cellular site of Au deposition in the bacterium *Shewanella algae* ATCC 51181. Gericke and Pinches (2006) have reported that variations in pH could affect in a very peculiar way the size, shape and number of AuNPs produced by the fungus *Verticillium luteoalbum* DSM 63545. In particular, changes in the pH value from 3 to 9 have produced: (i) a greater heterogeneity in NPs size (ii) a decrease numbers of NPs produced per cell and (iii) a greater heterogeneity in shape. In particular, the spherical shape has been predominant at pH 3, while triangular-, hexagonal-, spherical-, and rod-shaped NPs have been produced at pH 5 NPs and small spherical NPs and bigger NPs with irregular, undefined shapes have been observed at pH 7 and 9. These results have suggested that pH would play an important role in the tuning of nanoparticles morphology produced by *V. luteoalbum* DSM 63545. The variation of temperature seemed to exert a similar tuning effect on the morphology of AgNPs synthesized by *Morganella psychrotolerans* (Ramanathan et al. (2011). These results have shown that small spherical-shaped AgNPs were produced at 20 °C, the optimal temperature for the growth and physiological activities of *M. psychrotolerans*, while triangular and hexagonal nanoplates along with spherical nanoparticles have been observed at 5 °C above and below 20 °C. At 4 °C, the number of nanoplates was greater than the number of spherical NPs, but larger compared to those obtained at 20 °C. Other recent examples include the temperature dependence of size and monodispersity observed in AgNPs biosynthesis by mold *Trichoderma viride* (Mohammed Fayaz et al. 2009) and the effect of temperature on the release of Au nanostructures from the yeast *Yarrowia lipolytica* NCIM 3589 cell wall into the aqueous phase. Changes in type and size of Au nanostructures have been also observed though modification in the proportion of cell concentration and in the concentration of the precursor gold salt (Pimprikar et al. 2009).

The photoautotrophic metabolism of microalgae and cyanobacteria is based on carbon dioxide (as carbon source), light (as energy source), inorganic nutrients, and water. This condition generally reduces the costs of culture media (compared to the costs for culture media used for the growth of bacteria, yeasts, and molds) thus promoting the future scaling-up from the laboratory to the industrial scale, also through the design and the development of atmospheric CO_2 mitigation microbial-based technologies. One example has been provided by a recent study, where a potentiation in AgNPs biosynthesis has been observed for high CO_2 acclimated strain green algae *Desmodesmus abundans* and *Spirulina platensis* (Mora-Godínez et al. 2020). The effects of light on the ability to produce nanoparticles have been also reported in the literature for photosynthetic microorganisms like cyanobacteria and green algae. For instance, the influence of light/dark condition on the AgNPs biosynthesis by different selected cyanobacterial and green algae strains has been investigated by Patel et al. (2015).

Among abiotic parameters, intensity and spectral characteristics of light play a primary role for diatoms cultivation and can strongly influence frustule characteristics. Su et al. (2015) have found that in *Coscinodiscus granii,* the use of red, yellow, green, and blue light at 300 μmol photons $m^{-2}s^{-1}$ can affect valve thickness, Si concentration, frustule diameter, foramen characteristics. All these morphological changes influence photonic properties of frustules (Su et al. 2015, 2017, 2018b). Although the controlling of biomineralization process remains a challenge, these results could have important implication on the development of light-based modification methods of diatom frustules. The effect on diatom frustules properties (e.g., pore sizes and pore density) has been also explored by changing operational parameters of experimental setup and abiotic factors as pH, salinity, pH, temperature, nutrient concentration, and precursor $Si(OH)_4$ concentration (Su et al. 2018a). Townley et al. (2007) have reported alteration in pore sizes of *Coscinodiscus wailesii* frustules when exposed to sublethal concentration of Ni. In diatoms, the possibility of in vivo doping of frustules has been described through the addition of sublethal concentrations of given precursors to the culture medium. For instance, the doping of diatom *Pinnularia* sp. frustules or diatoms *Thalassiosira weissflogii* frustules with titania (TiO_2) has been reported (Jeffryes et al. 2008a; Lang et al. 2013) as well as the nanobiosynthesis of frustules containing Si–Ge oxides nanocomb in diatoms *Nitzschia* and *Pinnularia* by adding $Ge(OH)_4$ or GeO_2 in the culture medium (Rorrer et al. 2005; Qin et al. 2008; Jeffryes et al. 2008b, c).

The costs of culture media for microbial growth should be seriously considered not to limit the future large-scale productions of microbial biosynthetic nanomaterials. A notable example is bacterial nanocellulose whose application in biomedical field is still limited, mainly because of culture medium costs (Jacek et al. 2019a) and low bacterial nanocellulose production yield. Chandrasekaran et al. (2017) have reported a yield enhancement of bacterial nanocellulose synthesis in *Gluconobacter xylinus* cultured in Super Optimal Broth, with a 50% conversion of the carbon source to bacterial nanocellulose, compared to a 7% conversion in the traditional Hestrin-Schramm medium. Several researches have attempted to produce bacterial nanocellulose from various alternative low-cost substrates using agro-wastes industrial by-products. This is a case in point of the use of a white biotechnology approach in microbial nanobiosynthesis, i.e., the microbial conversion of agro-wastes into a value-added product like bacterial nanocellulose. The results obtained using agro-wastes-based low-cost media for the production of bacterial nanocellulose have been described in some recent works: a low-cost coconut water medium containing 1% (v/v) acetic acid and 2% (v/v) ethanol for the thermotolerant bacterium *K. xylinus* MSKU 12 (Naloka et al. 2020), vinasse 40% culture medium for *K. xylinus* PTCC 1734 (Barshan et al. 2019), and sugarcane molasses for *K. rhaeticus* (Machado et al. 2018).

Taken together, these results have provided interesting insight about how bacterial nanocellulose could be synthesized with good production yields using low-cost media. The production of bacterial magnetosomes is characterized by some drawbacks, including specific microaerophilic growth conditions required, low yield (Yan et al. (2017) and time-consuming purification steps (Xu et al. 2014). For different batch, fed-batch, and semi-continuous microbial growth strategies reported in literature, O_2 concentration seemed to be one of the most important abiotic parameters for bacterial magnetosomes production (Heyen and Schüler 2003; Sun et al. 2008;

Liu et al. 2010; Zhang et al. 2011). In the genus *Magnetospirillum*, the most widely exploited and studied magnetotactic bacteria genus, microaerobic (O_2 tension < 10%) or anaerobic conditions have enhanced magnetosomes production yield. The composition of culture medium has also shown to be an important factor. The use of nitrate and Fe(III) citrate as nitrogen and iron sources, respectively, has improved magnetosomes formation in *Magnetospirillum* strains (Heyen and Schüler 2003). As reported by Alphandéry et al. (2012), the presence of iron-chelating agents in culture medium with a concentration range between 0.4 and 40 μM (i) stimulated the cellular growth (ii) it enhanced the production of magnetosomes and (iii) it modified the size and chain length of magnetosomes in *Magnetospirillum magneticum* AMB-1. In addition, remarkable changes in these magnetosome properties have resulted in improved heating properties of magnetosomes. The heating properties of magnetosomes are important for possible application in cancer focal therapies like magnetic hyperthermia.

3 The Biochemistry Behind Microbial Nanobiosynthesis

The understanding of the involved biochemical mechanisms of nanomaterials synthesis by microorganisms is a key aspect of nanobiotechnology research in order to define more reliable and reproducible nanobiosynthetic methods and to spur future developments through the wide opportunities offered by genetic engineering tools (see Sect. 4). The main results of research concerning the identification of biochemical mediators involved in microbial nanobiosynthesis are summarized in Table 2. During the last years, the knowledge about the biosynthetic mechanisms of nanostructured materials of microbial origin like diatom frustules, magnetosomes, bacterial nanowires, and bacterial nanocellulose has evolved, providing significant insights into fundamental aspects of biochemical mechanisms and the molecular and enzymatic components involved. Concerning diatom cell wall silicification and frustules biogenesis, the complete understanding of cellular processes involved is still a challenge. A greater knowledge has been achieved about the role of proteins silaffins and long-chain polyamines in biosilicification process (Kröger and Poulsen 2008) and the possible involvement of cellular translocators (Yee et al. 2020). The magnetosome biogenesis in *Magnetospirillum* species is a very complex and still not completely elucidated process. To date, six different incomplete models have been proposed to elucidate magnetosome formation (Yan et al. 2017). Concerning bacterial nanowires, most of the studies have focused on *Geobacter sulfurreducens* (Lovley and Walker 2019), on other Gram-negative bacteria (Hospenthal et al. 2017) and some studies have been conducted on cyanobacteria *Microcystis* (Sure et al. 2015) and *Synechocystis* (Chen et al. 2020). The molecular mechanisms of bacterial nanocellulose biosynthesis by *Komagataeibacter* genus are very well characterized (Jacek et al. 2019a).

In addition to these aforementioned unique nanostructured materials of microbial origin, microorganisms have showed the ability to synthesize nanoparticles

Table 2 Biochemical mediators of microbial nanobiosynthesis processes

Microorganisms	Nanomaterials	Enzymes, reducing agents and stabilizing agents	References
Chlamydomonas reinhardtii	AgNPs	ATPase, sedoheptulose-1,7-bisphosphatase, carbonic anhydrase, ferredoxin $NADP^+$ reductase, superoxide dismutase, oxygen evolving enhancer protein, ribulose bisphosphate carboxylase, and H4 nuclear histone protein	Barwal et al. (2011)
Nitzschia	AuNPs	Catalase and peroxidase	Borase et al. (2017)
Neochloris oleoabundans	AgNPs	Chlorophylls	Bao and Lan (2018)
Amphora-46	AgNPs	Fucoxanthin	Jena et al. (2015)
Bacillus clausii	AgNPs	Nitrate reductase	Mukherjee et al. (2018)
Bacillus subtilis spizizenii NCIM-206	AgNPs	Nitrate reductase, endo-alpha-(1->5)-l-arabinanase, cyclic-di-AMP phosphodiesterase NADPH dehydrogenase	Sable et al. (2020)
Stenotrophomonas maltophilia SeITE02	SeNPs	Cytoplasmic thiol and/or peptides/proteins, alcohol dehydrogenase	Lampis et al. (2017)
Pseudomonas stutzeri TS44	SeNPs CdSe QDs	Glutathione reductase Reduced glutathione	Wang et al. (2019)
Pseudomonas aeruginosa ATCC 27853	AgNPs	Alkyl hydroperoxide reductase and azurin	Quinteros et al. (2019)
Shewanella oneidensis MR-1	UO_2NPs	Decaheme cytochrome MtrC	Marshall et al. (2006)
Geobacter sulfurreducens	AgNPs	c-cytochromes	Vasylevskyi et al. (2017)
Shewanella oneidensis MR-1	AuNPs	Proteins and co-enzyme complexes	Huang et al. (2019)
Shewanella oneidensis MR-1	AuNPs	Enzymes involved in nitrogen cycle, cell wall reductive enzymes and/or soluble secreted enzymes	Manivasagan et al. (2016), Manimaran and Kannabiran (2017)
Spirogyra insignis	AgNPs AuNPs	Pectins	Castro et al. (2013)

(continued)

Table 2 (continued)

Microorganisms	Nanomaterials	Enzymes, reducing agents and stabilizing agents	References
Aspergillus tubingensis AY876924	AgNPs	Glycoamilase, acid phosphatase, serine carboxypeptidase, and glucanosyltransferase	Ballottin et al. (2016)

either through extracellular or intracellular mechanisms. Bacteria-mediated nanoparticles synthesis mainly occurs through intracellular mechanisms while yeasts- and molds- and microalgae-mediated nanobiosynthesis can occur either through extracellular or intracellular mechanisms. Many of these biochemical mechanisms have been described as part of microbial resistance mechanisms for cellular detoxification that involves changes in solubility of inorganic ions by enzymatic reduction and/or precipitation of soluble toxic to insoluble non-toxic nanostructures.

The proposed biochemical mechanisms described in the literature have frequently suggested the possible role of enzymes in reduction reactions for the synthesis of nanoparticles, in particular oxidoreductase enzymes such as NADH-dependent nitrate reductase, NADPH-dependent sulfite reductase flavoprotein subunit α, cysteine desulfhydrase, electron transport chains, and other cellular components (Hulkoti and Taranath 2014; Shankar et al. 2016; Banerjee and Ravishankar Rai 2018). The enzymes may exert both a size-control on nanoparticles synthesis and act as stabilizing agents of nanoparticles. The role of microbial macromolecules and cellular secondary metabolites as reducing stabilizing and/or controlling agents has been also proposed in many works, as well as the involvement of physicochemical processes like complexation, nucleation, and capping to possibly mediate the biosynthesis and stabilization of nanoparticles. In particular, the chemical interactions between biocapping layer and nanoparticles could also likely contribute to the electrostatic and steric stabilizations of biosynthesized nanoparticles, generally characterized by a higher thermodynamic stability compared to chemically synthesized nanoparticles (Piacenza et al. 2018).

Future research will contribute to elucidate the stabilization mechanisms, also through a more complete characterization of biocapping layer agents and through the optimization of the required post-biosynthesis and/or purification treatments. The biochemical aspects of nanoparticles biosynthesis in photosynthetic microorganism have been also elucidated. Various proteins and enzymes possibly involved in the control of synthesis rate of AgNPs in *Chlamydomonas reinhardtii* have been identified by Barwal et al. (2011). The proteins and enzymes identified include ATPase, sedoheptulose-1,7-bisphosphatase, carbonic anhydrase, ferredoxin $NADP^+$ reductase, superoxide dismutase, oxygen evolving enhancer protein, ribulose bisphosphate carboxylase and H4 nuclear histone protein. In the diatom *Nitzschia,* roles of enzymes such as catalase and peroxidase as mediator in AuNPs biosynthesis have been outlined (Borase et al. 2017).

The involvement of photosynthetic pigments in nanoparticles assembly has been also described in literature. The essential role of chlorophylls in light-dependent

AgNPs synthesis in the microalga *Neochloris oleoabundans* has been demonstrated (Bao and Lan 2018). Another study has highlighted the active role of xanthophyll pigment fucoxanthin as photo-reducing agent for AgNPs formation through light-dependent reactions in diatom *Amphora*-46 (Jena et al. 2015). In cyanobacteria, the involvement in nanoparticles synthesis has been ascribed to several biochemical entities including polysaccharides, pigments, phytochelatins, NADPH-dependent reductase and nitrogen-fixing enzyme nitrogenase (Pathak et al. 2018). In bacteria belonging to the genus *Bacillus*, one of the most studied and promising bacterial genus for industrial application, two recent studies have confirmed the role of enzymes in the synthesis of AgNPs. Mukherjee et al. (2018) have adopted in silico approaches to profile the NADPH-dependent nitrate reductase enzyme involved in AgNPs using *Bacillus clausii*. The presence of reducing and capping agents in AgNPs biosynthesized by *Bacillus subtilis spizizenii* NCIM-2063 has been confirmed. The list of these reducing and capping agents include enzymes nitrate reductase, endo-alpha-(1->5)-l-arabinanase, cyclic-di-AMP phosphodiesterase NADPH dehydrogenase (Sable et al. 2020).

The bacterial genus *Pseudomonas* has been also widely studied for the production of primary and secondary metabolites for broader range of possible applications such as industrial, medical, and environmental with promising uses in microbial nanobiosynthesis. The intracellular ability to assembly and to extracellular release SeNPs has been studied in the bacterium *S. maltophilia* SeITE02. This bacterium has demonstrated a 100% reduction ability of 0.5 mM SeO_3^{2-} within 48 h growth. The involvement of cytoplasmic thiol containing molecules and/or peptides/proteins in the reduction of SeO_3^{2-} to Se^0 as well as the possible involvement of the identified alcohol dehydrogenase homolog in the extracellular SeNPs biogenesis process have been described (Lampis et al. 2017). The importance of glutathione reductase and reduced glutathione has been highlighted in *Pseudomonas stutzeri* TS44 for the formation of SeNPs CdSe QDs, respectively (Wang et al. 2019). Some capping agents in AgNPs biosynthesized by *P. aeruginosa* ATCC 27853 have been identified (Quinteros et al. 2019). Among the identified proteins, the involvement in AgNPs formation has been suggested for alkyl hydroperoxide reductase and azurin, while the role of stabilizing agent has been suggested for outer membrane protein OprG and glycine zipper 2 TM domain-containing protein.

Still on bacteria, under anaerobic conditions, dissimilatory metal-reducing bacteria display a remarkable respiratory versatility in the number of terminal electron acceptors, including insoluble metal electron acceptors. Such ability is largely due to the multihaeme c-type cytochromes in the respiratory electron transfer chain. The importance of the outer membrane decaheme cytochrome MtrC in extracellular synthesis UO_2 nanoparticles by the dissimilatory metal-reducing bacterium *S. oneidensis* MR-1 has been reported (Marshall et al. 2006). Vasylevskyi et al. (2017) have also described the role of the c-cytochromes of the electron transfer chain as biocatalytic component involved in AgNPs biosynthesis by the dissimilatory metal-reducing bacteria *G. sulfurreducens*. Recently, the involvement of extracellular electron transport pathway has been also revealed in photo-driven AuNPs biosynthesis by *S. oneidensis* MR-1 (Huang et al. 2019).

The filamentous bacteria actinomycetes can mediate both intracellular and extracellular synthesis of metal nanoparticles. It has been suggested that the first step of intracellular synthesis could involve the electrostatic binding of metal ions to the negatively charged carboxylate groups in enzymes present both on the cell wall and on the cytoplasmic membrane. Concerning extracellular synthesis, nitrogen cycle enzymes, cell wall reductive enzymes, and/or soluble secreted enzymes could exert a biocatalytic function; (Manimaran and Kannabiran 2017).

Compared to intracellular biosynthesis of nanoparticles, extracellular biosynthesis poses several undoubted advantages in terms of reduced nanoparticle downstream purification processes required as well as a possible reuse of cell cultures for new biosynthetic processes because lysis of cell to extract the nanoparticles is not required. The reducing hydroxyl groups of pectins located in the outer cell wall of the green alga *Spirogyra insignis* have been suggested to act as reducing agents to mediate AuNPs and AgNPs synthesis (Castro et al. 2013). The interesting role of the surface-layer in *D. radiodurans* as biological scaffold and biotemplate for the synthesis of nanomaterials, including noble metal and bimetallic nanoparticles, quantum dots, and nanopillars have been reviewed by Li et al. (2019).

The set of proteins and enzymes secreted by fungi, for either anchored to the cell surface or freed in the extracellular environment; i.e., fungal secretome components have been suggested to act as reducing and capping agents in nanoparticles biosynthesis (Kitching et al. 2015). In particular, molds belonging to *Penicillium, Fusarium, Trichoderma,* and *Aspergillus* genera have been extensively studied for nanosynthesis (Guilger-Casagrande and Lima 2019; Barabadi et al. 2019; Elegbede et al. 2018, 2019, 2020). However, the biochemical mechanisms involved in nanoparticles synthesis by fungal secretome have not yet been completely elucidated and characterized. Furthermore, fungal secretome possess a greater variety of extracellular enzymes in comparison with bacterial secretome and the identity and functions of many components of fungal secretome have not been completely elucidated (Bouws et al. 2008). To date, the promising potential of fungal secretome in nanobiosynthesis has yet to be explored. Ballottin et al. (2016) have studied the interactions between the secretome of *Aspergillus tubingensis* AY876924 and the biosynthesized AgNPs, and they have also characterized the components of AgNPs capping layer. In particular, mass spectrometry analyses have enabled the identification of eight proteins present in the capping layer of AgNPs, including glycoamilase, acid phosphatase, serine carboxypeptidase, and glucanosyltransferase, all involved in carbon, phosphorous and nitrogen uptake, and for the fungal growth.

Despite the current good level of knowledge about the biofabrication mechanisms of nanomaterials at cellular and molecular level, for the outlining of solutions concerning the control of monodispersity and nanoparticles size and in general to gain a greater control and tuning over all the biosynthetic processes steps, further biomolecular studies will be required.

4 Genetic Engineering and Synthetic Biology: From a Deeper Knowledge in Microbial Nanobiosynthesis to the Construction of Microbial 'Nanofactories'

For a proper scale-up of the microbial nanobiosynthesis processes that make microbial nanobiosynthesis industrially worthwhile, further steps to solve some key drawbacks will be required. One of these drawbacks is the inhomogeneity in the size and the shape of microbial-synthesized nanoparticles often reported in literature. Dispersity, i.e., the size distribution of the nanoparticle population as well as their shape of bionanoparticles of microbial origin, is key properties that strongly influence the nanoparticle's behavior in fluids but also their electronic and optical properties. Thus, the control over dispersity and shape is fundamental not to limit future applications of these fascinating materials. As described in Sect. 2, focused changes in culture conditions, optimization and standardization of microbial culture growth protocols can strongly contribute to the control, tuning, and to the improvement of biosynthesized microbial nanomaterials.

Today, genetic engineering techniques provide very powerful tools to enhance the performance of microbiological-based producing systems, while providing the identification of involved biochemical entities as well as a more complete knowledge of metabolic networks and controlling factors involved in microbial nanobiosynthesis. The main results of research concerning the application of genetic engineering to microbial nanobiosynthesis are summarized in Table 3. Wide arrays of genetic engineering toolbox for gene expression control are now available. Genetic techniques like gene silencing and gene deletion that enable the downregulation of gene expression can be useful for the identification of the specific role of a given gene(s) and gene(s)-encoded product(s) (proteins and/or enzymes) in microbial nanobiosynthesis. Thus, these genetic approaches can be very useful for the elucidation of metabolic pathways involved in microbial nanobiosythesis.

4.1 Microbial Synthesis of Metal Nanoparticles by Mutant and Engineered Cells

The silver-resistant strain and AgNPs producer *E. coli* 116AR have been isolated and studied by Lin et al. (2014). The deletion mutation of the gene encoding NapC (ΔnapC), the membrane-anchored tetra-heme c-type cytochrome subunit of the periplasmic nitrate reductase has resulted in a marked decrease in the cellular AgNPs accumulation, thus allowing to identify the molecular mechanism behind the AgNPs biosynthesis in *E. coli* 116AR (Lin et al. 2014). A novel aerobic selenite reductase CsrF able to form Se(0) and Cr(III) NPs in the bacterium *Alishewanella* WH16-1 has been identified through the *csrF* gene disruption (Xia et al. 2018). In addition to gene silencing and gene deletion, gene overexpression strategies are equally powerful genetic engineering tool to identify biochemical pathways underlying microbial

Table 3 Use of genetic engineering in microbial nanobiosynthesis

Microorganisms	Nanomaterials	Genetic tools	Results	References
Alishewanella WH16-1	Se(0) NPs Cr(III) NPs	Disruption of *csrF* gene	Identification of a novel aerobic selenite reductase	Xia et al. (2018)
Saccharomyces cerevisiae	CdSe QDs	Silencing of *met6* gene	Identification of the role of MET6 methionine synthase	Shao et al. (2018)
		Overexpression of *met6* gene	Increasing in CdSe QDs production yield	
Pichia pastoris	AgNPs SeNPs	Overexpression of *cyb5r* gene	Stable production of spherical NPs	Elahian et al. (2017)
	AuNPs PdNPs		Biosynthesis of monodispersed spherical NPs	Elahian et al. (2020)
Escherichia coli 116AR	AgNPs	Deletion mutation in *napC* gene	Identification of the role of NapC subunit of the periplasmic nitrate reductase	Lin et al. (2014)
Escherichia coli R189	CdS NPs	Co-overexpression of *pcs* gene from *Schizosaccharomyces pombe* and *gshi* gene	Biosynthesis of CdS NPs	Kang et al. (2008)
Escherichia coli DH5α	CdSeZn PrGd CdCs FeCoNPs	Co-expression of *atPCS* genes from *Arabidopsis thaliana* and *ppMT* gene from *Pseudomonas putida*	Biosynthesis of NPs never obtained before using chemical methods	Park et al. (2010)

(continued)

Table 3 (continued)

Microorganisms	Nanomaterials	Genetic tools	Results	References
	CdSeZn PrGd CdCs FeCo	Co-expression of *pcs* gene and/or *mt* gene		Choi et al. (2018)
	AgNPs	Transformation with a *mt* gene from *Candida albicans*	Enhancement of cellular growth and AgNPs production yield	Yuan et al. (2019)
Escherichia coli BL21	CdSe QDs	Introduction of *cds* gene	Biosynthesis of CdSe QDs	Mi et al. (2011)
*Escherichia coli*AG1		Overexpressing the *gshA* gene		Monrás et al. (2012)
Escherichia coli NEB 10-beta	Pilin-based electrically conductive protein nanowires	Co-expression genes encoding type IV pili biogenesis machinery from *Geobacter sulfurreducens* and a synthetic gene designed to yield a peptide monomer	Biosynthesis of electrically conductive protein nanowires with same diameter that those produced by *G. sulfurreducens*	Ueki et al. (2020)
Shewanella oneidensis MR-1	AgNPs Ag_2S NPs	Deletion mutation of *mtrC* and *omcA* genes	Control over size and properties of NPs	Ng et al. (2013)
	SeNPs CdSe NPs	Deletion mutation of *cymA* gene	Rapid and a high production of hexagonal, ultrafine and uniform-sized CdSe nanoparticles	Tian et al. (2017)
		Overexpression of *cymA* gene	High production of larger-sized spherical SeNPs	
Magnetospirillum gryphiswaldense	Magnetosomes	Bacterial conjugation with a donor strain of *E. coli*	Increase in magnetosomes production yield	Liu et al. (2008)

(continued)

Table 3 (continued)

Microorganisms	Nanomaterials	Genetic tools	Results	References
		Genomic amplification of single and multiple magnetosome gene clusters involved in magnetosomes biosynthesis	Control over magnetosome size and number	Lohße et al. (2016)
	Magnetosomes	Gene fusion of *mamA*, *mamG,* and *mamF* genes with different genes encoding enzymes beta-glucuronidase and glucose oxidase, mEGFP fluorophore and red fluorescent binding protein	Multiple surface functionalization of magnetosomes membrane	Mickoleit et al. (2020)
Desulfovibrio magneticus RS-1	Magnetosomes	Production and isolation of mutants using random mutagenesis and DNA sequencing	Identification of the role of *mad* genes in the morphology control of magnetosomes	Faivre and Baumgartner (2015)
Rhodospirillum rubrum *Magnetospirillum gryphiswaldense*	Organic-coated Fe_3O_4 nanoparticles	Heterologous expression of *mamAB*, *mamGFDC*, *mamXY*, and *mms6* genes from *Magnetospirillum gryphiswaldense*	Stable production for at least 40 cellular generations of *Rhodospirillum rubrum*	Kolinko et al. (2014)
Magnetospirillum magneticum AMB-1	Magnetosomes	Gene fusion of *mamC* gene encoding magnetosome membrane protein with *opd* gene encoding enzyme organophosphohydrolase from *Flavobacterium* ATCC 27551	Enzyme-labeled magnetosomes	Ginet et al. (2011)
		Gene fusion of *mamC* gene encoding magnetosome membrane protein with gene encoding RGD peptide	RGD-labeled magnetosomes	Boucher et al. (2017)

(continued)

Table 3 (continued)

Microorganisms	Nanomaterials	Genetic tools	Results	References
Thalassiosira pseudonana	Frustules	Expression of gene encoding immunoglobulin G-binding domain of protein G	Production of antibody-labeled genetically modified diatoms	Delalat et al. (2015)
		Multi-expression of gene encoding a single domain antibody against the EA1 S-layer protein of *Bacillus anthracis* Sterne strain and gene encoding a single chain variable fragment against the explosive TNT	Labeling of multiple antibodies on the frustules surface	Ford et al. (2016)
Komagataeibacter hansenii ATCC 23769	Bacterial nanocellulose	Overexpression of *motA* and *motB* genes	Increase in nanocellulose production yield	Jacek et al. (2019b)
Komagataeibacter oboediens MSKU 3 E3		Production of revertant strains from a *bcsC* loss-of-function mutant	Increase in nanocellulose production yield Production of bacterial nanocellulose with small average diameter and homogeneous fine fibrils with a narrower range of width	Taweecheep et al. (2019)
Geobacter sulfurreducens	Bacterial nanowires	Genetic modification of *pilA* gene	Production of nanowires with increased in electrical conductivity and reduced diameter	Tan et al. (2016)
Geobacter sulfurreducens MP *Geobacter metallireducens* GS15		Heterologous expression of *pilA* gene from *G. metallireducens* GS15	Production of nanowires with increased in electrical conductivity	Tan et al. (2017)

(continued)

Table 3 (continued)

Microorganisms	Nanomaterials	Genetic tools	Results	References
Geobacter sulfurreducens strain KN400		Genetic engineering for the production of *G. sulfurreducens* strain PilA-WT/PilA-6His	Addition of 9 peptides of to the amino acid monomer backbone of electrically conductive protein nanowires	Ueki et al. (2019)

nanobiosynthesis as well as very useful strategies to increase microbial nanomaterials production yield and/or to control the size and morphological features of nanomaterials. These gene overexpression strategies can include both homologous expression in the same original microorganism and heterologous expression in a different host microorganism.

The study of Shao et al. (2018) has shown how the biosynthesis CdSe QDs can be improved through the genetic modification of regulation mechanism in the Se metabolic flux of yeast *S. cerevisiae*. First, through gene silencing it has been demonstrated that *met6* gene encoding methionine synthase involved in the selenocysteine-to-selenomethionine pathway regulates the intracellular biosynthesis of CdSe QDs in yeast *S. cerevisiae*. Then, the overexpression of the *met6* gene in *S. cerevisiae* cells has increased CdSe QDs production yield compared to the wild type strain, 41.6 ± 3.1 nmol CdSe QDs per 50 mL of medium and 13.1 ± 3.6 nmol CdSe QDs per 50 mL, respectively (Shao et al. 2018). Elahian et al. (2017) have selected a strain of the yeast *Pichia pastoris* with Mut^+ phenotype and they have overexpressed the *cyb5r* gene, a metal-resistant gene encoding NADH-cytochrome b5 reductase. The recombinant *P. pastoris* has shown the production of stable 70–180 nm-sized spherical AgNPs and SeNPs. In a later paper, Elahian et al. (2020) have used the *cyb5r* gene overexpression to produce a recombinant strain of *P. pastoris* able to biosynthesize 100 nm-sized, monodispersed spherical AuNPs and PdNPs. The maximum 81.23 mg/g Au biosorption capacity was reached only after the first stages of fermentation and the maximum 493.35 mg/g Pd biosorption coincided with trophophase, i.e., the phase in the active growth of a culture in which primary metabolites are formed.

The rapid development of genomics in the past decade has broadened boundaries of biosynthetic and metabolic potential of engineered microorganisms. The heterologous expression using genetically engineered prokaryotic and eukaryotic microbial hosts has made available cellular biomachineries called 'microbial chassis' able to perform scalable and cost-effective biosynthesis of a wide variety of proteins, enzymes, pharmaceuticals, therapeutic and industrially valuable products. A 'microbial chassis' can be defined as host microorganism that supports both the expression of the inserted genetic components, like cluster of genes or operons and the function of the genetically encoded biochemical components involved in biosynthetic pathways of interest for given biotechnological application (Kent and Dixon 2020).

In order to produce improved nanomaterials of microbial origin, the adoption of suitable microbial chassis could have a remarkable future impact in the field of metabolic engineering and synthetic biology for microbial nanobiosynthesis. Microbial strains of the model bacterium *E. coli* with available full sets of genetic manipulation tools have been reported in many works as microbial chassis for the biosynthesis of nanoparticles. In the recombinant *E. coli* R189, Kang et al. (2008) have obtained the intracellular synthesis of fluorescent and water-soluble PC-coated CdSNPs through the overexpression of the gene coding phytochelatins synthase from the yeast *Schizosaccharomyces pombe* and the gene coding γ-glutamylcysteine synthetase, that synthesize the phytochelatins precursor glutathione. The yeast *S. pombe* has the intrinsic ability to form CdS nanocrystals in the vacuole as cadmium defense mechanism. Such multicompartment steps in *S. pombe* CdS biosynthesis make difficult the control and fine-tuning of the CdS nanocrystals properties. Through the described heterologous expression in *E. coli*, it has been by-passed the multicompartment steps that occur in *S. pombe* (Kang et al. 2008).

The in vivo synthesis of diverse metal NPs has been obtained using recombinant *E. coli* DH5α co-expressing genes encoding phytochelatin synthase from the plant *Arabidopsis thaliana* and metallothionein from the bacterium *Pseudomonas putida*. The products of biosynthesis mediated by the recombinant *E. coli* DH5α biosynthesis have included CdSeZn, PrGd, CdCs, and FeCoNPs, all never synthesized before using chemical methods (Park et al. 2010). Similarly, Choi et al. (2018) have demonstrated the in vivo synthesis of various metal NPs (e.g., CdSeZn, PrGd, CdCs, and FeCo) never synthesized before by chemical methods in recombinant *E. coli* DH5α cells co-expressing system for phytochelatin synthase and/or metallothionein. Yuan et al. (2019) have recently described AgNPs synthesis by *E. coli* DH5α transformed with a gene encoding copper binding protein metallothionein from the yeast *Candida albicans*. The expression of metallothionein from *C. albicans* enhanced both the *E. coli* cell growth and AgNPs production yield compared to the control. The biosyntheses of CdS QDs in genetically engineered *E. coli* have been also reported in literature. The biosynthesis of CdS QDs in genetically engineered *E. coli* BL21 has been obtained through the introduction of genes encoding CDS 7, a histidine-rich CdS binding peptide (Mi et al. 2011). Monrás et al. (2012) have exploited a wild type *E. coli* AG1 overexpressing the *gshA* gene encoding L-glutamate cysteine ligase, one of the principal enzymes involved in glutathione biosynthesis.

The genetic screening and the use of appropriate mutants for nanobiosynthesis are other fundamental biotechnological tools to investigate microbial nanobiosynthetic processes. As for other genetic tools described in this section, genetic studies on mutants can reveal the essential genetic requirements for the nanobiosynthetic process and related regulatory mechanisms and highlight the different roles of gene encoding products in nanomaterial assembly. In particular, gene overexpression can be employed in mutant genetic studies both for the identification of biochemical pathway components and to select the most suitable mutant phenotype for a given nanobiosynthetic process. Finally, comparative analyses of mutant strains can help to select the most suitable microbial strain for a given nanobiosynthetic product, according to specific requirements such as production yield, size, morphological and compositional features of the nanomaterial of interest.

The use of mutants for genes encoding outer membrane c-type cytochromes of the bacterium *S. oneidensis* have been investigated to take full advantage in nanoparticles synthesis mediated by these bacterial outer membrane proteins, casting light upon the possibilities of tuning morphology, size and composition of nanoparticles. The influences of outer membrane c-type cytochromes MtrC and OmcA on the size and activity of the extracellular silver AgNPs and Ag_2S NPs produced by the dissimilatory metal-reducing bacterium *S. oneidensis* MR-1 have been investigated by Ng et al. (2013). The *S. oneidensis* mutant Δ(*mtrC-omcA*) has been able to produce significantly smaller AgNPs and Ag_2S NPs comparing with the *S. oneidensis* MR-1, 24.4 ± 0.8 nm versus 40.9 ± 1.0 nm for AgNPs and 27.6 ± 6.4 versus 53.4 ± 12.4 for Ag_2S NPs, respectively. Furthermore, compared to NPs produced by *S. oneidensis* MR-1, the AgNPs and Ag_2S NPs from the *S. oneidensis* mutant Δ(*mtrC-omcA*) have shown higher antibacterial activity against *E. coli* and higher catalytic activity in methylviologen reduction, respectively. These results suggested that through the controlled expression of the genes encoding outer membrane c-type cytochromes in *S. oneidensis* MR-1, a possibility to control NPs size and properties of the extracellular biogenic NPs may be exerted (Ng et al. 2013).

The genetic modification of one component of extracellular electron transfer chain in *S. oneidensis* MR-1 has been also exploited by Tian et al. (2017) for the production of SeNPs or CdSe nanoparticles with fine-tuned composition, morphology, and size. In particular, the *S. oneidensis* mutant with the tetrahaem c-type cytochrome CymA-encoding gene deleted (ΔcymA) has exhibited a rapid and a high production of hexagonal, ultrafine, and uniform-sized CdSe nanoparticles (average diameter = 3.3 ± 0.6 nm). In the *S. oneidensis* mutant that overexpressed cymA gene, PYYDT-cymA, production of a larger-sized (average diameter = 104.6 ± 8.4 nm) spherical SeNPs has been observed (Tian et al. 2017).

4.2 *Microbial Synthesis of Magnetosomes, Frustules, Nanowires, and Nanocellulose*

Many studies using genetically engineered microbial strains for the biosynthesis of magnetosomes, frustules, bacterial nanowires, and bacterial nanocellulose have been extensively reported in the literature. The genomes of numerous magnetotactic bacteria strains of have been sequenced, but standardized protocols for genetic manipulation are only available for the bacteria *Magnetospirillum gryphiswaldense* and *M. magneticum.*

4.2.1 Magnetosomes

In particular, *M. magneticum* strain AMB-1 has been the most studied for the magnetite biomineralization process, also because *Magnetospirillum* species are easy to grow under laboratory conditions compared to other magnetotactic bacteria. The research on the other magnetotactic bacteria is still in progress. Liu et al. (2008) have obtained a mutant of *M. gryphiswaldense* named 'NPHB' from a conjugation experiment with a donor strain of *E. coli*. The *M. gryphiswaldense* NPHB mutant has showed a higher ATP hydrolyzing activity and a higher magnetosomes production yield 35% and 69% higher, respectively, compared to the wild type strain. Concerning magnetosomes production yield, *M. gryphiswaldense* NPHB mutant strain cultured in a 7.5 L bioreactor gave a maximum yield of 58.4 ± 6.4 mg magnetosomes per liter (Liu et al. 2008).

In another work, the combination of random mutagenesis and DNA sequencing has been applied to produce and isolate about 30 mutants of the magnetotactic bacterium *Desulfovibrio magneticus* RS-1. The role of unexpected genes called *mad* genes over the control of the magnetosomes morphology has been highlighted. In the near future, the combined use of this approach with physical and chemical analytical techniques could enable a more complete understanding of biomineralization process in different strains of magnetotactic bacteria (Faivre and Baumgartner 2015). The magnetic properties of magnetosomes are largely dependent on their shape, size, and degree of aggregation of magnetite nanoparticles forming the magnetic nanostructure. The key role of protein MamK in the nanoassembly of magnetite nanoparticles in magnetosomes stable chains has been demonstrated (Bennet et al. 2015).

A strategy for the genomic amplification of single and multiple magnetosome gene clusters involved in magnetosomes biosynthesis genes have been explored in *M. gryphiswaldense.* The results obtained from sequential chromosomal transposition have shown that the tuned expression of the *mam* and *mms* gene clusters can be employed as powerful strategy for the control of magnetosome size and number. More specifically, the duplication of *mamGFDC*, *mamAB*, *mms6*, and *mamXY* magnetosome operons has more than doubled the number of magnetosomes and the amplification of the *mms6* operon has caused an enlargement of magnetite crystals, compared to the wild type *M. gryphiswaldense* strain (Lohße et al. 2016). A successful heterologous expression in the photosynthetic model organism *Rhodospirillum rubrum* has been described for a minimal set of genes involved in the magnetosome biosynthesis in *M. gryphiswaldense.* This set of genes included *mamAB*, *mamGFDC*, *mamXY*, and *mms6* genes. The findings have shown the production of 24 nm-sized Fe_3O_4 nanoparticles surrounded by an external organic layer. The magnetic phenotype has remained stable for at least 40 generations under non-selective conditions (Kolinko et al. 2014).

The genetic modification of gene(s) encoding magnetosome surface protein(s) for the production of chimeric anchor gene transcripts has been described for focused surface functionalization of magnetosome membranes. Different works have reported the functionalization of magnetosome membranes with enzymes and biomolecules of interest exploiting the most abundant magnetosome surface proteins MamA, MamF, and MamG for the production of fusion or chimeric anchors to the magnetosome membrane. The genetic modification of *M. magneticum* AMB-1 magnetosomes functionalized with the bacterial enzyme organophosphohydrolase has been reported by Ginet et al. (2011). The functionalization has been carried out through the fusion of *mamC* gene encoding magnetosome membrane protein with *opd* gene encoding enzyme organophosphohydrolase from *Flavobacterium* ATCC 27551. The Opd-functionalized magnetosomes have been used for the hydrolysis of ethyl-paraoxon as a model organophosphate pesticide. The Opd-functionalized magnetosomes have retained a stable catalytic activity over repeated use for pesticide degradation. Such functionalized magnetic nanoparticles could be used for future applications in organophosphate pesticides bioremediation (Ginet et al. 2011).

Boucher et al. (2017) have described the functionalization of *M. magneticum* AMB-1 magnetosome membranes decorated with RGD peptide-MamC fusion proteins. The RGD-labeled magnetosomes have been used as magnetic resonance imaging contrast agent for in vivo molecular imaging of glioblastoma in a mouse model. Recently, Mickoleit et al. (2020) have confirmed in *M. gryphiswaldense* how the use of MamA, MamG, and MamF protein as anchor proteins can be useful for the making of genetically multifunctionalized magnetosomes. This research has produced a reusable magnetic biocomposite nanomaterial that simultaneously has displayed the enzymes beta-glucuronidase and glucose oxidase (GOx), a fluorophore (mEGFP) and a nanobody (red fluorescent binding protein) on the magnetosome membrane. This study could lay the groundwork for future generation of biohybrid nanomaterials with a great potential application in biotechnological and biomedical fields (Mickoleit et al. 2020).

4.2.2 Frustules

In recent years, the genetic and molecular details of diatoms frustule morphogenesis have been partially elucidated. To date, whole-genome sequencing has been generated for two marine diatom species *Thalassiosira pseudonana* and *Phaeodactylum tricornutum,* enabling their use as model for biotechnological studies in diatoms. In particular *P. tricornutum* is probably the best-characterized of all diatoms so far, with a small-size genome and easy to grow in laboratory (Butler et al. 2020). However, further studies on diatom genetics will be required for a complete knowledge of frustules biogenesis process (De Tommasi et al. 2017). Recently, genome editing approaches like Transcription Activator-Like Effector Nucleases (TALEN) endonucleases and the clustered regulatory interspaced short palindromic repeats (CRISPR)/Cas9 system are increasingly being used for the production of transgenic lines in some diatom species, useful in diatom functional genomic studies on frustule morphogenesis as well as for the development of future genome engineering approaches in diatoms. Both methods use engineered nucleases that catalyze a target gene-specific double-stranded DNA break, allowing a complete loss of a specific gene function and the generation of knockout strains. These approaches for gene down regulation are very promising in diatom research, especially for diploid species like *T. pseudonana* and *P. tricornutum* for which forward genetic methods based on random insertion and chemical mutagenesis cannot be not used (Kroth et al. 2018).

Antibody-functionalized frustules of *T. pseudonana* using genetic approaches have been described in literature. Delalat et al. (2015) have reported a genetically engineered diatom *T. pseudonana* with the incorporation of immunoglobulin G-binding domain of protein G onto the frustules surface. Such antibody-labeled genetically modified diatoms have enabled an in vitro selective cell targeting and selective killing of neuroblastoma and B-lymphoma cells (Delalat et al. 2015). Ford et al. (2016) have described a genetically modified *T. pseudonana* for the expression of biosilica-targeted fusion proteins comprising (i) a single domain antibody against the EA1 S-layer protein of *Bacillus anthracis* Sterne strain and (ii) a single chain variable fragment against the explosive TNT. This approach could be interesting for the development of future multi-functional sensing platforms (Ford et al. 2016).

4.2.3 Nanocellulose

The genetic engineering tool set has recently offered new perspectives also for the production of modified bacterial nanocellulose with improved features. Jacek et al. (2019b) have described the impact of *motA*, *motB* genes overexpression both on bacterial nanocellulose structure and production yield in *K. hansenii* ATCC 23769. More specifically, the analysis of two mutant phenotype *motA*+ and *motB*+ overexpressing *motA*, *motB* genes has revealed changes in cellular elongation, a threefold increase in speeding motility of mutant colonies and a 6% and 20% higher production yield, respectively, compared to the wild type *K. hansenii* ATCC 23769. Moreover, *motA*+ and *motB*+ overexpressing mutants have produced bacterial nanocellulose membranes with pores almost nine times larger and about two time thicker fibers. Although other studies will be required, these results suggested that the controlling of *K. hansenii* motility and cell size might tune the production of well-defined three-dimensional bacterial nanocellulose scaffolds (Jacek et al. 2019b).

Through repeated static cultures, Taweecheep et al. (2019) have obtained four bacterial nanocellulose-producing revertant strains from a *bcsC* loss-of-function mutant *Komagataeibacter oboediens* MSKU 3 E3. The revertant strains produced bacterial nanocellulose with different production yield and dimension. In particular, the R30-3 revertant strain produced the highest amount of bacterial nanocellulose (2.15 g dry weight l^{-1}) and the R37-9 revertant strain produced a bacterial nanocellulose with an exceptionally small average diameter of 34.58 nm and relatively homogeneous fine fibrils with a narrower range of width distribution (about 20 to 60 nm). The genome analysis of R37-9 revertant strain has revealed a one amino acid substitution in the *bcsCI* gene, the N713D mutation (Taweecheep et al. 2019).

4.2.4 Nanowires

Genetic engineering has proved to be useful also for functional and structural modification in bacteria nanowires. Tan et al. (2016) reported genetically engineered model microorganism *G. sulfurreducens* a 2000-fold increase in electrical conductivity and diameter of nanowires filaments has been reduced by half to 1.5 nm. In this case, genetic modification effected the modification of aminoacidic composition of the carboxyl end of PilA protein, the structural component of bacterial nanowires (Tan et al. 2016). In a later paper, Tan et al. (2017) performed the heterologous expression of the *pilA* gene of *G. metallireducens* GS15 in *G. sulfurreducens* MP. The individual pili prepared at physiologically relevant pH 7 showed a conductivity 5000-fold higher than the conductivity of *G. sulfurreducens* pili at pH 7 and nearly 1 million-fold higher than the conductivity of *G. uraniireducens* pili at the same pH. The functionalization of electrically conductive protein nanowires in *G. sulfurreducens* strain

KN400 to introduce new binding properties with surface-exposed peptides has been recently explored by Ueki et al. (2019). The constructed *G. sulfurreducens* strain PilA-WT/PilA-6His has demonstrated that peptides of up to 9 amino acids can be added to the 61 amino acid monomer backbone of *G. sulfurreducens* electrically conductive protein nanowires (Ueki et al. 2019).

A recent use of *E. coli* as microbial chassis for the biosynthesis of bacterial nanowires has been described by Ueki et al. (2020). In this work, a recombinant strain of *E. coli* NEB 10-beta has been produced for the biosynthesis of *G. sulfurreducens*' pilin-based electrically conductive protein nanowires. In particular, this heterologous co-expression concerned the genes encoding type IV pili biogenesis machinery and a synthetic gene designed to yield a peptide monomer that could be assembled into electrically conductive protein nanowires. This synthetic biology approach has allowed either to produce 3 nm diameter electrically conductive protein nanowires, the same diameter that produced by *G. sulfurreducens*, and to overcome technical complexity and costs required for anaerobic growth of *G. sulfurreducens*. These results could represent the basis for future opportunities in large-scale fabrication of novel electrically conductive protein nanowires (Ueki et al. 2020).

Altogether, these biotechnological approaches could strongly contribute to the future development of a tuneable control over microbial nanosynthetic processes on an industrial scale. As described in the literature reviewed in this section, a tight genetic-based control could indeed constitute a powerful synthetic biology tool able (i) to result in tailored implementations and suitable standardization of nanobiosynthetic processes, (ii) to develop engineered nanobiosynthetic pathways, and (iii) to construct synthetic metabolic pathways for the design and the build-up of efficient microbial cell 'nanofactories.'

5 Conclusion

In light of recent literature herein reported, fascinating developments toward fine-tuning and control over microbial nanobiosynthesis can be achieved in the next few years. Even if further research will be required to corroborate and complete the current knowledge about biochemical mechanisms behind microbial nanobiosynthesis, the application of microbiological and genetic methods for the tuning and control over microbial-mediated nanomaterials synthesis represents a concrete opportunity for future developments in nanomaterial synthesis. Starting from the isolation and screening of new microorganisms with potential for nanobiosynthesis and through the combined use of microbiological methods and the latest knowledge in the field of synthetic biology, breakthrough implementations in microbial nanotechnology could be achieved. In addition, future standardization and scale-up of microbial nanobiosynthesis protocols will spur the industrial application of sustainable microbial-based nanomanufacturing processes.

References

Abdul J, Salman S, Ibrahem KH, Ali FA (2014) Effect of culture media on biosynthesis of titanium dioxide nanoparticles using *Lactobacillus crispatus*. Int J Adv Res 2:1014–1021

Abo-Amer AE, El-Shanshoury AERR, Alzahrani OM (2015) Isolation and molecular characterization of heavy metal-resistant *Alcaligenes faecalis* from sewage wastewater and synthesis of silver nanoparticles. Geomicrobiol J 32:836–845. https://doi.org/10.1080/01490451.2015.1010754

Afzal B, Yasin D, Husain S, Zaki A, Srivastava P, Kumar R, Fatma T (2019) Screening of cyanobacterial strains for the selenium nanoparticles synthesis and their anti-oxidant activity. Biocatal Agric Biotechnol 21:101307. https://doi.org/10.1016/j.bcab.2019.101307

Alam MS, Huma Z, Shaikh R, Shaheen G, Mushtaq M (2017) Screening of different growth media for extracellular synthesis of silver nanoparticles using *Aspergillus niger*. Int J Biol Biotechnol 14:333–335. https://doi.org/10.1155/2011/546074

Alphandéry E, Amor M, Guyot F, Chebbi I (2012) The effect of iron-chelating agents on *Magnetospirillum magneticum* strain AMB-1: stimulated growth and magnetosome production and improved magnetosome heating properties. Appl Microbiol Biotechnol 96:663–670. https://doi.org/10.1007/s00253-012-4199-5

Ayano H, Kuroda M, Soda S, Ike M (2015) Effects of culture conditions of *Pseudomonas aeruginosa* strain RB on the synthesis of CdSe nanoparticles. J Biosci Bioeng 119:440–445. https://doi.org/10.1016/j.jbiosc.2014.09.021

Ballottin D, Fulaz S, Souza ML, Corio P, Rodrigues AG, Souza AO, Gaspari PM, Gomes AF, Gozzo F, Tasic L (2016) Elucidating protein involvement in the stabilization of the biogenic silver nanoparticles. Nanoscale Res Lett 11. https://doi.org/10.1186/s11671-016-1538-y

Banerjee K, Ravishankar Rai V (2018) A review on mycosynthesis, mechanism, and characterization of silver and gold nanoparticles. Bionanosci 8:17–31. https://doi.org/10.1007/s12668-017-0437-8

Bao Z, Lan CQ (2018) Mechanism of light-dependent biosynthesis of silver nanoparticles mediated by cell extract of *Neochloris oleoabundans*. Colloids Surf B Biointerf 170:251–257. https://doi.org/10.1016/j.colsurfb.2018.06.001

Barabadi H, Tajani B, Moradi M, Kamali KD, Meena R, Honary S, Mahjoub MA, Saravanan M (2019) *Penicillium* family as emerging nanofactory for biosynthesis of green nanomaterials: a journey into the world of microorganisms. J Clust Sci 30:843–856. https://doi.org/10.1007/s10876-019-01554-3

Barshan S, Rezazadeh-Bari M, Almasi H, Amiri S (2019) Optimization and characterization of bacterial cellulose produced by *Komagatacibacter xylinus* PTCC 1734 using vinasse as a cheap cultivation medium. Int J Biol Macromol 136:1188–1195. https://doi.org/10.1016/j.ijbiomac.2019.06.192

Barwal I, Ranjan P, Kateriya S, Yadav SC (2011) Cellular oxido-reductive proteins of *Chlamydomonas reinhardtii* control the biosynthesis of silver nanoparticles. J Nanobiotechnol 9:1–12. https://doi.org/10.1186/1477-3155-9-56

Bennet M, Bertinetti L, Neely RK, Scherte A, Kornig A, Flors C, Muller FD, Schuler D, Klumpp S, Faivre D (2015) Biologically controlled synthesis and assembly of magnetite nanoparticles. Faraday Discuss 181:71–83. https://doi.org/10.1039/c4fd00240g

Borase HP, Patil CD, Suryawanshi RK, Koli SH, Mohite BV, Benelli G, Pati SV (2017) Mechanistic approach for fabrication of gold nanoparticles by *Nitzschia* diatom and their antibacterial activity. Bioprocess Biosyst Eng 40:1437–1446. https://doi.org/10.1007/s00449-017-1801-3

Boucher M, Geffroy F, Préveral S (2017) Genetically tailored magnetosomes used as MRI probe for molecular imaging of brain tumor. Biomater 121:167–178. https://doi.org/10.1016/j.biomaterials.2016.12.013

Bouws H, Wattenberg A, Zorn H (2008) Fungal secretomes—nature's toolbox for white biotechnology. Appl Microbiol Biotechnol 80:381–388. https://doi.org/10.1007/s00253-008-1572-5

Brayner R, Barberousse H, Hemadi M, Djedjat C, Yéprémian C, Coradin T, Livage J, Flevet F, Couté A (2007) Cyanobacteria as bioreactors for the synthesis of Au, Ag, Pd, and Pt nanoparticles via an enzyme-mediated route. J Nanosci Nanotechnol 7:2696–2708. https://doi.org/10.1166/jnn.2007.600

Brooks J, Lefebvre DD (2017) Optimization of conditions for cadmium selenide quantum dot biosynthesis in *Saccharomyces cerevisiae*. Appl Microbiol Biotechnol 101:2735–2745. https://doi.org/10.1007/s00253-016-8056-9

Butler T, Kapoore RV, Vaidyanathan S (2020) *Phaeodactylum tricornutum*: a diatom cell factory. Trends Biotechnol 38:606–622. https://doi.org/10.1016/j.tibtech.2019.12.023

Castro L, Blázquez ML, Muñoz JA, González F, Ballester A (2013) Biological synthesis of metallic nanoparticles using algae. IET Nanobiotechnol 7:109–116. https://doi.org/10.1049/iet-nbt.2012.0041

Chandrasekaran PT, Bari NK, Sinha S (2017) Enhanced bacterial cellulose production from *Gluconobacter xylinus* using super optimal broth. Cellulose 24:4367–4381. https://doi.org/10.1007/s10570-017-1419-2

Chen A, Contreras LM, Keitz BK (2017) Imposed environmental stresses facilitate cell-free nanoparticle formation by *Deinococcus radiodurans*. Appl Environ Microbiol 83:1–14. https://doi.org/10.1128/AEM.00798-17

Chen Z, Li X, Tan X, Zhang Y, Wang B (2020) Recent advances in biological functions of thick pili in the Cyanobacterium *Synechocystis* sp. PCC 6803. Front Plant Sci 11:1–10. https://doi.org/10.3389/fpls.2020.00241

Choi Y, Park TJ, Lee DC, Lee SY (2018) Recombinant *Escherichia coli* as a biofactory for various single- and multi-element nanomaterials. Proc Natl Acad Sci USA 115:5944–5949. https://doi.org/10.1073/pnas.1804543115

Correa-Llantén DN, Muñoz-Ibacache SA, Castro ME, Muñoz PA, Blamey JM (2013) Gold nanoparticles synthesized by *Geobacillus* sp. strain ID17 a thermophilic bacterium isolated from deception island, Antarctica. Microb Cell Fact 12:2–7. https://doi.org/10.1186/1475-2859-12-75

Delalat B, Sheppard VC, Rasi Ghaemi S, Rao S, PrestidgeCA, McPhee G, Rogers M-L, Donoghue JF, Pillay V, Johns TG, Kroger N, Voelcker NH (2015) Targeted drug delivery using genetically engineered diatom biosilica. Nat Commun 6. https://doi.org/10.1038/ncomms9791

Deshpande P, Gaidhani S, Hitendra M, Shouche Y, Narhe R, Singh R, Wadhwani S, Shedbalkar U, Gond D, Chopade BA (2015) Biosynthesis of gold nanoparticles by human microbiota from healthy skins. J Nanomed Nanotechnol 06:10–12. https://doi.org/10.4172/2157-7439.1000300

De Tommasi E, Gielis J, Rogato A (2017) Diatom frustule morphogenesis and function: a multidisciplinary survey. Mar Genomics 35:1–18. https://doi.org/10.1016/j.margen.2017.07.001

Diko CS, Zhang H, Lian S, Fan S, Li Z, Qu Y (2020) Optimal synthesis conditions and characterization of selenium nanoparticles in *Trichoderma* sp. WL-Go culture broth. Mater Chem Phys 246:122583. https://doi.org/10.1016/j.matchemphys.2019.122583

Dragone R, Grasso G, Muccini M, Toffanin S (2017) Portable bio/chemosensoristic devices: innovative systems for environmental health and food safety diagnostics. Front Public Heal 5:1–6. https://doi.org/10.3389/FPUBH.2017.00080

Elahian F, Reiisi S, Shahidi A, Mirzaei SA (2017) High-throughput bioaccumulation, biotransformation, and production of silver and selenium nanoparticles using genetically engineered *Pichia pastoris*. Nanomed Nanotechnol Biol Med 13:853–861. https://doi.org/10.1016/j.nano.2016.10.009

Elahian F, Heidari R, Charghan VR (2020) Genetically modified *Pichia pastoris*, a powerful resistant factory for gold and palladium bioleaching and nanostructure heavy metal biosynthesis. Artif Cells Nanomed Biotechnol 48:259–265. https://doi.org/10.1080/21691401.2019.1699832

Elegbede JA, Lateef A, Azeez MA, Asafa TB, Yekeen TA, Oladipo IC, Adebayo EA, Beukes LS, Gueguim-Kana EB (2018) Fungal xylanases-mediated synthesis of silver nanoparticles for catalytic and biomedical applications. IET Nanobiotechnol 12(6):857–863. https://doi.org/10.1049/iet-nbt.2017.0299

Elegbede JA, Lateef A, Azeez MA, Asafa TB, Yekeen TA, Oladipo IC, Hakeem AS, Beukes LS, Gueguim-Kana EB (2019) Silver-gold alloy nanoparticles biofabricated by fungal xylanases exhibited potent biomedical and catalytic activities. Biotechnol Prog 35(5):e2829. https://doi.org/10.1002/btpr.2829

Elegbede JA, Lateef A, Azeez MA, Asafa TB, Yekeen TA, Oladipo IC, Aina DA, Beukes LS, Gueguim-Kana EB (2020) Biofabrication of gold nanoparticles using xylanases through valorization of corncob by *Aspergillus niger* and *Trichoderma longibrachiatum*: antimicrobial, antioxidant, anticoagulant and thrombolytic activities. Waste Biomass Valor 11(3):781–791. https://doi.org/10.1007/s12649-018-0540-2

Faivre D, Baumgartner J (2015) The combination of random mutagenesis and sequencing highlight the role of unexpected genes in an intractable organism. PLoS Genet 11:1–3. https://doi.org/10.1371/journal.pgen.1004895

Fernández J, Morena AG, Valenzuela SV, Pastor FJ, Díaz P, Martínez J (2019) Microbial cellulose from a *Komagataeibacter intermedius* strain isolated from commercial wine vinegar. J Polym Environ 27:956–967. https://doi.org/10.1007/s10924-019-01403-4

Ford NR, Hecht KA, Hu DH et al (2016) Antigen binding and site-directed labeling of biosilica-immobilized fusion proteins expressed in diatoms. ACS Synth Biol 5:193–199. https://doi.org/10.1021/acssynbio.5b00191

Gericke M, Pinches A (2006) Microbial production of gold nanoparticles. Gold Bull 39:22–28. https://doi.org/10.1007/BF03215529

Ginet N, Pardoux R, Adryanczyk G, Garcia D, Brutesco C, Pignol D (2011) Single-step production of a recyclable nanobiocatalyst for organophosphate pesticides biodegradation using functionalized bacterial magnetosomes. PLoS ONE 6:4–10. https://doi.org/10.1371/journal.pone.0021442

Grasso G, Zane D, Dragone R (2020) Microbial nanotechnology: challenges and prospects for green biocatalytic synthesis of nanoscale materials for sensoristic and biomedical applications. Nanomater 10. https://doi.org/10.3390/nano10010011

Guilger-Casagrande M, de Lima R (2019) Synthesis of silver nanoparticles mediated by fungi: a review. Front Bioeng Biotechnol 7:1–16. https://doi.org/10.3389/fbioe.2019.00287

Gurunathan S, Kalishwaralal K, Vaidyanathan R, Venkataraman D, Pandian SRK, Muniyandi J, Hariharan N, Eom SH (2009) Biosynthesis, purification and characterization of silver nanoparticles using *Escherichia coli*. Colloids Surf B Biointerf 74:328–335. https://doi.org/10.1016/j.colsurfb.2009.07.048

Hamimed S, Jebli N, Sellami H,Landoulsi A, Chatti A (2020) Dual valorization of olive mill wastewater by bio-nanosynthesis of magnesium oxide and *Yarrowia lipolytica* biomass production. Chem Biodivers 17. https://doi.org/10.1002/cbdv.201900608

Heyen U, Schüler D (2003) Growth and magnetosome formation by microaerophilic *Magnetospirillum* strains in an oxygen-controlled fermentor. Appl Microbiol Biotechnol 61:536–544. https://doi.org/10.1007/s00253-002-1219-x

Hospenthal MK, Costa TRD, Waksman G (2017) A comprehensive guide to pilus biogenesis in gram-negative bacteria. Nat Rev Microbiol 15:365–379. https://doi.org/10.1038/nrmicro.2017.40

Huang BC, Yi YC, Chang JS, Ng IS (2019) Mechanism study of photo-induced gold nanoparticles formation by *Shewanella oneidensis* MR-1. Sci Rep 9:1–11. https://doi.org/10.1038/s41598-019-44088-4

Hulkoti NI, Taranath TC (2014) Biosynthesis of nanoparticles using microbes-a review. Colloids Surf B Biointerf 121:474–483. https://doi.org/10.1016/j.colsurfb.2014.05.027

Ilshadsabah A, Suchithra TV (2019) Bacterial nanowires: an invigorating tale for future. In: Microbial nanobionics. Elsevier, p 109963

Jacek P, Dourado F, Gama M, Bielecki S (2019a) Molecular aspects of bacterial nanocellulose biosynthesis. Microb Biotechnol 12:633–649. https://doi.org/10.1111/1751-7915.13386

Jacek P, Ryngajłło M, Bielecki S (2019b) Structural changes of bacterial nanocellulose pellicles induced by genetic modification of *Komagataeibacter hansenii* ATCC 23769. Appl Microbiol Biotechnol 103:5339–5353. https://doi.org/10.1007/s00253-019-09846-4

Jeffryes C, Gutu T, Jiao J, Rorrer GL (2008a) Metabolic insertion of nanostructured TiO_2 into the patterned biosilica of the diatom *Pinnularia* sp. by a two-stage bioreactor cultivation process. ACS Nano 2:2103–2112. https://doi.org/10.1021/nn800470x

Jeffryes C, Gutu T, Jiao J, Rorrer GL (2008b) Two-stage photobioreactor process for the metabolic insertion of nanostructured germanium into the silica microstructure of the diatom *Pinnularia* sp. Mater Sci Eng C 28:107–118. https://doi.org/10.1016/j.msec.2007.01.002

Jeffryes C, Solanki R, Rangineni Y, Wang W, Chang CH, Rorrer GL (2008c) Electroluminescence and photoluminescence from nanostructured diatom frustules containing metabolically inserted germanium. Adv Mater 20:2633–2637. https://doi.org/10.1002/adma.200800292

Jena J, Pradhan N, Dash BP et al (2015) Pigment mediated biogenic synthesis of silver nanoparticles using diatom *Amphora* sp. and its antimicrobial activity. J Saudi Chem Soc 19:661–666. https://doi.org/10.1016/j.jscs.2014.06.005

Juibari MM, Abbasalizadeh S, Jouzani GS, Noruzi M (2011) Intensified biosynthesis of silver nanoparticles using a native extremophilic *Ureibacillus thermosphaericus* strain. Mater Lett 65:1014–1017. https://doi.org/10.1016/j.matlet.2010.12.056

Kalishwaralal K, Deepak V, Pandian SRK, Kottaisamy M, BarathManiKanth S, Kartikeyan B, Gurunathan S (2010) Biosynthesis of silver and gold nanoparticles using *Brevibacterium casei*. Colloids Surf B Biointerf 77:257–262. https://doi.org/10.1016/j.colsurfb.2010.02.007

Kang SH, Bozhilov KN, Myung NV, Mulchandani A, Chen W (2008) Microbial Synthesis of CdS nanocrystals in genetically engineered *E. coli*. Angew Chemie 120:5264–5267. https://doi.org/10.1002/ange.200705806

Kent R, Dixon N (2020) Contemporary tools for regulating gene expression in bacteria. Trends Biotechnol 38:316–333. https://doi.org/10.1016/j.tibtech.2019.09.007

Khan T, Abbas S, Fariq A, Yasmin A (2018) Microbes: nature's cell factories of nanoparticles synthesis. Exploring the realms of nature for nanosynthesis. Springer, Cham, pp 273–290

Kim HK, Jeong SW, Yang JE, Choi YJ (2019) Highly efficient and stable removal of arsenic by live cell fabricated magnetic nanoparticles. Int J Mol Sci 20. https://doi.org/10.3390/ijms20143566

Kitching M, Ramani M, Marsili E (2015) Fungal biosynthesis of gold nanoparticles: mechanism and scale up. Microb Biotechnol 8:904–917. https://doi.org/10.1111/1751-7915.12151

Kolinko I, Lohße A, Borg S, Raschdorf O, Jogler C, Tu Q, Posfai M, Tompa E, Plitzko JM, Brachmann A, Wanner G, Muller R, Zhang Y, Schuler D (2014) Biosynthesis of magnetic nanostructures in a foreign organism by transfer of bacterial magnetosome gene clusters. Nat Nanotechnol 9:193–197. https://doi.org/10.1038/nnano.2014.13

Konishi Y, Tsukiyama T, Tachimi T, Saitoh N, Nomura T, Nagamine S (2007) Microbial deposition of gold nanoparticles by the metal-reducing bacterium *Shewanella algae*. Electrochim Acta 53:186–192. https://doi.org/10.1016/j.electacta.2007.02.073

Kora AJ (2018) *Bacillus cereus*, selenite-reducing bacterium from contaminated lake of an industrial area: a renewable nanofactory for the synthesis of selenium nanoparticles. Bioresour Bioprocess 5. https://doi.org/10.1186/s40643-018-0217-5

Kröger N, Poulsen N (2008) Diatoms—from cell wall biogenesis to nanotechnology. Annu Rev Genet 42:83–107. https://doi.org/10.1146/annurev.genet.41.110306.130109

Kroth PG, Bones AM, Daboussi F, Ferrante MI, Jaubert M, Kolot M, Nymark M, Bartulos Rio, Ritter A, Russo MT, Serif M, Winge P, Falciatore A (2018) Genome editing in diatoms: achievements and goals. Plant Cell Rep 37:1401–1408. https://doi.org/10.1007/s00299-018-2334-1

Kulkarni RR, Shaiwale NS, Deobagkar DN, Deobagkar DD (2015) Synthesis and extracellular accumulation of silver nanoparticles by employing radiation-resistant *Deinococcus radiodurans*, their characterization, and determination of bioactivity. Int J Nanomed 10:963–974. https://doi.org/10.2147/IJN.S72888

Lampis S, Zonaro E, Bertolini C, Cecconi D, Monti F, Micaroni M, Turner RJ, Butker CS, Vallini G (2017) Selenite biotransformation and detoxification by *Stenotrophomonas maltophilia* SeITE02: novel clues on the route to bacterial biogenesis of selenium nanoparticles. J Hazard Mater 324:3–14. https://doi.org/10.1016/j.jhazmat.2016.02.035

Lang Y, Del Monte F, Rodriguez BJ, Dockery P, Finn DP, Pandit A (2013) Integration of TiO_2 into the diatom *Thalassiosira weissflogii* during frustule synthesis. Sci Rep 3:1–11. https://doi.org/10.1038/srep03205

Lateef A, Adelere IA, Gueguim-Kana EB, Asafa TB, Beukes LS (2015a) Green synthesis of silver nanoparticles using keratinase obtained from a strain of *Bacillus safensis* LAU 13. Int Nano Lett 5(1):29–35. https://doi.org/10.1007/s40089-014-0133-4

Lateef A, Ojo SA, Akinwale AS, Azeez L, Gueguim-Kana EB, Beukes LS (2015b) Biogenic synthesis of silver nanoparticles using cell-free extract of *Bacillus safensis* LAU 13: antimicrobial, free radical scavenging and larvicidal activities. Biologia 70(10):1295–1306. https://doi.org/10.1515/biolog-2015-0164

Lateef A, Ojo SA, Oladejo SM (2016) Anti-candida, anti-coagulant and thrombolytic activities of biosynthesized silver nanoparticles using cell-free extract of *Bacillus safensis* LAU 13. Process Biochem 51(10):1406–1412. https://doi.org/10.1016/j.procbio.2016.06.027

Li J, Li Q, Ma X, Tian B, Li T, Yu J, Dai S, Weng Y, Hua Y (2016a) Biosynthesis of gold nanoparticles by the extreme bacterium *Deinococcus radiodurans* and an evaluation of their antibacterial properties. Int J Nanomed 11:5931–5944. https://doi.org/10.2147/IJN.S119618

Li Q, Liu D, Jia Z, Csetenyi L, Gadd GM (2016b) Fungal biomineralization of manganese as a novel source of electrochemical materials. Curr Biol 26:950–955. https://doi.org/10.1016/j.cub.2016.01.068

Li J, Webster TJ, Tian B (2019) Functionalized nanomaterial assembling and biosynthesis using the extremophile *Deinococcus radiodurans* for multifunctional applications. Small 15:1–15. https://doi.org/10.1002/smll.201900600

Li X, Xu H, Chen ZS, Chen G (2011) Biosynthesis of nanoparticles by microorganisms and their applications. J Nanomater 2011. https://doi.org/10.1155/2011/270974

Lin IWS, Lok CN, Che CM (2014) Biosynthesis of silver nanoparticles from silver(i) reduction by the periplasmic nitrate reductase c-type cytochrome subunit NapC in a silver-resistant *E. coli*. Chem Sci 5:3144–3150. https://doi.org/10.1039/c4sc00138a

Liu J, Ding Y, Jiang W et al (2008) A mutation upstream of an ATPase gene significantly increases magnetosome production in *Magnetospirillum gryphiswaldense*. Appl Microbiol Biotechnol 81:551–558. https://doi.org/10.1007/s00253-008-1665-1

Liu Y, Li GR, Guo FF, Jiang W, Li Y, Li LJ (2010) Large-scale production of magnetosomes by chemostat culture of *Magnetospirillum gryphiswaldense* at high cell density. Microb Cell Fact 9:1–8. https://doi.org/10.1186/1475-2859-9-99

Lohße A, Kolinko I, Raschdorf O, Uebe R, BorgS Brachmann A, Plitzko JM, Müller R, Zhang Y, Schüler D (2016) Overproduction of magnetosomes by genomic amplification of biosynthesis-related gene clusters in a magnetotactic bacterium. Appl Environ Microbiol 82:3032–3041. https://doi.org/10.1128/AEM.03860-15

Lovley DR, Walker DJF (2019) Geobacter protein nanowires. Front Microbiol 10. https://doi.org/10.3389/fmicb.2019.02078

Luo K, Jung S, Park KH, Kim YR (2018) Microbial biosynthesis of silver nanoparticles in different culture media. J Agric Food Chem 66:957–962. https://doi.org/10.1021/acs.jafc.7b05092

Machado RTA, Meneguin AB, Sábio RM, Franco DF, Antonio SG, Gutierrez J, Tercjak A, Berretta AA, Ribeiro SJL, Lazarini SC, Lustri WR, Barud HS(2018) *Komagataeibacter rhaeticus* grown in sugarcane molasses-supplemented culture medium as a strategy for enhancing bacterial cellulose production. Ind Crops Prod 122:637–646. https://doi.org/10.1016/j.indcrop.2018.06.048

Manimaran M, Kannabiran K (2017) Actinomycetes-mediated biogenic synthesis of metal and metal oxide nanoparticles: progress and challenges. Lett Appl Microbiol 64:401–408. https://doi.org/10.1111/lam.12730

Manivasagan P, Venkatesan J, Sivakumar K, Kim SK (2016) Actinobacteria mediated synthesis of nanoparticles and their biological properties: a review. Crit Rev Microbiol 42:209–221. https://doi.org/10.3109/1040841X.2014.917069

Marshall MJ, Beliaev AS, Dohnalkova AC et al (2006) c-Type cytochrome-dependent formation of U(IV) nanoparticles by *Shewanella oneidensis*. PLoS Biol 4:1324–1333. https://doi.org/10.1371/journal.pbio.0040268

McCotter SW, Horianopoulos LC, Kronstad JW (2016) Regulation of the fungal secretome. Curr Genet 62:533–545. https://doi.org/10.1007/s00294-016-0578-2

Mi C, Wang Y, Zhang J, Huang H, Xu L, Wang S, Fang X, Fang J, Mao C, Xu S (2011) Biosynthesis and characterization of CdS quantum dots in genetically engineered *Escherichia coli.* J Biotechnol 153:125–132. https://doi.org/10.1016/j.jbiotec.2011.03.014

Mickoleit F, Lanzloth C, Schüler D (2020) A versatile toolkit for controllable and highly selective multifunctionalization of bacterial magnetic nanoparticles. Small 16. https://doi.org/10.1002/smll.201906922

Mohammed Fayaz A, Balaji K, Kalaichelvan PT, Venkatesan R (2009) Fungal based synthesis of silver nanoparticles-an effect of temperature on the size of particles. Colloids Surf B Biointerf 74:123–126. https://doi.org/10.1016/j.colsurfb.2009.07.002

Monrás JP, Díaz V, Bravo D, Montes RA, Chasteen TG, Osorio-Román IO, Vásquez CC, Pérez-Donoso JM (2012) Enhanced glutathione content allows the in vivo synthesis of fluorescent CdTe nanoparticles by *Escherichia coli.* PLoS ONE 7:1–10. https://doi.org/10.1371/journal.pone.0048657

Mora-Godínez S, Abril-Martínez F, Pacheco A (2020) Green synthesis of silver nanoparticles using microalgae acclimated to high CO_2. Mater Today Proc. https://doi.org/10.1016/j.matpr.2020.04.761

Mukherjee K, Gupta R, Kumar G, Kumari S, Biswas S, Padmanabhan P (2018) Synthesis of silver nanoparticles by *Bacillus clausii* and computational profiling of nitrate reductase enzyme involved in production. J Genet Eng Biotechnol 16:527–536. https://doi.org/10.1016/j.jgeb.2018.04.004

Naloka K, Matsushita K, Theeragool G (2020) Enhanced ultrafine nanofibril biosynthesis of bacterial nanocellulose using a low-cost material by the adapted strain of *Komagataeibacter xylinus* MSKU 12. Int J Biol Macromol 150:1113–1120. https://doi.org/10.1016/j.ijbiomac.2019.10.117

Nangia Y, Wangoo N, Goyal N, Shekhawat G, Suri CR (2009) A novel bacterial isolate *Stenotrophomonas maltophilia* as living factory for synthesis of gold nanoparticles. Microb Cell Fact 8:1–7. https://doi.org/10.1186/1475-2859-8-39

Ng CK, Sivakumar K, Liu X, Madhaiyan M, Ji L, Yang L, Tang C, Song H, Kjelleberg S, Cao B (2013) Influence of outer membrane c-type cytochromes on particle size and activity of extracellular nanoparticles produced by *Shewanella oneidensis.* Biotechnol Bioeng 110:1831–1837. https://doi.org/10.1002/bit.24856

Ojo SA, Lateef A, Azeez MA, Oladejo SM, Akinwale AS, Asafa TB, Yekeen TA, Akinboro A, Oladipo IC, Gueguim-Kana EB, Beukes LS (2016) Biomedical and catalytic applications of gold and silver-gold alloy nanoparticles biosynthesized using cell-free extract of *Bacillus safensis* LAU 13: antifungal, dye degradation, anti-coagulant and thrombolytic activities. IEEE Trans Nanobiosci 15(5):433–442. https://doi.org/10.1109/TNB.2016.2559161

Panwar V, Dutta T (2019) Diatom Biogenic Silica as a felicitous platform for biochemical engineering: expanding frontiers. ACS Appl Bio Mater 2:2295–2316. https://doi.org/10.1021/acsabm.9b00050

Park TJ, Lee SY, Heo NS, Seo TS (2010) In vivo synthesis of diverse metal nanoparticles by recombinant *Escherichia coli.* Angew Chemie—Int Ed 49:7019–7024. https://doi.org/10.1002/anie.201001524

Patel V, Berthold D, Puranik P, Gantar M (2015) Screening of cyanobacteria and microalgae for their ability to synthesize silver nanoparticles with antibacterial activity. Biotechnol Rep 5:112–119. https://doi.org/10.1016/j.btre.2014.12.001

Pathak J, Rajneesh, Ahmed H, Singh DK, Pandey A, Singh, SP, Sinha R P (2018) Recent developments in green synthesis of metal nanoparticles utilizing cyanobacterial cell factories. In: Nanomaterials in plants, algae and microorganisms. Academic Press, Elsevier Inc, pp 237–265

Piacenza E, Presentato A, Turner RJ (2018) Stability of biogenic metal(loid) nanomaterials related to the colloidal stabilization theory of chemical nanostructures. Crit Rev Biotechnol 38:1137–1156. https://doi.org/10.1080/07388551.2018.1440525

Pimprikar PS, Joshi SS, Kumar AR, Bravo D, Pérez-Donoso JM (2009) Influence of biomass and gold salt concentration on nanoparticle synthesis by the tropical marine yeast *Yarrowia lipolytica* NCIM 3589. Colloids Surf B Biointerf 74:309–316. https://doi.org/10.1016/j.colsurfb.2009.07.040

Plaza DO, Gallardo C, Straub YD et al (2016) Biological synthesis of fluorescent nanoparticles by cadmium and tellurite resistant Antarctic bacteria: exploring novel natural nanofactories. Microb Cell Fact 15:1–11. https://doi.org/10.1186/s12934-016-0477-8

Priyadarshini S, Gopinath V, Meera Priyadharsshini N, Mubarak Ali D, Velusamy P (2013) Synthesis of anisotropic silver nanoparticles using novel strain, *Bacillus flexus* and its biomedical application. Colloids Surf B Biointerf 102:232–237. https://doi.org/10.1016/j.colsurfb.2012.08.018

Prosa M, Bolognesi M, Fornasari L, Fornasari L, Grasso G, Lopez-Sanchez L, Marabelli F, Toffanin S (2020) Nanostructured organic/hybrid materials and components in miniaturized optical and chemical sensors. Nanomater 10(3):480. https://doi.org/10.3390/nano10030480

Qin T, Gutu T, Jiao J, Chang CH, Rorrer GL (2008) Biological fabrication of photoluminescent nanocomb structures by metabolic incorporation of germanium into the biosilica of the diatom *Nitzschia frustulum*. ACS Nano 2:1296–1304. https://doi.org/10.1021/nn800114q

Quinteros MA, Bonilla JO, Alborés S V, Villegas LB, Páez PL (2019) Biogenic nanoparticles: synthesis, stability and biocompatibility mediated by proteins of *Pseudomonas aeruginosa*. Colloids Surf B Biointerf 184:110517. https://doi.org/10.1016/j.colsurfb.2019.110517

Ramanathan R, O'Mullane AP, Parikh RY, Smooker PM, Bhargava SK, Bansal V (2011) Bacterial kinetics-controlled shape-directed biosynthesis of silver nanoplates using *Morganella psychrotolerans*. Langmuir 27:714–719. https://doi.org/10.1021/la1036162

Rorrer GL, Chang CH, Liu SH, Jeffryes C, Jiao J, Hedberg JA (2005) Biosynthesis of silicon-germanium oxide nanocomposites by the marine diatom *Nitzschia frustulum*. J Nanosci Nanotechnol 5:41–49. https://doi.org/10.1166/jnn.2005.005

Sable SV, Kawade S, Ranade S, Joshi S (2020) Bioreduction mechanism of silver nanoparticles. Mater Sci Eng C 107:110299. https://doi.org/10.1016/j.msec.2019.110299

Semjonovs P, Ruklisha M, Paegle L, Saka M, Treimane R, Skute M, Rozenberga L, Vikele L, Sabovics M, Cleenwerck I (2017) Cellulose synthesis by *Komagataeibacter rhaeticus* strain P 1463 isolated from Kombucha. Appl Microbiol Biotechnol 101:1003–1012. https://doi.org/10.1007/s00253-016-7761-8

Shankar PD, Shobana S, Karuppusamy I, Pugazhendhi A, Ramkumar VS, Arvindnarayan S, Kumar G (2016) A review on the biosynthesis of metallic nanoparticles (gold and silver) using bio-components of microalgae: formation mechanism and applications. Enzyme Microb Technol 95:28–44. https://doi.org/10.1016/j.enzmictec.2016.10.015

Shao M, Zhang R, Wang C, Hu B, Pang D, Xie Z (2018) Living cell synthesis of CdSe quantum dots: manipulation based on the transformation mechanism of intracellular Se-precursors. Nano Res 11:2498–2511. https://doi.org/10.1007/s12274-017-1873-z

Sharma C, Bhardwaj NK (2019) Bacterial nanocellulose: present status, biomedical applications and future perspectives. Mater Sci Eng C 104:109963. https://doi.org/10.1016/j.msec.2019.109963

Shivaji S, Madhu S, Singh S (2011) Extracellular synthesis of antibacterial silver nanoparticles using psychrophilic bacteria. Process Biochem 46:1800–1807. https://doi.org/10.1016/j.procbio.2011.06.008

Srivastava P, Bragança J, Ramanan SR, Kowshik M (2013) Synthesis of silver nanoparticles using haloarchaeal isolate *Halococcus salifodinae* BK3. Extremophiles 17:821–831. https://doi.org/10.1007/s00792-013-0563-3

Srivastava P, Nikhil EVR, Bragança JM, Kowshik M (2015) Anti-bacterial TeNPs biosynthesized by haloarcheaon *Halococcus salifodinae* BK3. Extremophiles 19:875–884. https://doi.org/10.1007/s00792-015-0767-9

Su Y, Lundholm N, Friis SMM, Ellegaard M (2015) Implications for photonic applications of diatom growth and frustule nanostructure changes in response to different light wavelengths. Nano Res 8:2363–2372. https://doi.org/10.1007/s12274-015-0746-6

Su Y, Lundholm N, Ellegaard M (2017) Long-term cultivation of the diatom *Coscinodiscus granii* at different light spectra: effects on frustule morphology. J Appl Phycol 29:1775–1779. https://doi.org/10.1007/s10811-017-1101-y

Su Y, Lundholm N, Ellegaard M (2018a) Effects of abiotic factors on the nanostructure of diatom frustules—ranges and variability. Appl Microbiol Biotechnol 102:5889–5899. https://doi.org/10.1007/s00253-018-9087-1

Su Y, Lundholm N, Ellegaard M (2018b) The effect of different light regimes on diatom frustule silicon concentration. Algal Res 29:36–40. https://doi.org/10.1016/j.algal.2017.11.014

Sun JB, Zhao F, Tang T, Jiang W, Tian JS, Li Y, Li JL (2008) High-yield growth and magnetosome formation by *Magnetospirillum gryphiswaldense* MSR-1 in an oxygen-controlled fermentor supplied solely with air. Appl Microbiol Biotechnol 79:389–397. https://doi.org/10.1007/s00253-008-1453-y

Sure S, Torriero AAJ, Gaur A, Li LH, ChenY Tripathi C, Adholeya A, Ackland ML, Kochar M (2015) Inquisition of *Microcystis aeruginosa* and *Synechocystis* nanowires: characterization and modelling. Antonie van Leeuwenhoek—Int J Gen Mol Microbiol 108:1213–1225. https://doi.org/10.1007/s10482-015-0576-2

Tan Y, Adhikari RY, Malvankar NS, Pi S, Ward JE, Woodard TL, Nevin KP, Xia Q, Tuominen MT, Lovley DR (2016) Synthetic biological protein nanowires with high conductivity. Small 12:4481–4485. https://doi.org/10.1002/smll.201601112

Tan Y, Adhikari RY, Malvankar NS, Ward, JE, Woodard, TL, Nevin, KP, Lovley, DR (2017) Expressing the *Geobacter metallireducens* PilA in *Geobacter sulfurreducens* yields Pili with exceptional conductivity. MBio 8:1–9. https://doi.org/10.1128/mBio.02203-16

Taweecheep P, Naloka K, Matsutani M, Yakushi T, Matsushita K, Theeragool G (2019) Superfine bacterial nanocellulose produced by reverse mutations in the *bcsC* gene during adaptive breeding of *Komagataeibacter oboediens*. Carbohydr Polym 226:115243. https://doi.org/10.1016/j.carbpol.2019.115243

Tian LJ, Li WW, Zhu TT, Chen JJ, Wang WK, An PF, Zhang L, Dong JC, GuanY, Orcid DFL, Zhou NQ, Liu G, Tian YC, Yu HQ (2017) Directed biofabrication of nanoparticles through regulating extracellular electron transfer. J Am Chem Soc 139:12149–12152. https://doi.org/10.1021/jacs.7b07460

Tiquia-Arashiro SM (2019) Fungi in extreme environments: ecological role and biotechnological significance. Fungi Extrem Environ Ecol Role Biotechnol Significance 383–394. https://doi.org/10.1007/978-3-030-19030-9

Tiquia-Arashiro S, Rodrigues DF (2016) Biotechnological applications of extremophiles in nanotechnology. In: Extremophiles: applications in nanotechnology. Springer International Publishing, Cham, Switzerland, p 193

Townley HE, Woon KL, Payne FP, White-Cooper H, Parker AR (2007) Modification of the physical and optical properties of the frustule of the diatom *Coscinodiscus wailesii* by nickel sulfate. Nanotechnology 18. https://doi.org/10.1088/0957-4484/18/29/295101

Ueki T, Walker DJF, Tremblay PL, Nevin KP, Ward JE, Woodard TL, Nonnenmann SS, Lovley DR (2019) Decorating the outer surface of microbially produced protein nanowires with peptides. ACS Synth Biol 8:1809–1817. https://doi.org/10.1021/acssynbio.9b00131

Ueki T, Walker DJF, Woodard TL, Nevin KP, Nonnenmann SS, Lovley DR (2020) An *Escherichia coli* chassis for production of electrically conductive protein nanowires. ACS Synth Biol 9:647–654. https://doi.org/10.1021/acssynbio.9b00506

Vasylevskyi SI, Kracht S, Corcosa P, Fromm KM, Giese B, Füeg M (2017) Formation of silver nanoparticles by electron transfer in peptides and c-cytochromes. Angew Chemie—Int Ed 56:5926–5930. https://doi.org/10.1002/anie.201702621

Voeikova TA, Zhuravliova OA, Gracheva TS, Bulushova NV, Ismagulova TT, Shaitan KV, Debabov VG (2018) Optimization of microbial synthesis of silver sulfide nanoparticles. Appl Biochem Microbiol 54:800–807. https://doi.org/10.1134/S0003683818080070

Wang D, Xia X, Wu S et al (2019) The essentialness of glutathione reductase GorA for biosynthesis of Se(0)-nanoparticles and GSH for CdSe quantum dot formation in *Pseudomonas stutzeri* TS44. J Hazard Mater 366:301–310. https://doi.org/10.1016/j.jhazmat.2018.11.092

Wong-Pinto L, Menzies A, Ordóñez JI (2020) Bionanomining: biotechnological synthesis of metal nanoparticles from mining waste—opportunity for sustainable management of mining environmental liabilities. Appl Microbiol Biotechnol 104:1859–1869. https://doi.org/10.1007/s00253-020-10353-0

Xia X, Wu S, Li N, Wang D, Zheng S, Wang G (2018) Novel bacterial selenite reductase CsrF responsible for Se(IV) and Cr(VI) reduction that produces nanoparticles in *Alishewanella* sp. WH16-1. J Hazard Mater 342:499–509. https://doi.org/10.1016/j.jhazmat.2017.08.051

Xu JK, Zhang FF, Sun JJ, Sheng J, Wang F, Sun M (2014) Bio and nanomaterials based on Fe_3O_4. Molecules 19:21506–21528. https://doi.org/10.3390/molecules191221506

Yan L, Da H, Zhang S, López VM, Wang W (2017) Bacterial magnetosome and its potential application. Microbiol Res 203:19–28. https://doi.org/10.1016/j.micres.2017.06.005

Yarzàbal LA (2016) Microbial models: from environmental to industrial sustainability. Springer, Singapore

Yee DP, Hildebrand M, Tresguerres M (2020) Dynamic subcellular translocation of V-type H^+-ATPase is essential for biomineralization of the diatom silica cell wall. New Phytol 225:2411–2422. https://doi.org/10.1111/nph.16329

Yin N, Gao R, Knowles B, Wang J, Wang P, Sun G, Cui Y (2019) Formation of silver nanoparticles by human gut microbiota. Sci Total Environ 651:1489–1494. https://doi.org/10.1016/j.scitotenv.2018.09.312

Yuan Q, Bomma M, Xiao Z (2019) Enhanced silver nanoparticle synthesis by *Escherichia coli* transformed with *Candida albicans* metallothionein gene. Mater (Basel) 12:1–14. https://doi.org/10.3390/MA12244180

Zhang Y, Zhang X, Jiang W, Li Y, Li J (2011) Semicontinuous culture of *Magnetospirillum gryphiswaldense* MSR-1 cells in an autofermentor by nutrient-balanced and isosmotic feeding strategies. Appl Environ Microbiol 77:5851–5856. https://doi.org/10.1128/AEM.05962-11

Zhong Y, Xia X, Deng S, Xie D, Shen S, Zhang K, Guo W, Wang X, Tu J (2018) Spore carbon from *Aspergillus oryzae* for advanced electrochemical energy storage. Adv Mater 30:1–9. https://doi.org/10.1002/adma.201805165

Current Advances in Fungal Nanobiotechnology: Mycofabrication and Applications

Th I. Shaheen, S. S. Salem, and A. Fouda

1 Introduction

Nanotechnology is the multidisciplinary science concerned with the creation and modification of materials in size range between 1 up to 100 nm. Nanotechnology science is greatly affected by various disciplines, especially physics, chemistry and biology (Elfeky et al. 2020; Hassan et al. 2019; Mohamed et al. 2019; Shaheen et al. 2019). This technology provides great opportunities by changing the significant properties of materials when they are in nano-size. When, the dimension of the materials is getting closer to nano-size the quantum physics play an important role instead of conventional laws of physics, which adds different and unique properties to the material. In other words, nano-science helps us to understand these new behaviors that emerge in nano-size due to quantum theories. Thus, nanotechnology lets us to design and synthesize nano-structures and provides us to use these extraordinary properties in order to generate new material forms and new application areas such as medicine, pharmacology, environment, textile, green approaches, agricultural, paper conservation and electronics (Fouda et al. 2019b, c; Mohamed et al. 2019; Mohmed et al. 2017a; Salem et al. 2020).

Nanobiotechnology is a new branch of nanotechnology, combining biological approaches with physical and chemical procedures to generate nano-scale size particles with structures and unique function. Nowadays, metal and metal oxide nanoparticles as silver, selenium, iron, platinum, gold, ZnO, TiO_2 and CuO have great popularity because of their many advantageous properties in several fields of applications (Aygün et al. 2020; El-Batal et al. 2020; El-Sayyad et al. 2019; Elegbede and Lateef

T. I. Shaheen (✉)
National Research Centre, Scopus affiliation ID 60014618, 33 El-Behouth St, Dokki, Giza 12622, Egypt
e-mail: shaheen_chem@yahoo.com

S. S. Salem · A. Fouda
Department of Botany and Microbiology, Faculty of Science, Al-Azhar University, Cairo, Egypt

A. Lateef et al. (eds.), *Microbial Nanobiotechnology*, Materials Horizons: From Nature to Nanomaterials, https://doi.org/10.1007/978-981-33-4777-9_4

2019; Elegbede et al. 2020; Fouda et al. 2018; Lateef et al. 2018; Maaroof and Mahmood 2019; Subramaniyan et al. 2018). Green procedure of NPs using various biological entities can overcome a lot of the destructive effects of physical and chemical techniques. Green assembly of nanomaterials includes synthesis via plants, actinomycetes, bacteria, algae, fungi and metabolites of arthropods that allow large-scale production of nanomaterials free from impurities (Adelere and Lateef 2016; Fouda et al. 2019c; Iqtedar et al. 2019; Lateef et al. 2016a, b; Lotha et al. 2019; Mohamed et al. 2019; Salem et al. 2019a). These living organisms, especially fungal cells are qualified for synthesis of active compounds and molecules that act as stabilizing and reducing agent for the fabrication of nanomaterials with assorted shapes, physiochemical and composition characteristics (Feroze et al. 2020; Fouda et al. 2019b; Salem et al. 2020).

2 Myconanotechnology

Myconanotechnology is defined as the science which interacts with mycology and nanotechnology (Rabeea et al. 2020). Myconanotechnology is rising up as a quickly improving filed with its usage in technology and science with the target goal of combining new smart materials at nano-scale for various uses (Mohamed et al. 2019). Myconanotechnology is an emerging field, where fungi can be harnessed for the synthesis of NPs or nano-structures with desirable shapes and sizes (Fouda et al. 2018; Subramaniyan et al. 2018). Potential applications of myconanotechnology have fascinated microbiologists and other investigators to contribute in providing incremental solutions through green chemistry approaches for targeted drug delivery. Nanoparticles have attracted worldwide attention due to their specific properties and applications in different fields, especially in biomedical sciences (Salem et al. 2020; Saravanakumar et al. 2019; Shaheen and Abd El Aty 2018). Mycosynthesis or mycofabrication is an unpretentious method for attaining easy and stable biological NPs formation. Most fungi consist of significant metabolites (proteins, carbohydrates and lipids) with greater bioaccumulation capacity and straightforward downstream handling are easy to cultivation for the efficient and low-cost formation of NPs (Fouda et al. 2019b; Hamedi et al. 2017).

Biological synthesis integrates biological rules (i.e., reduction/oxidation) by fungal enzymes with physical and chemical strategies to produce nano-sized particles. NPs are solid particles with all three external dimensions at the nano-scale (El-Sayed et al. 2020b; Elegbede et al. 2020) that can drastically amend physicochemical characteristic compared to the original bulk material. It can explicate actions relying on the chemical composition, biological actions, size and shape (Mohamed et al. 2019). Since metal and metal oxides NPs show considerable surface district to volume ratio; it courts wide range of implementations in industries (El-Sayyad et al. 2019; Mahanty et al. 2019). Many fungi have been explored for nanoparticles mycosynthesis as *Penicillium corylophilum, Aspergillus niger, Trichoderma longibrachiatum, Fusarium keratoplasticum, Penicillium chrysogenum, Saccharomyces cerevisiae*, *Alternaria tenuissima*, *Fusarium oxysporum*, *Penicillium pimiteouiense*, *Trichothecium* sp., *Cladosporium cladosporioides* and *Lentinus edodes* (Abd El Aty et al. 2020; Ahmed et al. 2018; El-Sayyad et al. 2019; Elegbede et al. 2020; Faramarzi et al. 2020; Fouda et al. 2019a; Joshi et al. 2017; Lateef and Adeeyo 2015; Mahanty et al. 2019; Mohamed et al. 2015; Mohmed et al. 2017a; Qu et al. 2018a; Rajput et al. 2016; Salem et al. 2020). Data in Table 1 showed sizes and applications of various NPs synthesized by different fungal species.

3 Current Advances on Green Mycosynthesis NPs

Fungi are classified as eukaryotic, heterotrophic and absorptive organisms having chitinous cell walls, which reproduce mostly asexually or sexually by producing spores and multiply by budding or even by cell elongation of the hyphal tip. They are characterized by a distinctive, multinucleate vegetative (somatic) called mycelium and survive by obtaining nourishment systemically or through rhizoids (Schumacher and Gorbushina 2020; Wurzbacher et al. 2020). Fungal cells are a premium origin of compounds valuable for the green mycosynthesis of NPs to be implemented in agricultural, textile, medical, wastewater treatment and environmental ambits (Abd El Aty et al. 2020; Mohamed et al. 2019; Parmar and Sharma 2020). Various products obtained from fungi cells are generated at trade scale, including proteins, organic acids, antibiotics and enzymes (Balaji et al. 2009; Eid et al. 2019; Fouda et al. 2015; Wierckx et al. 2020; Xiao et al. 2020). Moreover, fungi and their metabolites are significant biological factors for industrial implementation as bioremediation, bioaugmentation, biodegradation and biotransformation (Kalpana et al. 2018; Salem et al. 2019b; Salvadori et al. 2014; Sriramulu and Sumathi 2018).

The fungus-intermediated green formation of NPs has many advantages, including simple and easy scaling up, easy processing, economic viability, biomass processing and recovery of considerable surface distances with optimum outgrowth of mycelia (Abd El Aty et al. 2020; Salem et al. 2020; Subramaniyan et al. 2018). From economic overview, the uses of fungal biomass filtrate containing different metabolites as a green approach for fabrication of metallic NPs were advantages over other biological methods (Abdelhakim et al. 2020; Mohamed et al. 2019). The schematic explanation of the methodology for the mycosynthesis of NPs utilizing fungi is presented in Fig. 1. Moreover, various species of fungi grow quickly and formed enormous amount of mass fungal cells and safeguarding them in the definite laboratory are veritable easy (Elegbede et al. 2020; Fouda et al. 2018). Therefore, the extracellular fabrication of NPs is a favorable choice more than intracellular one which is appropriate for full-scale formation. Intracellular fabrication may be investigated for theoretic studies for academic. Basically, there are different species of fungi that can be utilized for preparation of gold and silver NPs as *Aspergillus* sp., *A. niger, Fusarium solani, Ganoderma lucidum, Trichoderma* sp. and *Rhodotorula glutinis* (Elegbede et al. 2020; Mohanta et al. 2018; Popli et al. 2018; Qu et al. 2018a, 2020; Ramos et al. 2020). Other species that included *Aspergillus terreus, A. niger, F.*

Table 1 Representative list of various types of NPs with different sizes and applications synthesized by fungal species

NPs	Synthesized by	Size (nm)	Applications	Reference
Au	*Aspergillus niger*	4.88–22.27	Antimicrobial, antioxidant, anticoagulant and thrombolytic activities	Elegbede et al. (2020)
	Trichoderma longibrachiatum	102.93–123.99		
	Flammulina velutipes	74	De-colorization of methylene blue	Rabeea et al. (2020)
	Aspergillus sp.	14	Uniform rod-shaped Au-NPs	Qu et al. (2020)
	Fusarium solani	40–45	Anticancer activities	Clarance et al. (2020)
	Trichoderma sp.	–	Azo dye de-colorization	Qu et al. (2017)
	Saccharomyces cerevisiae	18–15	Controllable mycosynthesis	Yang et al. (2016)
	Magnusiomyces ingens	20–28	Environmental pollution	Qu et al. (2018b)
	Trichoderma sp.	1–24	Catalytic activity on aromatic pollutants degradation	Qu et al. (2018a)
	Cladosporium cladosporioides	60	Antibacterial and antioxidant activity	Joshi et al. (2017)
	Trichoderma harzianum	32–44	Dye degradation, antibacterial activity	Tripathi et al. (2018)
	Pleurotus ostreatus	10–30	Antimicrobial, anticancer	El Domany et al. (2018)
	Aspergillus sp.	2.5–6.7	Reduction of nitrophenol compounds	Shen et al. (2017)
	Rhizopus oryzae	16–43	Hemocompatible activity	Kitching et al. (2016)
	Penicillium aurantiogriseum	153.3	Mycosynthesis NPs	Honary et al. (2013)
	Penicillium citrinum	172		
	Penicillium waksmanii	160.1		
Ag	*Penicillium chrysogenum*	5.2–9.7	Biomedical application	Abd El Aty et al. (2020)
	Ganoderma lucidum	12–22	Biochemical applications	Aygün et al. (2020)

(continued)

Table 1 (continued)

NPs	Synthesized by	Size (nm)	Applications	Reference
	Lentinus tuberregium	5–35	Antimicrobial and α-amylase inhibitory activity	Debnath et al. (2019)
	Penicillium chrysogenum	26.38–61.75	Biodeterioration	Fouda et al. (2019a)
	Fusarium keratoplasticum	6–36	Textile industry and biodeterioration of archeological manuscript	Fouda et al. (2019b); Mohmed et al. (2017a)
	Alternaria tenuissima	9.8	Antimicrobial, antioxidant property	Yousef et al. (2020)
	Phomopsis liquidambaris	18.7	Antimicrobial and larvicidal activity	Seetharaman et al. (2018)
	Trichoderma longibrachiatum	10	Antimicrobial against phytopathogen	Elamawi et al. (2018); Elegbede et al. (2018)
	Fusarium oxysporum	21.3–37	Antibacterial	Ahmed et al. (2018)
	Aspergillus terreus	10–18	Antifungal property	Ammar and El-Desouky (2016)
	Penicillium expansum	14–25		
	Ganoderma sessiliforme	45	Antibacterial, antioxidant and anticancer activities	Mohanta et al. (2018)
	Rhodotorula glutinis	15.45	Antifungal, cytotoxicity, dye degradation	Popli et al. (2018)
	Aspergillus sp	5–30	Antibacterial and cytotoxicity	Mohmed et al. (2017b)
	Arthroderma fulvum	15.5	Antifungal activity	Siddiqi and Husen (2016)
	Penicillium aculeatum	4–55	Antimicrobial activity, drug delivery	Ma et al. (2017)
	Penicillium chrysogenum	18–60	Anti-candida activity	Soliman et al. (2020)
	Fusarium oxysporum	10–50	Colloidal stability	Rajput et al. (2016)
	Fusarium oxysporum	34–44	Antibacterial activity	Hamedi et al. (2017)

(continued)

Table 1 (continued)

NPs	Synthesized by	Size (nm)	Applications	Reference
	Fusarium oxysporum	5–13	Antibacterial and antitumor activities	Husseiny et al. (2015)
	Trichoderma spp	–	Antibacterial activity	Ramos et al. (2020)
	Trichoderma harzianum	20–30	Antifungal activity	Guilger et al. (2017)
	Metarhizium anisopliae	28–38	Larvicidal activity	Amerasan et al. (2016)
	Rhodotorula glutinis	15.5	Antifungal activity and bioremediation	Ovais et al. (2018)
	Rhodotorula glutinis	15–220	Agricultural	Fernández et al. (2016)
	Saccharomyces cerevisiae	2–20	Mycosynthesis	Korbekandi et al. (2016)
	Rhodotorula mucilaginosa	11	Bioremediation	Salvadori (2019)
	Rhodotorula sp	8–21	Antimicrobial activity	Soliman et al. (2018)
	Cryptococcus laurentii	35–400	Agriculture	Fernández et al. (2016)
ZnO	*Periconium* sp	16–78	Antimicrobial and antioxidant activity	Ganesan et al. (2020)
	Alternaria tenuissima	15.45	Antimicrobial, antioxidant, anticancer and photo-catalytic activities	Abdelhakim et al. (2020)
	Aspergillus niger	80–130	Anticancer, antimicrobial and antioxidant activity	Gao et al. (2019)
	Penicillium chrysogenum	8.74–23.07	Biodeterioration	Fouda et al. (2019a)
	Aspergillus niger	8–38	Medical textile	Mohamed et al. (2019)
	Aspergillus niger	53–69	Dye degradation, antibacterial activity	Kalpana et al. (2018)
	Aspergillus terreus	10–45	Medical textile and UV protection	Fouda et al. (2018)
	Fusarium keratoplasticum	10–42	Textile industry	Mohamed et al. (2019)

(continued)

Table 1 (continued)

NPs	Synthesized by	Size (nm)	Applications	Reference
	Aspergillus fumigatus	22.4 ± 1.8	Agriculture	Raliya et al. (2016)
	Pichia fermentans	–	Antimicrobial property	Chauhan et al. (2015)
	Candida albicans	25	Synthesis of steroidal pyrazolines	Mashrai et al. (2017)
	Pichia kudriavzevii	10–61	Antibacterial and antioxidant activity	Moghaddam et al. (2017)
	Fusarium solani	60-95	Mycosynthesis	Venkatesh et al. (2013)
TiO_2	*Aspergillus flavus*	–	Immunogenic and hematologic effects	Maaroof and Mahmood (2019)
	Aspergillus flavus	12–15	Plant nutrient fertilizer	Raliya et al. (2015)
	Aspergillus tubingensis	<100	Improving crop production	Tarafdar et al. (2013)
	Yeast cells	10–20	Electrochemical	Li et al. (2015)
	Aspergillus niger	73.58–106.9	UV protection and bacterial resistance fabrication	Durairaj et al. (2014b)
			Larvicidal activity	Durairaj et al. (2014a)
CuO	*Trichoderma asperellum*	10–190	Development of anticancer nano-therapeutics	Saravanakumar et al. (2019)
	Penicillium chrysogenum	49.2	Agricultural area	El-Batal et al. (2020)
	Stereum hirsutum	5–20	Mycosynthesis	Cuevas et al. (2015)
	Penicillium aurantiogriseum	91–295	Downstream processing techniques	Honary et al. (2012)
	Penicillium citrinum	79–179		
	Penicillium waksmanii	89–250		
Cu	*Pleurotus ostreatus*	35	Wound pathogens	El-Batal et al. (2018)
	Stereum hirsutum	5–20	Mycosynthesis	Cuevas et al. (2015)

(continued)

Table 1 (continued)

NPs	Synthesized by	Size (nm)	Applications	Reference
	Trichoderma koningiopsis	87.5–328.27	Myco-remediation	Salvadori et al. (2014)
Fe	*Penicillium pimiteouiense*	2–16	Treatment of waste water	Mahanty et al. (2019)
	Alternaria alternata	5.4–12.1	Antimicrobial activities	Mohamed et al. (2015)
	Aspergillus oryzae	10–24.6	Mycosynthesis NPs	Raliya (2013)
Se	*Penicillium corylophilum*	29.1–48.9	Larvicidal, antibacterial and cytotoxicity activities	Salem et al. (2020)
	Fusarium semitectum	32.80–103.82	Anticancer and antimicrobial agents	Abbas and Abou Baker (2020)
	Monascus purpureus	46.58	Photo-catalytic and biological evaluation	El-Sayed et al. (2020b)
	Saccharomyces cerevisiae	75–709	Antioxidant activity	Faramarzi et al. (2020)
	Magnusiomyces ingens	70–90	Antibacterial	Lian et al. (2019)
	Penicillium chrysogenum	33.84	Biomedical and therapeutic applications	El-Sayyad et al. (2019)
	Aspergillus oryzae	55	Biomedical and food control	Mosallam et al. (2018)
	Aspergillus terreus	47	Mycosynthesis	Zare et al. (2013)
ZnS	*Aspergillus flavus*	12–24	Waste water treatment	Uddandarao et al. (2019)
ZnS–Gd		10–18		
Cobalt ferrite	*Monascus purpureus*	6.50	Anticancer, antioxidant and antimicrobial activities	El-Sayed et al. (2020a)
Te	*Aspergillus welwitschiae*	60	Antibacterial property against MRSA	Elsoud et al. (2018)
CdTe	*Saccharomyces cerevisiae*	2.0–3.6	Bio-imaging and bio-labeling	Luo et al. (2014)
Al_2O_3	*Colletotrichum* sp.	30–50	Antimicrobial activity	Suryavanshi et al. (2017)

(continued)

Table 1 (continued)

NPs	Synthesized by	Size (nm)	Applications	Reference
Pt	*Penicillium chrysogenum*	5–40	Cytotoxicity	Subramaniyan et al. (2018)
Pd	*Saccharomyces cerevisiae*	32	Dye degradations	Sriramulu and Sumathi (2018)
Ag–Au	*Aspergillus niger*	10.12–52.51	Biomedical and catalytic activities	Elegbede et al. (2019)
	Trichoderma longibrachiatum	6.98–25.20		
	Fusarium semitectum	10–35	Mycosynthesis	Sawle et al. (2008)
	Trichoderma harzianum	10–25	Catalytic property	Tripathi et al. (2015)
	Neurospora crassa	3–90, 3–110, 4–45	Mycosynthesis	Castro-Longoria et al. (2011)

keratoplasticum, *Aspergillus oryzae* and *Alternaria alternata* strains have also been reported to mycosynthesize zinc oxide and iron oxide NPs (Fouda et al. 2018; Gao et al. 2019; Mohamed et al. 2015, 2019; Raliya 2013).

Further, *Fusarium semitectum, Aspergillus flavus, P. corylophilum, Aspergillus tubingensis, P. chrysogenum* and yeast cells were employed for the synthesis of titanium oxide and selenium NPs (Abbas and Abou Baker 2020; El-Sayyad et al. 2019; Li et al. 2015; Maaroof and Mahmood 2019; Raliya et al. 2015; Salem et al. 2020). Additionally, *P. chrysogenum, Trichoderma asperellum, Penicillium aurantiogriseum, Penicillium waksmanii, Penicillium citrinum* and *Pleurotus ostreatus* were used for copper and copper oxide mycosynthesis (El-Batal et al. 2018, 2020; Honary et al. 2012; Saravanakumar et al. 2019), while *P. chrysogenum* and *S. cerevisiae* were utilized for the mycosynthesis of platinum and palladium NPs (Sriramulu and Sumathi 2018; Subramaniyan et al. 2018). *F. oxysporum* was also utilized for formation of cobalt ferrite, zinc sulfide (ZnS) and tellurium (Te) NPs, when the precursor salt solution was added to the fungal extracts (El-Sayed et al. 2020a; Elsoud et al. 2018; Uddandarao et al. 2019).

4 Mechanisms of Mycosynthesis of NPs

Fungi produce different types of extracellular metabolites that are used as the main factors for their survival when exposed to different environmental unfavorable conditions such as toxic materials, temperature and predators (Khan et al. 2016). During mycosynthesis of metal NPs, the fungal mycelia are directly in contact with metal salt solutions aiming to remediate them, which stimulate the secretion of fungal enzymes (proteins) and other metabolites in order to minimize their toxic effects. In

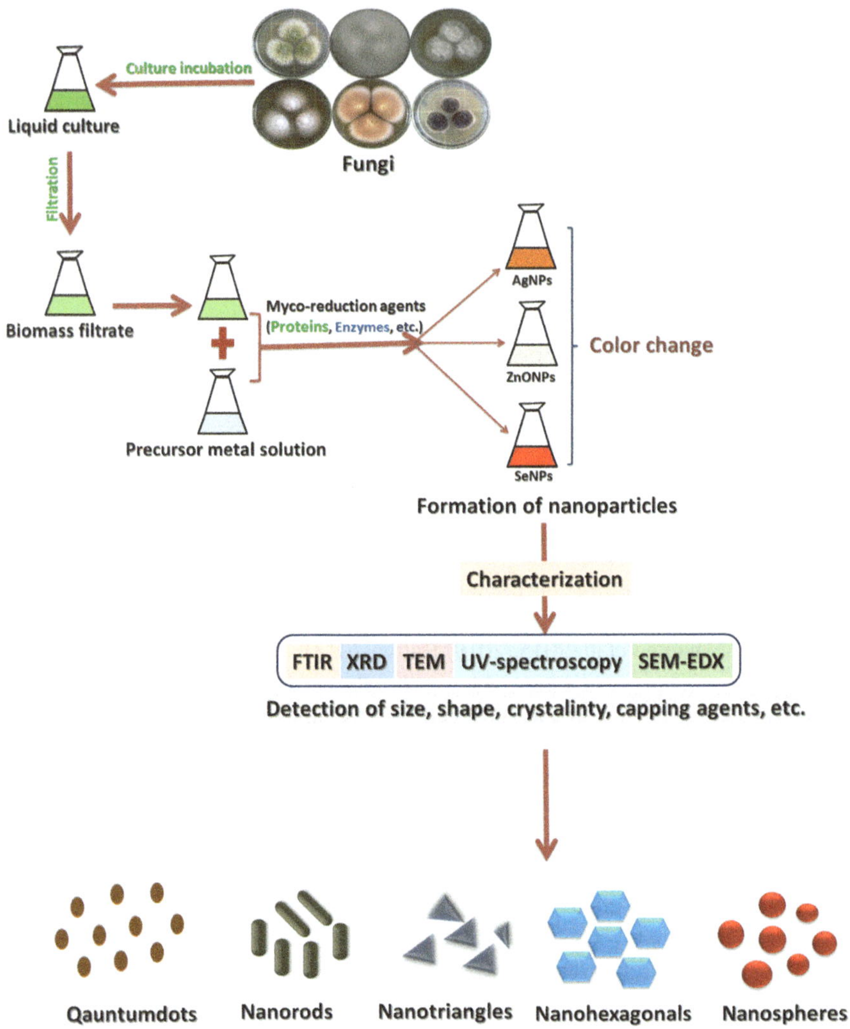

Fig. 1 Schematic explanation for the methodology used to mycosynthesize NPs

this approach, the venomous metal ions are reduced to non-venomous metallic static NPs as a catalytic impact of the metabolites and enzyme of the cell fungus (Salem and Fouda 2020; Zhao et al. 2018). The mechanism involves, for example, reduction of Ag^+ and Ag^0 via intracellular redox potential, and enzymes and/or proteins which is secreted by the fungal cell (Rahman et al. 2019). These reactions may occur either intracellularly or extracellularly (Fig. 2). For the extracellular mycosynthesis of NPs, a number of mechanisms have been proposed. Ahmad et al. (2003) for the first time proposed mechanism involved for the Ag-NPs mycosynthesis by *F. oxysporum*. The

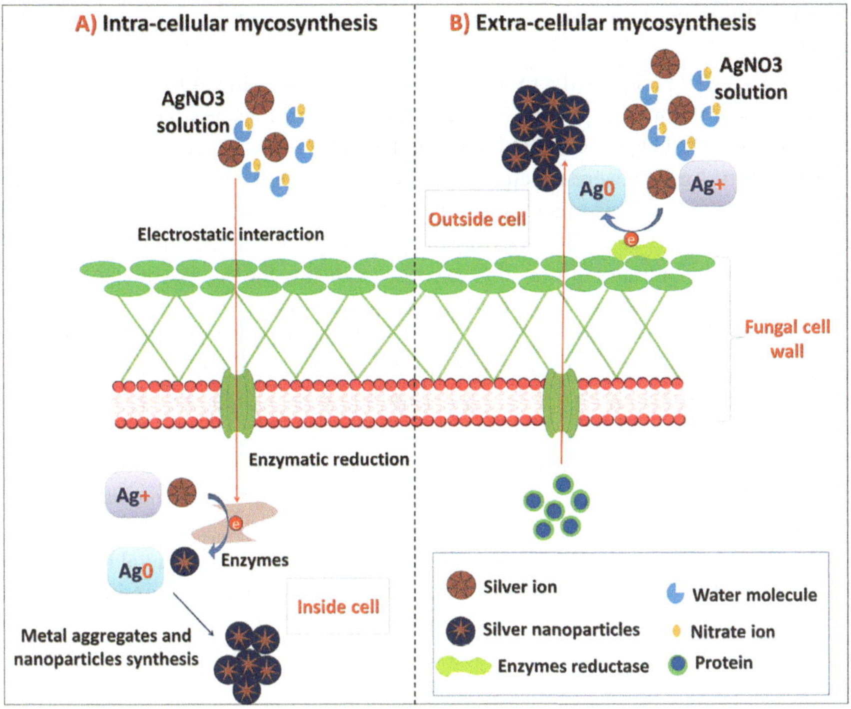

Fig. 2 Hypothetical mechanisms for Ag-NPs biosynthesis by fungal species

authors assayed the fungal protein to detect the presence of NADH-dependent reductase and reported that this enzyme was accountable for the Ag ions reduction and the subsequent formation of Ag-NPs. Durán et al. (2005) worked out the mechanism of mycosynthesis of Ag-NPs and reported that the mycosynthesis of Ag-NPs was achieved through anthraxquinone and NADPH-nitrate reductase. In this case, the electron required to fulfill the deficiency of aqueous silver ions (Ag^{+}) and convert it into Ag neutral (Ag^{0}) which was donated by both quinone and NADPH. Similarly, Ingle et al. (2008) proposed a mechanism for the mycosynthesis of Ag-NPs from *Fusarium acuminatum*. The authors supported the hypothetical mechanism where the co-factor NADH and nitrate reductase enzymes were accountable for the fabrication of Ag-NPs. They also confirmed the existence of nitrate enzyme reductase in the fungal cell filtrate. Mukherjee et al. (2008) also suggested Michaelis–Menten type of mechanism for the mycosynthesis of NPs where the reaction response initially offers pseudo-zero-order kinetics and subsequently follow up higher-order kinetics. Thus, at initial phase, when the concentration of silver nitrate was higher, the reaction was rather slow and as the reaction proceeded, the concentration of silver nitrate reduced considerably. The authors proposed that bioreduction of metal nanoparticles was brought about by protein extract including amino acid with -SH group and most likely cysteine subject de-hydrogenation on reaction with silver nitrate to form

Ag-NPs. While the free amino acid groups possibly served as a capping agent for Ag-NPs.

Jain et al. (2011) reported a two-step presumptive mechanism for mycosynthesis of Ag-NPs. In the first step, the reduction of bulk silver ions to silver nanoparticles was done by a 32 kDa protein, which might be a reductase secreted by *A. flavus*. In the latter step, Ag-NPs were covered by a 35 kDa protein of enzyme that capped NPs and confers stability. This phenomenon was confirmed by Durán and Seabra (2012) that reported the presence of two extracellular proteins with molecular weight of 21 and 24kDa in the biomass filtrate of *F. oxysporum* during the mycosynthesis of metal oxide NPs. The actual mechanism of mycosynthesis of NPs, however, is still not fully understood. According to Mukherjee et al. (2001), in intracellular synthesis, metal nanoparticles are synthesized below the cell surface, which is probably due to the metal ions reduction by enzymes present in the cell membrane. Mycosynthesis proceeds firstly by the metal ions entrapment on the fungal cell surface, which occurs due to the electrostatic interaction between metal ions and lysine residues (Riddin et al. 2006). The second step involved reduction of metal ions enzymatically, which leads to agglomeration and formation of NPs.

The fungal cell wall carbohydrates also play a major role in the metal ions reduction (Yadav et al. 2015). Some researchers suggested a linkage among the size of mycosynthesized NPs with sub-cellular spot, where the intracellular mycosynthesis NPs size was smaller than those gained by extracellular means (Ovais et al. 2018). Because the intracellular approach for NPs synthesis occurred on the fungal cell wall, the harvest techniques become harsh and hence the expense of NPs mycosynthesis rises (Molnár et al. 2018). Thus, the extracellular approach is believed to be an important design for NPs mycosynthesis (Qin et al. 2020). Finally, the mechanism of mycosynthesis of NPs utilizing fungi cells is necessarily needed to be better understood, as the NPs mycosynthesis could be affected by various factors and hence may lower the reproducibility of the mycosynthesis.

5 Factors That Affect Mycosynthesis of NPs

There are numerous factors that can influence the mycosynthesis process for NPs as fungal biomass, temperature, precursor concentrations and time contact, pH and the presence of certain enzymes (Fig. 3). Adjusting the size and shape of metal nanomaterials has been achieved either by controlling of environmental conditions or by shifting the functional molecules responsible for the reduction, capping and stabilizing NPs (Ma et al. 2017; Mohamed et al. 2019). Improving the reaction conditions for mycosynthesis of NPs, such as pH, incubation period, aeration, temperature, salt concentration, mixing ratio, redox conditions and irradiation have been extensively investigated (Fouda et al. 2018; Liang et al. 2019; Saxena et al. 2016). Size and shape of NPs depend on the chemical and physical factors during mycosynthesis which, in turn, influence the orientation of metal clusters during growth leading to form different shapes with different sizes. Of these factors are the optimum metal

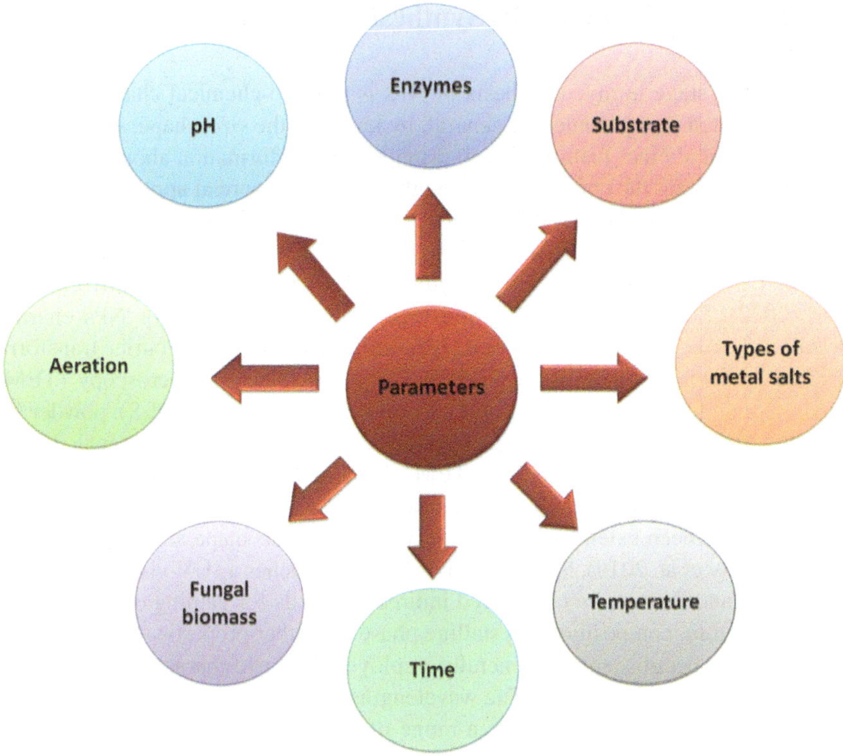

Fig. 3 Parameters affecting biosynthesis of nanoparticles by different fungal species

ions concentration, temperature and pH of reaction mixture which usually play a vital role in determining both size and shape of the formed nanoparticles. The rate of intracellular synthesis of NPs could, to an extent, be influenced by scheming parameters such as temperature, pH, substrate concentration and exposure period to substrate (Feroze et al. 2020; Fouda et al. 2017a, b; Mohamed et al. 2019).

Researchers have investigated the optimization of numerous parameters as temperature, inoculum of fungal biomass, pH and precursor concentration in the mycosynthesis of Ag-NPs, ZnO-NPs and Au-NPs from *Trichoderma* spp., *Sclerotinia sclerotiorum, A. terreus, A. niger* and *C. cladosporioides* (Fouda et al. 2017a, b, 2018; Joshi et al. 2017; Ramos et al. 2020; Saxena et al. 2016). They showed that optimizing the above factors will promote significantly the Ag-NPs, ZnO-NPs and Au-NPs mycosynthesis as well as its output. Moreover, the optimization of physical as well as cultural conditions during the mycosynthesis of Ag-NPs from *F. oxysporum* had a significant impact on their size as was investigated (Husseiny et al. 2015).

6 Characterization of NPs Synthesized by Fungi

An important stage in mycosynthesis of NPs is physico-chemical characterization of the fabricated nanoparticles. In general, by knowing the size, shape, surface area, homogeneity and other features, it will provide worth information about nano-scale systems facilitating the control of NPs synthesis for commercial applications. After the reaction, produced colloidal NPs can be separated by centrifugation at high speed or drying in oven and then can be observed via advanced nano-characterization systems (El-Batal et al. 2018; Elegbede et al. 2020; Mohamed et al. 2019; Salem et al. 2020). Some common tools and techniques can be used for NPs characterization involving color change test, UV-visible spectrometry, Fourier transform infrared spectroscopy (FT-IR), Raman spectroscopy, electron microscopy (TEM, SEM), atomic force microscopy (AFM), dynamic light scattering (DLS), powder X-ray diffraction (XRD) and energy dispersive spectroscopy (EDS) (Titus et al. 2019). Microscopy-based methods such as TEM, SEM and AFM are considered direct techniques for study the morphological characters of NPs. In particular, both TEM and SEM have been extensively utilized to determine the shape and sizes of metal NPs (Mahanty et al. 2019). Spectroscopy-based approaches as UV-vis, DLS, FT-IR, XRD, EDS and Raman are considered indirect methods for gaining data related to surface structure, composition, crystalline phase and other properties of metal NPs.

UV-visible spectroscopy is a generally employed system to characterize metal NPs (Ma et al. 2017; Qu et al. 2020). The wavelengths in the range between 200–700 nm are commonly used to characterize a range of metal and metal oxide NPs. Spectrophotometric measurements of absorbance in the different wavelength according to NPs determination, e.g., wavelength ranges between 500–550 nm, 300–400 nm and 400–460 nm are utilized to characterize gold, zinc oxide and silver NPs, respectively (Alsharif et al. 2020; Aygün et al. 2020; Fouda et al. 2018; Qu et al. 2020). FT-IR spectroscopy is suitable for characterizing the functional and surface chemistry of NPs. Biological functional groups (e.g., enzymes, carbonyls, proteins and others) attached to metal NPs surface and the other chemical residues that remain on the surface are detected via FT-IR spectroscopy (El-Sayyad et al. 2019; Fouda et al. 2019b; Mohamed et al. 2019). When the mycosynthesis is completed, the size average, shape and dispersion state of metal and metal oxide NPs are typically measured using AFM, SEM and TEM. The surface of NPs scanning by TEM and SEM is utilized for morphological properties on the nanometer to micrometer size scale. TEM has a 1000-fold higher resolution image than SEM (Gao et al. 2019; Kalpana et al. 2018; Molnár et al. 2018). The elemental composition of NPs is generally recognized using energy dispersive spectroscopy (EDS). AFM provides image in two and three dimensions that showing the surface of NPs. The resolution in the vertical, or *Z*-axis, is limited by the tool of vibration environment, whereas the resolution in the horizontal, or *X–Y*-axis, is limited by the diameter of the tip used for scanning (Singh et al. 2018). AFM affords surface characterization on the atomic scale size. XRD is applied for properties and identification of the crystallite structure of metal NPs. X-rays penetrate into the NPs and the resulting diffraction pattern is compared with standards to gain information of the structural properties (Mohamed et al. 2019). DLS is useful to determine the surface charge and distribution size of NPs suspended in a colloidal phase (Faramarzi et al. 2020; Mohamed et al. 2019).

7 Applications of Mycosynthesized NPs

Myconanotechnology is a quick-promoting industry which has effectiveness in the numerous fields of bionanotechnology. Fungus-based mycosynthesis of NPs has attracted the microbiologists and other researchers due to their enormous applications in several fields. Owing to their authentic properties, NPs have worthy implementation in many fields as catalysts, fuel additives, cosmetics, lubricants, paints, food packaging, agrochemicals, textile engineering, de-colorization of dyes, optics, electronics, environmental sensing, gene delivery and drug agents, biodetection of pathogens, nanomedicine, destruction of tumor, imaging, magnetic resonance, water treatments, agricultures studies and other applications as listed in Table 1 (El-Batal et al. 2020; El-Sayyad et al. 2019; Mohamed et al. 2019; Rabeea et al. 2020; Salvadori 2019; Uddandarao et al. 2019). In this context, fungus-mediated biosynthesis of NPs has recently gained much awareness due to their comprehensive application in multiple sectors. Applications of NPs in different fields may be summarized as follows:

7.1 Myconanotechnology in Agriculture Sector

Agriculture is the central backbone for providing human food, desirable products, animal feed and pivotal requisite materials for diverse industry as leathers, fibers, chemicals for industry as xylan, starch and sugars. Improving the agricultural system will have good effects on the other sectors. Nanotechnology has positive impacts on different sectors of agriculture. Agricultural knowledge and engineering scope have acquired momentum with the prospering implementations of myconanotechnology in agricultural field (Tarafdar et al. 2013). Mycosynthesis of nanomaterials as nanostructures, nanowires and nano-formulations are being utilized today to restrain plant pathogens as pesticide and insecticides (El-Batal et al. 2020; Fernández et al. 2016). The coming of myconanotechnology gave gateway to the evolution of nano-devices and nano-structures with foreseeable novel implementations in the agricultural sector (Raliya et al. 2015). Besides, NPs synthesized by different fungal species are used to inhibit plant microbes, and it can be used to protect plants from pests and insects infection (Raliya et al. 2016). Recently, pesticide-based NPs have been widely used in agricultural fields to retard the pest infections (Fernández et al. 2016; Rai and Ingle 2012). Metabolites secreted by fungal species are used as green methods for fabrication of new promising NPs for pests' control (Athanassiou et al. 2018; El-Batal et al. 2020).

The excessive chemical fertilizer usage and reduced water resources, as well as soil fertility are considered the most challenges for agricultural sector. To overcome these challenges, new formulation of nano-fertilizers is used which increased nutrient availability and reduced the excessive use of chemical compounds (Raliya et al. 2016). Therefore, NPs fabricated by fungal species can be used as promising tools for developing agricultural system, especially in growing countries. Recently, different formulations of nano-fertilizers such as nanoclays and nanocrystals have been used (Guo et al. 2018). Fungi are considered as the major plant pathogens and cause plant diseases more than those caused by other microorganisms and insects (Rai and Ingle 2012; Vijayabharathi et al. 2018). Fungal cells avail as a fundamental appliance for the fabrication of nanomaterials employed for plant cell disease control. Due to their high tolerance to metals and capability to pile metals, they are extensively exploited for the fabrication of NPs as platinum, silver, iron, gold among others. Ag-NPs have wide implementation as antifungal agent against phytopathogenic fungi (Hassan et al. 2018). Ag-NPs mycosynthesized by *Mucor hiemalis* were evaluated against pathogenic fungi *A. flavus, F. semitectum* and *Candida albicans* (Aziz et al. 2016). Nanoparticles with nano-carriers as silica-silver re-unite were scrutinized as a powerful biocide for plant cell diseases (Park et al. 2006). *F. oxysporum* has been discovered to be eligible of mycosynthesizing silica NPs (Bansal et al. 2005), a functional plant cell nutrient that enhances disease fighting in cells of plants. CuO-NPs mycosynthesized by the fungus cell *P. chrysogenum* were active against pathogenic fungi represented as *F. oxysporum, A. niger, Alternaria solani, P. citrinum* and *Erysiphe cichoracearum* (El-Batal et al. 2020).

7.2 Myconanotechnology and Wastewater Treatment

The progress and continuous development of nanotechnology have highlighted various methods that have a significant and effective impact on wastewater treatment. Myconanotechnology has provided a new strategy for solving most issues concerning water deficiency and quality (Shen et al. 2017; Uddandarao et al. 2019). The benefit of NPs in wastewater treatment is demonstrated in its antimicrobial and antioxidant properties as wastewater flows are rich in pathogenic microbes. Recently, nanotechnology-based wastewater treatment is able to provide high performance treated water containing less impurities, less toxic substances and able to remove heavy metals. Wide varieties of NPs are used to removal of organic and inorganic pollutants and toxic metals, ensure disinfection as well as detection of pathogens. At present, removal of heavy metals such as chromium, lead and cadmium through nanoparticles such as iron oxide is a new technique by myconanotechnology, where the results have proven the removal of chromium from water pollutants by 90% (Mahanty et al. 2019). According to economic vision, myconanotechnology is accepted as a new strategy for solving the challenges facing wastewater treatment (Khandel and Shahi 2018). Application of mycogenic nanomaterials leads to wastewater treatment as a growing field of research. This latest myconanotechnology can remarkably relief the contaminants of water in the environment and hazard to humans due to the employment of venomous solvents and chemicals.

7.3 *Myconanotechnology and Bioremediation*

Nano-remediation uses interactive nanomaterials to stimulate and reduce pollutants. Nowadays, NPs are suitable oxidizing factors to remedy the environmental contaminants. They are qualified for high permeation and reactivity versus hydrophobic-organic pollutants, redox contaminants and chlorinated hydrocarbons. The use of nanomaterials in environmental cleaning serves to maintain a healthy human life. Many research reports exhibited the use of NPs for bioremediation of heavy metals such as Cd and Cr from wastewater (El-Sayed et al. 2020b; Salvadori, 2019; Sriramulu and Sumathi 2018; Uddandarao et al. 2019). The role of nanomaterials such as Au-NPs, ZnO-NPs and Ag-NPs appears in the dye-removal process as it increases the reaction rate while reducing the time needed to complete the degradation process as its primary role as a catalyst for the process (Kalpana et al. 2018; Popli et al. 2018; Rabeea et al. 2020). Other reports hypothesized that the proteins surface of the Se-NPs and Au-NPs formed by the fungal cells of *Monascus purpureus* and *Cladosporium oxysporum* may improve the adsorption process of dyes (methylene blue and rhodamine B) as amino acids (enzymes) that are attached to rings of aromatic compounds to formation hydrophobic area that can improve the interaction between nano-catalyst and dye molecules (Bhargava et al. 2016; El-Sayed et al. 2020b). It is clear from the study of nano-scale silver, which demonstrated that the catalyst in the form of nano-metal has good effectiveness in removing pollutants, especially organic dyes, which leads to increased reaction rate and high efficiency, and thus the industrial role played by the nano-silver (Popli et al. 2018).

7.4 *Myconanotechnology and Textile Industry*

The textile industries are one of the most important industries in the world where obtaining a high-quality textile product is required by adding highly efficient materials, such as Ag-NPs, which in turn give the fabrics new properties such as antimicrobial activity and the prevention of harmful rays while giving them new performance characteristics for self-cleaning. Impregnating Ag-NPs in wound dressings and textile fabrics have been prosperous in controlling infection by microbial pathogens (El-Rafie et al. 2012; Mohmed et al. 2017a; Shaheen and Abd El Aty 2018). Also, some researchers have added biologically synthesized ZnO-NPs to textile fabrics to increase the performance properties of fabrics, such as blocking ultraviolet radiation and its antibacterial properties (Fouda et al. 2018; Mohamed et al. 2019). Recently, our published study concerned with various shapes of ZnO-NPs (hexagonal, nano-rod and spherical) were mycosynthesized by *F. keratoplasticum*, *A. niger* and *A. terreus*, respectively. The spherical, nano-rod and hexagonal ZnO-NPs shapes exhibit antimicrobial and cytotoxicity effects against cell lines (normal and cancer). The safe dose of three ZnO-NPs shapes was loaded on cotton fabrics to enhance their properties such as antibacterial activities against Gram negative and Gram positive bacteria and UV blocking property (Fouda et al. 2018; Mohamed et al. 2019).

7.5 Myconanotechnology in Food Industry

The applications of nanomaterials in the food industry represent a good development in terms of food safety from spoilage, packaging, safety and delivery systems. In the immediate future, the development of nanotechnology will serve two specific areas in food processing; namely food packaging and food additives (Sharaf et al. 2019). Myconanotechnology is used as a smart nanomaterial in food safety as the main trend is to reduce food spoilage. In order to improve the antimicrobial activity of packaging tissue, nano-gold was introduced into the packaging industry and gave it new properties (Ray et al. 2013). The nanomaterials were utilized in packaging method, taking into account food safety. Thus, the inclusion of ZnO-NPs in packing and packaging materials saves juices from deterioration without changing their properties and appearance as they become of better quality. Ag-NPs were also found to be slightly better than ZnO-NPs in its antimicrobial activity on mold and yeast cells. The juice packaged with NPs will work for a longer period if the process is done at moderate temperatures. Researchers have confirmed the mycosynthesis of ZnO-NPs by *Aspergillus fumigatus*, *A. alternata* and *C. albicans* (Mashrai et al. 2017; Sarkar et al. 2014). A study on Se-NPs showed that it has the ability to destroy pathogens during the cleaning process and remove contaminants that cause food spoilage (Mosallam et al. 2018). Ultimately, the availability of myconanotechnology in the food industry will be of benefit for the preservation of vegetables and fruits (Mohammed Fayaz et al. 2009).

7.6 Myconanotechnology in Biomedical Applications

The considerable confronts for medical practitioners are briefed in apparition of new drug-resistant microbes. Therefore, the developments of new drugs are necessary to overcome various diseases. The applications of NPs in medicine have various advantages as diagnosis using NP-based imaging, early detection systems and treatment of various diseases caused by drug-resistant microbes (Clarance et al. 2020; El-Sayed et al. 2020a; Elsoud et al. 2018; Yousef et al. 2020). Various types of NPs, including metals and metal oxides such as Ag, Se, Au, ZnO, TiO_2 and CuO have been developed by various researchers to utilize in medicinal applications (Abbas and Abou Baker 2020; Castillo et al. 2019; Dizaj et al. 2015; Durairaj et al. 2014b; Elegbede et al. 2020; Fouda et al. 2018; Gao et al. 2019; Hassan et al. 2018, 2019; Marquis et al.

2016; Mohamad Sukri et al. 2019; Mohanta et al. 2018; Saravanakumar et al. 2019; Thakur et al. 2019). Recently, mycosynthesized Se-NPs, ZnO-NPs, Ag-NPs, Cu-NPs and CuO-NPs showed antimicrobial activities against various pathogenic bacteria as *Escherichia coli, Pseudomonas aeruginosa, Klebsiella pneumoniae, Acinetobacter calcoaceticus, Enterobacter agglomerans, Enterobacter cloacae, Staphylococcus aureus, Bacillus subtilis, Erwinia amylovora, Citrobacter freundii* and *Ralstonia solanacearum* (El-Batal et al. 2018, 2020; Mohamed et al. 2019; Salem et al. 2020). The activities of NPs as antifungal agents for different pathogenic fungi have been widely evaluated (Guilger et al. 2017). Several studies have reported the activities of mycosynthesized Ag-NPs as antifungal agents against *Trichophyton mentagrophytes, A. flavus, Candida glabrata, A. fumigatus, Candida parapsilosis, Cryptococcus neoformans, Candida krusei, F. solani, Trichophyton rubrum, Cryptococcus gattii, Candida tropicalis, Sporothrix schenckii, Epidermophyton floccosum, C. albicans* and *M. hiemalis.* In the same regard, Ag-NPs showed activities against plant pathogenic fungi such as *A. niger*, *Colletotrichum* sp., *Fusarium* sp., *Culvularia lunata* and *Rhizoctonia solani* (Aziz et al. 2016; Parmar and Sharma 2020).

Damage of microbial cell wall, destroy the microbial cell components due to NPs activity or oxidative stress of NPs are considered three defined mechanisms to explain the activity of NPs against microbes (Fig. 4) (Qin et al. 2020). Due to difference in cell wall structure among bacteria, the metallic NPs exhibited more activity against

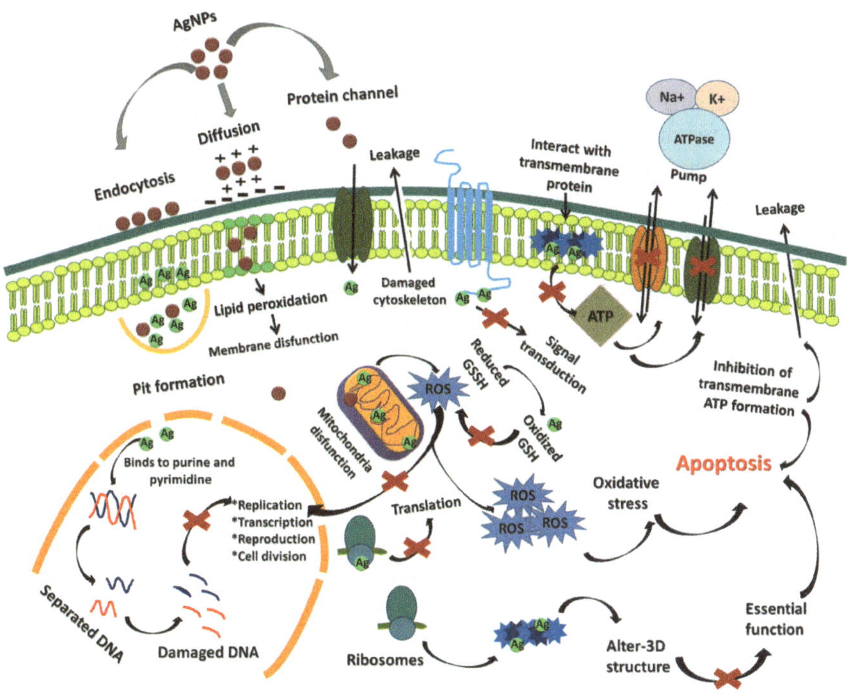

Fig. 4 Prospective mechanisms for antimicrobial activity of NPs

Gram negative bacteria than Gram positive bacteria (Roy et al. 2019). Also, this activity could be attributed to electrostatic attraction between lipopolysaccharides negative charge in Gram negative bacteria and positive charge of NPs which lead to disruption in selective permeability and hence cell death (Yun'an Qing et al. 2018). Due to change in membrane selective permeability, the essential components such as enzymes, minerals, proteins and nucleic acids are lost which have negative impacts on the cell metabolism. Therefore, we can conclude that the damage of cell wall and disruption in cell membrane selective permeability due to NPs activity is considered the main indicator for microbial death. Once metallic NPs enter inside the microbial cell, it reacts with their nucleic acids and then changes its normal shape and forfeit its replication activity (Qin et al. 2020). Moreover, microbial enzymes are converted to inactivated form due to NPs activity which combine with thiol (-SH) group present in amino acid as cysteine.

Recently, cancer diagnosis and treatment have received more attention. A large multiplicity in nanomaterials has been evaluated to improve its efficacy in cancer therapy as well as to reduce negative impacts compared with conventional therapies. The toxicity impacts of NPs synthesized by green methods are evaluated mainly by changes in viability and cell morphology, as well as metabolic activities (Fouda et al. 2018; Mohanta et al. 2018). NPs can be localized in mitochondria, inducing functional damage and structural as well as oxidative emphasis (Yan et al. 2018). Physico-chemical properties of NPs have critical important role in cytotoxicity effect. The nature and size of NPs, its surface area, and its surface functionalization (capping agents) are important factors that affect their toxicity (Golinska et al. 2017; Mohamed et al. 2019). Therefore, the construct new drug containing a different formulation of NPs are increasing day by day due to its safe, more effective, little side effect and high efficacy for targeting of cancer cells. Several types of NP-sized drugs have been evaluated in cancer therapy to reduce the negative impacts of conventional anticancer drugs and improve the antitumor drug efficacy target therapies. Recently, metal and metal oxide NPs are considered promising tools for applicable in biomedical fields such as gene delivery, cancer therapy, antibacterial agent, control of pathogenic fungi, anticoagulants, thrombolytics and cell imaging (Barabadi et al. 2017; Clarance et al. 2020; El-Batal et al. 2018; El-Sayed et al. 2020a; Fouda et al. 2018; Ma et al. 2017; Maaroof and Mahmood 2019; Mohamed et al. 2015; Elegbede et al. 2018, 2019, 2020; Salem et al. 2020).

8 Conclusion

Overall, the green synthesis of NPs using fungi in particular is highly preferred due to their huge secretion metabolites, simple and easy scaling up, easy processing, economic viability and biomass processing. Green biosynthesis of NPs encompasses myconanotechnology which highlights specifically the interaction between mycology and nanotechnology. Fungi, as a bio-factory, have the ability to easily fabricate NPs by remediating their ions in solution through either extracellular or intracellular approach. To improve mycosynthesis of NPs and control their size and shape, different environmental factors such as fungal biomass, temperature, precursor concentrations and time contact, pH and the being of certain enzyme were extensively investigated to optimize not only the formed NPs, but to maximize the production of NPs also. Moreover, physico-chemical characterization of mycosynthesized NPs is considered the critical factors for study the nature of NPs that clearly states the main characters of the prepared NPs. Due to the unique chemical, and physical characters of the mycosynthesized NPs, they have a wide rang of applications in many fields. Among NPs based applications, agriculture sector, wastewater treatment, heavy metal removal, textile industry, food packaging and safety, bioremediation of environmental hazard substances and biomedical applications are prospective areas for the utilization of mycosynthesized NPs. With no doubt, the biosynthesis of NPs using fungi opens up a new possibilities to make our daily life green as possible, however, a dense efforts have to be accomplished by both scientific communities and their governments for transporting such works to industrial world instead of chemical methods that threaten life.

References

Abbas H, Abou Baker D (2020) Biological evaluation of selenium nanoparticles biosynthesized by *Fusarium semitectum* as antimicrobial and anticancer agents. Egypt J Chem 63:18–19. https://doi.org/10.21608/ejchem.2019.15618.1945

Abd El Aty AA, Mohamed AA, Zohair MM, Soliman AA (2020) Statistically controlled biogenesis of silver nano-size by *Penicillium chrysogenum* MF318506 for biomedical application. Biocatal Agric Biotechnol 25:101592. https://doi.org/10.1016/j.bcab.2020.101592

Abdelhakim HK, El-Sayed ER, Rashidi FB (2020) Biosynthesis of zinc oxide nanoparticles with antimicrobial, anticancer, antioxidant and photocatalytic activities by the endophytic *Alternaria tenuissima*. J Appl Microbiol 128:1634–1646. https://doi.org/10.1111/jam.14581

Adelere I, Lateef A (2016) A novel approach to the green synthesis of metallic nanoparticles: the use of agro-wastes, enzymes, and pigments. Nanotechnol Rev 5:567–587. http://dx.doi.org/10.1515/ntrev-2016-0024

Ahmad A, Mukherjee P, Senapati S, Mandal D, Khan MI, Kumar R, Sastry M (2003) Extracellular biosynthesis of silver nanoparticles using the fungus *Fusarium oxysporum*. Colloids Surf B 28:313–318. https://doi.org/10.1016/S0927-7765(02)00174-1

Ahmed AA, Hamzah H, Maaroof M (2018) Analyzing formation of silver nanoparticles from the filamentous fungus *Fusarium oxysporum* and their antimicrobial activity. Turk J Biol 42:54–62. https://doi.org/10.3906/biy-1710-2

Alsharif MS, Salem SS, Abdel-Rahmanb AM, Fouda A, Eid MA, Hassan ES, Awad AM, Mohamed AA (2020) Multifunctional properties of spherical silver nanoparticles fabricated by different microbial taxa. Heliyon 6:e03943. https://doi.org/10.1016/j.heliyon.2020.e03943

Amerasan D, Nataraj T, Murugan K, Panneerselvam C, Madhiyazhagan P, Nicoletti M, Benelli G (2016) Myco-synthesis of silver nanoparticles using *Metarhizium anisopliae* against the rural malaria vector *Anopheles culicifacies* Giles (Diptera: Culicidae). J Pest Sci 89:249–256. https://doi.org/10.1007/s10340-015-0675-x

Ammar HAM, El-Desouky TA (2016) Green synthesis of nanosilver particles by *Aspergillus terreus* HA1N and *Penicillium expansum* HA2N and its antifungal activity against mycotoxigenic fungi. J Appl Microbiol 121:89–100. https://doi.org/10.1111/jam.13140

Athanassiou CG, Kavallieratos NG, Benelli G, Losic D, Usha Rani P, Desneux N (2018) Nanoparticles for pest control: current status and future perspectives. J Pest Sci 91:1–15. https://doi.org/10.1007/s10340-017-0898-0

Aygün A, Özdemir S, Gülcan M, Cellat K, Şen F (2020) Synthesis and characterization of *Reishi mushroom*-mediated green synthesis of silver nanoparticles for the biochemical applications. J Pharm Biomed Anal 178:112970. https://doi.org/10.1016/j.jpba.2019.112970

Aziz N, Pandey R, Barman I, Prasad R (2016) Leveraging the attributes of *Mucor hiemalis*-derived silver nanoparticles for a synergistic broad-spectrum antimicrobial platform. Front Microbiol 7:1984. https://doi.org/10.3389/fmicb.2016.01984

Balaji DS, Basavaraja S, Deshpande R, Mahesh DB, Prabhakar BK, Venkataraman A (2009) Extracellular biosynthesis of functionalized silver nanoparticles by strains of *Cladosporium cladosporioides* fungus. Colloids Surf B 68:88–92. https://doi.org/10.1016/j.colsurfb.2008.09.022

Barabadi H, Ovais M, Shinwari ZK, Saravanan M (2017) Anti-cancer green bionanomaterials: present status and future prospects. Green Chem Lett Rev 10:285–314. https://doi.org/10.1080/17518253.2017.1385856

Bhargava A, Jain N, Khan MA, Pareek V, Dilip RV, Panwar J (2016) Utilizing metal tolerance potential of soil fungus for efficient synthesis of gold nanoparticles with superior catalytic activity for degradation of rhodamine B. J Environ Manage 183:22–32. https://doi.org/10.1016/j.jenvman.2016.08.021

Castillo IF, De Matteis L, Marquina C, Guillén EG, Martínez de la Fuente J, Mitchell SG (2019) Protection of 18th century paper using antimicrobial nano-magnesium oxide. Int Biodeterior Biodegradation 141:79–86. https://doi.org/10.1016/j.ibiod.2018.04.004

Castro-Longoria E, Vilchis-Nestor AR, Avalos-Borja M (2011) Biosynthesis of silver, gold and bimetallic nanoparticles using the filamentous fungus *Neurospora crassa*. Colloids Surf B 83:42–48. https://doi.org/10.1016/j.colsurfb.2010.10.035

Chauhan R, Reddy A, Abraham J (2015) Biosynthesis of silver and zinc oxide nanoparticles using *Pichia fermentans* JA2 and their antimicrobial property. Appl Nanosci 5:63–71. https://doi.org/10.1007/s13204-014-0292-7

Clarance P, Luvankar B, Sales J, Khusro A, Agastian P, Tack JC, Al Khulaifi MM, Al-Shwaiman HA, Elgorban AM, Syed A, Kim HJ (2020) Green synthesis and characterization of gold nanoparticles using endophytic fungi *Fusarium solani* and its in-vitro anticancer and biomedical applications. Saudi J Biol Sci 27:706–712. https://doi.org/10.1016/j.sjbs.2019.12.026

Cuevas R, Durán N, Diez M, Tortella G, Rubilar O (2015) Extracellular biosynthesis of copper and copper oxide nanoparticles by *Stereum hirsutum*, a native white-rot fungus from chilean forests. J Nanomater 2015:1–7. https://doi.org/10.1155/2015/789089

Debnath G, Das P, Saha AK (2019) Green synthesis of silver nanoparticles using mushroom extract of *Pleurotus giganteus*: characterization, antimicrobial, and α-Amylase inhibitory activity. BioNanoScience 9:611–619. https://doi.org/10.1007/s12668-019-00650-y

Dizaj SM, Mennati A, JafariS, Khezri K, Adibkia K (2015) Antimicrobial activity of carbon-based nanoparticles. Adv Pharm Bull 5:19. https://doi.org/10.5681/apb.2015.003

Durairaj B, Xavier T, Muthu S (2014a) Fungal generated titanium dioxide nanoparticles: a potent mosquito (*Aedes aegypti*) larvicidal agent. Sch Acad J Biosci 2, 651–658

Durairaj B, Xavier T, Muthu S (2014b) Fungal generated titanium dioxide nanopartilces for UV Protective and bacterial resistant fabrication. Int. J. Eng. Sci. Technol. 6, 621

Durán N, Seabra AB (2012) Metallic oxide nanoparticles: state of the art in biogenic syntheses and their mechanisms. Appl Microbiol Biotechnol 95:275–288. https://doi.org/10.1007/s00253-012-4118-9

Durán N, Marcato PD, Alves OL, De Souza GIH, Esposito E (2005) Mechanistic aspects of biosynthesis of silver nanoparticles by several *Fusarium oxysporum* strains. J Nanobiotechnology 3:8. https://doi.org/10.1186/1477-3155-3-8

Eid AM, Salim SS, Hassan SED, Ismail MA, Fouda A (2019) Role of endophytes in plant health and abiotic stress management. In: Kumar V, Prasad R, Kumar M, Choudhary DK (eds) Microbiome in plant health and disease: challenges and opportunities. Springer, Singapore, pp 119–144. https://doi.org/10.1007/978-981-13-8495-0_6

Elamawi RM, Al-Harbi RE, Hendi AA (2018) Biosynthesis and characterization of silver nanoparticles using *Trichoderma longibrachiatum* and their effect on phytopathogenic fungi. Egypt J Biol Pest Control 28:28. https://doi.org/10.1186/s41938-018-0028-1

El-Batal AI, Al-Hazmi NE, Mosallam FM, El-Sayyad GS (2018) Biogenic synthesis of copper nanoparticles by natural polysaccharides and *Pleurotus ostreatus* fermented fenugreek using gamma rays with antioxidant and antimicrobial potential towards some wound pathogens. Microb Pathog 118:159–169. https://doi.org/10.1016/j.micpath.2018.03.013

El-Batal AI, El-Sayyad GS, Mosallam FM, Fathy RM (2020). *Penicillium chrysogenum*-mediated mycogenic synthesis of copper oxide nanoparticles using gamma rays for in vitro antimicrobial activity against some plant pathogens. J Clust Sci 31:79–90. https://doi.org/10.1007/s10876-019-01619-3

El Domany EB, Essam TM, Ahmed AE, Farghali AA (2018) Biosynthesis physico-chemical optimization of gold nanoparticles as anti-cancer and synergetic antimicrobial activity using *Pleurotus ostreatus* fungus. J Appl Pharm Sci 8:119–128. https://doi.org/10.7324/JAPS.2018.8516

Elegbede JA, Lateef A (2019) Green synthesis of silver (Ag), gold (Au), and silver–gold (Ag–Au) alloy nanoparticles: a review on recent advances, trends, and biomedical applications. In: Nanotechnology and nanomaterial applications in food, health, and biomedical sciences. Apple Academic Press, pp 3–89. https://doi.org/10.1201/9780429425660-1

Elegbede JA, Lateef A, Azeez MA, Asafa TB, Yekeen TA, Oladipo IC, Adebayo EA, Beukes LS, Gueguim-Kana EB (2018) Fungal xylanases-mediated synthesis of silver nanoparticles for catalytic and biomedical applications. IET Nanobiotechnol 12:857–863. https://doi.org/10.1049/iet-nbt.2017.0299

Elegbede JA, Lateef A, Azeez MA, Asafa TB, Yekeen TA, Oladipo IC, Hakeem AS, Beukes LS, Gueguim-Kana EB (2019) Silver-gold alloy nanoparticles biofabricated by fungal xylanases exhibited potent biomedical and catalytic activities. Biotechnol Prog 35:e2829. https://doi.org/10.1002/btpr.2829

Elegbede JA, Lateef A, Azeez MA, Asafa T, Yekeen TA, Oladipo I, Aina D, Beukes L, Gueguim-Kana E (2020). Biofabrication of gold nanoparticles using xylanases through valorization of corncob by *Aspergillus niger* and *Trichoderma longibrachiatum*: antimicrobial, antioxidant, anti-coagulant and thrombolytic activities. Waste Biomass Valor 11:781–791. https://doi.org/10.1007/s12649-018-0540-2

Elfeky AS, Salem SS, Elzaref AS, Owda ME, Eladawy HA, Saeed AM, Awad MA, Abou-Zeid RE, Fouda A (2020) Multifunctional cellulose nanocrystal/metal oxide hybrid, photo-degradation, antibacterial and larvicidal activities. Carbohydr Polym 230:115711. https://doi.org/10.1016/j.carbpol.2019.115711

El-Rafie MH, Shaheen TI, Mohamed AA, Hebeish A(2012). Bio-synthesis and applications of silver nanoparticles onto cotton fabrics. Carbohydr Polym 90:915–920. https://doi.org/10.1016/j.carbpol.2012.06.020

El-Sayed ESR, Abdelhakim HK, Zakaria Z (2020a) Extracellular biosynthesis of cobalt ferrite nanoparticles by *Monascus purpureus* and their antioxidant, anticancer and antimicrobial activities: yield enhancement by gamma irradiation. Mater Sci Eng C 107:110318. https://doi.org/10.1016/j.msec.2019.110318

El-Sayed ESR, Abdelhakim HK, Ahmed AS (2020b) Solid-state fermentation for enhanced production of selenium nanoparticles by gamma-irradiated *Monascus purpureus* and their biological evaluation and photocatalytic activities. Bioprocess Biosyst Eng 43:797–809. https://doi.org/10.1007/s00449-019-02275-7

El-Sayyad GS, El-Bastawisy HS, Gobara M, El-Batal AI (2019) Gentamicin-assisted mycogenic selenium nanoparticles synthesized under gamma irradiation for robust reluctance of resistant urinary tract infection-causing pathogens. Biol Trace Elem Res 1–20. https://doi.org/10.1007/s12011-019-01842-z

Elsoud MMA, Al-Hagar OE, Abdelkhalek ES, Sidkey N (2018) Synthesis and investigations on tellurium myconanoparticles. Biotechnol Rep 18:e00247. https://doi.org/10.1016/j.btre.2018.e00247

Faramarzi S, Anzabi Y, Jafarizadeh-Malmiri H (2020) Nanobiotechnology approach in intracellular selenium nanoparticle synthesis using *Saccharomyces cerevisiae*—fabrication and characterization. Arch Microbiol. https://doi.org/10.1007/s00203-020-01831-0

Fernández JG, Fernández-Baldo MA, Berni E, Camí G, Durán N, Raba J, Sanz MI (2016) Production of silver nanoparticles using yeasts and evaluation of their antifungal activity against phytopathogenic fungi. Process Biochem 51:1306–1313. https://doi.org/10.1016/j.procbio.2016.05.021

Feroze N, Arshad B, Younas M, Afridi MI, Saqib S, Ayaz A (2020) Fungal mediated synthesis of silver nanoparticles and evaluation of antibacterial activity. Microsc Res Tech 83:72–80. https://doi.org/10.1002/jemt.23390

Fouda AH, Hassan SED, Eid AM, Ewais EED (2015) Biotechnological applications of fungal endophytes associated with medicinal plant *Asclepias sinaica* (Bioss.). Ann Agric Sci 60:95–104. https://doi.org/10.1016/j.aoas.2015.04.001

Fouda A, Mohamed A, Elgamal MS, EL-Din Hassan S, Salem SS, Shaheen TI (2017) Facile approach towards medical textiles via myco-synthesis of silver nanoparticles. Der Pharma Chemica 9:11–18. https://www.derpharmachemica.com/abstract/facile-approach-towards-medical-textiles-via-mycosynthesis-of-silver-nanoparticles-12723.html

Fouda A, Saad E, Elgamal MS, Mohmed AA, Salem SS (2017) Optimal factors for biosynthesis of silver nanoparticles by *Aspergillus* sp. Al Azhar Bull Sci 9:161–172. https://www.academia.edu/33588039/OPTIMAL_FACTORS_FOR_BIOSYNTHESIS_OF_SILVER_NANOPARTICLES_BY_ASPERGILLUS_SP

Fouda A, El-Din Hassan S, Salem SS, Shaheen TI (2018) In-vitro cytotoxicity, antibacterial, and UV protection properties of the biosynthesized Zinc oxide nanoparticles for medical textile applications. Microb Pathog 125:252–261. https://doi.org/10.1016/j.micpath.2018.09.030

Fouda A, Abdel-Maksoud G, Abdel-Rahman MA, Eid AM, Barghoth MG, El-Sadany MAH (2019a) Monitoring the effect of biosynthesized nanoparticles against biodeterioration of cellulose-based materials by *Aspergillus niger*. Cellulose 26:6583–6597. https://doi.org/10.1007/s10570-019-02574-y

Fouda A, Abdel-Maksoud G, Abdel-Rahman MA, Salem SS, Hassan SED, El-Sadany MAH (2019b) Eco-friendly approach utilizing green synthesized nanoparticles for paper conservation against microbes involved in biodeterioration of archaeological manuscript. Int Biodeterior Biodegr 142:160–169. https://doi.org/10.1016/j.ibiod.2019.05.012

Fouda A, Hassan SED, Abdo AM, El-Gamal MS (2019c) Antimicrobial, antioxidant and larvicidal activities of spherical silver nanoparticles synthesized by endophytic *Streptomyces* spp. Biol Trace Elem Res 1–18. https://doi.org/10.1007/s12011-019-01883-4

Ganesan V, Hariram M, Vivekanandhan S, Muthuramkumar S (2020) *Periconium* sp. (endophytic fungi) extract mediated sol-gel synthesis of ZnO nanoparticles for antimicrobial and antioxidant applications. Mater Sci Semicond Process 105:104739. https://doi.org/10.1016/j.mssp.2019.104739

Gao Y, Arokia Vijaya Anand M, Ramachandran V, Karthikkumar V, Shalini V, Vijayalakshmi S, Ernest D (2019) Biofabrication of zinc oxide nanoparticles from *Aspergillus niger*, their antioxidant, antimicrobial and anticancer activity. J Clust Sci 30:937–946. https://doi.org/10.1007/s10876-019-01551-6

Golinska P, Rathod D, Wypij M, Gupta I, Składanowski M, Paralikar P, Dahm H, Rai M (2017) Mycoendophytes as efficient synthesizers of bionanoparticles: nanoantimicrobials, mechanism, and cytotoxicity. Crit Rev Biotechnol 37:765–778. https://doi.org/10.1080/07388551.2016.1235011

Guilger M, Pasquoto-Stigliani T, Bilesky-Jose N, Grillo R, Abhilash PC, Fraceto LF, Lima Rd (2017) Biogenic silver nanoparticles based on *trichoderma harzianum*: synthesis, characterization, toxicity evaluation and biological activity. Sci Rep 7:44421. https://doi.org/10.1038/srep44421

Guo H, White JC, Wang Z, Xing B (2018) Nano-enabled fertilizers to control the release and use efficiency of nutrients. Curr Opin Environ Sci Health 6:77–83. https://doi.org/10.1016/j.coesh.2018.07.009

Hamedi S, Ghaseminezhad M, Shokrollahzadeh S, Shojaosadati SA (2017) Controlled biosynthesis of silver nanoparticles using nitrate reductase enzyme induction of filamentous fungus and their antibacterial evaluation. Artif Cells Nanomed Biotechnol 45:1588–1596. https://doi.org/10.1080/21691401.2016.1267011

Hassan SELD, Salem SS, Fouda A, Awad MA, El-Gamal MS, Abdo AM (2018) New approach for antimicrobial activity and bio-control of various pathogens by biosynthesized copper nanoparticles using endophytic actinomycetes. J Radiat Res Appl Sci 11:262–270. https://doi.org/10.1016/j.jrras.2018.05.003

Hassan SED, Fouda A, Radwan AA, Salem SS, Barghoth MG, Awad MA, Abdo AM, El-Gamal MS (2019) Endophytic actinomycetes *Streptomyces* spp. mediated biosynthesis of copper oxide nanoparticles as a promising tool for biotechnological applications. J Biol Inorg Chem 24:377–393. https://doi.org/10.1007/s00775-019-01654-5

Honary S, Barabadi H, Gharaei-Fathabad E, Naghibi F (2012) Green synthesis of copper oxide nanoparticles using Penicillium aurantiogriseum, *Penicillium citrinum* and *Penicillium waksmanii*. Dig J Nanomater Bios 7:999–1005. http://www.chalcogen.ro/999_Honary.pdf

Honary S, Gharaei-Fathabad E, Barabadi H, Naghibi F (2013) Fungus-mediated synthesis of gold nanoparticles: a novel biological approach to nanoparticle synthesis. J Nanosci Nanotechnol 13:1427–1430. https://doi.org/10.1166/jnn.2013.5989

Husseiny SM, Salah TA, Anter HA (2015) Biosynthesis of size controlled silver nanoparticles by *Fusarium oxysporum*, their antibacterial and antitumor activities. Beni-Suef Uni J Appl Sci 4:225–231. https://doi.org/10.1016/j.bjbas.2015.07.004

Ingle A, Gade A, Pierrat S, Sonnichsen C, Rai M (2008) Mycosynthesis of silver nanoparticles using the fungus *Fusarium acuminatum* and its activity against some human pathogenic bacteria. Curr Nanosci 4:141–144. https://doi.org/10.2174/157341308784340804

Iqtedar M, Aslam M, Akhyar M, Shehzaad A, Abdullah R, Kaleem A (2019) Extracellular biosynthesis, characterization, optimization of silver nanoparticles (AgNPs) using *Bacillus mojavensis* BTCB15 and its antimicrobial activity against multidrug resistant pathogens. Prep Biochem Biotechnol 49:136–142. https://doi.org/10.1080/10826068.2018.1550654

Jain N, Bhargava A, Majumdar S, Tarafdar JC, Panwar J (2011) Extracellular biosynthesis and characterization of silver nanoparticles using *Aspergillus flavus* NJP08: a mechanism perspective. Nanoscale 3:635–641. https://doi.org/10.1039/C0NR00656D

Joshi CG, Danagoudar A, Poyya J, Kudva AK, Dhananjaya B (2017) Biogenic synthesis of gold nanoparticles by marine endophytic fungus-*Cladosporium cladosporioides* isolated from seaweed and evaluation of their antioxidant and antimicrobial properties. Process Biochem 63:137–144. https://doi.org/10.1016/j.procbio.2017.09.008

Kalpana VN, Kataru BAS, Sravani N, Vigneshwari T, Panneerselvam A, Devi Rajeswari V (2018) Biosynthesis of zinc oxide nanoparticles using culture filtrates of *Aspergillus niger*: antimicrobial textiles and dye degradation studies. OpenNano 3:48–55. https://doi.org/10.1016/j.onano.2018.06.001

Khan AL, Al-Harrasi A, Al-Rawahi A, Al-Farsi Z, Al-Mamari A, Waqas M, Asaf S, Elyassi A, Mabood F, Shin JH, Lee IJ (2016) Endophytic fungi from frankincense tree improves host growth and produces extracellular enzymes and indole acetic acid. PLoS ONE 11:e0158207–e0158207. https://doi.org/10.1371/journal.pone.0158207

Khandel P, Shahi SK (2018) Mycogenic nanoparticles and their bio-prospective applications: current status and future challenges. J Nanostructure Chem 8:369–391. https://doi.org/10.1007/s40097-018-0285-2

Kitching M, Choudhary P, Inguva S, Guo Y, Ramani M, Das SK, Marsili E (2016) Fungal surface protein mediated one-pot synthesis of stable and hemocompatible gold nanoparticles. Enzyme Microb Technol 95:76–84. https://doi.org/10.1016/j.enzmictec.2016.08.007

Korbekandi H, Mohseni S, Mardani Jouneghani R, Pourhossein M, Iravani S (2016) Biosynthesis of silver nanoparticles using *Saccharomyces cerevisiae*. Artif Cells Nanomed Biotechnol 44:235–239. https://doi.org/10.3109/21691401.2014.937870

Lateef, A, Ojo SA, Elegbede JA (2016b) The emerging roles of arthropods and their metabolites in the green synthesis of metallic nanoparticles. Nanotechnol Rev 5:601–622. http://dx.doi.org/10.1515/ntrev-2016-0049

Lateef A, Adeeyo AO (2015) Green synthesis and antibacterial activities of silver nanoparticles using extracellular laccase of *Lentinus edodes*. Not Sci Biol 7:405–411. https://doi.org/10.15835/nsb749643

Lateef A, Ojo S, Azeez M., Asafa T, Yekeen T, Akinboro A, Oladipo I, Gueguim-Kana E, Beukes L (2016a) Cobweb as novel biomaterial for the green and eco-friendly synthesis of silver nanoparticles. Appl Nanosci 6:863–874. https://doi.org/10.1007/s13204-015-0492-9

Lateef A, Ojo SA, Elegbede JA, Akinola PO, Akanni EO (2018) Nanomedical applications of nanoparticles for blood coagulation disorders. In: Dasgupta N, Ranjan S, Lichtfouse E (eds) Environmental nanotechnology. Springer, pp 243–277. https://doi.org/10.1007/978-3-319-76090-2_8

Li YN, Su J, Lv XY, Long YF, Wen YX (2015) Yeast bio-template synthesis of porous anatase TiO_2 and potential application as an anode for sodium-ion batteries. Electrochim Acta 182:596–603. https://doi.org/10.1016/j.electacta.2015.09.115

Lian S, Diko CS, Yan Y, Li Z, Zhang H, Ma Q, Qu Y(2019) Characterization of biogenic selenium nanoparticles derived from cell-free extracts of a novel yeast *Magnusiomyces ingens*. 3 Biotech 9:221. https://doi.org/10.1007/s13205-019-1748-y

Liang X, Perez MAMJ, Nwoko KC, Egbers P, Feldmann J, Csetenyi L, Gadd GM (2019) Fungal formation of selenium and tellurium nanoparticles. Appl Microbiol Biotechnol 103:7241–7259. https://doi.org/10.1007/s00253-019-09995-6

Lotha R, Shamprasad BR, Sundaramoorthy NS, Nagarajan S, Sivasubramanian A (2019) Biogenic phytochemicals (cassinopin and isoquercetin) capped copper nanoparticles (ISQ/CAS@ CuNPs) inhibits MRSA biofilms. Microb Pathog 132:178–187. https://doi.org/10.1016/j.micpath.2019.05.005

Luo QY, Lin Y, Li Y, Xiong LH, Cui R, Xie ZX, Pang DW (2014) Nanomechanical analysis of yeast cells in CdSe quantum dot biosynthesis. Small 10:699–704. https://doi.org/10.1002/smll.201301940

Ma L, Su W, Liu JX, Zeng XX, Huang Z, Li W, Liu ZC, Tang JX (2017) Optimization for extracellular biosynthesis of silver nanoparticles by *Penicillium aculeatum* Su1 and their antimicrobial activity and cytotoxic effect compared with silver ions. Mater Sci Eng C 77:963–971. https://doi.org/10.1016/j.msec.2017.03.294

Maaroof MN, Mahmood AE (2019) Determination of the immunogenic and hematologic effects of titanium nanoparticles manufactured from *Aspergillus flavus* in vivo. Jordan J Biol Sci 12:135–140. http://www.jjbs.hu.edu.jo/files/vol12/n2/Binder12n2.pdf#page=27

Mahanty S, Bakshi M, Ghosh S, Gaine T, Chatterjee S, Bhattacharyya S, Das S, Das P, Chaudhuri P (2019) Mycosynthesis of iron oxide nanoparticles using manglicolous fungi isolated from Indian sundarbans and its application for the treatment of chromium containing solution: Synthesis, adsorption isotherm, kinetics and thermodynamics study. Environ Nanotechnol Monit Manag 12:100276. https://doi.org/10.1016/j.enmm.2019.100276

Marquis G, Ramasamy B, Banwarilal S, Munusamy AP (2016) Evaluation of antibacterial activity of plant mediated CaO nanoparticles using *Cissus quadrangularis* extract. J Photochem Photobiol B 155:28–33. https://doi.org/10.1016/j.jphotobiol.2015.12.013

Mashrai A, Khanam H, Aljawfi RN (2017) Biological synthesis of ZnO nanoparticles using *C. albicans* and studying their catalytic performance in the synthesis of steroidal pyrazolines. Arab J Chem 10:S1530-S1536. https://doi.org/10.1016/j.arabjc.2013.05.004

Moghaddam AB, Moniri M, Azizi S, Rahim RA, Ariff AB, Saad WZ, Namvar F, Navaderi M, Mohamad R (2017) Biosynthesis of ZnO nanoparticles by a new *Pichia kudriavzevii* yeast strain and evaluation of their antimicrobial and antioxidant activities. Molecules 22:872. https://doi.org/10.3390/molecules22060872

Mohamad Sukri SNA, Shameli K, Mei-Theng Wong M, Teow SY, Chew J, Ismail NA (2019) Cytotoxicity and antibacterial activities of plant-mediated synthesized zinc oxide (ZnO) nanoparticles using *Punica granatum* (pomegranate) fruit peels extract. J Mol Struct 1189:57–65. https://doi.org/10.1016/j.molstruc.2019.04.026

Mohamed Y, Azzam A, Amin B, Safwat N (2015) Mycosynthesis of iron nanoparticles by *Alternaria alternata* and its antibacterial activity. Afr J Biotechnol 14:1234–1241. https://doi.org/10.5897/AJB2014.14286

Mohamed AA, Fouda A, Abdel-Rahman MA, Hassan SED, El-Gamal MS, Salem SS, Shaheen TI (2019) Fungal strain impacts the shape, bioactivity and multifunctional properties of green synthesized zinc oxide nanoparticles. Biocatal Agric Biotechnol 19:101103. https://doi.org/10.1016/j.bcab.2019.101103

Mohammed Fayaz A, Balaji K, Girilal M, Kalaichelvan PT, Venkatesan R (2009) Mycobased synthesis of silver nanoparticles and their incorporation into sodium alginate films for vegetable and fruit preservation. J Agric Food Chem 57:6246–6252. https://doi.org/10.1021/jf900337h

Mohanta YK, Nayak D, Biswas K, Singdevsachan SK, Abd_Allah EF, Hashem A, Alqarawi AA, Yadav D, Mohanta TK (2018) Silver nanoparticles synthesized using wild mushroom show potential antimicrobial activities against food borne pathogens. Molecules 23:655. https://doi.org/10.3390/molecules23030655

Mohmed AA, Fouda A, Elgamal MS, El-Din Hassan S, Shaheen TI, Salem SS (2017a) Enhancing of cotton fabric antibacterial properties by silver nanoparticles synthesized by new egyptian strain *fusarium keratoplasticum* A1-3. Egypt J Chem 60:63–71. https://doi.org/10.21608/ejchem.2017.1626.1137

Mohmed AA, Saad EL, Fouda A, Elgamal MS, Salem SS (2017b) Extracellular biosynthesis of silver nanoparticles using *Aspergillus* sp. and evaluation of their antibacterial and cytotoxicity. J Appl Life Sci 11:1–12. https://doi.org/10.9734/JALSI/2017/33491

Molnár Z, Bódai V, Szakacs G, Erdélyi B, Fogarassy Z, Sáfrán G, Varga T, Kónya Z, Tóth-Szeles E, Szűcs R, Lagzi I (2018) Green synthesis of gold nanoparticles by thermophilic filamentous fungi. Sci Rep 8:3943. https://doi.org/10.1038/s41598-018-22112-3

Mosallam FM, El-Sayyad GS, Fathy RM, El-Batal AI (2018) Biomolecules-mediated synthesis of selenium nanoparticles using *Aspergillus oryzae* fermented Lupin extract and gamma radiation for hindering the growth of some multidrug-resistant bacteria and pathogenic fungi. Microb Pathog 122:108–116. https://doi.org/10.1016/j.micpath.2018.06.013

Mukherjee P, Ahmad A, Mandal D, Senapati S, Sainkar SR, Khan MI, Parishcha R, Ajaykumar PV, Alam M, Kumar R, Sastry M (2001) Fungus-mediated synthesis of silver nanoparticles and their immobilization in the mycelial matrix: a novel biological approach to nanoparticle synthesis. Nano Lett 1:515–519. https://doi.org/10.1021/nl0155274

Mukherjee P, Roy M, Mandal BP, Dey GK, Mukherjee PK, Ghatak J, Tyagi AK, Kale SP (2008) Green synthesis of highly stabilized nanocrystalline silver particles by a non-pathogenic and agriculturally important fungus *T. asperellum*. Nanotechnol 19:075103. https://doi.org/10.1088/0957-4484/19/7/075103

Ovais M, Khalil AT, Ayaz M, Ahmad I, Nethi SK, Mukherjee S (2018) Biosynthesis of metal nanoparticles via microbial enzymes: a mechanistic approach. Int J Mol Sci 19:4100. https://doi.org/10.3390/ijms19124100

Park HJ, Kim SH, Kim H, Choi S (2006) A new composition of nanosized silica-silver for control of various plant diseases. Plant Pathol J 22:295–302

Parmar S, Sharma VK (2020) 13—Endophytic fungi mediated biofabrication of nanoparticles and their potential applications. In: Kumar A, Singh VK (eds) Microbial endophytes. Woodhead Publishing, pp 325–341. https://doi.org/10.1016/B978-0-12-818734-0.00013-9

Popli D, Anil V, Subramanyam AB, Namratha MN, Ranjitha VR, Rao SN, Rai RV, Govindappa M (2018) Endophyte fungi, *Cladosporium* species-mediated synthesis of silver nanoparticles possessing in vitro antioxidant, anti-diabetic and anti-alzheimer activity. Artif Cells Nanomed Biotechnol 46:676–683. https://doi.org/10.1080/21691401.2018.1434188

Qin W, Wang CY, Ma YX, Shen MJ, Li J, Jiao K, Tay FR, Niu LN (2020) Microbe-mediated extracellular and intracellular mineralization: environmental, industrial, and biotechnological applications. Adv Mater 32:1907833. https://doi.org/10.1002/adma.201907833

Qu Y, Shen W, Pei X, Ma F, You S, Li S, Wang J, Zhou J (2017) Biosynthesis of gold nanoparticles by *Trichoderma* sp. WL-Go for azo dyes decolorization. J Environ Sci 56:79–86. https://doi.org/10.1016/j.jes.2016.09.007

Qu Y, Li X, Lian S, Dai C, Jv Z, Zhao B, Zhou H (2018a) Biosynthesis of gold nanoparticles using fungus *Trichoderma* sp. WL-Go and their catalysis in degradation of aromatic pollutants. IET nanobiotechnol 13:12–17. https://doi.org/10.1049/iet-nbt.2018.5177

Qu Y, You S, Zhang X, Pei X, Shen W, Li Z, Li S, Zhang Z (2018b) Biosynthesis of gold nanoparticles using cell-free extracts of *Magnusiomyces ingens* LH-F1 for nitrophenols reduction. Bioprocess Biosyst Eng 41:359–367. https://doi.org/10.1007/s00449-017-1869-9

Qu Y, Lian S, Shen W, Li Z, Yang J, Zhang H (2020) Rod-shaped gold nanoparticles biosynthesized using Pb^{2+}-induced fungus *Aspergillus* sp. WL-Au. Bioprocess Biosyst Eng 43:123–131. https://doi.org/10.1007/s00449-019-02210-w

Rabeea MA, Owaid MN, Aziz AA, Jameel MS, Dheyab MA (2020) Mycosynthesis of gold nanoparticles using the extract of *Flammulina velutipes*, Physalacriaceae, and their efficacy for decolorization of methylene blue. J Environ Chem Eng, 103841. https://doi.org/10.1016/j.jece.2020.103841

Rahman S, Rahman L, Khalil AT, Ali N, Zia D, Ali M, Shinwari ZK (2019) Endophyte-mediated synthesis of silver nanoparticles and their biological applications. Appl Microbiol Biotechnol 103:2551–2569. https://doi.org/10.1007/s00253-019-09661-x

Rai M, Ingle A (2012) Role of nanotechnology in agriculture with special reference to management of insect pests. Appl Microbiol Biotechnol 94:287–293. https://doi.org/10.1007/s00253-012-3969-4

Rajput S, Werezuk R, Lange RM, McDermott MT (2016) Fungal isolate optimized for biogenesis of silver nanoparticles with enhanced colloidal stability. Langmuir 32:8688–8697. https://doi.org/10.1021/acs.langmuir.6b01813

Raliya R (2013) Rapid, low-cost, and ecofriendly approach for iron nanoparticle synthesis using *Aspergillus oryzae* TFR9. J Nanopart 2013:1–4. https://doi.org/10.1155/2013/141274

Raliya R, Biswas P, Tarafdar J (2015) TiO_2 nanoparticle biosynthesis and its physiological effect on mung bean (*Vigna radiata* L.). Biotechnol Rep 5:22–26. https://doi.org/10.1016/j.btre.2014.10.009

Raliya R, Tarafdar JC, Biswas P (2016) Enhancing the mobilization of native phosphorus in the mung bean rhizosphere using ZnO nanoparticles synthesized by soil fungi. J Agric Food Chem 64:3111–3118. https://doi.org/10.1021/acs.jafc.5b05224

Ramos MM, Morais ED, Sena ID, Lima AL, de Oliveira FR, de Freitas CM, Fernandes CP, de Carvalho JCT, Ferreira IM (2020) Silver nanoparticle from whole cells of the fungi *Trichoderma* spp. isolated from Brazilian Amazon. Biotechnol Lett 42:833–843. https://doi.org/10.1007/s10529-020-02819-y

Bansal V, Rautaray, D, Bharde A, Ahire K, Sanyal A, Ahmad A, Sastry M (2005) Fungus-mediated biosynthesis of silica and titania particles. J Mater Chem 15:2583–2589. https://doi.org/10.1039/B503008K

Ray PC, Khan SA, Fan Z, Senapati D (2013) Gold nanotechnology for targeted detection and killing of multiple drug resistant bacteria from food samples. Advances in applied nanotechnology for agriculture. J Am Chem Soc, 1–19. https://pubs.acs.org/doi/abs/10.1021/bk-2013-1143.ch001

Riddin TL, Gericke M, Whiteley CG (2006) Analysis of the inter- and extracellular formation of platinum nanoparticles by *Fusarium oxysporum f.* sp.*lycopersici* using response surface methodology. Nanotechnol 17:3482–3489. https://doi.org/10.1088/0957-4484/17/14/021

Roy A, Bulut O, Some S, Mandal AK, Yilmaz MD (2019) Green synthesis of silver nanoparticles: biomolecule-nanoparticle organizations targeting antimicrobial activity. RSC Adv 9:2673–2702. https://doi.org/10.1039/C8RA08982E

Salem SS, Fouda A (2020) Green synthesis of metallic nanoparticles and their prosective biotechnological applications: an overview. Biol Trace Elem Res. https://doi.org/10.1007/s12011-020-02138-3

Salem DM, Ismail MM, Aly-Eldeen MA (2019a) Biogenic synthesis and antimicrobial potency of iron oxide (Fe_3O_4) nanoparticles using algae harvested from the Mediterranean Sea, Egypt. Egypt J Aquat Res 45:197–204. https://doi.org/10.1016/j.ejar.2019.07.002

Salem SS, Mohamed AA, Gl-Gamal MS, Talat M, Fouda A (2019b) Biological decolorization and degradation of azo dyes from textile wastewater effluent by *Aspergillus niger*. Egypt J Chem 62:1799–1813. https://doi.org/10.21608/ejchem.2019.11720.1747

Salem, SS, Fouda, MMG, Fouda, A, Awad MA, Al-Olayan EM, Allam AA, Shaheen TI (2020) Antibacterial, cytotoxicity and larvicidal activity of green synthesized selenium nanoparticles using *Penicillium corylophilum*. J Clust Sci 2020. https://doi.org/10.1007/s10876-020-01794-8

Salvadori MR(2019). Processing of nanoparticles by biomatrices in a green approach. In: Prasad R (ed) Microbial nanobionics: volume 1, state-of-the-art. Springer International Publishing, Cham, pp 1–28. https://doi.org/10.1007/978-3-030-16383-9_1

Salvadori MR, Ando RA, Oller Do Nascimento CA, Correa B (2014) Bioremediation from wastewater and extracellular synthesis of copper nanoparticles by the fungus *Trichoderma koningiopsis*. J Environ Sci Health C 49:1286–1295. https://doi.org/10.1080/10934529.2014.910067

Saravanakumar K, ShanmugamS, Varukattu NB, MubarakAli D, Kathiresan K, Wang MH (2019) Biosynthesis and characterization of copper oxide nanoparticles from indigenous fungi and its effect of photothermolysis on human lung carcinoma. J Photochem Photobiol B 190:103–109. https://doi.org/10.1016/j.jphotobiol.2018.11.017

Sarkar J, Ghosh M, Mukherjee A, Chattopadhyay D, Acharya K (2014) Biosynthesis and safety evaluation of ZnO nanoparticles. Bioprocess Biosyst Eng 37:165–171. https://doi.org/10.1007/s00449-013-0982-7

Sawle BD, Salimath B, Deshpande R, Bedre MD, Prabhakar BK, Venkataraman A (2008) Biosynthesis and stabilization of Au and Au–Ag alloy nanoparticles by fungus, *Fusarium semitectum*. Sci Technol Adv Mater 9:035012. https://doi.org/10.1088/1468-6996/9/3/035012

Saxena J, Sharma PK, Sharma MM, Singh A (2016) Process optimization for green synthesis of silver nanoparticles by *Sclerotinia sclerotiorum* MTCC 8785 and evaluation of its antibacterial properties. Springer Plus 5:861. https://doi.org/10.1186/s40064-016-2558-x

Schumacher J, Gorbushina AA (2020) Light sensing in plant- and rock-associated black fungi. Fungal Biol 124:407–417. https://doi.org/10.1016/j.funbio.2020.01.004

Seetharaman PK, Chandrasekaran R, Gnanasekar S, Chandrakasan G, Gupta M, Manikandan DB, Sivaperumal S (2018) Antimicrobial and larvicidal activity of eco-friendly silver nanoparticles synthesized from endophytic fungi *Phomopsis liquidambaris*. Biocatal Agric Biotechnol 16:22–30. https://doi.org/10.1016/j.bcab.2018.07.006

Shaheen TI, Abd El Aty AA (2018) In-situ green myco-synthesis of silver nanoparticles onto cotton fabrics for broad spectrum antimicrobial activity. Int J Biol Macromol 118:2121–2130. https://doi.org/10.1016/j.ijbiomac.2018.07.062

Shaheen TI, Salem SS, Zaghloul S (2019) A new facile strategy for multifunctional textiles development through in situ deposition of SiO_2/TiO_2 nanosols hybrid. Ind Eng Chem Res 58:20203–20212. https://doi.org/10.1021/acs.iecr.9b04655

Sharaf OM, Al-Gamal MS, Ibrahim GA, Dabiza NM, Salem SS, El-ssayad MF, Youssef AM (2019) Evaluation and characterization of some protective culture metabolites in free and nano-chitosan-loaded forms against common contaminants of Egyptian cheese. Carbohydr Polym 223:115094. https://doi.org/10.1016/j.carbpol.2019.115094

Shen W, Qu Y, Pei X, Li S, You S, Wang J, Zhang Z, Zhou J (2017) Catalytic reduction of 4-nitrophenol using gold nanoparticles biosynthesized by cell-free extracts of *Aspergillus* sp. WL-Au. J Hazard Mater 321:299–306. https://doi.org/10.1016/j.jhazmat.2016.07.051

Siddiqi KS, Husen A (2016) Fabrication of metal nanoparticles from fungi and metal salts: scope and application. Nanoscale Res Lett 11:98. https://doi.org/10.1186/s11671-016-1311-2

Singh DK, Kumar J, Sharma VK, Verma SK, Singh A, Kumari P, Kharwar RN (2018) Mycosynthesis of bactericidal silver and polymorphic gold nanoparticles: physicochemical variation effects and mechanism. Nanomedicine 13:191–207. https://doi.org/10.2217/nnm-2017-0235

Soliman H, Elsayed A, Dyaa A (2018) Antimicrobial activity of silver nanoparticles biosynthesised by *Rhodotorula* sp. strain ATL72. Egypt J basic Appl Sci 5:228–233. https://doi.org/10.1016/j.ejbas.2018.05.005

Soliman AM, Abdel-Latif W, Shehata IH, Fouda A, Abdo AM, Ahmed YM (2020) Green approach to overcome the resistance pattern of *Candida* spp. using biosynthesized silver nanoparticles fabricated by *Penicillium chrysogenum* F9. Biol Trace Elem Res. https://doi.org/10.1007/s12011-020-02188-7

Sriramulu M, Sumathi S (2018) Biosynthesis of palladium nanoparticles using *Saccharomyces cerevisiae* extract and its photocatalytic degradation behaviour. Adv Nat Sci: Nanosci Nanotechnol 9:025018. https://doi.org/10.1088/2043-6254/aac506

Subramaniyan SA, Sheet S, Vinothkannan M, Yoo DJ, Lee YS, Belal SA, Shim KS (2018) One-pot facile synthesis of Pt nanoparticles using cultural filtrate of microgravity simulated grown *P. chrysogenum* and their activity on bacteria and cancer cells. J Nanosci Nanotechnol 18:3110–3125. https://doi.org/10.1166/jnn.2018.14661

Suryavanshi P, Pandit R, Gade A, Derita M, Zachino S, Rai M (2017) *Colletotrichum* sp. mediated synthesis of sulphur and aluminium oxide nanoparticles and its in vitro activity against selected food-borne pathogens. LWT—Food Sci Technol 81:188–194. https://doi.org/10.1016/j.lwt.2017.03.038

Tarafdar A, Raliya R, Wang WN, Biswas P, Tarafdar J (2013) Green synthesis of TiO_2 nanoparticle using *Aspergillus tubingensis*. Adv Sci Eng Med 5:943–949. https://doi.org/10.1166/asem.2013.1376

Thakur BK, Kumar A, Kumar D (2019) Green synthesis of titanium dioxide nanoparticles using *Azadirachta indica* leaf extract and evaluation of their antibacterial activity. S Afr J Bot 124:223–227. https://doi.org/10.1016/j.sajb.2019.05.024

Titus D, James Jebaseelan Samuel E, Roopan SM (2019) Nanoparticle characterization techniques. In: Shukla AK, Iravani S (eds) Green synthesis, characterization and applications of nanoparticles. Elsevier, pp 303–319. https://doi.org/10.1016/B978-0-08-102579-6.00012-5

Tripathi RM, Gupta RK, Bhadwal AS, Singh P, Shrivastav A, Shrivastav B (2015) Fungal biomolecules assisted biosynthesis of Au–Ag alloy nanoparticles and evaluation of their catalytic property. IET Nanobiotechnol 9:178–183. https://doi.org/10.1049/iet-nbt.2014.0043

Tripathi RM, Shrivastav BR, Shrivastav A (2018) Antibacterial and catalytic activity of biogenic gold nanoparticles synthesised by *Trichoderma harzianum*. IET nanobiotechnol 509–513. https://doi.org/10.1049/iet-nbt.2017.0105

Uddandarao P, Balakrishnan RM, Ashok A, Swarup S, Sinha P (2019) Bioinspired ZnS: Gd nanoparticles synthesized from an endophytic fungi *Aspergillus flavus* for fluorescence-based metal detection. Biomimetics 4:11. https://doi.org/10.3390/biomimetics4010011

Venkatesh K, Palani N, Krishnamoorthi S, Thirumal V, Ilangovan R (2013) Fungus mediated biosynthesis and characterization of zinc oxide nanorods. AIP Conf Proc 93–94. https://doi.org/10.1063/1.4810116

Vijayabharathi R, Sathya A, Gopalakrishnan S (2018) Extracellular biosynthesis of silver nanoparticles using *Streptomyces griseoplanus* SAI-25 and its antifungal activity against *Macrophomina phaseolina*, the charcoal rot pathogen of sorghum. Biocatal Agric Biotechnol 14:166–171. https://doi.org/10.1016/j.bcab.2018.03.006

Wierckx N, Agrimi G, Lübeck PS, Steiger MG, Mira NP, Punt PJ (2020) Metabolic specialization in itaconic acid production: a tale of two fungi. Curr Opin Biotechnol 62:153–159. https://doi.org/10.1016/j.copbio.2019.09.014

Wurzbacher C, Kreiling AK, Svantesson S, Van den Wyngaert S, Larsson E, Heeger F, Nilsson HR, Pálsson S (2020) Fungal communities in groundwater springs along the volcanic zone of Iceland. Inland Waters 1–10. https://doi.org/10.1080/20442041.2019.1689065

Xiao Z, Zhou W, Zhang Y (2020) Chapter ten—Fungal polysaccharides. In: Du G (ed), Adv pharmacology. Academic Press, pp 277–299. https://doi.org/10.1016/bs.apha.2019.08.003

Yadav A, Kon K, Kratosova G, Duran N, Ingle AP, Rai M (2015) Fungi as an efficient mycosystem for the synthesis of metal nanoparticles: progress and key aspects of research. Biotechnol Lett 37:2099–2120. https://doi.org/10.1007/s10529-015-1901-6

Yan X, He B, Liu L, Qu G, Shi J, Hu L, Jiang G (2018) Antibacterial mechanism of silver nanoparticles in *Pseudomonas aeruginosa*: proteomics approach. Metallomics 10, 557–564. https://doi.org/10.1039/C7MT00328E

Yang Z, Li Z, Lu X, He F, Zhu X, Ma Y, He R, Gao F, Ni W, Yi Y (2016) Controllable biosynthesis and properties of gold nanoplates using yeast extract. Nano-Micro Lett 9:5. https://doi.org/10.1007/s40820-016-0102-8

Yousef S, Ibrahim NA, Farag SS, El-mehalawy A, Ismaiel A, Ahmed A (2020) Mycosynthesis of silver nanoparticles by the endophytic fungus *Alternaria tenuissima* AUMC 13621 and evaluation of their antimicrobial, antioxidant effect. Egypt J Microbiol 54:63–76. https://doi.org/10.21608/ejm.2019.13564.1101

Yun'an Qing LC, Li R, Liu G, Zhang Y, Tang X, Wang J, Liu H, Qin Y (2018) Potential antibacterial mechanism of silver nanoparticles and the optimization of orthopedic implants by advanced modification technologies. Int J Nanomed 13:3311. https://doi.org/10.2147/IJN.S165125

Zare B, Babaie S, Setayesh N, Shahverdi AR (2013) Isolation and characterization of a fungus for extracellular synthesis of small selenium nanoparticles. Nanomed J. 1:13–19. https://doi.org/10.7508/NMJ.2013.01.002

Zhao X, Zhou L, Riaz Rajoka MS, Yan L, Jiang C, Shao D, Zhu J, Shi J, Huang Q, Yang H, Jin M (2018) Fungal silver nanoparticles: synthesis, application and challenges. Crit Rev Biotechnol 38:817–835. https://doi.org/10.1080/07388551.2017.1414141

Environmental Nanobiotechnology: Microbial-Mediated Nanoparticles for Sustainable Environment

O. M. Darwesh, M. F. Eida, and I. A. Matter

1 Introduction

It is clear that modern civilization, its technologies and related industries lead to many environmental changes and thus affect the style and quality of human life. Some of these changes and their impacts, directly or indirectly, include the depletion and disintegration of natural resources, thereby disrupting the "long-term environmental quality", i.e., "environmental sustainability" (Davis et al. 2017). To protect our environment and save its resources for the next generations, many efforts, studies, strategies and initiatives have been demonstrated. Most promising strategies for preserving the environment include using biotechnologies in general and microbial nanotechnology applications in particular to maximize and accelerate their environmental performance in many areas (Darwesh et al. 2015, 2020a). Microbial biotechnology is the field of science that deals with the utilization of microorganisms to obtain benefits and services. It includes directing of microorganisms and their enzymes for biodegradation of agricultural wastes, bioremediation of xenobiotics, production of biofuels, production of pharmaceuticals and food additives as well as the green synthesis of nanoparticles (Abou-Shanab et al. 2012; Darwesh et al. 2018a, 2019a, b).

Nanotechnology includes synthesizing and applications of nano-sized materials (1–100 nm); those materials could be nanoparticles, nanocomposites or nanotubes that have tremendous technological and biotechnological applications. The incorporation of nanotechnology into microbial and environmental biotechnology is a relatively new trend that has shown auspicious results in many areas and applications (Sangeetha et al. 2019; Besha et al. 2020). Due to the distinctive "size, shape, surface area and structure of particles", nanomaterials exhibit unique properties and behavior

O. M. Darwesh (✉) · M. F. Eida · I. A. Matter
Agricultural Microbiology Department, National Research Centre, Elbohouth St., Dokki, Cairo 12622, Egypt
e-mail: darweshosama@yahoo.com

A. Lateef et al. (eds.), *Microbial Nanobiotechnology*, Materials Horizons: From Nature to Nanomaterials, https://doi.org/10.1007/978-981-33-4777-9_5

compared to bulk materials, so their use often increases the efficiency and quality of different production systems. In addition, applying nanotechnologies to daily activities greatly increases resource sustainability either by reducing the quantities required from some or enabling the reusability and recyclability of others. Furthermore, it has been reported that the environmental toxicity of some nanomaterials is lower than that of similar non-nanoscale compounds, such as silver nanoparticles (Mourad et al. 2019; Abdel-Monem et al. 2020).

Consequently, a well-studied and good application of nanotechnology may be the best way to overcome many technical and environmental problems and promote economic development and environmental sustainability in many countries (Simate et al. 2013; Hadef 2018; Nasrollahzadeh et al. 2019). Here in this chapter, the potential role of microbial synthetic nanomaterials in different environmental applications and its impact on environmental sustainability are discussed.

2 The Need for Environmental Sustainability

Destroying the environment will not only "compromise the ability of future generations to meet their needs" but also could threaten the existence of humanity. Some of the global threats such as the pollution, the global warming and the erosion of the ozone layer consider as a critical and chronic problems for current generation and for the next generations too (Durgalakshmi et al. 2019). The reckless consumption behavior of humans as well as the steady increase of populations in addition to the competition between governments to obtain and sometimes take over resources which are already limited has resulted in the depletion of many resources over the years. Moreover, in order to increase productivity and efficiency of different production units such as agricultural farms and factories, humans have developed and used advanced technologies which in general only does not only considers the rights of next generations in the resources but also destroy it and pollute the environment. Moreover, humans have developed many methods and techniques to increase the productivity and efficiency of the various production units, but unfortunately, many of these technologies do not take into account the rights of future generations to resources and pollute the environment as well.

For instance, the dependence on the conventional resources energy such as fossil fuels has many downsides, which includes the possibility of depleting the fossil fuel and consequently depriving the next generation to benefit from it. In addition, the continuous combustion of such fuels will lead to increase the percentage of carbon dioxide gases and occurrence of the global warming. For these reasons, reliance on renewable energy such as solar and wind energies as well as biofuels became an urgent demand for energy and environment sustainability (El-Baz et al. 2016; Matter et al. 2018; Sohail et al. 2019; Darwesh et al. 2020b). Another example is the contamination of soils by different xenobiotics such as pesticides, aliphatic and aromatic hydrocarbons among other due to intensive and randomly applications of pesticides and using of low quality waters for irrigation in many places (Abou-Shanab et al. 2012; Hoballah et al. 2014; Saber et al. 2014). Monitoring of the pollution levels, find the ideal treatment to remove, detoxify, or reduce their negative impacts on environment considers as the minimum liability towards the environment. Pollution prevention from beginning will be more sustainable for humans and their environments as well. In addition, due to the depletion and limitations of many resources, it is an emergency demand to reserve and keep the limited resources as much as possible for next generations. In this regards, new technologies should be environmental friendly to prevent the destruction of the environment and keep it sustainable. Traditional technologies also should be revised and improved to fulfill the ethics for clean and sustainable environment.

3 Nanomaterials and Microbial Synthesis Procedures

The development of environmental-safe and ecofriendly technologies for nanomaterials synthesis and production superior substances is considered importance to increase their biological applications. The microbial-synthesized nanoparticles are distributed in a wide spectrum of potential areas from wastewater treatment, biofuel production, biofertilizers providing to food industry improvement (Li et al. 2011; Darwesh et al. 2018b). The nanomaterials are microbial-synthesized when the microbes targeted ions from their environment and convert these ions into element metal via enzymes system. Various microbes have different mechanisms for nanomaterials formation. Though, nanomaterials are usually manufactured by trapping the metal ions on the surface or inside the microbial cells. The trapped ions are reduced to nanosize by enzymes system. Generally, microorganisms influence the mineral formation in two ways. The first one is to modify the composition to become supersaturated or more supersaturated than it previously was with respect to a specific phase. A second means by which microorganisms can impact mineral formation is through the production of organic polymers, which can impact nucleation by favoring or inhibiting the stabilization of the very first mineral seeds (Fig. 1).

The NADH and NADH-dependent nitrate reductase enzymes are reported as important factors in microbial synthesis of metal nanomaterials. Nitrate reductase enzyme might be responsible for the bio-reduction of Ag^+ to Ag^0 and the subsequent formation of silver nanoparticles (Darwesh et al. 2019a). Microbial enzymes such as laccase, keratinase and xylanase among others have been used to synthesize metallic nanoparticles for different applications (Lateef et al. 2015a; Adelere and Lateef 2016; Lateef and Adeeyo 2015; Elegbede et al. 2020). Microorganisms have many metal resistance gene clusters allowing cell detoxification through a number of mechanisms like complexation, reductive precipitation or efflux. The metallophilic bacteria survive and thrive in environments containing high concentrations of heavy metals like mine waste rock piles, naturally mineralized zones and efflux streams of metal processing plants (Nies 1999). Many microbes were reported for nanomaterials manufacturing like genus of *Bacillus, Enterococcus, Pseudomonas, Fusarium,*

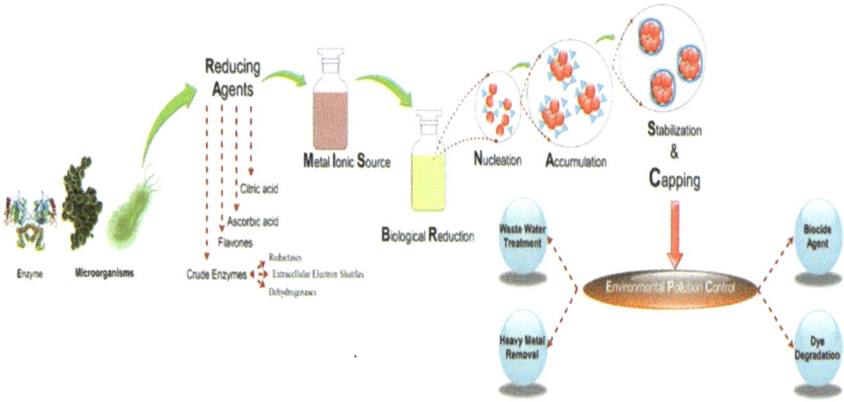

Fig. 1 Microbial synthesis of metal oxide nanoparticles

Trichoderma and *Aspergillus* (Ojo et al. 2016; Oladipo et al. 2017a, b; Elegbede et al. 2018, 2019; El-Shanshoury et al. 2020). These microbes were recorded for fabrication of silver, gold, silver–gold alloy, selenium, iron, zinc and copper nanoparticles.

4 Environmental Applications of Microbial Biosynthesized NPs

Due to the unique characteristics of NPs, their incorporation in different applications expected to influence the performance of the whole operation(s). Many of the operations that are related to environmental sustainability with the aid of NPs have been studied and reviewed. As mentioned before, the biosynthesized NPs are more environmental friendly and sustainable than those synthesized chemically, and hence, many researchers focused on using them indifferent environmental studies (Hasanin et al. 2018; Darwesh et al. 2019a; Hussein et al. 2019a). Among the most common topics related to environmental sustainability in which NPs could be used for are bioremediation (soil and water), biosensing, nano-pesticides and nano-fertilizers.

4.1 Applications of Microbial Nanoparticles in Bioremediation

Regarding the world scarcity of water, the water recycling and decontamination became one of the important pillars of environmental sustainability (Nguyen et al. 2019; Yang et al. 2014). Industrial and municipal wastewater distributes and increases daily, and this wastewater contains a lot of contaminants like heavy metals, organic

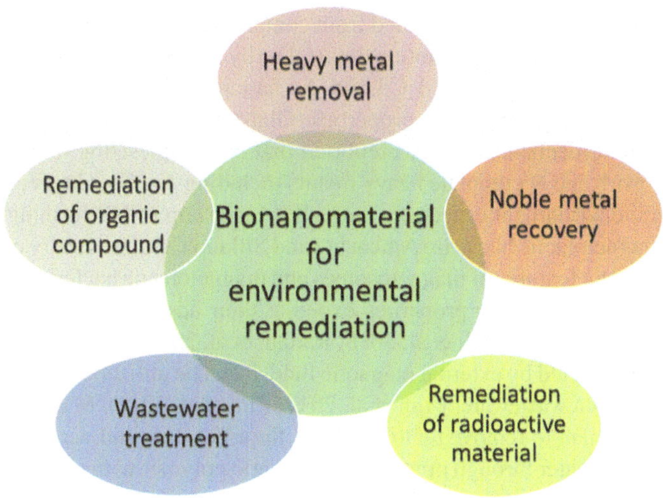

Fig. 2 Application of bionanomaterials in environmental remediation

pollutants (dyes and chemicals) and various pathogenic microbes. Several technologies were noted for treatment of wastewater to remove contaminants and/or detoxification of it like producing non-traditional materials. Among several technologies for wastewater treatments, the worldwide attention towards nanotechnology has vigorously increased in recent years, introducing an efficient and cheap remediation method for heavy metals, radioactive materials as well as organic pollutants (Singh et al. 2016; de Castro et al. 2019; Ranjan et al. 2019). As a result of unique physical and chemical properties of NPs and various ecofriendly applications concerning environmental cleanness as bioremediation, much attention from scientific society in diverse areas of environmental sciences was directed to nanomaterials as solution for environmental problems (Liu et al. 2014; Rizwan et al. 2014). Figure 2 illustrates some environmental application of biogenic NPs.

The potential utilization of nanotechnology in bioremediation practices is very promising as shown by Salvadori et al. (2017) where it was revealed that through application of fungal dead mycelium, wastes could be removed and NP can be synthesized. In addition, AuNPs manufactured by the *Rhizopus oryzae* displayed tremendous efficiency to minimize environmental impacts of organophosphorus pesticides contamination in both aquatic and terrestrial ecosystems (Das et al. 2009). As a tool in the bioremediation process, NPs that are microbial synthesized as a bioremediation approach introduces tremendous ecofriendly and economically feasible solution to decontaminate metal contaminated ecosystems. This technology increases the economic value of toxic metallic pollutants through the conversion to valuable nanostructures. In this regard, some microorganisms are applied for utilizing and recovering metals from contaminated water through synthesizing valuable nanomaterials. This topic is considered as a recently discovered field with high environmental and economic impact (Órdenes-Aenishanslins et al. 2014). Particularly, arsenic and cadmium were remediated from contaminated wastewater via producing its nano-sized structures (i.e., quantum dots (QDs) by Antarctic *Pseudomonas* strains. The importance of the produced Cd and As nanoparticles is due to its potential semiconductor and spectroscopic properties (Glatstein et al. 2018).

In a different mechanism, it was reported that NP synthesizing microorganisms itself can absorb and accumulate heavy metals including As, Co, Cu, Fe, Ni and Mg from aquatic environment especially through these group of NP forming microbes known as magnetotactic bacteria (Ambashta and Sillanpää 2010; Kabary et al. 2017). The characteristic features of magnetosomes and magnetotactic bacteria such as large surface areas and magnetic properties expedite their adsorption of pollutants and removal from wastewater (Ali et al. 2018; Ranjan et al. 2019). While these microbes can be simply affected by external magnetic field, it can be utilized for removing such heavy metals from wastewater (Yan et al. 2017; Zhu et al. 2012). Magnetosome was found to be efficient to remove up to 38 gkg^{-1} fluoride water and was able to adsorb Cu^{2+} from wastewater through preparing a magnetic polymer with polyvinyl acetate. The excellence of magnetosome as noble metal adsorbent is owing to high purity of extracted metals compared with utilization of the artificial magnetic particles of magnetosome (Das 2010; Yan et al. 2017). Thus, microbial magnetosome displays a vital role in wastewater treatment, heavy metal removal and resource recycling (Liu and Zhou 2008; Wang et al. 2020; Yan et al. 2017).

In addition to heavy metals, dye effluents discharged into the ecosystem are another dangerous source of environmental contamination that affects sustainability of natural resources (Darwesh et al. 2014). Degradation of dye could be achieved by harmless way utilizing NPs. Biogenic nanoparticles that synthesized through microorganisms can be utilized as a strong tool in this process (Sadhasivam et al. 2020). The application of biogenic iron NP for bioremediation of azo dye from the environment was confirmed by Sharma and Shirkot (2019). Furthermore, the CuNPs formed by a native *Escherichia* sp. were defined as photocatalysts for bioremediation of azo dye from textile effluents. The biogenic CuNPs decolorized 97.07, 90.55, 88.42 and 83.61% of congo red, malachite green, direct blue-1 and reactive black-5, respectively, after 5 h of application under sunlight. These results were obtained at a lower dyes concentration (25 mg L^{-1}), although the removal efficacy decreased by increasing dyes concentration to 100 mg L^{-1} (Noman et al. 2020). Similarly, the biogenic synthesized AuNPs using some non-pathogenic *Enterococcus* species have been reported to degrade 24.3–57.6%, and 88.85–97.36% of methylene blue and malachite green dyes, respectively, within 24 h (Oladipo et al. 2017b). Other reports have shown efficient degradation of dyes using microbe-mediated NPs (Ojo et al. 2016; Elegbede et al. 2018, 2019).

4.2 *Application of Microbial-Synthesized NPs in Bio-Sensing*

Knowing and assessing the problem is an important step to solve it; so the detection and monitoring of pollutants in an environment is an important step towards addressing, preserving and sustaining it for the current and next generations. Traditionally, the detection and quantification methods for pollutants are usually analyzed using conventional chromatography and spectroscopy techniques. Those conventional techniques are considered as time consuming, expensive and require highly trained persons and expensive equipment. Moreover, the environmental samples should be transferred into the lab for analysis. Therefore, the researchers focused on developing unconventional analysis techniques to overcome the problems of previously mentioned traditional methods, the most important of which are biosensors and nano-biosensors (Matter et al. 2020). The word "sensor" refers to any device that used to detect or measure an element or parameter as optical, electrical or mechanical signals produced by a specific "receptor" and transmitted through a "transducer" into a "detector" or read-out unit (Arora 2018; Sinha et al. 2020; Matter et al. 2020). Biosensors contain "bio-receptors" that could be specific enzymes, antibodies, nucleic acid, or even whole (living or dead) cells of microbial, plant or animal origin. Synthetic molecules such as synthetic antibodies, molecularly imprinted polymers and aptamers also could be used as bio-receptor elements and called "biomimetic receptors" (Ejeian et al. 2018; Zare and Shekari 2019; Sinha et al. 2020).

Merging of nanotechnology with the biosensors leads to the emergence of "Nano-biosensors", in which specific nanomaterials (nanoparticles, nanocomposites and nanotubes) are incorporated with the bio-receptors, and/or the transducer. The unique properties of nanomaterials are used to improve sensitivity and efficiency (by improving signal transformation) as well as reducing the overall size of the sensor (Ghaffar et al. 2020; Salouti and Derakhshan 2020). Classification of nano-biosensors could be based on the used receptors, signal transduction and applications. However, based on analysis mechanism of the transducer, nano-biosensors could be classified into optical, electrochemical and piezoelectric (mass-sensitive) biosensor (Saini et al. 2017). Biosensors and nano-biosensors could be designed to rapidly detect and monitor (with adequate accuracy) many of environmental pollutants in soil, water and air such as pesticides, herbicides and heavy metals. Many promising nanomaterials have been reported to be emerged successfully on the probes with conjugation to biomolecules for biochemical assays due to their unique properties such as large surface area as well as its highly luminescent, electrical, magnetic and plasmon resonance properties. Commonly, the nanomaterials that were used to manufacture nano-biosensors include magnetic NPs, gold NPs, nano-sized porous alumina, gold-platinum nanocomposites hybrids, multiwall carbon nanotubes, carbon-NPs and graphene (Chen et al. 2016; Matter et al. 2020).

Given that the use of nano-biosensors plays a useful role in environmental sustainability by detecting environmental hazardous materials, then the use of nanoparticles green-synthesized by microbes to manufacture these biosensors is considered more sustainable. Many efforts have been made to synthesize the nano constitute of nano-biosensors using the action of microorganisms. However, there is no available evidence according to our current knowledge if the efficiency of nanosensors can be affected by the way nanoparticles are manufactured, whether chemical, physical or biological. Magnetic NPs are known to be one of the most common nanoforms in various applications that include nanosensors as well, and in addition to their chemically manufactured possibility, they are efficiently biosynthesized by the action of many microorganisms. It is used to manufacture some nanosensors that can detect pathogenic bacteria (e.g., magnetic nanoparticle-urease nanosensors) (Sun et al. 2018), some antibiotics, e.g., Kanamycin (Chen et al. 2013), or some heavy metals, e.g., Hg (II) (Yang et al. 2012). To detect and measure herbicides, e.g., 2,4-dichlorophenoxyacetic acid (2,4-D), Fe_3O_4 magnetic NPs have been incorporated in alkaline phosphatase into a "sol gel/chitosan biosensor membrane" (Loh et al. 2008).

Gold NPs have been reported as an ingredient to manufacture nanosensors to detect the herbicide glyphosate. This nano-biosensor was prepared by "electrochemically depositing poly(2,5-dimethoxyaniline) (PDMA) doped with poly(4-styrenesulfonic acid) (PSS) onto the surface of a gold electrode followed by electrostatic attachment of the enzyme HRP onto the PDMA-PSS composite film" (Songa et al. 2009; Matter et al. 2020). Silver nanoparticles that have wide usage applications have been used to manufacturethe sensing surface of an optical fiber sensor with "a layer of tantalum(V) oxide nanoparticles sequestered in a nano–scaled matrix of reduced graphene oxide" to detect and determine fenitrothion (the organophosphate pesticide) (Kant 2020). Many nanocomposites have been used to manufacture nanosensors. For instance, the "amperometric response nano-biosensor" was prepared by Yang et al. (2013) to detect some pesticides based on AChE/SiO_2 nanosheet–Nafio nanocomposites with chitosan as a cross-linker. This nano-biosensor based on AChE inhibition by pesticides, i.e., methyl parathion, chlorpyrifos, and carbofuran detected the pesticides at low concentrations (1.0×10^{-12} to 1×10^{-8} M). Parallel to the developments in sensing and bio-sensing techniques, there is also a development in the biogenic synthesis of nanomaterials which led to improved efficacy, rapidity and accuracy of such nano-based sensors/biosensors. Consequently, this development in sensing techniques is a step towards the development of risk assessment followed by proper treatment for environmentally hazardous materials; hence, it is an important step towards environmental sustainability.

4.3 *Application of Microbial-Synthesized NPs in Wastes Recycling*

In addition to reducing waste production from various agricultural and industrial activities, the application of nanotechnology can assist in the recycling of harmful wastes. Waste recycling helps to introduce beneficial products (from low-value wastes) and thus reduces dependence on limited resources, which is an essential goal of environmental sustainability. As mentioned before, this sustainability could be maximized when the applied nanoparticles for waste recycling were sustainably synthesized, e.g., using microbial activities. The abundant agricultural wastes, for instance, mainly composed from lignocellulosic materials which could be a raw material to manufacture many useful products such as some furniture and paper.

Also, these cellulose-rich materials could be used as feedstock for the production of many beneficial products such as biofuel, single cell protein, chemicals, organic fertilizers and animal feeds (Hasanin et al. 2019). Successful degradation of cellulose-rich wastes using cellulases into fermentable sugars is the basis for further conversion into beneficial products (Darwesh et al. 2020b). However, production of celluloses could be a cost-effective process especially when the enzymes could not be reused for further degradation cycles. Given that, the immobilization of enzymes, including celluloses, enables the multiple utilization of enzymes in several cycles; so the improvement of immobilization process will enhance the overall process.

Nanomaterials especially magnetic NPS are considered as the most promising materials for enzymes immobilization due to enabling the recycling of enzymes, enhancement of enzyme stability and storage efficiency. Sustainability behind the second biofuel generation from lignocellulosic biomass could be increased with the sustainability of degrading enzymes stability and reusing with the aid of magnetic nanoparticles. Furthermore, nanoparticles could be applied as catalyst for pretreatment for enzyme reaction related to waste recycling, e.g., the application of iron oxide nanoparticle and acidic pretreatment for corn stover and subsequent degradation into fermentable sugars (Srivastava et al. 2017; Darwesh et al. 2019b).

4.4 Application of Biogenic NPs as Nanocatalysts

Catalysts accelerate the chemical reaction by forming bonds with the molecules of reactants to form the "product(s)" and detached from the final product to be available for another reaction cycle(s). Catalysts could be atoms, molecules and solid surfaces as well as biological enzymes. Catalysts are very important for industrial chemical transformations (considered as the workhorses in many industries) where about 85–90% of chemical industry products are produced by catalytic reactions. Without catalysts, many medical and essential chemical products "would not be feasible" while another products will cost more time, reaction temperature and reactants as well as generate more wastes (Singh and Tandon 2014; Chorkendorff and Niemantsverdriet 2017). This may mean, in other words, that "without catalysts, humanity and the environment together will cost more."

Nanocatalysts are catalytic materials that "have at least nanoscale dimension, either externally or in terms of internal structures" (Saoud 2018). Hence, application of nanotechnology in catalysts could be either by manufacturing nano-sized particles as "heterogeneous catalysis" or incorporation of synthetic nanoparticles with enzymes to enhance or keep their activities or enable its reuse (e.g., magnetic NPs). For instance, the esterification reaction to convert fatty acids using methanol into fatty acid methyl ester (biodiesel) could be performed in the presence of an homogeneous catalyst (either acidic, alkaline or enzyme) or with the aid of heterogeneous solid catalyst (El-Baz et al. 2016; Saoud 2018). Some magnetic nanocatalysts such as $CaO/KOHFe_3O_4$ and $KF/KOH\text{-}Fe_3O_4$ have been reported by Farrokheh et al. (2020) as promising heterogeneous nanocatalysts to produce biodiesel from

Chlorella vulgaris and *Spirulina platensis* microalgae by electrolysis. They obtained high process yield of 96.8%, at temperature of 25 °C without using of hazardous sodium hydroxide or sulfuric acids and without the side reaction of saponification which means high quality biodiesel product. Instead, in conventional esterification and transesterification, in addition to the required high heat, acid or base as catalysts are required, respectively, which are considered of less sustainability for overall process.

Moreover, in magnetic nanocatalysts, it is easy to recover (by external magnetic field), regenerate and reuse the catalyst for further production cycles. On the other hand, Darwesh et al. (2019b) used peroxidase enzymes immobilized on magnetic NPs as a nano-biocatalyst for bioremediation of textile wastewater dye in a reactor. They reported more stability, reusability of the biocatalyst due to the immobilization with the magnetic NPs as well as more bioremediation efficiency. It is could be concluded that, application of nanocatalysts enables the sustainability of catalyst, less hazardous waste discharges into the environment and more environmental sustainability.

4.5 Application of Biogenic NPs in Renewable Energy Production

Energy consumption is a manifestation of life on this planet as it is required for all life activities, and this consumption is linked to industrial development and population growth. Unfortunately, the biggest portion of consumed energy came from the limited carbon-based fossil fuels such as oil, coal and natural gas which in turn results in the greenhouse gas emission and global warming. In addition, the depletion of fossil fuels will threaten the current lifestyle through the outbreak of wars to compete for limited fuel in addition to the disruption of many factories and power plants around the world. Hence, the orientation towards relying on clean and renewable energy is an important and vital issue to keep the world sustainable. Solar and wind energy are among the cleanest and renewable energies on the planet. In addition, there are other important sources of clean, ecofriendly and renewable energies such as hydrothermal, marine, bio-mass and solar that are sustainable and leading to a sustainable environment. The application of nanotechnology had a clear imprint in improving the efficiency of production, transmission and consumption of new and renewable energy. As a result of advances in nanotechnology, solar and microbial fuel cells benefit from the advantages of unique properties of NPs such as titanium-based NPs (TiO_2), silicon dioxide (SiO_2) NPs, graphene and carbon nanotubes to efficiently produce energy (Sohail et al. 2019). It should be noted that the use of microbes to manufacture nanomaterials related to the production of clean energy is an environmentally friendly approach in addition to the main benefits arising from the use of clean and sustainable energy sources.

Biofuels could be defined as solid (e.g., biochar), liquid (e.g., bioethanol, biodiesel) or gaseous (e.g., biogas, biohydrogen) fuels that are predominantly produced from bio-renewable feedstocks. Biofuels are considered renewable because their raw materials result from photosynthesis either directly as in plants and algae or indirectly as in animal sources that feed on plants, so they are renewable and environmental friendly. The most common liquid biofuels are bioethanol and biodiesel. Bioethanol could be produced from the microbial degradation of lignocellulosic wastes into fermentable sugars (considered as second generation biofuel). While biodiesel is produced from the esterification or transesterification of plant, animal, microbial and algal lipids into fatty acid methyl ester. Incorporation of nanotechnology into the field of biofuel production offers many possibilities such working as growth enhancer, stress inducer to optimize lipid contents (in microalgae for instance), as well as for biomass harvesting (flocculation) (Matter et al. 2016, 2019) as well as improving the yield of biofuel (Sanusi et al. 2020). Furthermore, production of biofuels could be coupled with bioremediation of wastewater and municipal solid wastes which helps to gain environmental sustainability (Srivastava et al. 2017; Eida et al. 2018; Sangeetha et al. 2019).

5 Applications of Microbial Nanoparticles in Agriculture

As a fast developing field of science and technology, nanotechnology could be a very helpful tool to enhance agricultural controlling process. Nanotechnology has potential benefits that can improve sustainability in the agroecosystem including producing better and safe food, decreasing agricultural inputs, and implementation of nanoscale nutrients. Application of nanotechnology in plant nutrition and pest control could be with great impact on sustaining agro environment (Prasad et al. 2017). Advancing the application of nanotechnology increases the efficient use of agrochemicals (of fertilizers and pesticides) in agriculture that develops the smart characteristics in the agri-inputs. Improving agroprocesses as targeted delivery, controlled release, increasing solubility and long shelf-life through application of nanotechnology not only good for its efficacy but also minimizes the risk of environmental contamination (Chhipa 2019; Mishra et al. 2014).

5.1 Smart Nanofertilizers

One of the recent advancements in nanotechnology application in agriculture is nanofertilizers. Several studies revealed the higher efficacy of nanofertilizers compared to normal fertilizers which could be explained by its nanosize and high volume to surface area. Various trails have been reported from application of silver, iron, zinc, calcium, titanium, carbon nanotubes, molybdenum and silica as nanofertilizers for different crop systems. The findings of these studies have mentioned that nanomaterials were more efficient to increase root and shoot length, vegetative biomass, chlorophyll content, phytochemicals and seed germination at certain concentrations (Salama 2012; Sharma et al. 2012; Taha et al. 2016; Azeez et al. 2019a, b, 2020; Chhipa 2019; Hussein et al. 2019b). Some studies reported over a 50% increase in plant growth (e.g., maize and tobacco) due to the application CuO and carbon nanotubes, respectively, as nanofertilizers (Adhikari et al. 2016; Khodakovskaya et al. 2012). Moreover, the importance of nanotechnology for producing nanofertilizers in the agriculture rely not only on increasing the fertilization efficacy and bioavailability of such fertilizers but also on decreasing the contaminating effect of these fertilizers on surrounding environment (Shebl et al. 2019).

With regard to macronutrients, nano phosphorus was synthesized by *Aspergillius tubingensis* from tricalcium phosphate (Tarafdar et al. 2012). On the other hand, synthesis of several nano-micronutrient including Zn, Fe, Mn, Mg, Co among others that have vast application in agricultural sector (plant and/or animal production) is mediated by different microorganisms (Sohair et al. 2018; Yusof et al. 2019). Raliya et al. (2015) reported significant increase in plant vegetative parameters of mung bean (shoot length (17.02%), root length (49.6%), root area (43%), root nodule (67.5%), chlorophyll content (46.4%)) as well as total protein content (94%) when TiO_2 NPs mediated by *Aspergillus flavus* TFR 7 were sprayed on leaves after 14 days of cultivation. In addition, their results revealed an increase in rhizosphere microbiome (21.4–48.1%), acid and alkaline phosphatase (72 and 67.3%, respectively), phytase (64%) and dehydrogenase (108.7%) compared to control owing to application of TiO_2 NPs.

5.2 *Microbial Nano-Pest Control Agents*

Agricultural sustainability requires reducing the agricultural inputs including agrochemicals as well as preventing loss in production as a result of plant diseases. Thus, searching for more safe alternatives for chemical pesticides gains more attention through the last decades. Recently, nanoparticles have been introduced to the field of pest control as an ecofriendly alternative to chemical pesticides (Masum et al. 2019; Suman et al. 2013; Elshahawy et al. 2018). As reviewed by Ali et al. (2020), biologically mediated metallic NPs including Ag, Cu, Au and Zn displayed prevention effect against both Gram (+) and (−) bacteria such as *B. subtilis*, *E. coli* and *Staphylococcus aureus* as well as some pathogenic fungi, e.g., *A. niger*, *F. oxysporum*, *A. fumigatus* and other disease causing microbes. The green synthesis of silver NPs through plants, bacteria, fungi or yeasts increased the attention toward it (Rafique et al. 2017). In this respect, Ag-NPs mediated by *Streptomyces* showed great effect to control well-known plant pathogenic fungi, e.g., *Alternaria alternata*, *A. niger*, *F. oxysporum* and *Pythium ultimum* (Fouda et al. 2020). Also, a concentration of 200 μg/ml from AuNPs and Ag-AuNPs biosynthesized using cell-free extract of *Bacillus safensis* LAU 13 has been reported to display 66.67–90.78% growth inhibitions against strains of *A. fumigatus* and *A. niger*, respectively (Ojo et al. 2016). While, 100 μg/mL of Ag-NPs synthesized by the cell-free extract of the same bacterial strain exhibited growth inhibition for *Staphylococcus aureus*, *Escherichia coli*, *Klebsiella granulomatis* and *Pseudomonas aeruginosa* (Lateef et al. 2015b). Nanoparticles as pest control agents are applied for seed coating or as foliar spray. It could be utilized not only as pesticides but also as carriers to control the release of chemical pesticides from its formulation to reduce its environmental adverse impact (Ali et al. 2020; Khan and Rizvi 2014).

6 Conclusion and Future Perspective

This chapter reports the development of the microbial synthesis of nanoparticles and its role in environmental sustainability. The attention concerning green approach in nanotechnology has gained more attraction in the recent years regarding its wide range of advantages compared to chemical and physical processes such as environmental safety, cost efficient, applicability and easy scale-up. Thus, only the direction towards utilization of biogenic nanoparticles especially those that are mediated by microbes is a considerable sustainable environmental practice regarding to its higher safety of compared to chemical and physical synthesized nanomaterials. Several topics were addressed such as microbial synthesis of NPs and its advantages, application of microbial-mediated NPs in agriculture and bioremediation. When the energy issues have great importance from the sustainability point of view, the role of biogenic NPs for energy preservation and biofuel production cannot be overemphasized. As biosensors and biomonitoring are promising environmental applications of NPs, this chapter summarized the impact of microbial manufacturing of NPs on advancement of environmental biosensing. However, bridging the gap between bench studies and industrial level in the field of biogenic NPs for environmental applications still required a lot more research and pilot trials to improve the effectiveness of these bioprocesses. To effectively accomplish these targets, further combined research on screening for more efficient bionanofactories such as bacteria, actinomycetes, fungi, yeast, viruses, algae and even plants as well as more efforts to be dedicated to optimizing parameters and methodologies for the utilization/extraction of the biomatrices. Scaling up environmental utilization of NPs up today faces significant challenges such as controlling size, shape and crystallinity of produced nanomaterials as the actual bioformation mechanism is not fully understood yet. Developing production of biogenic NPs is so far a challenging issue that required more research. Although it looks less toxic, further researches to evaluate the toxicity behavior of microbial-mediated NPs are still necessary.

References

Abdel-Monem RA, Khalil AM, Darwesh OM, Hashim AI, Rabie ST (2020) Antibacterial properties of carboxymethyl chitosan Schiff-base nanocomposites loaded with silver nanoparticles. J Macromol Sci a 57:145–155. https://doi.org/10.1080/10601325.2019.1674666

Abou-Shanab RAI, Khalafallah MA, Emam NF, Aly MA, Abou-Sdera SA, Matter IA (2012) Characterisation and identification of carbofuran-utilising bacteria isolated from agricultural soil. Chem Ecol 28:193–203. https://doi.org/10.1080/02757540.2011.628317

Adelere IA, Lateef A (2016) A novel approach to the green synthesis of metallic nanoparticles: the use of agro-wastes, enzymes, and pigments. Nanotechnol Rev 5(6):567–587. https://doi.org/10.1515/ntrev-2016-0024

Adhikari T, Sarkar D, Mashayekhi H, Xing B (2016) Growth and enzymatic activity of maize (*Zea mays* L.) plant: solution culture test for copper dioxide nano particles. J Plant Nutr 39:99–115. https://doi.org/10.1080/01904167.2015.1044012

Ali I, Peng C, Khan ZM, Naz I, Sultan M (2018) An overview of heavy metal removal from wastewater using magnetotactic bacteria. J Chem Technol Biotechnol 93:2817–2832. https://doi.org/10.1002/jctb.5648

Ali M, Ahmed T, Wu W, Hossain A, Hafeez R, Masum IM, Wang Y, An Q, Sun G, Li B (2020) Advancements in plant and microbe-based synthesis of metallic nanoparticles and their antimicrobial activity against plant pathogens. Nanomater 10:1146. https://doi.org/10.3390/nano10061146

Ambashta RD, Sillanpää M (2010) Water purification using magnetic assistance: a review. J Hazard Mater 180:38–49. https://doi.org/10.1016/j.jhazmat.2010.04.105

Arora K (2018) Advances in nano based biosensors for food and agriculture. In: Nanotechnology, food security and water treatment. Springer, Cham, pp 1–52. https://doi.org/10.1007/978-3-319-70166-0_1

Azeez L, Adejumo AL, Lateef A, Adebisi SA, Adetoro RO, Adewuyi SO, Tijani KO, Olaoye S (2019) Zero-valent silver nanoparticles attenuate Cd and Pb toxicities on *Moringa oleifera* via immobilization and induction of phytochemicals. Plant Physiol Biochem 139:283–292. https://doi.org/10.1016/j.plaphy.2019.03.030

Azeez L, Lateef A, Wahab AA, Rufai MA, Salau AK, Ajayi IO, Ajayi EM, Maryam AK, Adebisi B (2019) Phytomodulatory effects of silver nanoparticles on *Corchorus olitorius*: its antiphytopathogenic and hepatoprotective potentials. Plant Physiol Biochem 136:109–117. https://doi.org/10.1016/j.plaphy.2018.12.006

Azeez L, Adejumo AL, Simiat OM, Lateef A (2020) Influence of calcium nanoparticles (CaNPs) on nutritional qualities, radical scavenging attributes of *Moringa oleifera* and risk assessments on human health. J Foo Meas Charact 14:2185–2195. https://doi.org/10.1007/s11694-020-00465-6

Besha AT, Liu Y, Bekele DN, Dong Z, Naidu R, Gebremariam GN (2020) Sustainability and environmental ethics for the application of engineered nanoparticles. Environ Sci Policy 103:85–98. https://doi.org/10.1016/j.envsci.2019.10.013

Chen YP, qiang Zou M, Qi C, Xie MX, Wang DN, Wang YF, Xue Q, Li JF, Chen Y (2013) Immunosensor based on magnetic relaxation switch and biotin–streptavidin system for the detection of kanamycin in milk. Biosens Bioelectron 39:112–117. https://doi.org/10.1016/j.bios.2012.06.056

Chen Y, Xianyu Y, Wu J, Yin B, Jiang X (2016) Click chemistry-mediated nanosensors for biochemical assays. Theranostics 6:969. https://doi.org/10.7150/thno.14856

Chhipa H (2019) Applications of nanotechnology in agriculture. In: Methods in microbiology, vol 46. Elsevier, Amsterdam, pp 115–142. https://doi.org/10.1016/bs.mim.2019.01.002

Chorkendorff I, Niemantsverdriet JW (2017) Concepts of modern catalysis and kinetics. John Wiley & Sons, Hoboken, p 524. https://doi.org/10.1002/3527602658

Darwesh OM, Moawad H, El-Rahim WM, Barakat OS, Sedik MZ (2014) Bioremediation of textile reactive blue (RB) azo dye residues in wastewater using experimental prototype bioreactor. Res J Pharm Biol Chem Sci 5:1203–1219

Darwesh OM, Moawad H, Barakat OS, Abd El- Rahim WM (2015) Bioremediation of textile reactive blue azo dye residues using nanobiotechnology approaches. Res J Pharm Biol Chem Sci 6:1202–1211
Darwesh OM, Eida MF, Matter IA (2018a) Isolation, screening and optimization of L-asparaginase producing bacterial strains inhabiting agricultural soils. Biosci Res 15:2802–2812
Darwesh OM, Sultan YY, Seif MM, Marrez DA (2018b) Bio-evaluation of crustacean and fungal nano-chitosan for applying as food ingredient. Toxicol Rep 5:348–356. https://doi.org/10.1016/j.toxrep.2018.03.002
Darwesh OM, Matter IA, Eida MF, Moawad H, Oh YK (2019a) Influence of nitrogen source and growth phase on extracellular biosynthesis of silver nanoparticles using cultural filtrates of *Scenedesmus obliquus*. Appl Sci 9:1465. https://doi.org/10.3390/app9071465
Darwesh OM, Matter IA, Eida MF (2019b) Development of peroxidase enzyme immobilized magnetic nanoparticles for bioremediation of textile wastewater dye. J Environ Chem Eng 7:102805. https://doi.org/10.1016/j.jece.2018.11.049
Darwesh OM, Ali SS, Matter IA, Elsamahy T, Mahmoud YA (2020a) Enzymes immobilization onto magnetic nanoparticles to improve industrial and environmental applications. In: Methods in enzymology, vol 630. Academic Press, Cambridge, pp 481–502. https://doi.org/10.1016/bs.mie.2019.11.006
Darwesh OM, El-Maraghy SH, Abdel-Rahman HM, Zaghloul RA (2020b) Improvement of paper wastes conversion to bioethanol using novel cellulose degrading fungal isolate. Fuel 262:116518. https://doi.org/10.1016/j.fuel.2019.116518
Das N (2010) Recovery of precious metals through biosorption—a review. Hydrometallurgy 103:180–189. https://doi.org/10.1016/j.hydromet.2010.03.016
Das SK, Das AR, Guha AK (2009) Gold nanoparticles: microbial synthesis and application in water hygiene management. Langmuir 25:8192–8199. https://doi.org/10.1016/j.hydromet.2010.03.016
Davis AS, Prakash P, Thamaraiselvi K (2017) Nanobioremediation technologies for sustainable environment. In: Bioremediation and sustainable technologies for cleaner environment. Springer, Cham. pp 13–33. https://doi.org/10.1007/978-3-319-48439-6_2
de Castro AL, Yáñez-Vilar S, Piñeiro-Redondo Y, Rivas J (2019) Novel magnetic nanostructured beads for cadmium (II) removal. Nanomaterials 9:356. https://doi.org/10.3390/nano9030356
Durgalakshmi D, Rajendran S, Naushad M (2019) Current role of nanomaterials in environmental remediation. In: Advanced nanostructured materials for environmental remediation. Springer, Cham, pp 1–20. https://doi.org/10.1007/978-3-030-04477-0_1
Eida MF, Darwesh OM, Matter IA (2018) Cultivation of oleaginous microalgae *Scenedesmus obliquus* on secondary treated municipal wastewater as growth medium for biodiesel production. J Ecol Eng 19:38–51. https://doi.org/10.12911/22998993/91274
Ejeian F, Etedali P, Mansouri-Tehrani HA, Soozanipour A, Low ZX, Asadnia M, Taheri-Kafrani A, Razmjou A (2018) Biosensors for wastewater monitoring: a review. Biosens Bioelectron 118:66–79. https://doi.org/10.1016/j.bios.2018.07.019
El-Baz FK, Gad MS, Abdo SM, Abed KA, Matter IA (2016) Performance and exhaust emissions of a diesel engine burning algal biodiesel blends. Int J Mechan Mechatr Eng 16:150–157
Elegbede JA, Lateef A, Azeez MA, Asafa TB, Yekeen TA, Oladipo IC, Adebayo EA, Beukes LS (2018) Fungal xylanases-mediated synthesis of silver nanoparticles for catalytic and biomedical applications. IET Nanobiotechnol 12:857–863. https://doi.org/10.1049/iet-nbt.2017.0299
Elegbede JA, Lateef A, Azeez MA, Asafa TB, Yekeen TA, Oladipo IC, Hakeem AS, Beukes LS, Gueguim-Kana EB (2019) Silver-gold alloy nanoparticles biofabricated by fungal xylanases exhibited potent biomedical and catalytic activities. Biotechnol Prog 35:2829. https://doi.org/10.1002/btpr.2829
Elegbede JA, Lateef A, Azeez MA, Asafa TB, Yekeen TA, Oladipo IC, Aina DA, Beukes LS, Gueguim-Kana EB (2020) Biofabrication of gold nanoparticles using xylanases through valorization of corncob by *Aspergillus niger* and *Trichoderma longibrachiatum*: antimicrobial, antioxidant, anticoagulant and thrombolytic activities. Waste Biomass Valoriz 11(3):781–791. https://doi.org/10.1007/s12649-018-0540-2

Elshahawy I, Abouelnasr HM, Lashin SM, Darwesh OM (2018) First report of *Pythiuma phanidermatum* infecting tomato in Egypt and its control using biogenic silver nanoparticles. J Plant Prot Res 58:137–151. https://doi.org/10.24425/122929

El-Shanshoury AE, Darwesh OM, Sabae SZ, Awadallah OA, Hassan SH (2020) Bio-manufacturing of selenium nanoparticles by *Bacillus subtilis* isolated from Qarun lake and evaluation their activity for water remediation. Biointerf Res Appl Chem 10:5834–5842. https://doi.org/10.33263/BRIAC104.834842

Farrokheh A, Tahvildari K, Nozari M (2020) Biodiesel production from the *Chlorella vulgaris* and *Spirulina platensis* microalgae by electrolysis using CaO/KOH-Fe_3O_4 and KF/KOH-Fe_3O_4 as magnetic nanocatalysts. Biomass Convers Biorefin 1–15. https://doi.org/10.1007/s13399-020-00688-z

Fouda A, Hassan SE, Abdo AM, El-Gamal MS (2020) Antimicrobial, Antioxidant and Larvicidal activities of spherical silver nanoparticles synthesized by endophytic *Streptomyces* spp. Biol Trace Elem Res 195:707–724. https://doi.org/10.1007/s12011-019-01883-4

Ghaffar N, Farrukh MA, Naz S. (2020) Applications of nanobiosensors in agriculture. In: Nanoagronomy. Springer, Cham, pp 179–196. https://doi.org/10.1007/978-3-030-41275-3_10

Glatstein DA, Bruna N, Gallardo-Benavente C, Bravo D, Carro Perez ME, Francisca FM, Pérez-Donoso JM (2018) Arsenic and cadmium bioremediation by antarctic bacteria capable of biosynthesizing CdS fluorescent nanoparticles. J Environ Eng 144:04017107. https://doi.org/10.1061/(ASCE)EE.1943-7870.0001293

Hadef F (2018) An introduction to nanomaterials. In: Environmental nanotechnology. Environmental chemistry for a sustainable world, vol 14. Springer, Cham, pp 1–58. https://doi.org/10.1007/978-3-319-76090-2_1

Hasanin MS, Mostafa AM, Mwafy EA, Darwesh OM (2018) Eco-friendly cellulose nano fibers via first reported Egyptian *Humicola fuscoatra* Egyptia X4: isolation and characterization. Environ Nanotechnol Monit Manag 10:409–418. https://doi.org/10.1016/j.enmm.2018.10.004

Hasanin MS, Darwesh OM, Matter IA, El-Saied H (2019) Isolation and characterization of non-cellulolytic *Aspergillus flavus* EGYPTA5 exhibiting selective ligninolytic potential. Biocatal Agric Biotechnol 17:160–167. https://doi.org/10.1016/j.bcab.2018.11.012

Hoballah E, Saber M, Matter I, Zaghloul A (2014) Bioremediation of polychlorinated biphenyl (PCBs) in a sewaged soil by certain remediative amendments followed by phytoremediation. Res J Pharm Biol Chem Sci 5:91–103

Hussein HA, Darwesh OM, Mekki BB (2019) Environmentally friendly nano-selenium to improve antioxidant system and growth of groundnut cultivars under sandy soil conditions. Biocatal Agric Biotechnol 18:101080. https://doi.org/10.1016/j.bcab.2019.101080

Hussein HA, Darwesh OM, Mekki BB, El-Hallouty SM (2019) Evaluation of cytotoxicity, biochemical profile and yield components of groundnut plants treated with nano-selenium. Biotechnol Rep 24:00377. https://doi.org/10.1016/j.btre.2019.e00377

Kabary H, Eida MF, Attia MM, Awad N, Easa SM (2017) Magnetotactic characterization and environmental application *P. aeruginosa* kb1 isolate. Annu Res Rev Biol 20:1–10. https://doi.org/10.9734/arrb/2017/37737

Kant R (2020) Surface plasmon resonance based fiber–optic nanosensor for the pesticide fenitrothion utilizing Ta_2O_5 nanostructures sequestered onto a reduced graphene oxide matrix. Microchim Acta 187:8. https://doi.org/10.1007/s00604-019-4002-8

Khan MR, Rizvi TF (2014) Nanotechnology: scope and application in plant disease management. Plant Pathol J 13:214–231. https://doi.org/10.3923/ppj.2014.214.231

Khodakovskaya MV, De Silva K, Biris AS, Dervishi E, Villagarcia H (2012) Carbon nanotubes induce growth enhancement of tobacco cells. ACS Nano 6:2128–2135. https://doi.org/10.1021/nn204643g

Lateef A, Adeeyo AO (2015) Green synthesis and antibacterial activities of silver nanoparticles using extracellular laccase of *Lentinus edodes*. Not Sci Biol 7(4):405–411. https://doi.org/10.15835/nsb749643

Lateef A, Adelere IA, Gueguim-Kana EB, Asafa TB, Beukes LS (2015) Green synthesis of silver nanoparticles using keratinase obtained from a strain of *Bacillus safensis* LAU 13. Int Nano Lett 5(1):29–35. https://doi.org/10.1007/s40089-014-0133-4

Lateef A, Ojo SA, Akinwale AS, Azeez L, Gueguim-Kana EB, Beukes LS (2015b) Biogenic synthesis of silver nanoparticles using cell-free extract of *Bacillus safensis* LAU 13: antimicrobial, free radical scavenging and larvicidal activities. Biologia 70:1295–306. http://dx.doi.org.https://doi.org/10.1515/biolog-2015-0164

Li X, Xu H, Chen ZS, Chen G (2011) Biosynthesis of nanoparticles by microorganisms and their applications. J Nanomater 1:2011. https://doi.org/10.1155/2011/270974

Liu J, Zhou PG (2008) Application and development of magnetotactic bacteria in heavy metal wastewater treatment. J Environ Sci Technol 21:60–66

Liu M, Wang Z, Zong S, Chen H, Zhu D, Wu L, Hu G, Cui Y (2014) SERS detection and removal of mercury (II)/silver (I) using oligonucleotide-functionalized core/shell magnetic silica sphere@ Au nanoparticles. ACS Appl Mater Interf 6:7371–7379. https://doi.org/10.1021/am5006282

Loh KS, Lee YH, Musa A, Salmah AA, Zamri I (2008) Use of Fe_3O_4 nanoparticles for enhancement of biosensor response to the herbicide 2, 4-dichlorophenoxyacetic acid. Sensors 8:5775–5791. https://doi.org/10.3390/s8095775

Masum M, Islam M, Siddiqa M, Ali KA, Zhang Y, Abdallah Y, Ibrahim E, Qiu W, Yan C, Li B (2019) Biogenic synthesis of silver nanoparticles using *Phyllanthus emblica* fruit extract and its inhibitory action against the pathogen *Acidovorax oryzae* strain RS-2 of rice bacterial brown stripe. Front Microbiol 10:820. https://doi.org/10.3389/fmicb.2019.00820

Matter IA, Darwesh OM, El-baz FK (2016) Using the natural polymer chitosan in harvesting *Scenedesmus* species under different concentrations and cultural pH values. Int J Pharm Biol Sci 7:254–260. https://doi.org/10.22376/ijpbs.2016.7.4.b254-260

Matter IA, Darwesh OM, Eida MF (2018) Harvesting of microalgae *Scenedesmus obliquus* using chitosan-alginate dual flocculation system. Biosci. Res. 15:540–548

Matter IA, Bui VK, Jung M, Seo JY, Kim YE, Lee YC, Oh YK (2019) Flocculation harvesting techniques for microalgae: a review. Appl Sci 9:3069. https://doi.org/10.3390/app9153069

Matter IA, Darwesh OM, Matter HAB (2020) Nano-sensors for herbicides monitoring in soil. Springer, Cham (in press)

Mishra S, Singh A, Keswani C, Singh HB (2014) Nanotechnology: exploring potential application in agriculture and its opportunities and constraints. Biotech Today. 4:9–14. https://doi.org/10.5958/2322-0996.2014.00011.8

Mourad R, Helaly F, Darwesh OM, Sawy SE (2019) Antimicrobial and physicomechanical natures of silver nanoparticles incorporated into silicone- hydrogel films. Contact Lens Anter Eye 42:325–333. https://doi.org/10.1016/j.clae.2019.02.007

Nasrollahzadeh M, Sajadi SM, Sajjadi M, Issaabadi Z (2019) Applications of nanotechnology in daily life. In: Interface science and technology, vol 28. Elsevier, Amsterdam, pp 113–143. https://doi.org/10.1016/b978-0-12-813586-0.00004-3

Nguyen CC, Hugie CN, Kile ML, Navab-Daneshmand T (2019) Association between heavy metals and antibiotic-resistant human pathogens in environmental reservoirs: a review. Front Environ Sci Eng 13:46. https://doi.org/10.1007/s11783-019-1129-0

Nies DH (1999) Microbial heavy-metal resistance. Appl Microbiol Biotechnol 51(6):730–750. https://doi.org/10.1007/s002530051457

Noman M, Shahid M, Ahmed T, Niazi MB, Hussain S, Song F, Manzoor I (2020) Use of biogenic copper nanoparticles synthesized from a native *Escherichia* sp. as photocatalysts for azo dye degradation and treatment of textile effluents. Environ Pollut 257:113514. https://doi.org/10.1016/j.envpol.2019.113514

Ojo SA, Lateef A, Azeez MA, Oladejo SM, Akinwale AS, Asafa TB, Yekeen TA, Akinboro A, Oladipo IC, Gueguim-Kana EB, Beukes LS (2016) Biomedical and catalytic applications of gold and silver-gold alloy nanoparticles biosynthesized using cell-free extract of *Bacillus safensis* LAU 13: antifungal, dye degradation, anti-coagulant and thrombolytic activities. IEEE Trans Nanobiosci 15:433–442. https://doi.org/10.1109/TNB.2016.2559161

Oladipo IC, Lateef A, Azeez MA, Asafa TB, Yekeen TA, Akinboro A, Akinwale AS, Gueguim-Kana E.B, Beukes LS (2017a) Green synthesis and antimicrobial activities of silver nanoparticles using cell-free extracts of *Enterococcus* species. Not Sci Biol 9(2): 196–203. http://dx.doi.org/https://doi.org/10.15835/nsb929938

Oladipo IC, Lateef A, Elegbede JA, Azeez MA, Asafa TB, Yekeen TA, Akinboro A, Gueguim-Kana EB, Beukes LS, Oluyide TO, Atanda OR (2017) *Enterococcus* species for the one-pot biofabrication of gold nanoparticles: characterization and nanobiotechnological applications. J Photochem Photobiol B Biol 173:250–257. https://doi.org/10.1016/j.jphotobiol.2017.06.003

Órdenes-Aenishanslins NA, Saona LA, Durán-Toro VM, Monrás JP, Bravo DM, Pérez-Donoso JM (2014) Use of titanium dioxide nanoparticles biosynthesized by *Bacillus mycoides* in quantum dot sensitized solar cells. Microb Cell Fact 13:90. https://doi.org/10.1186/s12934-014-0090-7

Prasad R, Bhattacharyya A, Nguyen QD (2017) Nanotechnology in sustainable agriculture: recent developments, challenges, and perspectives. Front Microbiol 8:1014. https://doi.org/10.3389/fmicb.2017.01014

Rafique M, Sadaf I, Rafique MS, Tahir MB (2017) A review on green synthesis of silver nanoparticles and their applications. Artif Cells Nanomed Biotechnol 45:1272–1291. https://doi.org/10.1080/21691401.2016.1241792

Raliya R, Biswas P, Tarafdar JC (2015) TiO_2 nanoparticle biosynthesis and its physiological effect on mung bean (*Vigna radiata* L.). Biotechnol Rep 5:22–26. https://doi.org/10.1016/j.btre.2014.10.009

Ranjan B, Pillai S, Permaul K, Singh S (2019) Simultaneous removal of heavy metals and cyanate in a wastewater sample using immobilized cyanatehydratase on magnetic-multiwall carbon nanotubes. J Hazard Mater 363:73–80. https://doi.org/10.1016/j.jhazmat.2018.07.116

Rizwan M, Singh M, Mitra CK, Morve RK (2014) Ecofriendly application of nanomaterials: nanobioremediation. J Nanopart 2014:431787. https://doi.org/10.1155/2014/431787

Saber M, Hoballah E, Matter I, Zaghloul A (2014) Bioremediation of chlorinated hydrocarbons in a sewaged soil by certain remediative amendments followed by phytoremediation. Res J Pharm Biol Chem Sci 5:83–90. https://rjpbcs.com/pdf/2014_5(2)/[11].pdf

Sadhasivam S, Vinayagam V, Balasubramaniyan M (2020) Recent advancement in biogenic synthesis of iron nanoparticles. J Mol Struct 1217:128372. https://doi.org/10.1016/j.molstruc.2020.128372

Saini RK, Bagri LP, Bajpai AK (2017) Smart nanosensors for pesticide detection. In: New pesticides and soil sensors. Academic Press, Cambridge, pp 519–559. https://doi.org/10.1016/B978-0-12-804299-1.00015-1

Salama HM (2012) Effects of silver nanoparticles in some crop plants, common bean (*Phaseolus vulgaris* L.) and corn (*Zea mays* L.). Int Res J Biotechnol 3:190–197

Salouti M, Derakhshan FK (2020) Biosensors and nanobiosensors in environmental applications. In: Biogenic nano-particles and their use in agro-ecosystems. Springer, Singapore, pp 515–591. https://doi.org/10.1007/978-981-15-2985-6_26

Salvadori MR, Ando RA, Nascimento CA, Corrêa B (2017) Dead biomass of Amazon yeast: a new insight into bioremediation and recovery of silver by intracellular synthesis of nanoparticles. J Environ Sci Health Part A 52:1112–1120. https://doi.org/10.1080/10934529.2017.1340754

Sangeetha J, Mundaragi A, Thangadurai D, Maxim SS, Pandhari RM, Alabhai JM (2019) Nanobiotechnology for agricultural productivity, food security and environmental sustainability. In: Nanotechnology for agriculture: crop production & protection. Springer, Singapore, pp 1–23. https://doi.org/10.1007/978-981-32-9374-8_1

Sanusi IA, Suinyuy TN, Lateef A, Kana GE (2020) Effect of nickel oxide nanoparticles on bioethanol production: process optimization, kinetic and metabolic studies. Process Biochem 92:386–400. https://doi.org/10.1016/j.procbio.2020.01.029

Saoud K (2018) Nanocatalyst for biofuel production: a Review. In: Green nanotechnology for biofuel production. Springer, Cham, pp 39–62. https://doi.org/10.1007/978-3-319-75052-1_4

Sharma H, Shirkot P (2019) Bioremediation of azo dyes using biogenic iron nanoparticles. J Microbiol Exp 7:12–15. https://doi.org/10.15406/jmen.2019.07.00232

Sharma P, Bhatt D, Zaidi MG, Saradhi PP, Khanna PK, Arora S (2012) Silver nanoparticle-mediated enhancement in growth and antioxidant status of *Brassica juncea*. Appl Biochem Biotechnol 167:2225–2233. https://doi.org/10.1007/s12010-012-9759-8

Shebl A, Hassan AA, Salama DM, El-Aziz A, AbdElwahed MS (2019) Green synthesis of nanofertilizers and their application as a foliar for *Cucurbita pepo* L. J Nanomat 2019:3476347. https://doi.org/10.3762/bxiv.2019.45.v1

Simate GS, Ndlovu S, Iyuke SE, Walubita LF (2013) Biotechnology and nanotechnology: a means for sustainable development in Africa. In: Chemistry for sustainable development in Africa. Springer, Berlin, Heidelberg, pp 159–191. https://doi.org/10.1007/978-3-642-29642-0_9

Singh SB, Tandon PK (2014) Catalysis: a brief review on nano-catalyst. J Energ Chem Eng 2:106–115

Singh P, Kim YJ, Zhang D, Yang DC (2016) Biological synthesis of nanoparticles from plants and microorganisms. Trends Biotechnol 34:588–599. https://doi.org/10.1016/j.tibtech.2016.02.006

Sinha R, Dwivedi S, Kumar A, Srivastava P (2020) Materials in bio-sensing of water pollutants. In: Sensors in water pollutants monitoring: role of material. Springer, Singapore, pp 187–211. https://doi.org/10.1007/978-981-15-0671-0_11

Sohail MI, Waris AA, Ayub MA, Usman M, urRehman MZ, Sabir M, Faiz T (2019) Environmental application of nanomaterials: a promise to sustainable future. In: Comprehensive analytical chemistry, vol 87. Elsevier, Amsterdam, pp 1–54. https://doi.org/10.1016/bs.coac.2019.10.002

Sohair EED, Abdall AA, Amany AM, Hossain MF, HoudaRA (2018) Evaluation of nitrogen, phosphorus and potassium nano-fertilizers on yield, yield components and fiber properties of Egyptian cotton (*Gossypium barbadense* L.). J Plant Sci Crop Prot 1:208

Songa EA, Arotiba OA, Owino JH, Jahed N, Baker PG, Iwuoha EI (2009) Electrochemical detection of glyphosate herbicide using horseradish peroxidase immobilized on sulfonated polymer matrix. Bioelectrochemistry 75:117–123. https://doi.org/10.1016/j.bioelechem.2009.02.007

Srivastava N, Srivastava M, Manikanta A, Singh P, Ramteke PW, Mishra PK (2017) Nanomaterials for biofuel production using lignocellulosic waste. Environ Chem Lett 15:179–184. https://doi.org/10.1007/s10311-017-0622-6

Suman TY, Rajasree SR, Kanchana A, Elizabeth SB (2013) Biosynthesis, characterization and cytotoxic effect of plant mediated silver nanoparticles using *Morinda citrifolia* root extract. Colloids Surf B 106:74–78. https://doi.org/10.1016/j.colsurfb.2013.01.037

Sun Y, Fang L, Wan Y, Gu Z (2018) Pathogenic detection and phenotype using magnetic nanoparticle-urease nanosensor. Sens Actuat B 259:428–432. https://doi.org/10.1016/j.snb.2017.12.095

Taha RA, Hassan MM, Ibrahim EA, Abou Baker NH, Shaaban EA (2016) Carbon nanotubes impact on date palm in vitro cultures. Plant Cell Tissue Organ Cult 127:525–534. https://doi.org/10.1007/s11240-016-1058-6

Tarafdar JC, Raliya R, Rathore I (2012) Microbial synthesis of phosphorous nanoparticle from tri-calcium phosphate using *Aspergillus tubingensis* TFR-5. J Bionanosci 6:84–89. https://doi.org/10.1166/jbns.2012.1077

Wang X, Li Y, Zhao J, Yao H, Chu S, Song Z, He Z, Zhang W (2020) Magnetotactic bacteria: characteristics and environmental applications. Front Environ Sci Eng 14:1–14. https://doi.org/10.1007/s11783-020-1235-z

Yan L, Da H, Zhang S, López VM, Wang W (2017) Bacterial magnetosome and its potential application. Microbiol Res 203:19–28. https://doi.org/10.1016/j.micres.2017.06.005

Yang H, Tian Z, Wang J, Yang S (2012) A magnetic resonance imaging nanosensor for Hg (II) based on thymidine-functionalized supermagnetic iron oxide nanoparticles. Sens Actuat B Chem 161:429–433. https://doi.org/10.1016/j.snb.2011.10.055

Yang L, Wang GC, Liu YJ, An JJ, Wang M (2013) Development of a stable biosensor based on a SiO_2 nanosheet–Nafion–modified glassy carbon electrode for sensitive detection of pesticides. Anal Bioanal Chem 405:2545–2552. https://doi.org/10.1007/s00216-012-6634-y

Yang J, Lei M, Chen T, Gao D, Zheng G, Guo G, Lee D (2014) Current status and developing trends of the contents of heavy metals in sewage sludges in China. Front Environ Sci Eng 8:719–728. https://doi.org/10.1007/s11783-013-0600-6

Yusof HM, Mohamad R, Zaidan UH (2019) Microbial synthesis of zinc oxide nanoparticles and their potential application as an antimicrobial agent and a feed supplement in animal industry: a review. J Anim Sci Biotechnol 10:57. https://doi.org/10.1186/s40104-019-0368-z

Zare HR, Shekari Z (2019) Types of monitoring biosensor signals. In: Electrochemical biosensors. Elsevier, Amsterdam, pp 135–166. https://doi.org/10.1016/b978-0-12-816491-4.00006-1

Zhu J, Li Q, Jiao W, Jiang H, Sand W, Xia J, Liu X, Qin W, Qiu G, Hu Y, Chai L (2012) Adhesion forces between cells of *Acidithiobacillus ferrooxidans*, *Acidithiobacillus thiooxidans* or *Leptospirillum ferrooxidans* and chalcopyrite. Colloids Surf B 94:95–100. https://doi.org/10.1016/j.colsurfb.2012.01.022

Nanotechnology in Bioprocess Development: Applications of Nanoparticles in the Generation of Biofuels

I. A. Sanusi, Y. Sewsynker-Sukai, and E. B. Gueguim-Kana

1 Introduction

The global energy demand is rapidly increasing due to the growing human population, with the current energy sources unable to meet the staggering daily fuel consumption. More specifically, the major reliance on fossil fuels as an energy source has led to its depletion (Hirsch et al. 2005; Ge et al. 2018). Fossil fuels such as coal and oil are non-renewable energy sources that are most exploited globally and have been predicted to run out in the next 115 and 50 years, respectively (BP 2016). In addition to their depletion, the consumption of fossil fuels are harmful to the environment, since its combustion releases greenhouse gases such as CO_2, nitrous oxide and fluorinated oxide (Fazal et al. 2018; Ge et al. 2018). Consequently, the development of alternative energy sources such as biofuels that are renewable, cost-effective and environmentally friendly is attracting unprecedented research interest. Biofuels have emerged as sustainable energy sources which could be produced from lignocellulosic waste, vegetable oils, animal fat and industrial wastewaters, among others (Sekoai et al. 2019). These fuels are considered 'green' since it is non-toxic, easily biodegradable, environmentally benign and sustainable in nature (Moodley and Gueguim-Kana 2019).

Biofuels can be classified based on the carbohydrate sources used for their production. For example, first-generation biofuels involve the use of edible feedstocks to produce biofuels (bioethanol, biodiesel and biogas) (Khawla et al. 2014; Baeyens et al. 2015). Even though first-generation biofuels generate high yields, the utilization of staple crops such as wheat, sorghum, corn and sugarcane debilitate food security.

I. A. Sanusi · Y. Sewsynker-Sukai · E. B. Gueguim-Kana (✉)
School of Life Sciences, University of KwaZulu-Natal, Pietermaritzburg, South Africa
e-mail: kanag@ukzn.ac.za

Y. Sewsynker-Sukai
SMRI/NRFSARChI Research Sugarcane Biorefining Group, Discipline of Chemical Engineering, University of KwaZulu-Natal, Durban, South Africa

A. Lateef et al. (eds.), *Microbial Nanobiotechnology*, Materials Horizons: From Nature to Nanomaterials, https://doi.org/10.1007/978-981-33-4777-9_6

Alternatively, second-generation biofuels consist of non-food crops as substrate, commonly lignocellulosic materials which include rice bran, sawdust, potato waste, corncob, bagasse and sugarcane leaves (Gueuguim-Kana et al. 2012; Baeyens et al. 2015; Sebayang et al. 2017; Sewsynker-Sukai and Gueguim-Kana 2018; Chohan et al. 2020). And, the third-generation biofuels use various types of microalgal biomass as feedstock for the production of biofuels (Sekoai et al. 2019). In the recent time, second- and third-generation biofuels have gained interest as advanced energy sources that could be harnessed from waste materials such as potato peels, cassava peels, sugarcane leaves, bagasse and corn cobs (Khawla et al. 2014; Sekoai et al. 2019). These wastes are abundantly available via agricultural activities (Lateef et al. 2008; Adeoye et al. 2015; Elegbede and Lateef 2018).

Currently, the bio-energy market is expanding (Renewable Fuels Association 2018). For instance, the global biodiesel production is expanding with an estimated annual increase of 7.3%, to a total of 54.8 billion USD by 2025, while, the production of bioethanol is projected to have an annual growth of 3–7% (Sekoai et al. 2019). The global bioethanol production was estimated at around 100 billion litres in 2017 and is envisaged to double by 2027 (Sekoai et al. 2019). The International Energy Agency (IEA) predicts that biofuels production is expected to increase 15% over the next five years, reaching 165 billion litres.

The most commonly researched biofuels attracting global attention include bioethanol, biodiesel, biohydrogen and biogas. Particularly, bioethanol from lignocellulosic feedstock. The production of lignocellulosic-based bioethanol involves four phases. These include pretreatment, hydrolysis, fermentation and distillation processes (Deniz et al. 2015; Kim et al. 2016; Sewsynker-Sukai and Gueguim-Kana 2018; Moodley and Gueguim-Kana 2019). Bioethanol has been characterised as an environmentally friendly, efficient fuel with higher oxygen content and higher octane number. In the same vein, fermentative biohydrogen production has unlimited potential, as its production is commonly based on the use of wastes generated from the food, municipal and industrial sectors (Zhao et al. 2013). Biohydrogen is particularly remarkable as a fuel source for the near future, due to its high energy yield (Zhang and Shen 2007; Han et al. 2011; Gueguim-Kana et al. 2013). Additionally, the combustion of hydrogen produces water as its major end-product (Zhang and Shen 2007; Zhao et al. 2013). Biodiesel production using microalgae, animal and vegetable oils represents another potential fuel source. Biodiesel has similar combustion properties to petroleum-based diesel and therefore can be used in current diesel engines (Ma and Hanna 1999). In addition, it is less toxic, renewable and accounts for lower carbon emissions compared to traditional diesel (Schenk et al. 2008).

Despite the merits associated with microbial biofuel production processes from lignocellulosic biomass, microalgae biomass and industrial wastewaters, several challenges plague its application at large scale. Major limitations of biofuel production may include (1) high cost and energy requirement, (2) lack of a suitable substrate that can be utilized by the microbes and (3) low product yield (Aruwajoye et al. 2017; Sewsynker-Sukai and Gueguim-Kana 2018). In an attempt to overcome these challenges, several bioprocess optimization strategies are being investigated such as more cost-effective substrates, low-cost pretreatment procedures, low energy input and the application of biocatalysts such as nano-sized biocatalysts (Kim et al. 2016; Faloye et al. 2014; Sewsynker et al. 2015; Sewsynker-Sukai and Gueguim-Kana 2018).

The use of nano-sized materials has attracted significant attention in the recent decades due to intrinsic properties that promote its application in several biotechnological fields such as bioprocessing, agriculture, biosensor, biopharmaceuticals and medicine (Lateef et al. 2018; Elegbede and Lateef 2019a, b; Sekoai et al. 2019). More specifically, nanotechnology has been employed in various bioprocesses to improve the microbial metabolic activities by their integration with metabolic intermediates and key enzyme activities. Consequently, increased glycolytic rates, cell-substrate affinity, growth rate, mass transfer rate, modulation of oxidation–reduction potential (ORP) enhanced process performance and ultimately high product yields. This chapter focuses on the recent progress made in the application of nanotechnology for the enhancement of biofuel production processes. Furthermore, the impacts of nanoparticles as biocatalytic additives in biodiesel, bioethanol, biohydrogen, biomethane and biogas production are reviewed. Finally, the current challenges and future outlook on the application of nanotechnology in biofuel production processes are highlighted.

2 The Use of Nanotechnology in Biofuel Production

In the recent years, nanoparticles have attracted significant attention due to its distinctive physical and chemical nature that has shown to stimulate microbial and enzymatic biochemical activities in biofuel production. Previous studies have indicated that the stimulatory and catalytic properties of nanoparticles are majorly influenced by their shape, size, concentration, surface coating and operating conditions (Resham and Priyabrata 2008). The use of biocompatible and bioactive nanomeric additives such as NiO, Fe_3O_4 and AgO nanoparticles in biofuel production processes that could significantly improve bioprocess performance and productivity is being implemented (Abdelsalam et al. 2016). Nanoparticle surface properties can regulate stability, solubility and targeting of specific cellular receptors (Howarth et al. 2008). Moreover, the suitability of nanoparticle additives is strongly dependent on other factors such as operating parameters, the type of substrate and additive type (Abdelsalam et al. 2016). For instance, monovalent and functionalized nanomaterial might be used to stimulate or regulate the activities of individual proteins or enzymes (Fu et al. 2004; Howarth et al. 2008). More specifically, for biological applications, the surface property is generally polar, which gives high aqueous solubility that prevents nanoparticles aggregation and ultimately, their performance (Prime and Whitesides 1991; Liu et al. 2010).

Nanomaterials can also play a vital role to improve the thermal and pH stability of enzymes (Pandurangan and Kim 2015). Furthermore, a coated nanoparticle that is multivalent or polymeric confers high stability. Multivalent nanoparticles, bearing multiple targeting groups, can cluster receptors, which could activate cellular signalling pathways and result in stronger attachment and reactivity. Other significant properties of nanoparticles that could improve bioprocess performance include large surface-to-volume ratio, high surface reaction activity, high catalytic efficiency, strong adsorption ability and redox potential that is normally high due to small atomic size (Willner et al. 2006; Ansari and Husain 2012; Abdelsalam et al. 2016). Nanotechnology has been applied in the production of different biofuels such as biodiesel, bioethanol, biohydrogen and biogas, and these are subsequently discussed.

2.1 Biodiesel Production

Biodiesels are alkyl esters of both short- and long-chain fatty acids from either animal fats or vegetable oils. Production of biodiesel from microalgal lipids, vegetable oils and animal oils has attracted interest due to the many benefits of biodiesel; (i) feed stocks are highly abundant since they are regarded as waste, (ii) food security-wise, (iii) reduced production cost, (iv) decreased CO_2 emissions and (v) its degradability (Sekoai et al. 2019). Biodiesel is one clean energy source that is considered a suitable substitute for the conventional petroleum diesel. This is due to its higher energy density, enhanced lubricating property, environmental friendliness, and they can be produced using non-edible oils (Sekoai et al. 2019). Nanotechnology has been employed in biodiesel production to achieve high product yields (Lee et al. 2015). Chen et al. (2018) reported on the impact of supplementing Fe_3O_4/ZnMg(Al)O nanoparticles in biodiesel production using microalgal oil (Table 1). The incorporation of Fe_3O_4/ZnMg(Al)O NPs favoured the biodiesel production, resulting in high yield of 94% (Chen et al. 2018). Similarly, Tahvildari et al. (2015) evaluated the catalytic and synergistic potential of CaO and MgO nanocatalysts on biodiesel

Table 1 Nano-additives employed in biodiesel production processes

NPs	Feedstock	NPs (wt%)	Yield (%)	Cycle	References
Fe_3O_4/ZnMg(Al)O	Microalgal oil	ND	94	7	Chen et al. (2018)
CaO	Microalgal oil	1.7	86	ND	Pandit and Fulekar (2017)
ZnO	Waste cooking oil	1.5	96	ND	Varghese et al. (2017)
$SO4^{2-}/ZrO^2$	Waste cooking oil	2.9	94		Vahida et al. (2018)
Ni-ZnO	Castor oil	11.1	95	ND	Baskar et al. (2018)
CaO	*Bombax ceiba* oil	1.5	96	ND	Hebbar et al. (2018)
Calcite-Au	Sunflower	0.3	98	10	Bet-Moushoul et al. (2016)
sulfamic silica-Fe/Fe_3O_4	Glyceryl trioleate	ND	>95	5	Wang et al. (2015)

ND, Not determined

production from waste cooking oil. The combination of both NPs showed an excellent catalytic efficiency, resulting in biodiesel yield of 98.95%, which was attained with 0.7 g of CaO and 0.5 g of MgO nanoparticles (Tahvildari et al. 2015).

In another study, Dantas et al. (2017) reported on the influence of copper-magnetic nanoferrites on methyl transesterification of soybeans oil and demonstrated up to 85% enhancement in the biodiesel yield. Furthermore, acid-functionalized magnetic nanocatalyst was evaluated for catalytic potential in the transesterification of glyceryl trioleate to biodiesel (Dantas et al. 2017). The acid-functionalized nanoparticles (sulfamic silica-coated crystalline Fe/Fe_3O_4 core–shell magnetic nanoparticles) additives showed notable catalytic activity, with high biodiesel conversion of more than 95% (Wang et al. 2015). Also, Chiang et al. (2015) used functionalized nanoparticles (Fe_3O_4@silica core–shell nanoparticles) for microalgae oil conversion to biodiesel and obtained a high percentage yield (97.1%). The use of calcite-Au nanoparticles for biodiesel production has been evaluated by Bet-Moushoul et al. (2016). These authors recorded a conversion value of 97.5% at 3% calcite-Au nanoparticles catalyst loading. The application of nanoparticles in biodiesel production has showed an enhanced substrate conversion, increased productivity, catalyst recovery and reusability. Various nanoparticles have been employed as a biocatalyst for the enhancement of biodiesel production (Table 1).

2.2 *Bioethanol Production*

Bioethanol is produced when microbes such as *Saccharomyces cerevisiae* or *Zymomonas mobilis* that metabolize fermentable sugars under microaerophilic or anaerobic conditions to produce ethanol and CO_2 (Baeyens et al. 2015). Different attempts have been made to improve the bioethanol fermentation process (Kim et al. 2016). Metallic nanoparticles in fermentation process nutrient formulation have recently been identified as advantageous in stimulating and promoting the bioactivity of ethanol-producing microorganisms and fermentation productivity (Demirel and Scherer 2011; Miazek et al. 2015; Pádrová et al. 2015; Kim et al. 2016). Kim et al. (2014) supplemented six different nanoparticles to enhance bioethanol production in syngas fermentation using *Clostridium ljungdahlii*. The aforementioned study revealed a 34.5, 166.1, and 29.1% increase in the levels of biomass, ethanol, and acetic acid production, respectively due to supplementation with nanoparticles (Kim et al. 2014). These enhancements were ascribed to enhanced gas–liquid mass transfer by methyl and isopropyl hydrophobic surface modification on the silica NPs (Kim et al. 2014). The effects could also be attributed to improved metabolic and enzymatic activities, buffering capacity and oxidation–reduction potential (ORP) of the nano-system.

In addition, Sanusi et al. (2019) assessed the potential of nine different metallic oxide nanoparticles to improve bioethanol production using *S. cerevisiae*. Of these metallic nanooxides, NiO NPs, Fe_3O_4 NPs, CuO NPs, CoO and ZnO NPs inclusions showed desirable catalytic potentials, with Fe_3O_4 NPs inclusion producing the

Table 2 Nanoparticles as biostimulatory catalysts in bioethanol production

	Studies				
Sources	Kim et al. (2014)	Kim and Lee (2016)	Kim and Lee (2016)	Sanusi et al. (2019)	Sanusi et al. (2020)
Strain	*C. ijungdahlii*	*C. ijungdahlii*	*C. ijungdahlii*	*S. cerevisiae*	*S. cerevisiae*
Temperature (°C)	30	37	37	30	37
Nano supplement	0.3 wt% SiO_2–CH_3	0.3 wt% SiO_2–CH_3	0.3 wt% $CoFe_2O_4$@SiO_2–CH_3	0.01 wt% Fe_3O_4	0.05 wt% NiO
Time (h)	24	60	60	24	24
Substrate	0.9 g Fructose	0.9 g Fructose	0.9 g Fructose	20 g Glucose	49 g Glucose
Ethanol (g/L)	–	0.354	0.489	5.21	31.58
Ethanol yield (% improvement)	166.1%	126.9%	213.5%	0.26 g/g	0.66 g/g
Repeated cycle	ND	5-Batch	5-Batch	ND	ND
pH	6.8	6.8	6.8	5	5
Productivity (g/L/h)	ND	0.020	0.027	0.22	1.97

ND, Not determined

highest bioethanol yield of 0.26 g/g (13% improvement). Similarly, the impact of CuO NPs gave 11% increment in biomass proliferation and accumulation, while ZnO NPs inclusion enhanced the process buffering capacity (Sanusi et al. 2019). In another study by Sanusi et al. (2020), NiO nanoparticle inclusion was optimal for bioethanol production, with 18% improvement in bioethanol yield for the glucose fermentation at optimum setpoints of 0.05 wt%, 10 g/L, 4.86 and 32.25 °C for NiO nanoparticles, substrate concentration, pH, and temperature, respectively. These metallic oxides are vital ingredient for the formation of cytochromes and ferroxins (Fd) which are crucial for cell energy metabolism, hence product formation during fermentation. Various reports on the use of nano biocatalysts such as SiO_2–CH_3, $CoFe_2O_4$@SiO_2–CH_3, and metal oxides in bioethanol production are presented in Table 2.

2.3 Biohydrogen Production

Biohydrogen is generated during the microbial fermentation of suitable substrates, and it involves diverse groups of microorganisms (Han et al. 2011; Faloye et al.

2014). These microorganisms are able to utilize organic matter such as lignocellulosic wastes, food wastes, municipal wastes, and animal manure during dark fermentation (Han et al. 2011). The application of nanoparticles for the improvement of dark fermentative biohydrogen production has been reported in several studies as shown in Table 3. Many of these efforts have yielded positive and desirable results. A plausible explanation for increased biohydrogen yields is due to the ability of nanoparticles to improve the process buffering capacity, which in turn stimulates and enhances the activity of hydrogenase enzymes and substrate hydrolysis (Han et al. 2011). The addition of nanoparticles has shown to enhance the hydrogen-producing metabolic pathways such as acetate and butyrate reactions, hydrolysis and acidification processes. For instance, the study by Han et al. (2011) supplemented haematite nanoparticles at 200 mg/L as a bioactive to a bacterial mixed culture (pH 6.0, at 35 °C), and this resulted in a 30% improvement in the hydrogen yield (Table 3). The authors attributed this increase to enhanced metabolic activities that favour hydrogen formation pathways (Han et al. 2011).

Furthermore, Wimonsong and Nitisoravut (2015) investigated the activity of nanoporous activated carbon (NAC) in batch fermentative biohydrogen production (using sucrose-fed anaerobic mixed bacteria culture, at 37 °C). The nanoporous activated carbon resulted in low concentration of butyric acid with 77% absorption capacity, thereby increasing the buffering capacity of the system (Wimonsong and Nitisoravut 2015). This invariably improves the physiological state and fermentative activities of biohydrogen producing microbes that lead to high hydrogen yield in the system (Wimonsong and Nitisoravut 2015). Moreover, the effects of silver nanoparticles concentration (0–200 nmol/L) on glucose-fed and pre-treated mixed bacteria culture in an anaerobic batch reactor were investigated and revealed a 61.45% improvement in fermentative hydrogen production at 20 nmol/L silver nanoparticles (Zhao et al. 2013). In another study, MCM41 nanoparticles with or without a functional group influenced syngas fermentation in a system for biohydrogen production (Haiyang et al. 2010). Findings from the aforementioned study showed that biohydrogen yield was enhanced by a twofold in the present of 0.6 wt% of the MCM41 nanoparticles functionalized with 5% molar ratio of mercaptopropyl group (Haiyang et al. 2010). The enhanced hydrogen yield was due to improved CO-water mass transfer (water–gas shift was biologically and effectively mediated) through the addition of the functionalized MCM41 nanoparticles (Haiyang et al. 2010). Similarly, in a recent study, Vi et al. (2017) optimized fermentative biohydrogen-producing conditions of substrate concentration, pH and $FeSO_4$ nanoparticle concentration. Cumulative biohydrogen yield of 3.50 g/L was achieved at optimized setpoints of 27.63 g/L, 6.10, and 0.063 g/L, for substrate concentration, pH, and $FeSO_4$ NP concentration, respectively. Additional studies on the influence of different nanoparticles on biohydrogen production are summarized in Table 3 (Hydrogen yield: Highest H_2 yield in mol H_2/mol substrate, Nanoporous activated carbon: NAC, Nickel-graphene: Ni-C).

Table 3 Effect of nanobiocatalysts on fermentative biohydrogen yield

Techniques	Substrates	Inoculum	Additives	Highest H2 yield	References
Batch	Starch	*Enterobacter aerogenes*	Fe_2O_3 NPs	192.40	Lin et al. (2016)
Batch	Glucose	Sewage sludge	Gold NPs	2.48	Zhao et al. (2013)
Batch	Sucrose	Activated sludge	NAC	2.60	Wimonsong and Nitisoravut (2015)
Batch	Sucrose	Digested sludge	Fe^0 NPs	4.20	Heguang et al. (2014)
Batch	Glucose	*E. cloacae*	Iron NPs	1.90	Nath et al. (2015)
Batch	Glucose	Activated sludge	Nickel NPs	2.54	Mullai et al. (2013)
Batch	Wastewater	Sewage sludge	Nickel NPs	24.73 H2/g COD	Elreedy et al. (2017)
Batch	Wastewater	Sewage sludge	Ni-C NPs	41.28 H2/g COD	El reedy et al. (2017)
Batch	Glucose	*C. butyricum*	Silver NPs	0.97	Sekoai et al. (2019)
Batch	Glucose	*E. cloacae*	Copper NPs	1.39	Mohanraj et al. (2016)
Batch	Glucose	*C. butyricum*	Copper NPs	1.01	Sekoai et al. (2019)
Batch	Glucose	*C. butyricum*	Pd NPs	0.97	Sekoai et al. (2019)
Batch	Distillery wastewater	Activated sludge	Fe_2O_3 + NiO NPs	8.83 mmol/g-COD	Gadhe et al. (2015a)
Batch	Dairy wastewater	Activated sludge	Fe_2O_3 + NiO NPs	17.2 mmol/g-COD	Gadhe et al. (2015b)
Batch	Glucose	*C. pasteurianum* CH5	TiO_2 NPs	2.1	Sekoai et al. (2019)
Batch	Distillery wastewater	Activated sludge	NiO NPs	6.73 mmol/g-COD	Gadhe et al. (2015a)
Batch	Dairy wastewater	Activated sludge	NiO NPs	15.7 mmol/g-COD	Gadhe et al. (2015b)
Batch	Glucose and starch	Activated sludge	NiO NPs	1.3	Sekoai et al. (2019)
Batch	Palm oil effluent	*Bacillus anthracis*	NiO NPs	560 mL/g-COD	Mishra et al. (2018)
Batch	Glucose	Activated sludge	Fe_3O_4 NPs	1.53	Sekoai et al. (2019)
Continuous	Sucrose wastewater	Sewage sludge	Fe_2O_3 NPs	300 mL/g-sucrose	Salem et al. (2017)

(continued)

Table 3 (continued)

Techniques	Substrates	Inoculum	Additives	Highest H2 yield	References
Batch	Distillery wastewater	Activated sludge	Fe_2O_3 NPs	7.85 mmol/g-COD	Gadhe et al. (2015a)
Batch	Dairy wastewater	Activated sludge	Fe_2O_3 NPs	16.75 mmol/g-COD	Gadhe et al. (2015b)
Batch	Glucose and starch	Activated sludge	Fe_2O_3 NPs	1.92	Sekoai et al. (2019)
Batch	Cassava starch	*E. aerogenes* ATCC13408	Fe_2O_3 NPs	124.3 mL/g-starch	Lin et al. (2016)
Batch	Glucose	*E. aerogenes* ATCC13408	Fe_2O_3 NPs	192.4 mL/g-glucose	Lin et al. (2016)
Batch	Palm oil effluent	*B. anthracis*	CoO NPs	0.487 L/g-COD	Mishra et al. (2018)

2.4 Biogas Production

Anaerobic digestion is one of the most important techniques used to convert organic waste biomass into renewable energy in the form of biogas (Abdelsalam et al. 2016). The anaerobic digestion process is relatively slow and is carried out by a mixed consortium of microorganisms. Anaerobic digestion depends on various process parameters such as pH, temperature, hydraulic retention time and carbon/nitrogen (C/N) ratio, among others (Abdelsalam et al. 2016). This process consists of a series of microbial processes that convert organic matter to biogas, which could take place under psychrophilic (<20 °C), mesophilic (25–40 °C) or thermophilic (50–65 °C) conditions (Abdelsalam et al. 2017a). Biogas production from organic matter mainly depends on the contents of the substrates that can be degraded to CH_4 and CO_2. Substrate composition, biodegradability and nutrients are key factors for biogas yield.

The use of catalytic, stimulatory and biochemical nanoparticles additives in anaerobic digestion processes could improve biogas production significantly and have previously recorded promising results (Table 4). These positive outcomes have been related to effective electron transfer (oxidation–reduction potential), cofactor of key enzymes and improved hydrolysis of organic matter. For example, the influence of zero valent iron (ZVI) and Fe_2O_3 nanoparticles on biogas production using activated waste sludge was reported by Wang et al. (2016). These authors indicted that ZVI NPs (10 mg/g TSS) and Fe_2O_3 NPs (100 mg/g TSS) enhanced the biogas production by 2.20-fold and 2.17-fold, respectively (Wang et al. 2016). Their results demonstrate that nanoparticles inclusion has a positive effect on the activity of methanogenic archaea. Similarly, Su et al. (2013) assessed the effects of 0.1 wt% ZVI NPs on biogas production and methane production using activated waste sludge, and these resulted in 30% and 13.2% increase in concentration, respectively. Furthermore, Abdelsalam et al. (2017a) reported that the addition of Co NPs (1 mg/L) notably increased biogas and methane volume by 1.64 and 1.86 times, respectively. The same authors also observed substantial improvements in the biogas and methane volume by 1.74 and 2.01 times, respectively, when 2 mg/L Ni NPs were included in the anaerobic digestion of livestock slurry. On the other hand, the study by Abdelsalam et al. (2016) varied Co, Ni, Fe and Fe_3O_4 nanoparticle concentrations (1, 2, 20 and 20 mg/L) to assess their impacts on biogas and methane production from anaerobic digestion of livestock slurry. The aforementioned study revealed that these NPs enhanced the biogas and methane production.

(Abdelsalam et al. 2016). Similarly, a study by Gonzalez-Estrella et al. (2013) revealed that Fe_3O_4 and ZVI nanoparticles enhanced biogas production by 66% and 45%, respectively. This improved process productivity of nano-base anaerobic digestion can be ascribed to the proliferation of methanogens resulting from the promotion of direct interspecies electron transfer by nanoparticles (Park et al. 2018). The inclusion of nanometric materials also enhances the formation of essential biogas pathway intermediates such as acetate, butyrate, formate and hydrogen, while reducing others like ethanol (Sekoai et al. 2019). In addition to the above-mentioned impacts of NPs in anaerobic digestion, other catalytic effects include

Table 4 Nanocatalysts in anaerobic digestion for biogas production

NPs	Inoculum	Biogas (mL gas g^{-1}VS)	Methane[a]	References
Ni	Livestock slurry	615	362	Abdelsalam et al. (2017a)
Co	Livestock slurry	579	333	Abdelsalam et al. (2017a)
Ni	Livestock manure	512	304	Abdelsalam et al. (2016)
Fe_3O_4	Livestock manure	496	303	Abdelsalam et al. (2016)
Fe	Livestock manure	424	234	Abdelsalam et al. (2016)
Co	Livestock manure	491	281	Abdelsalam et al. (2016)
Nzvi	Livestock manure	513	286	Abdelsalam et al. (2017b)
Fe_3O_4	Livestock manure	584	352	Abdelsalam et al. (2017b)
Nzvi	*Dehalococcoides* sp.	ND	275 μmol	Xiu et al. (2010)
Nzvi	Waste-activated sludge	ND	217	Wang et al. (2016)
Fe_2O_3	Waste-activated sludge	ND	212	Wang et al. (2016)
CuO	Anaerobicgranular sludge	ND	6 g COD CH_4 L^{-1} d^{-1}	Otero-Gonzalez et al. (2014)

ND, Not determined; [a]mL gas g^{-1}VS except otherwise stated

increased substrate conversion, which can be attributed to the large surface area-to-volume ratio provided by nanoparticles for microbes and enzymes to bind in active sites, thereby promoting their biochemical (such as complexation aggregation) and metabolic processes. Furthermore, nano-based anaerobic digestion provides an interdependent process condition that permits the oxidation state of nanoparticles to be altered by microorganisms acting as catalytic agents. A few studies showing the impacts of different nanoparticles on the biogas production is summarized in Table 4.

3 Nanoparticles as Immobilization Matrix for Biofuel Production

Common immobilization approaches (cells or enzymes) employ materials with sizes above the nanometre (10^{-9}) range (Nikzad et al. 2015; Wiboon et al. 2012; Charlimagne et al. 2015; Martins et al. 2013; Zhaohui et al. 2016). Nevertheless, nanoparticles have been well established as an immobilization matrix and have desirable biotechnological advantages. These advantages include large surface to volume ratio, high loading of the specific molecules targeted, easy separation from the reaction using an external magnetic field and strong adsorption ability (Han et al. 2011). In addition, multivalent nanoparticles bearing multiple targeting groups can give stronger anchoring for cells and enzymes (Willner et al. 2006; Han et al. 2011; Ansari and Husain 2012). Nanoparticles can also play a vital role for the improvement of thermal and pH stability of microbial cells and enzymes (Pandurangan and Kim 2015). For instance, magnetic nanoparticles are suitable for lipase immobilization, due to their ability to form nanocrystals. This phenomenon increases thermal stability and reusability of the enzyme. Furthermore, the formation of nanocrystals tends to clump together, increasing the biocatalyst surface area.

The incorporation of nano-immobilized enzymes/microbes is beneficial for bioprocessing, since they enhance process costs, conversion efficiency and the overall process performance (Sekoai et al. 2019). Limited studies exist on the application of nanoparticles as an immobilization matrix in biofuel production (Table 5).

The study by Tran et al. (2012) immobilized lipase on magnetic nanoparticles (Fe_3O_4) coated with silica (Fe_3O_4–SiO_2) and demonstrated an increased tolerance to high methanol to oil ratio (67:93) in an *in-situ* transesterification of *Chlorella vulgaris* lipid. Furthermore, the immobilized lipase could withstand high water content of 71% with a biodiesel conversion value of 97.3 wt% oil (Tran et al. 2012). Likewise, Ivanova et al. (2011) assessed the effect of immobilized *S. cerevisiae* on bioethanol productivity. The immobilized cells substantially enhanced bioethanol productivity. Moreover, the entrapped cells were employed successfully over 42 days without a significant loss of bioethanol productivity. Similarly, the study by Cherian et al. (2015) reported on cellulase immobilized on MnO_2 nanoparticles and found the immobilized enzyme to be more thermostable at 70 °C. Also, the reusability (after five cycles, retained 60% activity) of cellulase was significantly increased after immobilization. Studies showing the influence of various nanoparticles as immobilization agents for biofuel production are summarized in Table 5.

4 Present Challenges and Future Perspectives on Biofuel Production

Fermentative production of biofuel is achieved by the bioconversion fermentable sugars contained in organic waste through pre-treatment and fermentation. However,

Table 5 Nanoparticles immobilization matrix employed in biofuel production

NPs matrix	Cell/enzymes	Process	Biofuel produced	References
MnO_2	Cellulase	Sugar hydrolysis	Bioethanol	Cherian et al. (2015)
SiO_2	β-galactosidase	Sugar hydrolysis	Bioethanol	Beniwal et al. (2018)
Nanofiber	β-Glucosidase	Sugar hydrolysis	Bioethanol	Lee et al. (2010)
Fe_3O_4	α-amylase	Starch liquefaction	Bioethanol	Ivanova et al. (2011)
Fe_3O_4	Glucoamylase	Sugar saccharification	Bioethanol	Ivanova et al. (2011)
Fe_3O_4	*S. cerevisiae*	Fermentation	Bioethanol	Ivanova et al. (2011)
Fe_3O_4–SiO_2	Lipase	Lipid transesterification	Biodiesel	Tran et al. (2012)
SiO_2	Lipase	Lipid transesterification	Biodiesel	Babaki et al. (2016)
Fe_3O_4–SiO_2	Lipase	Lipid transesterification	Biodiesel	Thangaraj et al. (2016)
Modified Fe_3O_4	Lipase	Lipid transesterification	Biodiesel	Raita et al. (2015)
Modified Fe_3O_4	Lipase	Lipid transesterification	Biodiesel	Zhang et al. (2016)
Magnetic NPs	*Candida antarctica* lipase	Lipid transesterification	Biodiesel	Mehrasbi et al. (2017)
carbon nanotubes	Lipase	Lipid transesterification	Biodiesel	Fan et al. (2016)

the wide application of fermentation has always been limited due to the need for suitable feedstock, inefficient pre-treatment regimes and low process yield. To address some of these challenges in biofuel production, catalytic nano-sized materials are been employed (Han et al. 2011; Ban and Paul 2014; Abdelsalam et al. 2016, 2017a; Sanusi et al. 2019, 2020). In relation to microbial fermentation, there are several processes that nanoparticles have shown to influence: (i) it could serve as an essential trace element for fermenting microbes and enhance metabolic activities leading to improved cell growth and productivity; (ii) activities of liquefaction, saccharification and fermenting metalloenzymes could be enhanced by nanoparticles; (iii) reduction in the oxidation–reduction potential of the process; (iv) improvement of buffering capacity; (v) reduction in organic inhibitors accumulation through chelating potential; and (vi) efficient immobilization matrix due to large surface area and modifiable surfaces (Han et al. 2011; Ban and Paul 2014; Abdelsalam et al. 2016, 2017a; Sekoai et al. 2019).

Although the application of nanoparticles in biofuel processing has demonstrated promise, further research is necessary to determine the most efficient method of application. Despite the merits with nanoparticle supplementation in biofuel processes, several challenges still plague its implementation. Some of these challenges include the cost of synthesizing the nanoparticles, biocompatibility of nanomaterials and their mechanism of action, reusability and its environmental effects. Additionally, achieving a cost-effective biofuel production process will require thorough considerations of the economic implication involved in the synthesis of nanomaterials. Similarly, nanoparticles with precise biocatalytic properties should be synthesized in order to improve specific biofuel production process. Also, biogenic synthesis of nanoparticles using retinue of biological resources can lead to low-cost production of environmentally benign nanoparticles that may be useful for applications in biofuel research and development (Adelere and Lateef 2016; Lateef et al. 2016).

5 Conclusion

Nanomaterials could play a crucial role towards the advancement of biofuel production owing to their distinctive and beneficial properties as described in this chapter. The inclusion of nanoparticles in biofuel process development has the potential to ensure an economically viable biofuel production by enhancing various stages of product formation, thus reducing the negative impact of fossil fuel consumption. Additionally, these biocatalysts are environmentally friendly as they can be recovered and reused. However, in order to fast-track their application for biofuel production, many hindrances needs to be addressed. These include (i) synthesizing nanoparticles that are less expensive, (ii) using nanoparticles that are biocompatible or non-toxic to microorganisms, (iii) use of nanoparticles of distinctive and specific function and (iv) using nanoparticles that are eco-friendly.

References

Abdelsalam E, Samer M, Attia YA, Abdel-Hadi MA, Hassan HE, Badr Y (2016) Comparison of nanoparticles effects on biogas and methane production from anaerobic digestion of cattle dung slurry. Renew Energy 87:592–598. https://doi.org/10.1016/j.renene.2015.10.053

Abdelsalam E, Samer M, Attia YA, Abdel-Hadi MA, Hassan HE, Badr Y (2017a) Effects of Co and Ni nanoparticles on biogas and methane production from anaerobic digestion of slurry. Energy Convers Manag 141:108–119. https://doi.org/10.1016/j.enconman.2016.05.051

Abdelsalam E, Samer M, Attia YA, Abdel-Hadi MA, Hassan HE, Badr Y (2017b) Influence of zero valent iron nanoparticles and magnetic iron oxide nanoparticles on biogas and methane production from anaerobic digestion of manure. Energy 120:842–853. https://doi.org/10.1016/j.energy.2016.11.137

Adelere IA, Lateef A (2016) A novel approach to the green synthesis of metallic nanoparticles: the use of agro-wastes, enzymes and pigments. Nanotechnol Rev 5(6):567–587. https://doi.org/10.1515/ntrev-2016-0024

Adeoye AO, Lateef A, Gueguim-Kana EB (2015) Optimization of citric acid production using a mutant strain of *Aspergillus niger* on cassava peel substrate. Biocatal Agric Biotechnol 4(4):568–574. https://doi.org/10.1016/j.bcab.2015.08.004

Ansari SA, Husain Q (2012) Potential applications of enzymes immobilized on/in nano materials: a review. Biotechnol Adv 30:512–523. https://doi.org/10.1016/j.biotechadv.2011.09.005

Aruwajoye GS, Faloye FD, Gueguim Kana EB (2017) Soaking assisted thermal pretreatment of cassava peels wastes for fermentable sugar production: process modelling and optimization. Energy Convers Manage 150:558–566. https://doi.org/10.1016/j.ijhydene.2014.01.163

Babaki M, Yousefi M, Habibi Z, Mohammadi M, Yousefi P, Mohammadi J (2016) Enzymatic production of biodiesel using lipases immobilized on silica nanoparticlesas as highly reusable biocatalysts: effect of water, t-butanol and blue silicagel contents. Renew Energy 91:196–206. https://doi.org/10.1016/j.renene.2016.01.053

Baeyens J, Kang Q, Appels L, Dewil R, Lv Y, Tan T (2015) Challenges and opportunities in improving the production of bio-ethanol. Prog Energy Combust 47:60–88. https://doi.org/10.1016/j.pecs.2014.10.003

Ban DK, Paul S (2014) Zinc oxide nanoparticles modulates the production of β-glucosidase and protects its functional state under alcoholic condition in *Saccharomyces cerevisiae*. Appl Biochem Biotechnol 173:155–166. https://doi.org/10.1007/s12010-014-0825-2

Baskar G, Selvakumari A, Aiswarya R (2018) Biodiesel production from castor oil using heterogeneous Ni doped ZnO nanocatalyst. Bioresour Technol 250:793–798. https://doi.org/10.1016/j.biortech.2017.12.010

Beniwal A, Saini P, Kokkiligadda A, Vij S (2018) Use of silicon dioxide nanoparticles for β-galactosidase immobilization and modulated ethanol production by co-immobilized *K. marxianus* and *S. cerevisiae* in deproteinized cheese whey. LWT Food Sci Technol 87:553–561. https://doi.org/10.1016/j.lwt.2017.09.028

Bet-Moushoul E, Farhadi K, Mansourpanah Y, Nikbakht A, Molaei R, Forough M (2016) Application of CaO-based/Au nanoparticles as heterogeneous nanocatalysts inbiodiesel production. Fuel 164:119–127. https://doi.org/10.1016/j.fuel.2015.09.067

BP (2016) British petroleum (BP) statistical review of world energy 2016.

Charlimagne MM, Emerson RD, Lawrence VS, Jay RTA, Rizalinda LD (2015) Continuous bioethanol production using *Saccharomyces cerevisiae* cells immobilized in nata de coco (Biocellulose). In: 2nd International conference on environment and industrial innovation IPCBEE (2012), vol 35

Chen Y, Liu T, He H, Liang H (2018) Fe3O4/ZnMg(Al)O magnetic nanoparticles for efficient biodiesel production. Appl Organo met Chem 32:1–10. https://doi.org/10.1002/aoc.4330

Cherian E, Dharmendirakumar M, Baskar G (2015). Immobilization of cellulase onto MnO_2 nanoparticles for bioethanol production by enhanced hydrolysis of agricultural waste. Chinese J Catal 36:1223–1229. https://doi.org/10.1016/S1872-2067(15)60906-8.

Chiang Y, Dutta S, Chen C, Huang Y, Lin K, Wu J (2015). Functionalized Fe3O4@silica core-shell nanoparticles as microalgae harvester and catalyst for biodiesel production. ChemSusChem 8:789–94. https://doi.org/10.1002/cssc.201402996

Chohan NA, Aruwajoye GS, Sewsynker-Sukai Y, Gueguim Kana EB (2020) Valorisation of potato peel wastes for bioethanol production using simultaneous saccharification and fermentation: process optimization and kinetic assessment. Renew Energy 146:1031–1040. https://doi.org/10.1016/j.renene.2019.07.042

Dantas J, Leal E, Mapossa A, Cornejo D, Costa A (2017) Magnetic nanocatalysts of Ni0.5Zn0.5Fe2O4 doped with Cu and performance evaluation in transesterification reaction for biodiesel production. Fuel 191:463–471. https://doi.org/10.1016/j.fuel.2016.11.107

Demirel and Scherer (2011) Trace element requirements of agricultural biogas digesters during biological conversion of renewable biomass to methane. Biomass Bioenerg 35:992–998. https://doi.org/10.1016/j.biombioe.2010.12.022

Deniz I, Imamoglu E, Vardar Sukan F (2015) Evaluation of scale-up parameters of bioethanol production from *Escherichia coli* KO11. Turkish J Biochem 40(1):74–80. https://doi.org/10.5505/tjb.2015.33603

Elegbede JA, Lateef A (2018) Valorization of corn-cob by fungal isolates for production of xylanase in submerged and solid state fermentation media and potential biotechnological applications. Waste Biomass Valor 9(8):1273-1287. https://doi.org/10.1007/s12649-017-9932-y

Elegbede JA, Lateef A (2019a) Green synthesis of silver (Ag), gold (Au) and silver-gold (Ag-Au) alloy nanoparticles: a review on recent advances, trends and biomedical applications. In: Verma DK, Goyal MR, Suleria HAR (eds) Nanotechnology and nanomaterial applications in food, health and biomedical sciences. Apple Academic Press Inc./CRC Press, Taylor and Francis Group, Oakville, Ontario, Canada, pp 3-89. ISBN 978-1-77188-764-9. https://doi.org/10.1201/9780429425660-1

Elegbede JA, Lateef A (2019) Green nanotechnology in Nigeria: the research landscape, challenges and prospects. Ann Sci Technol 2:6–38. https://doi.org/10.2478/ast-2019-0008

Elreedy A, Ibrahim E, Hassan N, El-Dissouky A, Fujii M, Yoshimura C (2017) Nickel graphene nanocomposite as a novel supplement for enhancement of biohydrogen production from industrial wastewater containing mono-ethylene glycol. Energy Convers Manage 140:133–144. https://doi.org/10.1016/j.enconman.2017.02.080

Faloye FD, Gueguim-Kana EG, Schmidt S (2014) Optimization of biohydrogen inoculum development via a hybrid pH and microwave treatment technique—semi pilot scale production assessment. Int J Hydrogen Energy 39:5607–5616. https://doi.org/10.1016/j.enconman.2017.08.046

Fan Y, Wu G, Su F, Li K, Xu L, Han X (2016) Lipase oriented-immobilized on dendrimer-coated magnetic multi-walled carbon nanotubes toward catalysing biodiesel production from waste vegetable oil. Fuel 178:172–178. https://doi.org/10.1016/j.fuel.2016.03.071

Fazal T, Mushtaq A, Rehman F, Khan AU, Rashid N, Farooq W, Rehman MSU, Xu J (2018) Bioremediation of textile wastewater and successive biodiesel production using microalgae. Renew Sust Energy Rev 82:3107–3126. https://doi.org/10.1016/j.rser.2017.10.029

Fu A, Micheel CM, Cha J, Chang H, Yang H, Alivisatos AP (2004) Discrete nanostructures of quantum dots/Au with DNA. J Am Chem Soc 126(35):10832–10833. https://doi.org/10.1021/ja046747x

Gadhe A, Sonawane SS, Varma MN (2015a) Influence of nickel and hematite nanoparticle powder on the production of biohydrogen from complex distillery wastewater in batch fermentation. Int J Hydrogen Energy 40(34):10734–10743. https://doi.org/10.1016/j.ijhydene.2015.05.198

Gadhe A, Sonawane SS, Varma MN (2015b) Enhancement effect of hematite and nickel nanoparticles on biohydrogen production from dairy wastewater. Int J Hydrogen Energy 40(13):4502–4511. https://doi.org/10.1016/j.ijhydene.2015.02.046

Ge S, Qiu S, Tremblay D, Viner K, Champagne P, Jessop PG (2018) Centrate wastewater treatment with *Chlorella vulgaris*: simultaneous enhancement of nutrient removal, biomass and lipid production. Chem Eng J 342:310–320. https://doi.org/10.1016/j.cej.2018.02.058

Gonzalez-Estrella J, Sierra-Alvarez R, Field J (2013) Toxicity assessment of inorganic nanoparticles to acetoclastic and hydrogenotrophic methanogenic activity in anaerobic granular sludge. J Hazard Mater 260:278–285. https://doi.org/10.1016/j.jhazmat.2013.05.029

Gueguim-Kana EB, Oloke JK, Lateef A, Adesiyan MO (2012) Modeling and optimization of biogas production on saw dust and other co-substrates using artificial neural network and genetic algorithm. Renew Energy 46:276–281. https://doi.org/10.1016/j.renene.2012.03.027

Gueguim-Kana EB, Schmidt S, Azanfack Kenfack RHA (2013) Web-enabled software for real-time biogas fermentation monitoring-assessment of dark fermentations for correlations between medium conductivity and biohydrogen evolution. Int J Hydrogen Energy 38:10235–10244. https://doi.org/10.1016/j.ijhydene.2013.06.019

Haiyang Z, Brent HS, Dong WC, Theodore JH (2010) Effect of functionalized MCM41 nanoparticles on syngas fermentation. Biomass Bioenerg 34:1624–1627. https://doi.org/10.1016/j.biombioe.2010.06.008

Han H, Cui M, Wei L, Yang H, Shen J (2011) Enhancement effect of hematite nanoparticles on fermentative hydrogen production. Bioresour Technol 102:7903–7909. https://doi.org/10.1016/j.biortech.2011.05.089

Hebbar H, Math M, Yatish K (2018) Optimization and kinetic study of CaO nano-particles catalyzed biodiesel production from *Bombax ceiba* oil. Energy 143:25–34. https://doi.org/10.1016/j.energy.2017.10.118

Heguang Z, Peter S, Parker W (2014) Enhanced dark fermentative hydrogen production under the effect of zero-valent iron shavings. Int J Hydrogen Energy 39:19331–19336. https://doi.org/10.1016/j.ijhydene.2014.06.055

Hirsch RL, Bezdek RH, Wendling RM (2005) Peaking oil production: sooner rather than later? Issues Sci Technol 21(3):25–30. https://www.jstor.org/stable/43312533/

Howarth M, Liu W, Puthenveetil S, Zheng Y, Marshall LF, Schmidt MM, Wittrup KD, Bawendi MG, Ting AY (2008) Monovalent, reduced-size quantum dots for imaging receptors on living cells. Nat Method 5(5):397–399. https://doi.org/10.1038/nmeth.1206

Ivanova V, Petrova P, Hristov J (2011) Application in the ethanol fermentation of immobilized yeast cells in matrix of alginate/magnetic nanoparticles, on chitosan magnetite microparticles and cellulose-coated magnetic nanoparticles. Int Rev Chem Eng 3:289–299. https://arxiv.org/ftp/arxiv/papers/1105/1105.0619

Khawla BJ, Sameh M, Imen G, Donyes F, Dhouha G, Raoudha EG, Oumèma N-E (2014) Potato peel as feedstock for bioethanol production: a comparison of acidic and enzymatic hydrolysis. Ind Crops Prod 52:144–149. https://doi.org/10.1016/j.indcrop.2013.10.025

Kim Y, Lee H (2016) Use of magnetic nanoparticles to enhance bioethanol production in syngas fermentation. Bioresour Technol 204:139–144. https://doi.org/10.1007/s10562-019-02796-6

Kim Y, Park SE, Lee H, Yun JY (2014) Enhancement of bioethanol production in syngas fermentation with *Clostridium ljungdahii* using nanoparticles. Bioresour Technol 159:446–450. https://doi.org/10.1016/j.biortech.2014.03.046

Lateef A, Oloke JK, Gueguim-Kana EB, Oyeniyi SO, Onifade OR, Oyeleye AO, Oladosu OC, Oyelami AO (2008) Improving the quality of agro-wastes by solid state fermentation: enhanced antioxidant activities and nutritional qualities. World J Microbiol Biotechnol 24(10):2369–2374. https://doi.org/10.1007/s11274-008-9749-8

Lateef A, Ojo SA, Elegbede JA (2016) The emerging roles of arthropods and their metabolites in the green synthesis of metallic nanoparticles. Nanotechnol Rev 5(6):601–622. https://doi.org/10.1515/ntrev-2016-0049

Lateef A, Ojo SA, Elegbede JA, Akinola PO, Akanni EO (2018) Nanomedical applications of nanoparticles for blood coagulation disorders. In: Dasgupta N, Ranjan S, Lichtfouse E (eds) Environmental nanotechnology, vol 1. Springer International Publishing AG, Cham, Switzerland, pp 243-277. ISBN 978-3-319-76089-6. https://doi.org/10.1007/978-3-319-76090-2_8

Lee S, Jin L, Kim J, Han S, Na H, Hyeon T (2010) β-Glucosidase coating on polymer nanofibers for improved cellulosic ethanol production. Bioprocess Biosyst Eng 33:141–147. https://doi.org/10.1007/s00449-009-0386-x

Lee Y, Lee K, Oh Y (2015) Recent nanoparticle engineering advances in microalgal cultivation and harvesting processes of biodiesel production: a review. Bioresour Technol 184:63–72. https://doi.org/10.1016/j.biortech.2014.10.145

Lin R, Cheng J, Ding L, Song W, Liu M, Zhou J (2016) Enhanced dark hydrogen fermentation by addition of ferric oxide nanoparticles using *Enterobacter aerogenes*. Bioresour Technol 207:213–219. https://doi.org/10.1016/j.biortech.2016.02.009

Liu W, Greytak AB, Lee J, Wong CR, Park J, Marshall LF, Jiang W, Curtin PN, Ting AY, Nocera DG, Fukumura D, Jain RK, Bawendi MG (2010) Compact biocompatible quantum dots via RAFT-mediated synthesis of imidazole-based random copolymer ligand. J Am Chem Soc 132(2):472–478. https://doi.org/10.1021/ja908137d

Ma F, Hanna MA (1999) Biodiesel production: a review. Bioresour Technol 70:1–15. https://doi.org/10.1016/S0960-8524(99)00025-5

Martins SCS, Martins CM, Fiúza LM, Cidrão G, Santaella ST (2013) Immobilization of microbial cells: a promising tool for treatment of toxic pollutants in industrial wastewater. Afr J Biotechnol 2(28):4412–4418. https://doi.org/10.5897/AJB12.2677

Mehrasbi M, Mohammadi J, Peyda M, Mohammad M (2017) Covalent immobilization of *Candida antarctica* lipase on core-shell magnetic nanoparticles for production of biodiesel from waste cooking oil. Renew Energ 101:596–602. https://doi.org/10.1016/j.renene.2016.09.022

Miazek K, Iwanek W, Remacle C, Richel A, Goffin D (2015) Effect of metals, metalloids and metallic nanoparticles on microalgae growth and industrial product biosynthesis: a review. Int J Mol Sci 16:23929–23969. https://doi.org/10.3390/ijms161023929

Mishra P, Thakur S, Mahapatra DM, AbWahid Z, Liu H, Singh L (2018) Impacts of nano-metal oxides on hydrogen production in anaerobic digestion of palm oil mill effluent—a novel approach. Int J Hydrogen Energ 43(5):2666–2676. https://doi.org/10.1016/j.ijhydene.2017.12.108

Mohanraj S, Anbalagan K, Rajaguru P, Pugalenthi V (2016) Effects of phytogenic copper nanoparticles on fermentative hydrogen production by *Enterobacter cloacae* and *Clostridium acetobutylicum*. Int J Hydrogen Energ 41:10639–10645. https://doi.org/10.1016/j.ijhydene.2016.04.197

Moodley P, Gueguim-Kana EB (2019) Bioethanol production from sugarcane leaf waste: effect of various optimized pretreatments and fermentation conditions on process kinetics. Biotechnol Rep 22:e00329. https://doi.org/10.1016/j.btre.2019.e00329

Mullai P, Yogeswari MK, Sridevi K (2013) Optimisation and enhancement of biohydrogen production using nickel nanoparticles—a novel approach. Bioresour Technol 141:212–219. https://doi.org/10.1016/j.biortech.2013.03.082

Nath D, Manhar A, Gupta K, Saikia D, Das S, Mandal M (2015) Phytosynthesized iron nanoparticles: effects on fermentative hydrogen production by *Enterobacter cloacae* DH-89. Bull Mater Sci 38:1533–1538. https://doi.org/10.1007/s12034-015-0974-0

Nikzad M, Movaghernajed K, Talebnia F, Mighani M (2015) Modeling of alkali pretreatment of rice husk using response surface methodology and artificial neural network. Chem Eng Commu 202(6):728–738. https://doi.org/10.1080/00986445.2013.871707

Otero-Gonzalez L, Field J, Sierra-Alvarez R (2014) Inhibition of anaerobic wastewater treatment after long-term exposure to low levels of CuO nanoparticles. Water Res 58:160–168. https://doi.org/10.1016/j.watres.2014.03.067

Pádrová K, Lukavský J, Nedbalová L, Čejková A, Cajthaml T, Sigler K, Vítová M, Řezanka T (2015) Trace concentrations of iron nanoparticles cause over production of biomass and lipids during cultivation of cyanobacteria and microalgae. J Appl Phycol 27:1443–1451. https://doi.org/10.1007/s10811-014-0477-1

Pandit P, Fulekar M (2017) Egg shell waste as heterogeneous nanocatalyst for biodiesel production: optimized by response surface methodology. J Environ Manage 198:319–329. https://doi.org/10.1016/j.jenvman.2017.04.100

Pandurangan M, Kim DH (2015) ZnO nanoparticles augment ALT, AST, ALP and LDH expressions in C2C12 cells. Saudi J Biol Sci 22:679–684. https://doi.org/10.1016/j.sjbs.2015.03.013

Park J-H, Kang H-J, Park K-H, Park H-D (2018) Direct interspecies electron transfer via conductive materials: a perspective for anaerobic digestion applications. Bioresour Technol 254:300–311. https://doi.org/10.1016/j.biortech.2018.01.095

Prime KL, Whitesides GM (1991) Self-assembled organic monolayers: model systems for studying adsorption of proteins at surfaces. Science 252(5009):1164–1167. https://www.jstor.org/stable/2876292

Raita M, Arnthong J, Champreda V, Laosiripojana N (2015) Modification of magnetic nanoparticle lipase designs for biodiesel production from palm oil. Fuel Process Technol 134:189–197. https://doi.org/10.1016/j.fuproc.2015.01.032

Renewable Fuels Association (2018)

Resham B, Priyabrata M (2008) Biological properties of "naked" metal nanoparticles. Adv Drug Deliv Rev 60:1289–1306. https://doi.org/10.1016/j.addr.2008.03.013

Salem AH, Mietzel T, Brunstermann R, Widmann R (2017) Effect of cell immobilization, hematite nanoparticles and formation of hydrogen-producing granules onbiohydrogen production from sucrose wastewater. Int J Hydrogen Energy 42(40):25225–25233. https://doi.org/10.1016/j.ijhydene.2017.08.060

Sanusi AI, Faloye FD, Gueguim-Kana EB (2019) Impact of various metallic oxide nanoparticles on ethanol production by *Saccharomyces cerevisiae* BY4743: screening, kinetic study and validation on potato waste. Catal Lett 149(7):2015–2031. https://doi.org/10.1007/s10562-019-02796-6

Sanusi IA, Suinyuy TN, Lateef A, Gueguim Kana EB (2020) Effect of nickel oxide nanoparticles on bioethanol production: process optimization, kinetic and metabolic studies. Process Biochem 92:386–400. https://doi.org/10.1007/s10562-019-02796-6

Schenk PM, Thomas-Hall SR, Stephens E, Marx UC, Mussgnug JH, Posten C, Kruse O, Hankamer B (2008) Second generation biofules: high-efficiency microalgae for biodiesel production. Bioenerg Res 1:20–43. https://doi.org/10.1007/s12155-008-9008-8

Sebayang AH, Masjuki HH, Ong HC DS, Silitonga AS, Kusumo F, Milano J (2017) Optimization of bioethanol production from sorghum grains using artificial neural networks integrated with ant colony. Ind Crops Prod 97:146–155. https://doi.org/10.1016/j.indcrop.2016.11.064

Sekoai PT, Ouma CNM, du Preez SP, Modisha P, Engelbrecht N, Bessarabov DG, Ghimire A (2019) Application of nanoparticles in biofuels: an overview. Fuel 237:380–397. https://doi.org/10.1016/j.fuel.2018.10.030

Sewsynker Y, Gueguim-Kana EB, Lateef A (2015) Modelling of biohydrogen generation in microbial electrolysis cells (MECs) using a committee of artificial neural networks (ANNs). Biotechnol Biotechnol Equip 29(6):1208–1215. https://doi.org/10.1080/13102818.2015.1062732

Sewsynker-Sukai Y, Gueguim-Kana EB (2018) Simultaneous saccharification and bioethanol production from corn cobs: process optimization and kinetic studies. Bioresour Technol 262:32–41. https://doi.org/10.1016/j.biortech.2018.04.056

Su L, Shi X, Guo G, Zhao A, Zhao Y (2013) Stabilization of sewage sludge in the presence of nanoscale zero-valent iron (nZVI): abatement of odor and improvement of biogas production. J Mater Cycles Waste Manage 15:461–468. https://doi.org/10.1007/s10163-013-0150-9

Tahvildari K, Anaraki Y, Fazaeli R, Mirpanji S, Delrish E (2015) The study of CaO and MgO heterogenic nano-catalyst coupling on transesterification reaction efficacy in the production of biodiesel from recycled cooking oil. J Environ Health Sci Eng 13:73–81. https://doi.org/10.1186/s40201-015-0226-7

Thangaraj B, Jia Z, Dai L, Liu D, Du W (2016) Effect of silica coating on Fe_3O_4 magnetic nanoparticles for lipase immobilization and their application for biodiesel production. Arab J Chem https://doi.org/10.1016/j.arabjc.2016.09.004

Tran D, Chen C, Chang J (2012) Immobilization of *Burkholderia* sp. lipase on a ferric silica nanocomposite for biodiesel production. J Biotechnol 158:112–119. https://doi.org/10.1016/j.jbiotec.2012.01.018

Vahida B, Saghatoleslami N, Nayebzadeh H, Toghiani J (2018) Effect of alumina loading on the properties and activity of SO_4^{2-}/ZrO_2 for biodiesel production: process optimization via response surface methodology. J Taiwan Inst Chem Eng 83:115–123. https://doi.org/10.1016/j.jtice.2017.12.007

Varghese R, Henry J, Irudayaraj J (2017) Ultrasonication-assisted transesterification for biodiesel production by using heterogeneous ZnO nanocatalyst. Environ Prog Sustain Energy 37:1176–1182. https://doi.org/10.1007/978-981-15-3761-5_2

Vi L, Salakkam A, Reungsang M (2017) Optimization of key factors affecting bio-hydrogen production from sweet potato starch. Energy Procedia 41:973–978. https://doi.org/10.1016/j.egypro.2017.10.092

Wang H, Covarrubias J, Prock H, Wu X, Wang D, Bossmann S (2015) Acid-functionalized magnetic nanoparticle as heterogeneous catalysts for biodiesel synthesis. J Phys Chem C 119:26020–26028. https://doi.org/10.1021/acs.jpcc.5b08743

Wang T, Zhang D, Dai L, Chen Y, Dai X (2016) Effects of metal nanoparticles on methane production from waste-activated sludge and microorganism community shift in anaerobic granular sludge. Sci Rep 6:1–10

Wiboon R, Maneewan S, Poonsuk P (2012) Application of palm pressed fiber as a carrier for ethanol production by *Candida shehatae* TISTR5843. Electron J Biotechnol 15(6) ISSN: 0717–3458https://doi.org/10.2225/vol15-issue6-fulltext-1

Willner I, Baron R, Willer B (2006) Growing metal nanoparticles by enzymes. Adv Mater 18:1109–1120. https://doi.org/10.1002/adma.200501865

Wimonsong P, Nitisoravut R (2015) Comparison of different catalysts for fermentative hydrogen production. J Clean Energ Technol 3(2):128–131. https://doi.org/10.7763/JOCET.2015.V3.181

Xiu Z, Jin Z, Li T, Mahendra S, Lowry G, Alvarez P (2010) Effects of nano-scale zerovalent iron particles on a mixed culture dechlorinating trichloroethylene. Bioresour Technol 101:1141–1146. https://doi.org/10.1016/j.biortech.2009.09.057

Zhang Y, Shen J (2007) Enhancement effect of gold nanoparticles on biohydrogen production from artificial wastewater. Int J Hydr Energy 32(1):17–23. https://doi.org/10.1016/j.ijhydene.2006.06.004

Zhang Q, Zheng Z, Liu C, Tan T (2016) Biodiesel production using lipase immobilized onepoxychloropropane-modified Fe_3O_4 sub-microspheres. Colloids Surf B 140:446–451. https://doi.org/10.1016/j.colsurfb.2016.01.013

Zhao W, Zhang Y, Du B, Wei D, Wei Q, Zhao Y (2013) Enhancement effect of silver nanoparticles on fermentative biohydrogen production using mixed bacteria. Bioresour Technol 142:240–245. https://doi.org/10.1016/j.biortech.2013.05.042

Zhaohui X, Qi Z, Xueyi S, Yan L (2016) Saccharomyces cerevisiae immobilized in alginate for continuous fermentation. J Clean Energy Technol 4(1):48–51. https://doi.org/10.7763/JOCET.2016.V4.252

Microbial Enzymes in Nanotechnology and Fabrication of Nanozymes: A Perspective

J. A. Elegbede and A. Lateef

1 Introduction

Nanotechnology is all about strategies for synthesis and manipulation of structure of particles with sizes that range from approximately 1–100 nm. Within this range of size, all the characteristics such as chemical, biological, physical and electrical are tailored in elementary ways of both individual atoms/molecules and consequently their bulk (Ahmed et al. 2016). Nanotechnology is constantly generating great attention due to its of widespread applications in numerous facets life endeavors. It is a major subject that traverses many disciplines of science and technology (Elegbede and Lateef 2019a; Lateef et al. 2019). Nanobiotechnology well described as a subset of nanotechnology is a multidisciplinary subject which aggregates various spheres of knowledge which include nanotechnology, biotechnology, physical methodology, chemical processing and system engineering, molecular motors, nanobiomaterials and nanocrystals (Lateef et al. 2015a). Nanobiotechnology is currently speedily receiving attentions because it improves the method of recreating biocompatible nanomaterials that is applicable in several fields such as biomedicine and healthcare, feed and food, environment, cosmetics, chemical industries, single electron transistors, catalysis, drug–gene delivery, electronics, mechanics, light emitters, energy science, space industries, nonlinear optical devices, in addition to photoelectrochemical applications (Elegbede and Lateef 2019b).

The green synthesis approach is a highly prospective aspect of nanobiotechnology wherein materials that are environmentally friendly are used in the biofabrication of nanomaterials. Green synthesis is favorable over chemical techniques and physical methods because it is uncomplicated, safe, reasonably reproducible, cost-effective and regularly generates more stable nanomaterials. Biological process deployed in

J. A. Elegbede (✉) · A. Lateef
Laboratory of Industrial Microbiology and Nanobiotechnology, Department of Pure and Applied Biology, Ladoke Akintola University of Technology, Ogbomoso, Nigeria
e-mail: josephadetunji18@gmail.com

A. Lateef et al. (eds.), *Microbial Nanobiotechnology*, Materials Horizons: From Nature to Nanomaterials, https://doi.org/10.1007/978-981-33-4777-9_7

the production of metallic nanoparticles (MeNPs) with the desired morphology and sizes is of great significance in nanoscience. The method does not include toxic chemicals mostly used in the synthetic processes, which produces some undesirable environmental consequences (Adelere and Lateef 2016). The pursuit for obtaining nanomaterials that are environmentally compatible and cost-effective having desired features has resulted in various advances in green (biological) synthesis. In our laboratories, numerous biological agents like microorganism (fungi and bacteria), plants, extracts, some arthropods and their metabolites, including enzymes have been applied to mediate the synthesis of MeNPs (Lateef and Adeeyo 2015; Lateef et al. 2015a, b, c, 2016a, b, c, d, e, f, g, h 2017; Azeez et al. 2017; Ojo et al. 2016; Oladipo et al. 2017a, b; Elegbede et al. 2018, 2019, 2020; Adebayo et al. 2019a, 2019b; Aina et al. 2019). There are also comprehensive reviews that elucidated the impacts of greenly synthesized nanoparticles in various spheres of life (Adelere and Lateef 2016; Lateef et al. 2016e, 2018, 2019; Ovais et al. 2018a; Elegbede and Lateef 2019b; Azeez et al. 2020; Elegbede and Lateef 2020).

The search for novel biocompatible nanomaterials capable of being applied in diverse spheres particularly the biomedical domain has impacted on the progressive exploration of various biological resources including various enzymes of microbial origin in the fabrication of nanomaterials which in turn have expanded the scale of applications of greenly synthesized nanomaterials. Generally, proteinaceous compounds have been broadly applied in the biofabrication of nanomaterials. Thus, since enzymes are proteinous in nature, preset to carry out a unique task and function like a key in a lock, which permits each enzyme to fix together with a definite substrate in a specifying way (Sabu 2003), and they have potentials of being a chief source of specialized biomolecules in nanoparticles production. Majority of technical enzymes are applied in their bulk form for manufacture of detergents, leather, textiles, paper, pulp and biofuels among others. Over five hundred (500) commercial products have been prepared using enzymes (Johannes and Zhao 2006), and in year 2017, the global enzyme market size was estimated at USD 6.3 billion and it was projected to rise to about 6.8% compound annual growth rate (CAGR) up to the year 2024 (Sharma and Upadhyay 2020). Other ways in which enzymes are employed include animal feed, household care, foods, pharmaceuticals and fine chemicals (Sanchez and Demain 2017).

Enzymes are biomolecules that facilitate necessary chemical reactions that sustain life and all living organisms produce them. Native and recombinant enzymes in biopharmaceutical, chemical and agricultural industries act in various areas of diagnostics, detergents, feed, leather and textiles, paper and pulp and in plastics. Enzymes obtained from edible tissues of both plants and animals are also generated by microorganisms (archaea, bacteria, fungi and yeasts) and are being employed for centuries in food production (Poeta et al. 2018). Enzymes are very much specific in action and hasten the rate of a unique reaction by decreasing the activation energy without it going through any permanent change; therefore, they are biomolecules essential for life support (Aldridge 2013). They require characteristically milder conditions of temperature and pressure for facilitating reactions and are applied as substitute to hazardous chemicals because of their nontoxic and biodegradable nature (Choi et al. 2015). Other exclusive characteristics of enzymes that include better yields and reduction in waste generation offer flexibility in relation to operating conditions required in production reaction. Enzymes mostly replace chemical processes if they effectively compete on a cost basis in the chemical industry. They mostly necessitate less energy, produce high yield titer with improved catalytic effectiveness, catalyst waste and by-products are minimized and lower volumes of wastewater streams are generated (Sanchez and Demain 2017).

2 Microbial Enzymes: Sources, Nature and Production

2.1 Sources and Nature of Microbial Enzymes

Since the entire living organisms generate enzymes, therefore three diverse resources of enzymes can be proposed: microbes, animals and plants. However, the expenditure of production increases with increased complexity of living systems from microbes to plants to animals. Microbes are usually utilized for production of enzymes being the most primitive of living systems. Also, they are popular for many industrial processes because of factors including low cost, low energy, nontoxic and eco-friendliness (Illanes et al. 2012; Choi et al. 2015). Many microbes like actinomycetes, fungi, yeast and bacteria intracellularly or extracellularly generate resourceful and attractive enzymes having vast diversity of structures and thus commercial applications. Microbial enzymes, for instance, proteases, amylases, pectinases, xylanases, lipases, laccases and cellulases, are extracellularly synthesized while catalases among others have been produced from *Saccharomyces cerevisiae* and *Aspergillus niger* as intracellular enzymes (Liu and Kokare 2017). In relation with their rankings, the highest percentage of enzymes applicable in industries are synthesized by yeast and fungi (about 50%), followed by bacteria (35%) and plants (15%) (Saranraj and Naidu 2014). Microorganisms are the foremost source of enzymes because they synthesize superior concentrations of extracellular enzymes. Microorganisms utilized for enzyme production include more or less fifty generally regarded as safe (50 GRAS) bacteria and fungi species among which bacteria are largely characterized by *Bacillus licheniformis*, *Bacillus subtilis* and various *Streptomyces* species while fungi are typically embodied by *Aspergillus*, *Rhizopus* and *Mucor* (Sanchez and Demain 2017).

Microorganisms of marine origin have a different range of enzymatic activity and they can possibly catalyze various biochemical reactions with new enzymes. Thus, there is massive scope for the researches exploring the prospects of deriving new products of economic significance from potential marine microorganisms (Sabu 2003). In comparison to enzymes of plant and animal sources, enzymes from microbial sources present several advantages. Enzymes from microbial sources are more active and stable and this is because preferred strains are able to synthesize well-characterized and purified enzymes on a big scale through the improvement of fermentation processes. Also, microbial enzymes exhibit high yield, present ease for modification of product and optimization owing to the openness to gene manipulation, and biochemical diversity in microorganisms. The process of enzyme production can be improved by engineering techniques and the characteristics of enzymes can be altered by protein engineering (Gurung et al. 2013). Moreover, microbes represent a plenteous source for the novel microbial enzymes discovery via contemporary methods such as genome mining, metagenome screening and investigating the biodiversity of extremophiles (Sabu 2003; Adrio and Demain 2014).

Microbial enzymes are widely employed in diverse sectors including food, agriculture, medicine, chemicals and energy (Singh et al. 2016). Moreover, protein engineering and DNA recombinant techniques have made exploitation of microbial strains achievable in meeting the ever-rising need for enzymes with novel characteristics (Liu et al. 2015). Furthermore, the microorganism exhibits metabolic dynamism, ease of large scale cultivation by fermentation, need of non-complex nutrients and is unaffected by seasonal conditions (Sharma and Upadhyay 2020). It will be true to assume the impracticability of any biological process with no enzymes. Enzymes are biocatalysts, which boost the reaction speed to several folds. All enzymes are protein in nature excluding catalytic RNA molecules, and all are required for all living organisms. Usually, the catalytic activity of enzymes results from the integrity of protein conformation and this is lost if the enzyme is denatured. The active or functional site of enzymes is the specific site which is the area of connection of the enzyme to its specific substrate to produce the product. Typically, the functional site is cleft or groove on the surface of enzyme that creates a non-polar environment that helps in proficient attachment to a substrate (Sharma and Upadhyay 2020).

Enzymes are comprised of several amino acid molecules with each residue of amino acid covalently joined to its neighbor; the amide linkage between them is described as a peptide bond. They are large macromolecules ranging between kilodaltons to megadaltons in terms of molecular mass (Singh et al. 2016). Generally, enzymes contain 20 different amino acids types, and depending on the kind of reaction they facilitate, according to the International Union of Biochemistry (IUB), they can be grouped into six classes: hydrolases, transferases, lyases, isomerases, oxidoreductases and ligases (Sanchez and Demain 2017). Currently, according to Li et al. (2012), just about two hundred (200) kinds of enzymes from microbial resources of about four thousand (4000) recognized enzymes are commercially applied and just about twenty (20) are produced on an actual industrial scale.

2.2 *Production of Microbial Enzymes*

Enzyme production is a field that is very important in biotechnology (Couto and Sanromán 2005). Lignocellulosic biomass is utilized as a low-cost raw material which is used for cost-effective manufacture of enzymes as a result of their cheap commercial value. The application of lignocellulosic resources in enzymes production has resulted in decrease in environmental pollution caused by the unfavorable production of massive amounts of lignocellulosic resources. Microbes (fungi and bacteria) flourish on natural lignocelluloses in the normal ecosystems. Among the most proficient decomposers of natural lignocelluloses and wood are white-rot fungi and other mushrooms. Mostly, enzyme production at industrial scale involves the exploitation of lignocellulose-degrading fungi. Many microorganisms, including yeast, fungi, bacteria and actinomycetes, can produce enzymes using various lignocellulosic substances. Filamentous fungi are able to produce enzymes to a large degree compared with other microorganisms. Therefore, fungi are imperative for the enzymes production for commercial purposes. Various strains of fungi that are being widely involved in the production of enzymes are the genera *Trichoderma* and *Aspergillus* (Abdeshahian et al. 2020). Furthermore, strain improvement is most central to the production of enzyme of microbial sources. Strain improvement involves microbial strain engineering which entails inserting a specific gene into a host (production strain) for the synthesis of a specific enzyme and this host must have a safe (GRAS) status (Illanes 2008). Achieving a GRAS status for a new organism is difficult; therefore, cloning an enzyme structural gene into a host of GRAS category is preferred (Sharma and Upadhyay 2020).

Natural enzymes are often inapt for industrial biocatalysis and they require modifications for most industrial applications. The production strains or hosts are frequently modified via genetic manipulation to achieve enhanced properties and elevated levels of production. The establishment of recombinant DNA technology has made cloning of genes that encode for an enzyme from microbes possible, which can express them at higher levels as much as hundreds times superior than those generated by original microorganisms. The enzyme industry has rapidly adopted the technology and moved enzyme production from industrially unsuitable strains into industrial strains (Sanchez and Demain 2017). Metagenomics, genomics, recombinant DNA technology and proteomics are used to assist the novel enzymes discovery from microbes in environment and to generate or develop enhanced enzymes. Via metagenomics, many novel and functional enzymes have been acquired (Ferrer et al. 2007). It is implied that the enzyme market will maintain growth due to enhanced production efficiency which results in inexpensive enzymes, new enzymes via various selection techniques and by engineering characteristics of traditional enzymes, and new application fields. Tailoring enzymes for precise and specific applications, more comprehension of structure–function relationships and improved searching for enzymes from unusual and exotic environments will be a potential inclination with endlessly improving tools (Sanchez and Demain 2017).

Microbial enzyme production is cost-effective on a great scale because of inexpensive culture media and short fermentation cycles (Sanchez and Demain 2017). There are diverse methods of fermentation by which these important enzymes may be produced. Liquid cultures contained in huge bioreactors are preferable for the bulk production of enzymes on commercial scale. Processes such as solid state fermentation (SSF), fermentation on inert solid supports and immobilization are also broadly utilized for enzyme production. Submerged fermentation (SmF) which involves the cultivation of microbial cells in liquid media kept under controlled conditions in bioreactors for the synthesis of desired metabolites. Monitoring of process parameters via online platform and automation is some of the merits of SmF. On the other hand, SSF involves culturing microorganisms on solid substrates with slightly raised moisture content with total or near deficiency of free water. This is described as a fermentation process on semisolid or solid substrate or such that occurs on a nutritionally static solid support. SSF presents some benefits such as access to nutrients for microorganisms and high purity of product. Moreover, immobilization of cells is the binding of cells or cellular inclusion on a distinct solid phase which permits exchange of substrates, product, in addition to inhibitors, etc. and simultaneously separates the catalytic cell biomass from the bulk phase containing substrates and products (Sabu 2003).

2.3 Applications of Microbial Enzymes in the Synthesis of Nanomterials

The compass of biosynthesis of nanomaterials is on a continuous expansion in recent times by reason of the vast accessibility of bioresources (Ojo et al. 2016; Oladipo et al. 2017a; Adelere and Lateef 2016; Lateef et al. 2016e; Elegebede and Lateef 2020). Interests in applications of biological systems including microorganisms as cell factories in fabrication of nanoparticles having novel biological activities keep increasing dramatically. Although there are a lot of other biomaterials employed in nanobiotechnology, plants and microorganisms (fungi and bacteria) are at the first line of investigations in nanoparticles biosynthesis via both extracellular and intracellular routes (Ogi et al. 2010; Pandian et al. 2010). The biosynthetic processes are less labor-intensive, nontoxic, low-cost technique. Diverse types of biogenic nanomaterials including cadmium sulfide, zinc, copper, titanium, gold, magnesium, silver and platinum have been biosynthesized using biological resources (Gholami-Shabani et al. 2015).

Microorganisms such as fungi, bacteria and algae and their metabolites contribute a noteworthy function in nanoparticles biosynthesis. Enzymes, extracellular polysaccharides, DNA and rhamnolipids are all involved in biosynthesis of nanoparticles which would lead to the fabrication of polydispersed nanoparticles as a result of the attendance of multiple organic constituents in cell-free extract or cell suspension. Proteins mostly cover the face of such bionanoparticles and may confer stability to it (Durán et al. 2011; Gholami-Shabani et al. 2015; Kitching et al. 2015). Also, it was proposed that proteins, organic acids and polysaccharides released by different fungi were capable of distinguishing different crystal shapes and growth was directed into spherical crystals (Shedbalkar et al. 2014). Proteins and peptides which are biomolecules that can also be established in microbial enzymes have been described to be responsible for biosynthesis nanoparticles. The existence of peptide groups with proteins molecules (amino acid groups) in enzymes opens up a probability of them being directly implicated in the bioreduction of metals in the biofabrication of nanoparticles. Functional groups indicating the occurrence of protein moieties have been implicated in various studies in the synthesis of nanoparticles (Adelere and Lateef 2016; Lateef et al. 2016e; Elegbede and Lateef 2020). As such, many microbial enzymes have been produced and deployed for the synthesis of various metallic nanoparticles (Table 1, Fig. 1).

Studies have shown that the synthesized proteins or enzymes and reducing agents like organic acids, amino acids and peptides in biological constituents are accountable for nanoparticles bioproduction. These compounds contain functional groups that get involved in the reduction and consequently the synthesis in addition to the stabilization of nanoparticles (Safaepour et al. 2009). Also, it has been reported from FTIR analysis in several studies that biomolecules loaded with amine (N–H) and hydroxyl (O–H) groups were liable for the reduction, capping as well as stabilization of nanoparticles; thus, these demonstrate the contribution of proteins in the biosynthetic process (Elegbede et al. 2019; Shankar et al. 2014).

These proteins are believed to create electrostatic attraction of negatively charged carboxylic groups, thus stabilizing the nanoparticles by capping to avoid their aggregation through the creation of repulsive forces (El-Batal et al. 2015). Durán et al. (2015) described in a review article that residues of peptides and amino acid such as arginine, cysteine, lysine and methionine have been engaged in the biosynthesis AgNPs. Also, Slocik et al. (2005) found out that tyrosine residues on viral capsid surface keenly activated the reduction $AuCl_4^-$ which resulted in the fabrication of AuNPs. Proteins responsible for PtNP production from *Acinetobacter calcoaceticus* have also been purified according to Gaidhani et al. (2014). Likewise, gold nanoplates were biosynthesized utilizing 28 KDa purified protein obtained from *Chlorella vulgaris* (Xie et al. 2007a). URAK purified from *Bacillus cereus* NK1 was employed for the biosynthesis of AgNPs (60 nm) and AuNPs (20 nm) which occurred at 5 min and 12 h after addition of NaOH, respectively (Deepak et al. 2011).

In recent times, majority of the biofabricated nanoparticles are made by using fungal and bacterial cells but not purified enzymes and this may lead to several problems especially in scaling up for applications industrially. The foremost difficulty in this cell-based procedure is binding of the biosynthesized nanoparticles to the microbial biomass and this necessitates further steps of microbial cells breakdown and then nanoparticles isolation for purification of nanoparticles. Although culture supernatants are preferable to microbial biomass in nanoparticles biosynthesis on account of the lower costs, maintenance and simplicity for application, they also suffer from drawbacks such as limited numbers of microbes possessing vital secretory proteins for biosynthesis and the requirement for downstream processes for purification. Also, both microbial biomass and culture supernatant cannot be reused making the biosynthetic process non-economic particularly for applications in industries. However, enzymatically synthesized nanoparticles have some advantages over those synthesized by microbial biomasses and culture supernatants. There is considerable decrease of the downstream purification steps, and they have superior prospects for use in industries as they can be immobilized for recycling (Gholami-Shabani et al. 2015).

Enzyme technology has successfully gained access into nanobiotechnology, as result of biosynthesis via enzymes being proteinous in nature and thus can operate as bioreducing agents of bulk materials (Elegbede et al. 2020). Enzymes can as well

Table 1 Biosynthesis of nanomaterials using microbial enzymes and their applications

Microorganism	Enzyme produced	Nanoparticles synthesized	Activities/Applications	References
Lentinus edodes	Laccase	AgNPs	Antibacterial	Lateef and Adeeyo (2015)
Trametes versicolor	Laccase	AgNPs	–	Durán et al. (2014)
Stenotrophomonas maltophilia R13	Keratinase	AgNPs	Antimicrobial, antioxidant (DPPH and ABTS), anticollagenase	Jang et al. (2018)
Bacillus safensis LAU 13	Keratinase	AgNPs	Antibacterial	Lateef et al. (2015a)
Aspergillus niger L3	Xylanase	AgNPs	Antibacterial, antifungal, antioxidant (DPPH and H_2O_2), dye degradation (malachite green and methylene blue), anticoagulation and thrombolytic	Elegbede et al. (2018)
Trichoderma longibrachiatum L2	Xylanase	AgNPs	Antibacterial, antifungal, antioxidant (DPPH and H_2O_2), dye degradation (malachite green and methylene blue), anticoagulation and thrombolytic	Elegbede et al. (2018)
Trichoderma ressei ATCC 26,921	Cellulase	AgNPs	–	Rai and Panda (2015)
Fusarium oxysporum	Nitrate reductase	AgNPs	Antibacterial, antifungal	Gholami-Shabani et al. (2014)
Acinetobacter sp.	Lignin peroxidase	AgNPs	–	Singh et al. (2017)
Aspergillus oryzae	Alpha-amylase	AgNPs	–	Mishra and Sardar (2012)
Bacillus cereus NK1	Fibrinolytic enzyme (URAK)	AgNPs	–	Deepak et al. (2013)
Pleurotus ostreatus	Laccase	AuNPs	–	El-Batal et al. (2015)
Paraconiothyrium variabile	Laccase	AuNPs	–	Faramarzi and forootanfar (2011)
Phanerochaete chrysosporium	Laccase and ligninase	AuNPs	–	Sanghi et al. (2011)

(continued)

Table 1 (continued)

Microorganism	Enzyme produced	Nanoparticles synthesized	Activities/Applications	References
Acinetobacter sp. SW30	Lignin peroxidase	AuNPs	–	Wadhwani et al. (2018)
Aspergillus niger L3	Xylanase	AuNPs	Antibacterial, antifungal, antioxidant (DPPH and H_2O_2), anticoagulation and thrombolytic	Elegbede et al. (2020)
Trichoderma longibrachiatum L2	Xylanase	AuNPs	Antibacterial, antifungal, antioxidant (DPPH and H_2O_2), anticoagulation and thrombolytic	Elegbede et al. (2020)
A. saitoi	Protease	AuNPs	–	Bharde et al. (2007)
A. oryzae	Amylase	AuNPs	Biodegradation of nitroaromatic pollutants (*p*-nitrophenol and *p*-nitroaniline)	Mishra and Sardar (2014)
Bacillus cereus NK1	Fibrinolytic enzyme	AuNPs	–	Deepak et al. (2013)
Bacillus subtilis RSE163	keratinase and disulfide reductase	AuNPs	Nontoxic interaction with *Escherichia coli* (ATCC11103)	Gupta et al. (2015)
Escherichia coli	Sulfite reductase	AuNPs	Antifungal	Gholami-Shabani et al. 2015
Acinetobacter sp. SW30	Lignin peroxidase	SeNPs	–	Wadhwani et al. (2018)
Bacterial strain JS-11	Disulfide reductase and fumarate reductase	SeNPs	–	Dwivedi et al. (2013)
Bacillus methylotrophicus	Alpha-amylase	SeNPs	–	Dizaj et al. (2013)
Aspergillus niger L3	Xylanase	Ag–AuNPs	Antibacterial, antifungal, antioxidant (DPPH and H_2O_2), dye degradation (malachite green and methylene blue), anticoagulation and thrombolytic	Elegbede et al. (2019)
Trichoderma longibrachiatum L2	Xylanase	Ag–AuNPs	Antibacterial, antifungal, antioxidant (DPPH and H_2O_2), dye degradation (malachite green and methylene blue), anticoagulation and thrombolytic	Elegbede et al. (2019)

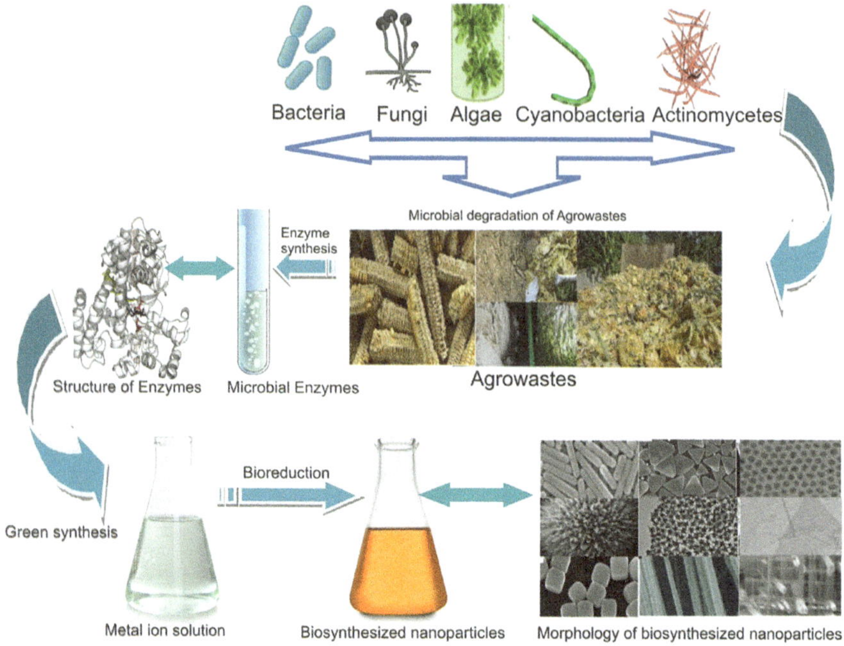

Fig. 1 Schematic representative of production of microbial enzymes and application in biosynthesis of nanomaterials

be attached to nanomaterials as an immobilized base and carrier for the enzyme. Many enzymes that originated from microbes such as protease, xylanase, phytase and amylases have been immobilized on a variety of nanoparticles for example 1,3,5-triazine-functionalized $Fe_3O_4@SiO_2$ nanoparticles, Fe_3O_4-coated chitosan, gold nanoparticles, carbon nanoparticles among others. The enzyme activity, storage stability, thermostability, pH stability, in addition to reusability of these nanoparticles of microbial enzymes have shown noteworthy superiority compared with free microbial enzymes. According to Soozanipour et al. (2015), xylanase immobilized metallic nanoparticles showed quite impressive stability after nine reaction cycles with about 65% of its original activity retained. Enzymatically produced nanoparticles are valuable for homogeneous (soluble-free enzymes) in addition to heterogeneous (immobilized enzymes) catalysis with relevance in chemical reactions in human body and chemical industrial processes by acting as catalytic site or as support for catalytic processes (Kalimuthu et al. 2008).

2.4 Biosynthetic Mechanism of Microbial Enzymes in the Nanomaterials Fabrication

By their universal characteristics, enzymes catalyze the synthesis but not used up in the reactions, but sometimes, may function as agents causing reduction and stabilization. The process may be catalyzed by the whole enzyme or amino acids liberated after denaturation of the enzyme as a result of the reaction conditions (Adelere and Lateef 2016). Microbial enzymes can facilitate the biosynthesis of nanomaterials either by intracellularly or extracellularly produced enzymes. Microbes such as fungi, bacteria, viruses, actinomycetes and yeasts serve as biofactories for the bioreduction of silver, gold–silver alloy, gold, cadmium, silica, platinum, selenium, titania, magnetite, palladium and other metals to their nanoparticles (Narayanan and Sakthivel 2010) by employing their intracellularly produced enzymes. In the intracellular method, enzymes with ability to bioreduce the metal ions to their particular nanoparticles are secreted in the cell wall of the microbe. The inter-relations of intracellular enzymes and positively charged groups are exploited in enthralling of metallic ions from the solution leading to consequent reduction within the cell (Thakkar et al. 2010; Dauthal and Mukhopadhyay 2016). Microscopic examinations have revealed that nanoparticles are stored in the cytoplasmic membrane, periplasmic space and cell wall and this was as a consequence of the diffusion of metal ions across these membranes and enzymatic reduction (Ovais et al. 2018b).

Alkalo-tolerant and alkalo-thermophilic actinomycetes species, *Rhodococcus* sp. and *Thermomonospora* sp. respectively were employed for the intracellular synthesis of AuNPs (Ahmad et al. 2003a, b), and the Au^{3+} reduction was observed to be mediated by enzymes located at the cytoplasmic membrane and mycelia surface. According to Mukherjee et al. (2001), *Verticillium* biomass was reacted with $AgNO_3$ solution which brought about the intracellular reduction and subsequent development of AgNPs. Electron microscopy illustrated that the AgNPs formed were located under the cell wall surface owing to enzymatic bioreduction, which is nontoxic to the fungi. *Brevibacterium casei* reportedly biosynthesize AuNPs and AgNPs via intracellular enzymes which were spherically shaped (Kalishwaralal et al. 2010). Likewise, Konishi et al. (2007) reported *Shewanella algae* as an efficient bioreducer of $AuCl_4$ ions to elemental gold. AuNPs biosynthesized were located in the periplasmic space which was facilitated via intracellular enzymes.

In extracellular scheme, reductase enzymes are secreted by microbial cell and are applied in the reduction of metal ions (Hulkoti and Taranath 2014). Studies have suggested that cofactors like both nicotinamide adenine dinucleotide (NADH) and reduced form of nicotinamide adenine dinucleotide phosphate (NADPH)-dependent enzymes play important responsibilities as reducing agents through the electron transfer from NADH by NADH-contingent enzymes that functions as electron carriers (Bose and Chatterjee 2016). He et al. (2007) emphasized that the bioreduction of gold was instigated via electron transfer from NADH by NADH-contingent reductase enzymes that are at hand in bacterium *Rhodopseudomonas capsulata* in

AuNPs biosynthesis. Nitrate reductase is an attractive enzyme that mediates the translation of nitrate to nitrite in the nitrogen cycle by the appropriate microbial enzymes (Moteshafi et al. 2012). It is as well a multidomain enzyme having the combination of the prosthetic groups; Fe-heme, molybdopterin and flavin adenine dinucleotide (FAD) in a 1:1:1 stoichiometry which facilitates shift of electron from NAD(P)H to nitrate. In fungi, extracellular biosynthesis of nanoparticles has been said to be the outcome of nitrate reductase through bioreduction of metal ions (Ingle et al. 2008). It was also emphasized that NADH-dependent reductases play important function in the reduction of gold chloride salt in fungi (Rai et al. 2009).

Studies have elucidated the active participation of nitrate reductase in the biosynthesis of metallic nanoparticles extracellularly (Kumar et al. 2007). Commercially available nitrate reductase disks have revealed that NADH-reliant reductases were implicated in the bioreduction of Ag^+ ions to Ag^0 and thus formation of AgNPs (Ingle et al. 2008). It was reported by Senapati et al. (2005) that *Fusarium oxysporum* produced extracellular reductases which caused the bioreduction of Au^{3+} and Ag^+ to Au–Ag alloy NPs. In another study, nitrate reductase obtained from *F. oxysporum* was in addition applied in a condition free of oxygen in the presence of NADPH, phytochelatin (stabilizer protein) and 4-hydroxyquinoline (electron carrier) for the biosynthesis of AgNPs (Karbasian et al. 2008). *Penicillium brevicompactum* was also described to release NADH-reliant enzyme nitrate reductases which caused the bioreduction of Ag^+ ions (Shaligram et al. 2009). Moreover, several fungi have produced some extracellular enzymes for instance acetyl xylan esterase, cellobiohydrolase, D-glucosidase cellulose, xylanase, laccase, keratinase, peroxidases, proteases, reductases and fibrinolytic enzyme (URAK) which have been known to take a key function in the eco-friendly synthesis of metallic nanoparticles (Deepak et al. 2011; Gholami-Shabani et al. 2015; El-Batal et al. 2015; Lateef and Adeeyo 2015; Lateef et al. 2015a; Elegbede et al. 2018, 2019, 2020; Wadhwani et al. 2018). Reports have also revealed that protein moieties from these enzymes are mostly accountable for eco-friendly synthesis of nanoparticles where they could be functioning as non-enzymatic catalytic reaction. Durán et al. (2015) described the contribution of disulfide bridge moieties and thiol groups of enzymes as the reaction sites of nanoparticles formation in a review article. The disulfide and thiols protein moieties were proposed as the likely catalytic site for most chemical oxido-reductions resulting in metallic nanoparticles biosynthesis. Correspondingly, the S–H and S–S moieties of enzymes which have been denatured could reduce metallic ions to nanoparticles.

2.5 *Some Enzyme-Mediated Biosynthesized Nanomaterials*

2.5.1 Silver Nanoparticles (AgNPs)

Silver nanoparticles are at present commercially available as antimicrobial agents used in places such as railway stations, elevators etc. across the world. They are moreover employed as antimicrobial substances implanted in surgical catheters so as to lessen possibilities of post surgery infections and they are suggested to possess anti-fungal, anti-angiogenic, anticancer, anti-inflammatory, and antipermeability activity against gases such as oxygen which is essential for microbial growth (Kalishwaralal et al. 2009). The relevance of microbial enzymes in the eco-friendly synthesis of AgNPs is still an emerging and growing field having great prospects. The application of laccase from *Lentinus edodes* in the biogenic synthesis of AgNPs, with the creation of walnut-shaped nanoparticles with sizes ranging from 50–100 nm, was described by Lateef and Adeeyo (2015). The AgNPs absorbed at 430 nm in the UV–Vis spectrum and displayed antibacterial potentials of 11–20 mm against strains of *E. coli*, *K. pneumoniae* and *P. aeruginosa*. Also, Durán et al. (2014) reported that semipurified laccase obtained from *Trametes versicolor* was employed in the biogenic synthesis of AgNPs. This led to the fabrication of spherical-shaped AgNPs with size of the particles below 100 nm according to TEM micrograph and it was reported that the fabrication of AgNPs by laccase was mostly likely on account of the existence of free cysteine as a reducing molecules.

Jang et al. (2018) reported that keratinase produced by *Stenotrophomonas maltophilia* R13 was utilized as a bioreducing means for the biogenic synthesis of AgNPs. The crystalline AgNPs were revealed to be spherical in shapes having mean diameter of 8.4 nm by dynamic light scattering and electron microscopy analysis. FTIR analysis showed that AgNPs were stabled by proteins molecules at hand in the crude enzyme. Also, the AgNPs displayed a wide spectrum antimicrobial effect against many pathogenic microorganisms, causing structural distortion of cells leading to leakage of membrane and consequent lysis was proposed as the mechanism for the antimicrobial activities. Additionally, the AgNPs displayed potent antioxidant properties in DPPH and ABTS radical mitigating activities with IC_{50} of 0.0112 and 0.0243 mg/ml, respectively, in addition to anticollagenase activity with IC_{50} of 23.5 mg/ml. Moreover, Lateef et al. (2015a) reported the biogenic synthesis of AgNPs using crude keratinase obtained from *Bacillus safensis* LAU 13 which was isolated from feather waste dumpsite (Lateef et al. 2015b). The spherical formed AgNPs synthesized absorbed at 409 nm in UV–vis spectrum and TEM showed that their sizes ranged from 5–30 nm. The AgNPs displayed good antibacterial potentials against clinical *E. coli* strain in the tune of 8.6–12.5 mm. Additionally, Elegbede et al. (2018) reported the novel use of xylanases obtained from *Aspergillus niger* L3 (NEA) and *Trichoderma longibrachiatum* L2 (TEA) in the biogenic synthesis of AgNPs. The occurrence of proteineous molecules in the fungal xylanases produced in both SmF and SSF (Elegbede and Lateef 2018) and successfully optimized (Elegbede and Lateef 2019c) was reported to be accountable for the reduction of the Ag ions and subsequent stabilization of the AgNPs. The brownish AgNPs was revealed to be spherically shaped with sizes which ranged from 15.21–77.49 nm as made known by TEM was having surface plasmon resonance at 410 and 402.5 nm for TEA-AgNPs and NEA-AgNPs, respectively. The AgNPs were described to possess antimicrobial (against clinical fungal and bacterial isolates), antioxidant (DPPH and H_2O_2), dye degradation (malachite green and methylene blue), thrombolytic and anticoagulation activities.

Rai and Panda (2015) studied the reaction of commercial cellulase produced by *Trichoderma ressei* ATCC 26,921 with silver ions in both aqueous and non-aqueous media which produced stable narrow-sized AgNPs at room temperature. The redox potentials of residues of tyrosine and protein backbone reportedly played an active function in the bioreduction and cellulase was said to function as the capping ligand that enabled stabilization of the AgNPs even in methanol. The probable mechanism for the reduction was proposed as resulting from interaction of the residues of amino acid present in cellulase (composed of 344 amino acid) and silver ions which probably when coordinated to groups having residues of amino acid such as -NH, -SH and -OH groups were subsequently reduced by them. Meanwhile, Xie et al. (2007b) reported that Au(III) or Ag(I) ions could be reduced by Tyr residues; thus, cellulase containing 14 Tyr residues and other residues with reduction properties can also act as a robust reducing agent. Moreover, Gholami-Shabani et al. (2014) utilized purified nitrate reductase produced by *Fusarium oxysporum* for the biogenic synthesis of AgNPs which used gelatin as a capping agent. The purified enzyme with a molecular weight of 214 KDa as projected by gel filtration on Sephacryl S-300 biosynthesized the AgNPs in a NADPH-dependent mode. The AgNPs synthesized were spherically natured with an mean size of 50 nm and were studied for antimicrobial activities using the disk-diffusion method which displayed strong growth inhibitory activity against all tested human pathogenic bacteria and fungi as evident from inhibition zones that ranged from 14 to 25 mm. Also, lignin peroxidase produced by *Acinetobacter* sp. was purified via diethyl aminoethyl cellulose ion exchange, then by Biogel P-150 gel filtration column chromatography and employed for the biogenic synthesis of AgNPs. The purified lignin peroxidase which had molecular weight of 99 kDa and consisted of dimers of two polypetides of 23.9 and 24.6 kDa as proven by native and SDS-PAGE upon reduction of Ag ions was reported to produced spherical polydispersed AgNPs of 50 nm in dimension as observed under TEM (Singh et al. 2017). Furthermore, Mishra and Sardar (2012) reported the eco-friendly fabrication of AgNPs using pure alpha-amylase obtained from *Aspergillus oryzae*. Results obtained displayed that the AgNPs having a peak of 422 nm in UV–Vis spectrum was characterized by TEM to be monodispersed, hexagonally and triangularly fashioned with sizes ranging from 22–44 nm. Deepak et al. (2011) reported that purified fibrinolytic enzyme (URAK) generated by *Bacillus cereus* NK1 was employed for biosynthesizing AgNPs within 24 h which was significantly reduced to 5 min upon the addition of NaOH solution. The AgNPs biosynthesized was reported to absorb maximally at 440 nm under the UV–Vis spectrum. Also, the AgNPs were said to be crystalline, spherically shaped with mean size of 60 nm.

2.5.2 Gold Nanoparticles (AuNPs)

El-Batal et al. (2015) described the synthesis of laccase generated by *Pleurotus ostreatus* by means of SSF. The production was optimized using factorial design and the laccase produced was partially purified, characterized and applied for the biofabrication of AuNPs. The biosynthesis led to the development of a violet-colored solution that maximally absorbed at 550 nm in the UV–Vis spectrum and TEM showed the AuNPs to be highly monodispersed having size variation of 22–39 nm. Laccase was observed to perform the reduction course as a protein and not as an active enzyme because laccase in its active form only catalyzed the oxidation and lost its activity once exposed to increased temperature or gamma radiation as the nature of its protein structure was broken down, thus exposing its various amino acids. According to Faramarzi and Forootanfar (2011), purified laccase generated by *Paraconiothyrium variabile* was employed for biogenic synthesis of AuNPs, and the nature of the AuNPs was characterized. UV–vis spectrum showed peak at 530 nm related to surface plasmon absorbance of AuNPs, and TEM image of AuNPs revealed well-dispersed AuNPs with sizes ranging from 71–266 nm as also displayed by the laser light scattering technique. Sanghi et al. (2011) reported the extracellular and intracellular biosynthesis of AuNPs using laccase and ligninase obtained from *Phanerochaete chrysosporium*, respectively. Characterization via atomic force microscopy (AFM) displayed that the biosynthesized AuNPs was spherical having size varying from 10 to 100 nm. Wadhwani et al. (2018) reported the biogenic synthesis of AuNPs using lignin peroxidase, purified by DEAE-cellulose anion exchange and Biogel P-150 gel filtration chromatography from the cell supernatant of *Acinetobacter* sp. SW30. The lignin peroxidase was a monomeric enzyme with molecular weight of 97.4 KDa, and the resultant nanoparticles biosynthesized were spherical crystalline AuNPs with mean size 10 ± 2 nm. Elegbede et al. (2020) reported the novel relevance of xylanases produced by strains of *Aspergillus niger* L3 (NEA) and *Trichoderma longibrachiatum* L2 (TEA) in the ecological-friendly synthesis of AuNPs for biomedical applications. The spherical and flower-shaped AuNPs with sizes varying from 4.88–123.99 nm as shown by TEM were a purple color solution which displayed absorbance at 545 and 560 nm for NEA-AuNPs and TEA-AuNPs, respectively, attributable to their surface plasmon resonance. The biosynthesized AuNPs had good antimicrobial against clinical fungal and bacterial isolates, antioxidant (DPPH and H_2O_2), thrombolytic and anticoagulation activities.

Bharde et al. (2007) revealed the important role of protease in the shape controlled environmental-friendly synthesis of gold nanoparticles. The results obtained indicated the probable responsibility of protease induced with bovine serum albumin (BSA) for the reduction of gold ions and consequent formation of AuNPs. The function of protease in the synthesis of AuNPs was further clarified by utilizing commercial protease from fungus *A. Saitoi* for the synthesis resulting in color development to pink from pale-yellow after about 8 h of reaction which indicated the creation of AuNPs which were established to be hexagonal and triangular structured using TEM. However, reacting the culture supernatant containing BSA with 1 mM $HAuCl_4$ in the attendance of pepstatin (protease inhibitor) did not lead to formation of AuNPs even after 12 h of reaction; thus, establishing that protease was accountable for the bioreduction. Moreover, the synthesis of the anisotrophic AuNPs within 8 h reported was observed to be a noteworthy development over synthesis of anisotrophic AuNPs which have been well reported to take about 24 h. The existence of the enzyme was observed to speed up the reaction rate which can make them applicable catalyst even for the synthesis of nanomaterials. Mishra and Sardar

(2014) described the environmental benign synthesis and characterization of AuNPs using purified amylase from *A. oryzae*. The SPR band at 525 nm confirmed the existence of AuNPs, which were revealed to be spherically structured with sizes varying from 2–20 nm via TEM microscopy. The nanoparticles were examined for the chemocatalysis of *p*-nitrophenol and *p*-nitroaniline in aqueous solutions, and the results obtained could be applied in the biomanagement of nitroaromatic pollutants in solution. Also, purified fibrinolytic enzyme obtained from *Bacillus cereus* NK1 was reported by Deepak et al. (2011) to mediate the biogenic synthesis of AuNPs within 60 h which was considerably reduced to 12 h upon the addition of NaOH solution. The spherical-shaped AuNPs obtained were revealed to be crystalline in nature with an approximate size of 20 nm.

Moreover, Babu Maddinedi et al. (2015) reported the efficiency of diastase as both reducing and stabilizing mediator for the size controlled biosynthesis of AuNPs, and it was observed that varying the quantity of the diastase used can successfully tune the size variation of the AuNPs. Also, the existence of thiol groups of enzyme was proposed to play foremost role in the shape and size tuning. Cytotoxicity study carried out in vitro via MTT assay on A549 and HCT116 cancer cells revealed that cytotoxicity of AuNPs relies on its dose and size. Gupta et al. (2015) reported the effect of combined interaction of keratinase and reductase for simplistic biogenic synthesis of AuNPs. Keratinase and disulfide reductase generated by *Bacillus subtilis* RSE163 in chicken feather degradation was utilized to reduce Au^{3+} to synthesize AuNPs. The AuNPs were absorbed around 540 nm under the UV–Vis spectrum attributed to SPR absorbance of gold. HRTEM showed that the AuNPs are well dispersed, spherically shaped with sizes varying from 50–80 nm. Interaction with *Escherichia coli* ATCC 11,103 established the nontoxic nature of biogenic AuNPs, where the growth rate improvement of *E. coli* observed in the log phase indicated their nontoxicity. Also, Gholami-Shabani et al. (2015) biosynthesized AuNPs utilizing cell-free purified α-NADPH-dependent sulfite reductase produced by *E. coli*. The AuNPs were described to be spherically structured with an approximate size of 10 nm. Also, the biosynthesized AuNPs demonstrated good antifungal activity against some human fungal pathogens.

2.5.3 Other Metal-Based Nanoparticles Synthesized Using Microbial Enzymes

Selenium nanoparticles have been described to be biosynthesized using enzymes from microbial sources. Wadhwani et al. (2018) reported the biogenic synthesis of SeNPs using purified lignin peroxidase produced by *Acinetobacter* sp. SW30. The biosynthesized SeNPs were revealed to be amorphous and crystalline with approximate size of 100 ± 10 nm. NADH, NADH-dependent reductases, disulfide reductase and fumarate reductase were reported to play noteworthy roles in biosynthesis of SeNPs involving bacterial strain JS-11 (Dwivedi et al. 2013). Also, the environmental-friendly synthesis of seleniun nanoparticles (SeNPs) was reported by Dizaj et al. (2013) using purified α-amylase produced by *B. methylotrophicus* which resulted in formation of light red-colored SeNPs after 24 h. The environmental-friendly synthesis of titanium dioxide nanoparticles was also reported by Ahmad et al. (2015) using α-amylase as the main reducing and capping agent. The biosynthesized nanoparticles were described by XRD, TEM and FTIR, and the particles displayed very effective antimicrobial potency against gram positive and gram negative bacteria. Also, palladium nanoparticles have been biosynthesized using enzymes from microbial sources. Pereira et al. (2011) reported the eco-benign synthesis of palladium nanoparticles associated to glucose oxidase (GOx-PdNPs) in a supramolecular arrangement. GOx molecules interact with Pd salt leading to metal ion and FAD reduction most likely through the thiol group of cysteine 521 residues and formic acid acted as a co-adjuvant reducing mediator. GOx had a role in the redox processes and was also essential for the stability of nanoparticles. The synthesis occurred in two stages, the Pd bioreduction and the development of the 80 nm sized supramolecular clusters having manifold units of GOx connected to 3.5 nm sized PdNPs.

Moreover, bimetallic nanoparticles such as silver–gold alloy (Ag–AuNPs) have been described to be synthesized using microbial enzymes. Elegbede et al. (2019) reported the novel relevance of xylanases produced by strains of *Trichoderma longibrachiatum* L2 (TE) and *Aspergillus niger* L3 (NE) in the eco-benign synthesis of Ag–AuNPs. The biosynthesized Ag–AuNPs were light purple and ruby red with surface plasmon resonance at 534 and 520 nm for TEAg–AuNPs and NEAg–AuNPs, respectively. The anisotropic oval, spherical and irregular fashioned Ag–AuNPs had sizes that ranged from 6.98 to 52.51 nm as made known by TEM. The biosynthesized Ag–AuNPs were reported to possess antimicrobial against clinical fungal and bacterial isolates, antioxidant (DPPH and H_2O_2), dye degradation (malachite green and methylene blue), thrombolytic and anticoagulation activities.

3 Nanozymes: Synthesis and Applications

Operational nanomaterials having characteristics like enzymes are termed nanozymes (Cheng et al. 2017; Wang et al. 2017a; Lin et al. 2018; Yao et al. 2018), and recently, nanozymes have become known as innovative kind of artificial or synthetic enzymes (Wu et al. 2018). In recent years, researches are concentrated on leveraging on the catalytic capacities of chemical molecules, such as metal complexes, cyclodextrins, porphyrins, polymeric and supramolecules, as advantage to use them as alternatives to natural enzymes (Raynal et al. 2014). However, their biocompatibility and overall catalytic efficiency became a few of the concerns with usage of these molecules. Moreover, the intrinsic drawbacks (such as high cost, low stability and storage difficulty) of natural enzymes have inspired the surfacing and advances of a variety of enzyme mimics termed artificial enzymes (Wu et al. 2019). Current expansions in nanotechnology have resulted in an exponential increase in fabrication of natural enzyme-like nanomaterials (Nanozymes), which possess quite a lot of intrinsic worth (Wei and Wang 2013) and as a sphere of research in the subject of artificial or synthetic or non-natural enzymes which is still emerging (Table 2). Nanozymes have attracted enormous research interests on account of their exceptional properties when related with natural and classic artificial enzymes (Wei and Wang 2013; Chen et al. 2014). Generally, natural enzymes have need of stringent physiological circumstances and conditions in carrying out catalytic functions. These stringent conditions include their restricted stability in severe environmental circumstances, the expensive cost of production, isolation, in addition to purification. But unlike natural enzymes, nanozymes present undaunted biocatalytic potentials even under the extreme conditions of temperatures, pH and forbearance to the digestion by proteases (Singh 2019). Additionally, nanozymes have other advantages, such as robustness, ease and lower cost of mass production. Moreover, nanozymes are proficient in displaying some unique properties with regard to their responses toward external stimuli, shape, size and composition-dependent catalytic activities, large surface area, in addition to displaying multiple functionalities (Wu et al. 2019).

Till now, nanomaterials have been investigated to mimic a series of natural enzymes, like oxidase, peroxidase, catalase, superoxide dismutase (SOD), nuclease, esterase, phosphatase, ferroxidase and protease (He et al. 2011; Singh et al. 2011; Li et al. 2014; Vernekar et al. 2014; Janoš et al. 2015; Wang et al. 2016). The inbuilt catalytic abilities of nanozymes have guaranteed their exploitation as artificial enzymes to either reinstate or improve abnormal enzymatic activities in patients (Bhushan and Gopinath 2015). These nanozymes are being expansively analyzed to ascertain a wide range of use in disease diagnosis and therapy, biosensing, immunoassays, theranostics, cell/tissue growth and proliferation, protection from oxidative stress and pollutants removal (Xie et al. 2012; Lin et al. 2014; Zhou et al. 2017). According to Singh (2019), the nanozymes that have been revealed so far can be broadly divided into two categories; antioxidants (superoxidase dismutase and catalase mimetic nanomaterials) and pro-oxidants (peroxidase and oxidase mimetic nanomaterials), considering their functions either as free radicals scavenger or free radicals generator during the catalytic reaction, respectively (Fig. 2).

Moreover, some nanomaterials are referred to as multifunctional nanozymes because they combine the enzyme-mimicking abilities with the added characteristics like electrical, optical and magnetic properties. Hence, the multifunctional nanozymes designed would not just incorporate multiple processes such as catalysis and separation more competently and cost-effectively but also present high-performance opportunities in practical applications for instance ultrasensitive sensing and in situ monitoring. Some of these nanomaterials that have been expansively examined for various relevancies in biosensor development, diagnosis and therapeutics, and environmental remediation, include iron oxide nanozymes having magnetic properties and metal nanozymes displaying SPR (Wu et al. 2018). Magnetic iron oxide nanomaterials which are also denoted as magnetic nanoparticles (MNPs) are exceptional on account of their interactions with both the magnetic fields and field gradients. These facilitate easy magnetic separation, enrichment, and recycling, tracking and visualization of the local environment of cells labeled with MNPs via magnetic resonance imaging (MRI) (Lee et al. 2015). MNPs are regarded as being catalytically inert; therefore, they are being broadly applied as a support which binds enzymes

Table 2 Nanozymes and their applications

Nanozyme	Enzyme mimicked	Activities/Applications	References
Carbon nanomaterials	Peroxidase	Hydrogen peroxide degradation/ colorimetric and UV–Vis absorption detection of hydrogen peroxide	Zeng et al. (2017)
Single-wall carbon nanotubes (SWNT)	Peroxidase	Neuroprotective effects against β-amyloid (Aβ) fibrillation-induced neurotoxicity in SH-SY5Y human neuroblastoma cells	Shin et al. (2017)
Cobalt-doped magnetite/graphene nanocomposites (Co-MGNCs)	Peroxidase	Dopamine detection in human serum samples	Hosseini et al. (2017)
Graphene dots	Peroxidase	H_2O_2, reduced glutathione (GSH) and glucose detection	Zheng et al. (2013)
Hemin/reduced graphene oxide (hemin/rGO) nanocomposite	Peroxidase	Catalytic oxidation of indole-3-acetic acid (IAA)	Liu et al. (2016)
AgNPs, bimetallic nanoparticles (Ag–Au/Au–Ag) and AgNPs-decorated Prussian blue nanocomposite (PB–AgNP)	Peroxidase	Bioanalytical applications	Pandey et al. (2015)
AuNPs-decorated nanodiamonds	Peroxidase	Catalytic oxidation of o-phenylenediamine	Kim et al. (2016)
Copper nanomaterials	Peroxidase	Analysis of blood glucose concentration	Fang et al. (2018)
Copper nanoparticles (CuNPs)	Peroxidase	Oxidation of 3,3′,5,5′-tetramethylbenzidine (TMB), colorimetric detection of glucose	Wang et al. (2015)
Ionic liquid-coated nanoparticles (IL-NPs)	Peroxidase	Colorimetric assay for hydrogen peroxide	Zarif et al. (2018)
Pd/Fe_3O_4-PEI-RGO nanohybrid	Peroxidase	Colorimetric detection of H_2O_2 in solution	Li et al. (2016a; b)
Platinum nanoclusters (PtNCs)	Peroxidase	Colorimetric detection of glucose	Jin et al. (2017)
Glutathione-capped platinum nanoparticles (GSH-capped PtNPs)	Peroxidase	Colorimetric detection of Hg^{2+} in both drinking water and biological fluid	Li et al. (2015)

(continued)

Table 2 (continued)

Nanozyme	Enzyme mimicked	Activities/Applications	References
AuNPs-Hg	Peroxidase	Determination of metallothioneins in (spiked) human urine	Li et al. (2016a; b)
platinum–gold bimetal nanoparticles	Peroxidase	Quantitative and visual detection of *E. coli* O157:H7	Jiang et al. (2016)
Cerium dioxide nanoparticle	Peroxidase	Glucose detection	Liu et al. (2017a; b)
Platinum nanoparticles (PtNPs)	Catechol oxidase	Oxidization and detection of polyphenols	Liu et al. (2015)
Nanoceria	Oxidase	Colorimetric assays for dopamine and catechol detection	Hayat et al. (2015)
Manganese dioxide nanoflakes (MnO_2 NFs)	Glucose oxidase and peroxidase	Colorimetric detection of glucose and H_2O_2	Han et al. (2018)
Platinum nanoparticles (PtNPs)	Ferroxidase	–	Sennuga et al. (2011)
AuNPs/Cadmium sulfide quantum dots	Glucose oxidase	Sensitive detection of DNA methyltransferase (MTase)	Zhou et al. (2016)
Copper-containing carbon dots (Cu-CDs)	Laccase	Oxidation of *p*-phenylenediamine	Ren et al. (2015)
Platinum nanoparticles	Laccase	Oxidation of catechol, dopamine, *p*-phenylenediamine and hydroquinone	Wang et al. (2017a; b)
Nanoceria	Oxidase	Colorimetric assay for single-nucleotide polymorphism (SNP) typing	Xu et al. (2014)
Molybdenum trioxide (MoO_3) nanoparticles	Sulfite oxidase	Stimulation of sulfite oxidase deficiency cells	Ragg et al. (2014)
Gold nanoclusters (AuNCs)	Oxidase	Colorimetric assay to sense trypsin	Wang et al. (2015)
Palladium-based nanostructures (Pd nanosheets, Pd@Au and Pd@Pt nanoplates)	Oxidase, peroxidase and catalase	Colorimetric detection of glucose and H_2O_2, and electrocatalytic reduction of H_2O_2	Wei et al. (2015)
AuNPs	Catalase and superoxide dismutase	Rapid decomposition of H_2O_2, and superoxide	He et al. (2013)
Graphene oxide quantum dots (GOQDs)	Catalase	Reduction reactive oxygen species (ROS) and H_2O_2, and repair of MPP^+-induced neurotoxicity	Ren et al. (2018)

(continued)

Table 2 (continued)

Nanozyme	Enzyme mimicked	Activities/Applications	References
Polyamidoamine (PAMAM) dendrimer-encapsulated gold nanoclusters (AuNCs–NH_2)	Catalase	Anticancer treatment	Liu et al. (2017a; b)
CuO nanozyme	Catalase	Ultrasensitive electrochemical detection of circulating tumor cells (CTCs)	Tian et al. (2018)
Hematite nanozyme	Catalase	Wound healing support	Hu et al. (2017a; b)
MoS_2 nanosheets	Catalase, superoxide dismutase and peroxidase	Protection of bacteria and cells against oxidative injury that could be caused by H_2O_2	Chen et al. (2018)
Silicon nanowire arrays (SiNWAs)	Catalase and peroxidase	Decomposition of H_2O_2 and oxidation of o-phenylenediamine	Wang et al. (2013)
Platinum nanoparticles (PtNPs)	Superoxide dismutase, catalase and peroxidase	Reduction of ROS levels and restoration of physiological homeostasis in a cellular model of an oxidative stress-related disorder	Moglianetti et al. (2016)
Carbon cluster nanoparticles	Superoxide dismutase	conversion of superoxide to hydrogen peroxide and dioxygen	Samuel et al. (2015)
Manganese oxide (MnO) nanoparticles	Superoxide dismutase	Responsive magnetic resonance imaging contrast agent for the imaging and treatment of cancer cells	Ragg et al. (2016)
Ce-doped titanate nanosheets (Ce-TNS)	Superoxide dismutase	Protection of DNA molecules from oxidative damage induced by ultraviolet light	Kamada and Soh, (2015)
Ceria/Polyoxometalates hybrid (CeONP@POMs)	Superoxide dismutase and protease	Degradation of amyloid-β peptides (Aβ) aggregates and reduction of intracellular ROS	Guan et al. (2016)
Graphene oxide (GO)-peptide nanofiber hybrid hydrogels (GO–PNFs)	β-glycosyl hydrolase	Degradation of cellopentaose and cellobiose	He et al. (2017)
AuNPs	Phosphodiesterases	–	Salvio and Cincotti (2014)
AgNPs, AuNPs and bimetallic Ag–AuNPs	Phosphodiesterases	Cleavage and detoxification of organophosphates	Silva et al. (2017)

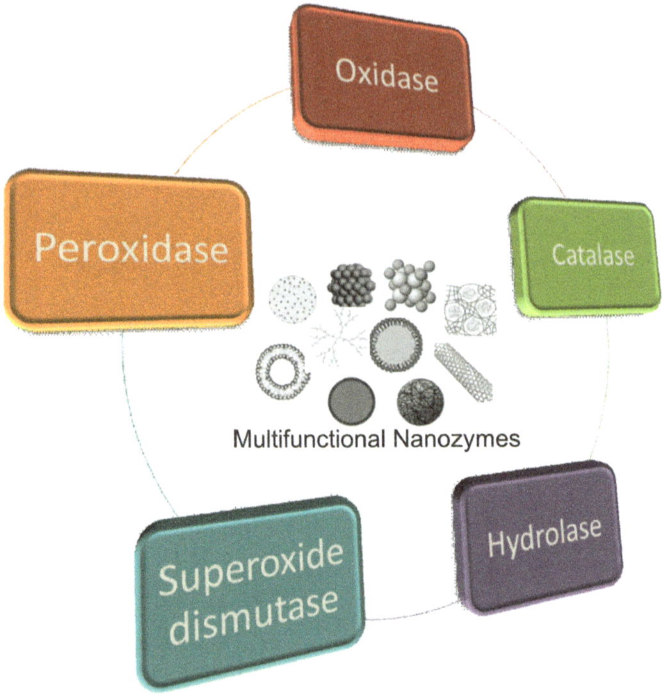

Fig. 2 Multifunctional nanozymes displaying nanozymatic antioxidant system (NAS)-mimicking enzymatic antioxidant system (EAS)

or catalysts for further functionalization (Gawande et al. 2013). Ease of separation and control in an external magnetic field is a major benefit of MNPs and this is attributable to their intrinsic magnetic properties (Wu et al. 2018). MNPs have been described for glucose and hydrogen peroxide (H_2O_2) detection on a the account of their peroxidase-mimicking potentials, as well as applied for single-nucleotide polymorphism (SNP) genotyping in an inexpensive colorimetric assay and as a nanozyme strip for identifying Ebola virus glycoprotein (EBOV-GP) among others (Gao et al. 2007, 2013; Wei and Wang 2008; Duan et al. 2015).

Moreover, metal nanomaterials have apparent optical properties which include surface enhanced Raman scattering (SERS) and localized surface plasmon resonance (LSPR) (Wei et al. 2011; Wu et al. 2014; Guo et al. 2017a). These characteristics are robustly subjective to their inter-particle distance, size and morphology. Immense interest is being displayed by researchers in recent times to develop sensitive assays with AgNPs as SERS substrates and AuNPs providing colorimetric signals (Guo and Wang 2011; Tan et al. 2014; Guo et al. 2017b). Although metallic nanomaterials are being employed as catalysts, hidden talents such as glucose oxidase-mimicking activity in AuNPs (Luo et al. 2010), and peroxidase-mimicking properties displayed in AuNPs, Ag@carbon dots (CDs), Au/FeS (Au/Co_3O_4) composite, and rGO/CuS/Au composite nanosheets were deemed to be exceptional (Jin et al. 2016; Ma et al. 2016; Song et al. 2016; Hu et al. 2017a).

3.1 Fabrication of Nanozymes and Their Enzyme-Like Activities

3.1.1 Peroxidase-Like Nanomaterials

Carbon nanotubes (CNTs) and carbon nanomaterials have become known as a leading nanomaterial in biomedical applications as a consequence of their unusual properties, which make them valuable as delivery vehicles for proteins, drugs, in addition to DNA into cells (Shin et al. 2017). Zeng et al. (2017) reported that an economical and easy synthesis of water-soluble carbon nanomaterials obtained from graphite and polypropylene carbon fibers. Commercially obtainable graphite powder or pulverized carbon fiber powder was added into a mix of concentrated H_2SO_4 (95%) and HNO_3 (65%), after which this was sonicated and refluxed. The pH of the solution was adjusted to 8 using Na_2CO_3, and then the solution was filtered and dialyzed in a dialysis bag for seven days. The carbon nanomaterials powder was hereafter obtained via freeze drying. It was detected that the prepared carbon nanomaterials (with sizes varying from 1–8 nm) could instigate the production of hydroxyl radicals during hydrogen peroxide degradation, thus own inherent peroxidase-like potential for both colorimetric and UV–Vis absorption detection of hydrogen peroxide. These carbon nanomaterials exhibited outstanding hydrogen peroxide sensitivity with the threshold of detection reported as low as 0.014 mM. Shin et al. (2017) demonstrated a self-degradation means for single-wall carbon nanotubes (SWNT) facilitated by the in-built peroxidase-like potentials of biocompatible bacterial magnetic nanoparticles (BMPs) which was biosynthesized using *Magnetospirillum* sp. AMB-1. The resulting SWNT–BMP hybrid was established to demonstrate highly synergetic peroxidase-like activity, and BMPs acted as a proficient intrinsic peroxidase in effecting the self-degradation of BMP-coated SWNTs. Moreover, the result obtained reported that the SWNT–BMP hybrid could offer an innovative approach for prevention and treatment of neurodegenerative diseases as it was revealed that they exercised neuroprotective effects against β-amyloid (Aβ) fibrillation-induced neurotoxicity in SH-SY5Y human neuroblastoma cells by decreasing the growth of β-amyloid (Aβ) fibrils considered as a key element in Alzheimer's disease.

Graphene-based nanomaterials have also been said to have peroxidase-like potentials. As a result of their high diffusion, high catalytic potential and outstanding biocompatibility, graphene dots are probable to be applied in various fields, such as biotechnology, environmental monitoring and medical diagnostics (Zheng et al. 2013). According to Hosseini et al. (2017), cobalt-doped magnetite/graphene nanocomposites (Co-MGNCs) were synthesized in a simplistic one-pot electrochemical process. Characterization of the Co-MGNCs via X-ray photoelectron spectroscopy and field emission scanning electron microscopy techniques revealed that cobalt element was greatly dispersed in the graphene scaffold. It was reported that the nanocomposite could effectively catalyze the reaction between H_2O_2 and 3,3,5,5-tetramethylbenzidine (TMB). A simple, visual and sensitive colorimetric method based on Co-MGNCs nanocomposite as peroxidase mimetic was proven for detecting dopamine which was productively applied to dopamine detection in samples of human serum. Zheng et al. (2013) depicted that graphene dots (GDs) possessed very proficient peroxidase-like catalytic potentials which was much superior than activity of graphene oxide (GO) having larger sizes. GDs were synthesized when Vulcan XC-72 carbon black was dispersed in 7 M nitric acid, then sonicated for 1 h, and refluxed for 24 h at 130 °C; eventually, a reddish-brown solid was obtained. The GDs synthesized was reported to mediate the oxidation of 3,3,5,5-tetramethylbenzidine (TMB) (peroxidase substrate) with H_2O_2 in attendance to generate a blue product, which can be employed for H_2O_2 detection by examining the absorbance change. This catalytic reaction was said to be capable of being used for detection of some other analytes (such as reduced glutathione (GSH) and glucose) by examining the generation or consumption of H_2O_2. This GDs-based system permitted detection of as little as 10 nM of H_2O_2, which is a lot lesser than that of other nanomaterial-catalyzed schemes. However, the limit of detection of this system was 0.5 μM for glucose and GSH detection. Also, hemin/reduced graphene oxide (hemin/rGO) nanocomposite created by a easy and facile hydrothermal method not including addition of any reducing agent was reported by Liu et al. (2016). The synthesized hemin/rGO nanocomposite displayed peroxidase-like activity by catalyzing the reaction of indole-3-acetic acid (IAA) in the presence of oxygen. The oxygen consumption has a linear correlation with IAA concentration in the range from 0.1 to 43 μM and from 43 to 183 μM with the limit of detection as small as 0.074 μM.

Moreover, a simplistic technique to synthesize processable AgNPs, bimetallic nanoparticles (Ag–Au/Au–Ag) and a decorated Prussian blue nanocomposite (PB–AgNP) was reported by Pandey et al. (2015). The synthesis was facilitated by 3-aminopropyltrimethoxysilane (APTMS) and cyclohexanone as reducing agents. The peroxidase mimectic potentials of the nanomaterials were spectrophotometrically determined by quantifying the development of the oxidized product of odianisidine at 430 nm. The nanocomposite displayed produced better catalytic activity than that recorded with natural enzyme and the peroxidase mimetic capacity was observed to correspond to the concentration of 3-APTMS which revealed the prospective potential of functional AgNPs in bioanalytical purposes. According to Kim et al. (2016), AuNPs were deposited on carboxylated nanodiamonds (NDs) by in situ chemical reduction of Au^+ ions to fabricate a novel nanodiamond–gold nanocomposites (NDAus), which displayed potentials to mediate o-phenylenediamine oxidation carried out in hydrogen peroxide presence comparable to peroxidase. The remarkable catalytic activity was displayed only by the AuNPs-decorated nanodiamonds and not observed in either AuNPs or nanodiamond individually. The studies on the kinetics of the oxidative catalysis of the process showed that NDAus produced a ping-pong system having activation energy of 93.3 kJ/mol, with the concentration of substrate being relative to the oxidation reaction rate. Also, the NDAus maintained substantial activities even after a number of reuse and are well matched with a natural enzyme.

Also, Fang et al. (2018) made a needle-type microelectrode which comprises of copper nanoflowers, nafion, glucose oxidase and polyurethane membranes successfully synthesized via layer-by-layer deposition. The copper nanomaterials created electrocatalytic activity and a large specific surface area typical of nanomaterials for glucose detection. The polyurethane layers functioned as mass-transport membranes which appreciably improved the stability and linearity of sensors. The resulting biosensor displayed a wide linear range of 0–20 mM, with a high-quality sensitivity of 42.38 nA mM^{-1} and a rapid response time of below 15 s. In vivo experiments using anesthetized rats displayed outstanding real-time response to the variation of blood glucose concentration. And the affinity of variation of the sensor output was in agreement with that of glucose meter. The research brought to fore a prospective approach in applying implantable device for diabetes management in addition to other medical diagnosis. Wang et al. (2015) reported the synthesis of copper nanoparticles (CuNPs) produced through a simple and easy annealing process employing humic acid as both the reducing and stabilizing mediator. The synthesized CuNPs displayed remarkable fundamental peroxidase-like activity, which speedily catalyzed the peroxidase substrate oxidation, 3,3′,5,5′-tetramethylbenzidine (TMB), in the attendance of H_2O_2 to create a blue-colored reaction. The H_2O_2 detection limit by CuNPs can be as small as 1.32×10^{-7} M. More significantly, the CuNPs displayed exceptional stability which was hardly oxidized even after six months. Also, a simple, selective and rapid colorimetric method for discovering the presence of glucose was developed with limit of detection of 6.86×10^{-7} M.

Ionic liquid-coated nanoparticles (IL-NPs) which consisted of zero-valent iron were reported by Zarif et al. (2018) to present fundamental peroxidase-like activity. Iron nanoparticles were produced by adding 140 mM sodium borohydride in dropwise manner to 20 mM ferrous sulfate through titration followed leading to formation of black particles, after which the ionic liquid-coated nanoparticles were prepared. The IL-NPs produced improved the prospect to mediate the oxidation of the chromogenic substrate 3,3′,5,5′-tetramethylbenzidine (TMB) in the presence of H_2O_2 which led to the development of a blue green-colored product that was detectable with bare eyes and quantified by photometry at 652 nm. The IL-NPs were additionally coupled with bismuth to boost its catalytic properties; this also led to the creation of an easy, responsive and selective colorimetric assay for hydrogen peroxide. The response obtained was linear in the 30–300 μM H_2O_2 concentration range with limit of detection of 0.15 μM. Li et al. (2016a) reported that Pd/Fe_3O_4-PEI-RGO nanohybrid was synthesized by distribution of PdNPs and Fe_3O_4 NPs on polyetheylenimine (PEI) modified graphene oxide sheets. This nanohybrid was observed to display superior peroxidase-like potential, and could competently oxidize 3,3′,5,5′-Tetramethylbenzidine (TMB) by hydrogen peroxide (H_2O_2) to develop a color reaction. This led to the detection of H_2O_2 in solution via colorimetric analysis with linear range obtained for H_2O_2 being from 0.5 to 150 μM and the limit of detection as small as 0.1 μM. Moreover, simplistic preservation and quick separation of the nanohybrids were observed to be advantageous, and while comparing with other catalysts (Fe_3O_4-PEI-RGO, PEI-RGO and Pd-PEI-RGO), the nanohybrid exhibited enhanced peroxidase catalytic activity.

Platinum nanoclusters (PtNCs) were synthesized via an easy one-pot technique using yeast extract as both reductant and stabilizer as reported by Jin et al. (2017). Asides from their outstanding water solubility, the yeast extract-stabilized PtNCs also possess attractive peroxidase-mimicking property. The PtNCs efficiently catalyzed 3,3,5,5-tetramethylbenzidine (TMB) oxidation in the coexistence of H_2O_2. Catalytic mechanism analysis proposed that the peroxidase-mimicking potential of the PtNCs was like on account of their characteristic of speeding up electron transfer between TMB and H_2O_2, and their enzymatic kinetics was in accordance with the typical Michaelis–Menten theory. A highly responsive colorimetric system for detection of glucose was produced, and the limit of detection was estimated as little as 0.28 μM. The system was further applied for discovery of glucose in human blood serum and was successfully demonstrated, which suggested its promising use as peroxidase imitators in pharmaceuticals, clinical diagnosis and environmental chemistry. Li et al. (2015) described the production of glutathione (GSH)-capped platinum nanoparticles by dissolving different concentrations of GSH (1–300 mM) in 10 mM PBS buffer (pH 5.0), then mixed with 300 mM K_2PtCl4 and incubated for 2 h at 25 °C. Then freshly prepared dimethylamine borane (DMAB) aqueous solution was mixed to the above mixture for a 12 h-reduction at 25 °C. The GSH-capped PtNPs displayed activity as peroxidase imitator which ranged from 2.3 to 3.6 nm. The limit of detection was as little as 0.25 nM by using GSH-Pt having mean diameter of 3.6 nm. GSH-Pt was reiterated as a capable component for colorimetric detection of Hg^{2+} in both drinkable water and biological fluids.

Some bimetallic nanomaterials have also been described to display peroxidase mimetic potentials. Li et al. (2016b) found that cysteine-rich metallothioneins (MTs) could improve the peroxidase mimic activity of citrate-capped AuNPs coated with traces of mercury (AuNPs-Hg). The AuNPs with mean size of 13 nm were fabricated by the citrate reduction method using 1 mM $HAuCl_4$ solution as a precausor. The AuNPs accumulated by the cysteine-rich metallothioneins (AuNP-Hg-MTs) successfully catalyzed H_2O_2-mediated oxidation of 2,2′-azino-bis(3-ethylbenzothiazoline6-sulfonate) (ABTS). This led to a color change and subsequent absorbance at 416 nm which linearly corresponded to the concentration of MTs in the 4.3–49 nM range, and the limit of detection was 1.3 nM. Thus, the technique was effectively applied to determine MTs in (spiked) human urine. Jiang et al. (2016) produced a new immunochromatographic assay (ICA) with improved sensitivity for both quantitative and visual exposure of *E. coli* O157:H7. Sandwich-type immunoreactions were carried out on the immunochromatographic assay, and platinum–gold bimetal nanoparticles (Pt–Au NPs) were aggregated on the test area. The Pt–AuNPs was fabricated by ultrasonically dissolving Pluronic F127 (10 mg) into 1.0 ml of a solution containing K_2PtCl_4 (20 mM), $HAuCl_4$ (20 mM) and HCl (6 M), and using ascorbic acid as reducing substance while further washed with bovine serum albumin. The signal amplification was based on Pt–AuNPs having high peroxidase activity toward 3,3′,5,5′-tetramethylbenzidine, which produced characteristic-colored signals and thus facilitated visual recognition of presence of *E. coli* O157:H7 even without instrumentation. By reason of the outstanding peroxidase activity of Pt–AuNPs, they emit powerful visible color intensity in under 1 min for visual examination even at low concentration range of *E. coli* O157:H7. Estimation carried out by means of a commercial assay meter recorded that the susceptibility was better and above 1000 folds relative to the normal test strip founded on colored gold colloids. It was opined that this technique could effortlessly be developed to be a general signal amplification method and used for foodborne pathogens detection in addition to protein biomarkers.

Some metallic oxide-based nanomaterials have also been described to have peroxidase mimetic potentials. Liu et al. (2017a) reported that ceria or cerium dioxide nanoparticle functionalized by porphyrin (Por-Ceria) was fabricated by a one-step technique by reacting 0.04 M cerium (III) nitrate with 0.125 mM 20-tetrakis(4-carboxyl pheyl)-porphyrin (H_2TCPP), after which 200 μL H_2O_2 was injected, and then heat was applied. The porphyrin-functionalized ceria nanoparticles were exploited as a probe in colorimetric detection for H_2O_2. Por-Ceria nanoparticles exhibited powerful inherent peroxidase potential against 3,3′,5,5′-tetramethylbenzidine (TMB), in the coexistence of H_2O_2. It was observed that improved peroxidase-like potential of Por-Ceria was in consequent of the synergistic result of ceria and porphyrin, thus establishing the elevated performance of Por-Ceria as a synthetic peroxidase mimic. Also, when Por-Ceria was integrated with glucose oxidase, glucose detection was achieved and a limit of detection of 1.9×10^{-2} mM glucose with a linear range up to 0.15 mM was obtained. According to Deng et al. (2017), CuO nanoparticles were synthesized using quick-precipitation method with 0.02 M aqueous copper acetate solution as precausor. It was observed that inherent alkaline peroxidase-like potency was demonstrated by CuO nanoparticles when 3-(4-hydroxyphenyl) propionic acid was utilized as a substrate. From the result, a fluorometric assay with a limit of detection as little as 0.81 μM was authenticated for H_2O_2 under alkaline conditions. However, ammonia was observed to hinder the alkaline peroxidase-like potentials of the CuO nanoparticles. Thus, a sensing package for ascertaining of urease and urea was successfully constructed, with the detection limits for urea and urease being 27 μM and 2.6 U/l, correspondingly. The sensing technique was subsequently applied for identification of urea and urease in human urine and soil, respectively, which yielded acceptable results.

Furthermore, Guo et al. (2017b) reported that Hemin–MnO_2 nanocomposite was used as the label for human immunoglobulin G in an enzyme-linked immunosorbent assay (ELISA). Hemin-MnO_2 nanocomposite was synthesized by using hemin treated with 10 mM of aqueous phosphate buffer, and 50 μl of aqueous ammonia which was reacted with 100 mM manganese acetate. Improved sensitivity was recorded attributable to the better catalytic activity of the hemin-MnO_2 nanocomposite to 3,3′,5,5′-tetramethylbenzidine when related to MnO_2 and hemin separately. After assembly of the sandwich-type immunoassay in the 96 wells of the plate, the hemin–MnO_2-based label catalyzed 3,3′,5,5′-tetramethylbenzidine into blue compounds that were monitored by a plate reader. The absorbance correspondingly rise with human immunoglobulin G concentration. The immunoassay was also utilized for ascertaining human immunoglobulin G in serum with positive results. Also, a facile and low-cost electrochemical biosensor for glucose detection was established using TiO_2 nanotube arrays produced by means of potentiostatic anodization (Zhang et al. 2013). The TiO_2 nanotube arrays were observed to have an inherent peroxidase-like activity. The electrochemical and colorimetric assays both revealed their outstanding catalytic potential toward H_2O_2 reduction.

Metal sulfide nanoparticles have also been revealed to portray peroxidase activities. Dutta et al. (2012) reported the production of iron sulfide (FeS) and iron selenide (FeSe) nanoparticles (NPs) via solvothermal decomposition of precursor complex $[Fe_3(\mu_3\text{-}O)(\mu_2\text{-}O_2CCH_2Cl)_6(H_2O)_3]NO_3{\cdot}H_2O$ in the co-presence of thiourea and sodium selenite, correspondingly. Structural analyzes of the fabricated nanoparticles via TEM showed that the FeSNPs and FeSeNPs were composed of needle-like and spherical-shaped particles, respectively. It was observed that the FeSNPs and FeSeNPs displayed photocatalytic activity for the degradation of both rose bengal and methylene blue dyes under illumination of white light. They also showed outstanding catalytic potency in the oxidation of 3,3′,5,5′-tetramethylbenzidine (TMB) in the coexistence with H_2O_2 which was in line with the Michaelis–Menten kinetics. Additionally, both FeSNPs and FeSeNPs displayed electrocatalytic potency in the reduction of H_2O_2, which on immobilized on glass-carbon (GC) electrodes operated as amperometric sensors for H_2O_2 detection. Also, the catalytic action of uncapped nanobranch-based copper sulfide (CuS) clews synthesized via a simple one-pot hydrothermal technique as a peroxidase was studied by Niu et al. (2016). The synthesized CuS catalytically oxidized 3,3′,5,5′-tetramethylbenzidine by H_2O_2 to create a visible color reaction with quick response (having maximum change within 5 min). The CuS nanozyme exhibited catalytic kinetics preferred over normal horseradish peroxidase (HRP). Also, when incorporated with glucose oxidase (GOD), the system displayed capability to monitor glucose in blood samples with exceptional performance.

3.1.2 Oxidase-Like Nanomaterials

Liu et al. (2015) demonstrated that platinum nanoparticles (PtNPs) displayed catechol oxidase-like potentials while oxidizing polyphenols resulting in o-quinones. Four unique approaches were employed to confirm the catechol oxidase-like potential of the PtNPs. Firstly, UV–vis spectroscopy was firstly used to observe polyphenols oxidation catalyzed by PtNPs. Also, the oxidized polyphenols products were detected via ultrahigh-performance liquid chromatography (UHPLC) separation, after which identification by high-resolution mass spectrometry (HRMS). In the third method, the O_2 consumption during the oxidation was confirmed using the electron spin resonance (ESR) oximetry. And finally, semiquinone radicals (an intermediate product) produced during the polyphenols oxidation were identified via ESR using spin stabilization. The oxidizing potential of polyphenols by the PtNPs was found very useful because polyphenols and related bioactives have been examined as potent antioxidants that could be of great use in the impediment of cardiovascular diseases and cancer. Also, Hayat et al. (2015) studied the potential oxidase-like characteristics of nanoceria particles in the creation of colorimetric assays for the detection of dopamine and catechol. The limits of detection of 1.5 and 0.2 μM were observed for dopamine and catechol respectively and similar assay was utilized as a sensing system for the detection of other phenolics. However, the sensitivity of the system was significantly dependent on the type of particle, composition of buffer, pH and the configuration of the phenolic compounds.

Sennuga et al. (2011) reported the fabrication of platinum nanoparticles (PtNPs) with ferroxidase-like using horse spleen apoferritin (HSAF) mixed with diverse concentrations of K_2PtCl_4 followed by a 20-fold concentration of sodium borohydride leading to a change in color from colorless or whitish yellow to brown. TEM examination revealed that PtNPs with mean size of 2–6 nm were generated, while FTIR confirmed that the PtNPs were attached to protein. The upshot of PtNPs on the ferroxidase activity of HSAF displayed a precise activity of 360 ρmol min^{-1} mg^{-1}, (ninefold enhancement over the control). Also, according to Zhou et al. (2016), an electrochemiluminescence (ECL) biosensor without background for the responsive detection of DNA methyltransferase (MTase) potential was established dependent on incorporating the catalytic utility of gold nanoparticles (AuNPs) for glucose and the ECL enhancement of AuNPs for cadmium sulfide quantum dots (CdSQDs). The glucose oxidase-mimicking AuNPs was synthesized via rapid stirring of 1% w/v $HAuCl_4$ solution and heating at 100 °C, after which 1 wt% sodium citrate solution was added up quickly, and the solution was kept boiling till the color changed from light yellow to deep red. While the Cds QDs was synthesized by heating $Cd(NO_3)_2 \cdot 4H_2O$ suspended in ultrapure water at 70 °C under stirring, after which aqueous solution of Na_2S was slowly injected resulting in an orange-yellow solution.

Moreover, Han et al. (2018) reported the bovine serum albumin-directed method for the fabrication of Manganese dioxide nanoflakes (MnO_2 NFs) whose morphology could be tuned by adjusting the synthesis conditions. The MnO_2 NFs displayed twin enzyme potentials (GO*x*-like and peroxidase-like activity) at comparable pH range. Furthermore, the colorimetric detection of glucose, and H_2O_2 were simultaneously conducted under the catalysis of the single nanozyme (MnO_2 NFs). The method used displayed low limit of detection, high sensitivity and short detection time, as a result of the in situ reaction and proximity effect. Copper-containing carbon dots (Cu-CDs) were produced via a nontoxic and single-pot hydrothermal technique (Ren et al. 2015). The Cu-CDs possessed inherent laccase-like activities which catalyzed the oxidation of *p*-phenylenediamine (PPD) (laccase substrate) to develop a characteristic color change from colorless to brown. Polymethacrylic acid sodium salt (PMAA) was utilized as the carbon source, reducing agent and also assisted flocculation between Cu-CDs and polyacrylamide by providing carboxyl groups, which aided the exclusion of PPD. Significantly, based on laccase mimics and fluorescence quenching, the intrinsic fluorescence of the Cu-CDs could point out the occurrence of hydroquinone.

Platinum nanoparticles were produced by employing different oligonucleotides (A_{10}, T_{10}, C_{10} and G_{10},) as stabilizing agents and their catalytic activities were assessed in laccase substrates oxidation (Wang et al. 2017b). This synthesized platinum nanozyme demonstrated good stability in the temperature range of 20–90 °C and pH 3–9, beyond the range for normal laccase. The most proficient was the C_{10}-templated Pt nanozyme (4.6 nm sized) which displayed three times superior affinity with regard to 2,4-dichlorophenol than natural laccase. DNA-stabilized Pt nanoparticles produced excellent catalytic potentials in substrates oxidation which include catechol, dopamine, *p*-phenylenediamine and hydroquinone. In addition, Xu et al. (2014) described the fabrication of nanoceria with oxidase-like activity wherein a solution containing a mixture of 1 M cerium (III) nitrate solution and 1.0 M dextran were added in a dropwise manner to 30.0 ml ammonium hydroxide solution (25%) which was stirred continuously for 24 h at ambient temperature, then centrifuged and precipitated. Nucleoside triphosphates (NTPs) were applied to augment the oxidase-like potential of nanoceria, and it was observed that the improvement directly corresponded with the specific type of NTP applied. The enhancement effect was said to be in consequence of coupling of the oxidative reaction of nanoceria with the hydrolysis reactions of NTP. And founded on the effectiveness of the NTP-improved oxidase-like potential of nanoceria, series effective colorimetric assays for single-nucleotide polymorphism (SNP) typing were generated.

The detoxification processes in the mitochondria involve sulfite oxidase, where defects in the enzyme lead to harsh infant disease resulting in early death with no capable or cost-effective therapy in view. Ragg et al. (2014) reported that molybdenum trioxide (MoO_3) nanoparticles demonstrated a fundamental biomimetic sulfite oxidase (SuOx) activity. The molybdenum trioxide nanoparticles was synthesized by dissolving 10.4 mmol of molybdenum powder and H_2O_2 in distilled water and cooled to 0 °C. Acetic acid was added dropwise under rigorous stirring eventually leading to creation of a yellow precipitate washed with distilled water and ethanol. The nanoparticles produced were then functionalized with a bifunctional ligand (composed of both dopamine and triphenylphosphonium ion, as anchor group and targeting agent respectively), which selectively targeted the mitochondria. The surface-functionalized MoO_3 nanoparticles demonstrated a built-in biomimetic SuOx activity which permitted the intracellular oxidation of sulfite to sulfate. When chemically stimulated sulfite oxidase overwhelmed cells were treated with MoO_3 nanoparticles, they recuperated their sulfite oxidase potential in vitro, and this makes MoO_3 nanoparticles a prospective curative agent for sulfite oxidase deficiency. Additionally, Wang et al. (2015) found that the bovine serum albumin (BSA)-stabilized red- or blue-emitting gold nanoclusters (AuNCs) displayed enzyme-like activity under visible light irradiation. Red-emitting AuNCs were produced by mixing aqueous 10 mM $HAuCl_4$ solution to BSA while vigorous stirring, after which 1.0 mol/l NaOH solution was introduced and incubated at 37 °C while vigorous stirring for 24 h which resulted in a change in color from light yellow to deep brown while for the fabrication of blue-emitting AuNCs, 1.0 mol/l NaOH solution was replaced by 50 μL of ascorbic acid in process described for red-emitting AuNCs. The BSA–AuNCs produced better stability in severe conditions when related with natural enzyme. Photoactivated BSA–AuNCs were designed to fashion a simplistic, inexpensive and rapid colorimetric assay to sense trypsin via trypsin digestion of the protein model of BSA-stabilized AuNCs. The limit of detection for trypsin was 0.6 μg/ml much beneath the average level of trypsin in urine or serum of patient.

Wei et al. (2015) investigated the mimetic enzyme activities of two-dimensional (2D) palladium-based nanostructures (Pd nanosheets, Pd@Au and Pd@Pt nanoplates) and found out that they had inherent oxidase-, peroxidase-, and catalase-like activities. For production of Pd nanosheets (PdNSs), a known weight of $Pd(acac)_2$, PVP and tetrabutylammonium bromide (TBAB) were mixed together with N, N-dimethylformamide (DMF) and distilled water in a glass pressure vessel which was then stimulated with CO to 1 bar and heated from ambient temperature to 90 °C in 1 h, then maintained for an added 3 h before being cooled to ambient temperature resulting in a dark blue product precipitated by acetone. While for synthesis of Pd@Au bimetallic nanoplates, the synthesized PdNSs stock solution (1.2 mg/ml) and varying mass of $AuPPh_3Cl$ (2 mg/ml prepared in DMF) were mixed in DMF, then 100 μL hydrazine was added dropwise while the mixture was vigorously stirred. The solution was left untouched at ambient temperature in excess of 12 h, then the products obtained were precipitated by acetone. For Pd@Pt bimetallic nanoplates, the similar procedure was applied except that $Pt(acac)_2$ (2 mg/ml prepared in DMF) was employed to replace $AuPPh_3Cl$. The nanostructures synthesized were capable of activating H_2O_2 or dissolved oxygen (DO) to enable catalysis of the oxidation of organic substrates and breakdown hydrogen peroxide to produce oxygen. The Pd-based nanostructures were employed for detection of glucose and H_2O_2 via colorimetry, in addition to electrocatalytic reduction of H_2O_2.

3.1.3 Catalase-Like Nanomaterials

He et al. (2013) investigated the catalytic activity of AuNPs in biologically reactions through the electron spin resonance spectroscopy (ESR). The AuNPs with spherical morphology had sizes which ranged from 10–100 nm and were coated with either tannic acid or polyvinylpyrrolidone (PVP). It was reported that the AuNPs facilitated the rapid decomposition of H_2O_2 which occurred in conjunction with the development of oxygen at higher pH and hydroxyl radicals at lower pH. In addition, it was noticed that the AuNPs proficiently catalyzed the breakdown of superoxide, thus mimicking superoxide dismutase. These results revealed that AuNPs can function as catalase and superoxide dismutase mimetics. Ren et al. (2018) described the effectiveness of graphene oxide quantum dots (GOQDs) as nanozymes which competently diminish reactive oxygen species (ROS) and H_2O_2 in 1-methyl-4-phenyl-pyridinium ion (MPP^+)-induced PC12 cells. Besides, the GOQDs exercised neuroprotective impacts in a neuronal cell model by reducing α-synuclein and apoptosis. GOQDs also proficiently reduce ROS, mitochondrial impairment and apoptosis in zebrafish treated with MPP^+. Furthermore, the GOQDs-pretreated zebrafish displayed improved locomotive activity in Nissl bodies in the brain, thus establishing that GOQDs can amend MPP^+-induced neurotoxicity. The Fenton reactions and H_2O_2 decomposition suggested the catalase-like potential of the GOQDs.

It was concluded that GOQDs could translocate into zebrafish brains and exercise catalase-mimicking activity to prevent oxidation intracellularly.

Also, Liu et al. (2017b) reported that amine-terminated polyamidoamine (PAMAM) dendrimer-encapsulated gold nanoclusters (AuNCs–NH_2) produced O_2 for photodynamic therapy (PDT) through their in-built catalase-like activity. The AuNCs–NH_2 was produced by mixing 200 μL of 150 × 10^{-3} M $HAuCl_4$ with deionized water containing 94.9 μL of 5 μmol G_4NH_2 (20 wt% methanol solution). The mixture was incubated at 4 °C overnight before being irradiated by microwave for 30 min. The AuNCs–NH_2 synthesized displayed optimal H_2O_2 consumption through its catalase-like activity in the physiological pH range (pH 4.8–7.4), and also extended such activity to acidic pH conditions. The prospect to release O_2 through the catalase-like potential of AuNCs–NH_2 for PDT against hypoxia of cancer cells was studied also which presented the potentials of the AuNCs–NH_2 enabling effective anticancer treatments.

Metal oxide-based nanoparticles have also been reported to produce inherent catalase mimetic activities. Tian et al. (2018) illustrated the fabrication of the CuO nanozyme by adding 10 mM copper nitrate solution to 5 mmol oxalic acid, the mixture was magnetically stirred, and the suspension colloids obtained were washed with distilled water and ethanol several times to purify the product, followed by drying at 50 °C for 12 h. The CuO nanozyme was utilized as a catalyst in the ultrasensitive electrochemical detection of circulating tumor cells (CTCs) by employing reduced graphene oxide/gold nanoparticles composites (rGO/AuNPs nanocomposites) as a support substance. The MCF-7 circulating tumor cells were sensed by an electrochemical cytosensor having efficient surface recognition between MUC-1 aptamer and specific mucin 1 protein (MUC-1) over-expressed on the cell membranes of MCF-7. The CuO nanozyme was utilized as a signal-amplifying nanoprobe in an ultrasensitive electrochemical cytosensor. Also, Hu et al. (2017a; b) describe the fabrication of hematite nanozyme by dissolving 20 mmol $FeCl_3$ and 0.4 mmol KH_2PO_4 in distilled water to produce a clear yellowish solution that was stirred magnetically and heated under reflux for 72 h leading to a brown turbid dispersion which was filtered, precipitated and centrifuged to achieve a brown powder. The hematite nanozyme were incorporated into electrospun and then cross-linked polyvinyl alcohol membranes, where the polymeric mesh created a porous scaffold with improved water permeability, and the nanozyme functioned as a catalyst with catalase-resembling potential that competently converted H_2O_2 into O_2. The hematite nanozyme was utilized to support wound healing, and the results obtained showed that the development of fibroblasts at an H_2O_2 concentration of 50 μM was principally improved when the nanozyme dressing was applied. Thus, the application of the nanozyme dressing can appreciably decrease the harmful effects of H_2O_2 at higher concentrations.

Moreover, Chen et al. (2018) demonstrated MoS_2 nanosheets as a multifunctional nanozyme which possessed fundamental potentials resembling enzymes like catalases, superoxide dismutases and peroxidases at certain physiological conditions (pH 7.4, 25 °C). The MoS_2 nanosheets were fabricated by dissolving MoS_2 powder in dispersion solvent containing a combination of ethanol and distilled water. This was sonicated for 8 h, then centrifuged, after which the sample was dehydrated at −50 °C in a vacuum. The results actualized in the experiments showed that the MoS_2 nanosheets displayed peroxidase-like activity by transferring electrons as against generating ROS. Nanozymatic antioxidant system (NAS) (which is similar to enzymatic antioxidant system (EAS)) was developed by MoS_2 nanosheets for oxidative stress management. Interestingly, the NAS could efficiently scavenge other free radicals which include nitrogen-centered free radicals (DPPH), hydroxyl radicals (OH) and nitric oxide (NO). Applying the MoS_2-based NAS in vivo, in *Staphylococcus aureus* and *Escherichia coli*, and A549 cell models showed that MoS_2 nanosheets superiorly protected these bacteria and cells against oxidative injury that could be instigated by H_2O_2, and this reinforces the NAS unique characteristics of the MoS_2 nanosheets. In a study reported by Wang et al. (2013), silicon nanowire arrays (SiNWAs) synthesized via chemical etching technique which involved cleaning of silicon wafers in a freshly piranha solution (7:3 v/v H_2SO_4/ H_2O_2) at 90 °C for 30 min, then rinsing with deionized water and dried in nitrogen stream. The cleaned silicon wafers were submerged in an etching solution (5.0 mol/l hydrogen floride and 20 mmol/l $AgNO_3$) at 50 °C for 5, 10 and 30 min resulting in silicon nanowire arrays with 4, 9 and 24 μm nanowire lengths, respectively. The SiNWAs were observed to produce catalytic activities comparable to those of biological enzymes (catalase and peroxidase). It was reported that the SiNWAs facilitated H_2O_2 decomposition reaction into oxygen and water, in addition to the oxidation of o-phenylenediamine (OPD), a general substrate for peroxidases. The morphology of the SiNWAs and existence of Si–H bonds were said to be crucial to the incidence of such catalytic activity.

3.1.4 Superoxide Dismutase-Like Nanomaterials

Superoxide radicals are connected with the incident of many severe diseases, cancer inclusive. In non-pathogenic conditions, the natural enzyme superoxide dismutase (SOD) controls and normalizes the concentrations of superoxide in intracellular environments; in fact, nearly all tumor tissues display reduced SOD levels (Ragg et al. 2016). Moglianetti et al. (2016) showed the potentials of biocompatible platinum nanoparticles (PtNPs) as antioxidant nanozymes with strong and broad properties which include acting as catalase, superoxide dismutase and peroxidase enzymes, which showed similar and even superior activities than natural enzymes, with higher adaptability to changing environmental conditions. Two varieties of PtNPs were synthesized; Pt5 and Pt20. Pt5 was fabricated by mixing 0.5 M H_2PtCl_6 (160 μl) to 79.65 ml of MilliQ water together with 192 μl of 0.5 M sodium citrate. A solution 0.06 M of $NaBH_4$ was added dropwisely while vigorously stirred; the heat was then raised to 75 °C and sustained for 30 min. Pt20 was synthesized by utilizing the seed-growth method (Bigall et al. 2008). The PtNPs synthesized were revealed to be monodispersed, with Pt5 being quasi-spherical-shaped having mean size of 5 nm while Pt20 was flower shaped with mean size of 20 nm. The activities of the PtNPs as radical scavenging agents in a cellular prototype of an oxidative stress-associated disorder, mainly human Cerebral Cavernous Malformation (CCM) disease, which is connected to a noteworthy rise in intracellular ROS levels were examined, and note worthily, it was observed that Pt nanozymes proficiently reduced ROS levels and completely restored the physiological homeostasis of the cell.

Samuel et al. (2015) reported that nontoxic hydrophilic carbon cluster nanoparticles displayed the capacity to directly convert of superoxide to hydrogen peroxide and dioxygen. This was achieved faster than in most natural enzymes, and this is exactly what dificiency in dioxygen need in an injured state where ROS particularly superoxide, overpower the natural enzymes required to eliminate superoxide. It was confirmed that the hydrophilic carbon clusters produced no reaction against nitric oxide radical, a strong vasodilator that has a noteworthy function in cytoprotection and neurotransmission. Also, Ragg et al. (2016) demonstrated that manganese oxide (MnO) nanoparticles (NPs) produced a basic superoxide dismutase (SOD)-like activity, which was superior to that of the natural Mn-dependent SOD. The MnONPs were synthesized according to an approach described by Schladt et al. (2009) with 40 mM manganese (II) chloride tetrahydrate as precausor. Along with the SOD-like property displayed, the MnONPs appreciably improved the magnetic resonance imaging (MRI) contrast on exposure to superoxide radicals, which made them responsive MRI contrast agents for the imaging and treatment of cancer cells with diminished SOD levels. Lastly, Kamada and Soh (2015) reported that colloidal solutions of Ce-doped titanate nanosheets (Ce-TNS) with tiny dimensions of around 10 nm were fabricated through hydrolysis reaction of titanium tetra isopropoxide and $Ce(NO_3)_3$, and their potentials for annihilation of reactive oxygen species (ROS) were investigated. The results obtained indicated that Ce-TNS possessed inherent SOD mimetic potentials to obliterate of superoxide anion radicals (O_2^-) as target ROS. Moreover, the Ce-TNS was revealed to protect DNA molecules from oxidative damage induced by ultraviolet light.

3.1.5 Hydrolase-Like Nanomaterials

Some nanomaterials have been described in recent times to act as nanozymes presenting hydrolase-like potentials which spans through activities as proteases, amidases, nucleases and phosphesterases among others. Protein misfolding to amyloid aggregates is a major feature for neurodegenerative diseases, and recently, a lot of concentration has been to inspect natural proteases with competence of degrading amyloid-β peptides (Aβ), but majority of the screened ones are difficult to apply in the clinics having intractable problem of immunogenicity in living organisms; thus, the need arose for the exploitation of artificial enzymes that can overcome this challenge. Guan et al. (2016) described the prototype of a proteolytic enzyme by coating Ceria nanoparticles (CeONP) with Polyoxometalates (POMs) resulting in Ceria/Polyoxometalates hybrid (CeONP@POMs). CeONP was fabricated by reacting 1 M cerium (III) nitrate to 30.0 ml ammonium hydroxide which resulted in a color development from deep brown to light yellow. Then CeONP was redissolved in distilled water and reacted with a solution containing POMs (10 mg/mL) which brought about the fabrication of CeONP@POMs. The CeONP@POMs displayed both superoxide dismutase (SOD) and proteolytic activities. The results achieved in the study showed that CeONP@POMs proficiently degraded amyloid-β peptides (Aβ) aggregates and also reduced intracellular ROS. In vivo studies using mice further proved CeONP@POMs nanozyme was biocompatibile which was revealed by a thorough examination of their biodistribution, in vivo toxicology and body weight change.

In a study presented by He et al. (2017), homogenous graphene oxide (GO)-peptide nanofiber hybrid hydrogels (GO-PNFs) were fabricated to mimic β-glycosyl hydrolase for proficient degradation of cellopentaose and cellobiose. Free peptides were employed to generate nanofiber hydrogels after self-assembling for 24 h at environmental temperature. Then, GO-PNFs were fabricated by treating the peptide nanofiber hydrogels with 0.2 mg/ml Graphene oxide solution in citrate–phosphate buffer (pH 5.0). TEM microscopy demonstrated that the single-layered GO nanosheets was about 1 nm thick while the PNF has entwined nanofibrous network with diameters of fiber which ranged from 10 to 40 nm. The graphene oxide reacted with free peptides, free peptides and self-assembled peptide nanofibers (PNFs) was also examined for hydrolase mimetics potentials in cellobiose degradation but it was GO-PNFs that displayed the uppermost hydrolysis activity. The study suggested that the high catalytic activities of GO-PNFs for hydrolysis of cellopentaose and cellobiose may be credited to the development of nanofiber structures of peptides, less steric hindrance to access the substrate, in addition to optimum molecular conformation. The results obtained were deemed insightful enough in opening up the potentials of peptides and GO hybrid composite nanoenzymes for application in efficient cellulose hydrolysis. Silva et al. (2017) described the chemical synthesis Ag, Au and bimetallic Ag–Au nanoparticles (NPs) sustained on SiO_2 which were functionalized with methimazole (MTZ) (comprising of imidazole and thiol groups). The synthesized nanomaterials were utilized as nanocatalysts for the cleavage of organophosphate and detoxifying them in a Phosphesterase-like manner. The MTZ-developed nanocatalysts were reported to be efficient in cleavage of organophosphates. The enhancement rate of 10^8 fold was achieved for toxic pesticide (Paraoxon), when related with the uncatalyzed reaction. Moreover, Au-derived nanocatalysts were appreciably more efficient because their imidazole group is free to react with the organophosphate, which was not feasible in Ag–N interactions.

4 Future Prospects and Conclusion

The report has presented an ample review touching the applications enzymes of microbial sources in the biofabrication of nanomaterials in addition to also elucidating the creation and potentials of nanomaterials producing enzyme-like activities (nanozymes). With their extensively range of potential applications in catalysis and biomedical nanotechnology, there is a great future for microbial enzyme-mediated synthesis of nanomaterials when related with nanomaterials synthesized through various chemical and physical means owing to their inexpensive and eco-friendliness of biosynthetic processes and biocompatibility of the nanomaterials. We have elucidated that the concept of green synthesis using enzymes of microbial origin can be accomplished by simply mixing the precursor salt solution and the microbial enzymes, thus eliminating the utilization of chemicals that could have impact on the toxicity of the nanoparticles to the environment and organisms. If the vast pool of microbes with capacities to synthesize microbial enzymes is properly utilized, it could lead to the biofabrication of nanomaterials that can turn out to be a prospective game changer in the very nearest future. However, many challenges must be handled prior to industrial scale production and subsequent application. There is massive potential for microbial enzyme-mediated nanomaterials especially in therapeutic applications as they exhibit low-cost fabrication, less toxicity in addition to high degradability potentials.

Moreover, nanozymes are reported as a budding technology having huge prospect of applications especially in the biomedical fields. Certainly, investigations on nanozymes will progressively increase at the interface in between many fields such as enzymology, nanomedicines, animal biotechnology, including materials science in the coming years. Nanozymes with antioxidant properties have been demonstrated to shield cells from oxidative stress while nanozymes having pro-oxidant properties are studied for exploitations in biosensing including other immunoassays. This current literature has mostly encompass the mimicking potentials of five types of biological enzymes (peroxidase, oxidase, catalase, SOD and hydrolase), showing a need for the fabrication of nanomaterials with other enzyme-like potentials. Additionally, majority of the existing literatures on nanozyme study their in vitro catalytic activity, and use in immunoassays, there is the need for detailed exploration and characterization of nanozymes when administered in vivo conditions.

It has been reiterated and established by literature that nanozymes have several distinct improvements over natural enzymes including some artificial enzymes. However, they still face some limitations which require being examined and overcomed in the nearest future. For instance, biological enzymes are known to be highly selective to their targets and substrates, but nanozymes show partial substrate and target selectivity. Also, the catalytic effectiveness of a lot of nanozymes is still not as good as that of natural enzymes including other organic catalysts; thus, efforts must be concentrated on fabricating nanozymes exhibiting high performance in the nearest future. Additionally, the catalytic activities of nanozymes when dispersed in appropriate buffers require being investigated to expand the scope of their biological applications. Finally, safety concerns of nanomaterials are presently receiving substantial attentions in consequence of their probable effects on human and environment health at large; therefore, there is also a necessity to evaluate the benignity of synthesized nanozymes within cells and also to guarantee that they do not constitute greater environmental concerns in the long run. In the light of this, a safe-by-design method for synthesis of nanozymes could be employed to fabricate biocompatible nanomaterials. Advances in green synthesis should be seen as leverage to realize the biofabrication of high-performance nanozymes of potential wide applications.

Acknowledgements AL is grateful to Tertiary Education Trust Fund (TETFund), Abuja, Nigeria and Ladoke Akintola University of Technology (LAUTECH), Ogbomoso, Nigeria, for supporting some of the research activities mentioned in this review. JAE appreciates LAUTECH, Ogbomoso, Nigeria, for the award of honorary lectureship during which this work was undertaken.

References

Abdeshahian P, Kadier A, Rai PK, da Silva SS (2020) Lignocellulose as a renewable carbon source for microbial synthesis of different enzymes. Lignocellulosic Biorefin Technol 23:185–202. https://doi.org/10.1002/9781119568858.ch9

Adebayo EA, Ibikunle JB, Oke AM, Lateef A, Azeez MA, Oluwatoyin AO, Ajala VA, Olowoporoku TB, Okunlola OC, Ogundele OA, Badmus JA, Asafa TB, Beukes LS, Gueguim-Kana EB, Abbas SH (2019) Antimicrobial and antioxidant activity of silver, gold and silver-gold alloy nanoparticles phytosynthesized using extract of *Opuntia ficus-indica*. Rev Adv Mater Sci 58(1):313–326. https://doi.org/10.1515/rams-2019-0039

Adebayo EA, Oke AM, Lateef A, Oyatokun AA, Abisoye OD, Adiji IP, Fagbenro DO, Amusan TV, Badmus JA, Asafa TB, Beukes LS, Gueguim-Kana EB, Abbas HS (2019) Biosynthesis of silver, gold and silver-gold alloy nanoparticles using *Persea americana* fruit peel aqueous extract for their biomedical properties. Nanotechnol Environ Eng 4:13. https://doi.org/10.1007/s41204-019-0060-8

Adelere IA, Lateef A (2016) A novel approach to the green synthesis of metallic nanoparticles: the use of agro-wastes, enzymes, and pigments. Nanotechnol Rev 5(6):567–587. https://doi.org/10.1515/ntrev-2016-0024

Adrio JL, Demain AL (2014) Microbial enzymes: tools for biotechnological processes. Biomolecules 4(1):117–139. https://doi.org/10.3390/biom4010117

Ahmad A, Senapati S, Khan MI, Kumar R, Ramani R, Srinivas V, Sastry M (2003) Intracellular synthesis of gold nanoparticles by a novel alkalotolerant actinomycete *Rhodococcus* Species. Nanotechnology 14(7):824. https://doi.org/10.1088/0957-4484/14/7/323

Ahmad A, Senapati S, Khan MI, Kumar R, Sastry M (2003) Extracellular biosynthesis of monodisperse gold nanoparticles by a novel extremophilic actinomycete *Thermomonospora* Sp. Langmuir 19(8):3550–3553. https://doi.org/10.1021/la0267721

Ahmad R, Mohsin M, Ahmad T, Sardar M (2015) Alpha amylase assisted synthesis of TiO_2 nanoparticles: structural characterization and application as antibacterial agents. J Hazard Mater 283:171–177. https://doi.org/10.1016/j.jhazmat.2014.08.073

Ahmed S, Ahmad M, Swami BL, Ikram S (2016) A review on plants extract mediated synthesis of silver nanoparticles for antimicrobial applications: a green expertise. J Adv Res 7(1):17–28. https://doi.org/10.1016/j.jare.2015.02.007

Aina DA, Owolo O, Lateef A, Aina FO, Hakeem AS, Adeoye-Isijola M, Okon V, Asafa TB, Elegbede JA, Olukanni OD, Adediji I (2019) Biomedical applications of *Chasmanthera dependens* stem extract mediated silver nanoparticles as antimicrobial, antioxidant, anticoagulant, thrombolytic, and larvicidal agents. Karbala Int J Mod Sci 5(2):2. https://doi.org/10.33640/2405-609X.1018

Aldridge S (2013) Industry backs biocatalysis for greener manufacturing. Nat Biotechnol 31:95–96. https://doi.org/10.1038/nbt0213-95

Azeez MA, Lateef A, Asafa TB, Yekeen TA, Akinboro A, Oladipo IC, Gueguim-Kana EB, Beukes LS (2017) Biomedical applications of cocoa bean extract-mediated silver nanoparticles as antimicrobial, larvicidal and anticoagulant agents. J Clust Sci 28(1):149–164. https://doi.org/10.1007/s10876-016-1055-2

Azeez L, Adejumo AL, Ogunbode SM, Lateef A, Adetoro RO (2020) Influence of calcium nanoparticles (CaNPs) on nutritional qualities, radical scavenging attributes of *Moringa oleifera* and risk assessments on human health. J Food Measur Charact. https://doi.org/10.1007/s11694-020-00465-6

Babu Maddinedi S, Mandal BK, Ranjan S, Dasgupta N (2015) Diastase assisted green synthesis of size-controllable gold nanoparticles. RSC Adv 5(34):26727–26733. https://doi.org/10.1039/C5RA03117F

Bharde A, Kulkarni A, Rao M, Prabhune A, Sastry M (2007) Bacterial enzyme mediated biosynthesis of gold nanoparticles. J Nanosci Nanotechnol 7(12):4369–4377. https://doi.org/10.1166/jnn.2007.891

Bhushan B, Gopinath P (2015) Antioxidant nanozyme: a facile synthesis and evaluation of the reactive oxygen species scavenging potential of nanoceria encapsulated albumin nanoparticles. J Mater Chem B 3(24):4843–4852. https://doi.org/10.1039/C5TB00572H

Bigall NC, Härtling T, Klose M, Simon P, Eng LM, Eychmüller A (2008) Monodisperse platinum nanospheres with adjustable diameters from 10 to 100 nm: synthesis and distinct optical properties. Nano Lett 8(12):4588–4592. https://doi.org/10.1021/nl802901t

Bose D, Chatterjee S (2016) Biogenic synthesis of silver nanoparticles using guava (*Psidium guajava*) leaf extract and its antibacterial activity against *Pseudomonas aeruginosa*. Appl Nanosci 6(6):895–901. https://doi.org/10.1007/s13204-015-0496-5

Chen X, Ramström O, Yan M (2014) Glyconanomaterials: emerging applications in biomedical research. Nano Res 7(10):1381–1403. https://doi.org/10.1007/s12274-014-0507-y

Chen T, Zou H, Wu X, Liu C, Situ B, Zheng L, Yang G (2018) Nanozymatic antioxidant system based on MoS_2 nanosheets. ACS Appl Mater Interf 10(15):12453–12462. https://doi.org/10.1021/acsami.8b01245

Cheng H, Liu Y, Hu Y, Ding Y, Lin S, Cao W, Wang Q, Wu J, Muhammad F, Zhao X, Zhao D (2017) Monitoring of heparin activity in live rats using metal–organic framework nanosheets as peroxidase mimics. Anal Chem 89(21):11552–11559. https://doi.org/10.1021/acs.analchem.7b02895

Choi JM, Han SS, Kim HS (2015) Industrial applications of enzyme biocatalysis: current status and future aspect. Biotechnol Adv 33:1443–1454. https://doi.org/10.1016/j.biotechadv.2015.02.014

Couto SR, Sanromán MA (2005) Application of solid-state fermentation to ligninolytic enzyme production. Biochem Eng J 22(3):211–219. https://doi.org/10.1016/j.bej.2004.09.013

Dauthal P, Mukhopadhyay M (2016) Noble metal nanoparticles: plant-mediated synthesis, mechanistic aspects of synthesis, and applications. Ind Eng Chem Res 55(36):9557–9577. https://doi.org/10.1021/acs.iecr.6b00861

Deepak V, Umamaheshwaran PS, Guhan K, Nanthini RA, Krithiga B, Jaithoon NM, Gurunathan S (2011) Synthesis of gold and silver nanoparticles using purified URAK. Colloids Surf B 86(2):353–358. https://doi.org/10.1016/j.colsurfb.2011.04.019

Deng HH, Zheng XQ, Wu YY, Shi XQ, Lin XL, Xia XH, Peng HP, Chen W, Hong GL (2017) Alkaline peroxidase activity of cupric oxide nanoparticles and its modulation by ammonia. Analyst 142(20):3986–3992. https://doi.org/10.1039/C7AN01293D

Dizaj RR, Gharib-Bibalan F, Mollania N (2013) Biological method for selenium nanoparticles synthesis assisted by α-amylase enzyme from *Bacillus methylotrophicus*. In: 1st Tabriz international life science conference and 12th Iran biophysical chemistry conference 2013 May 22. Tabriz University of Medical Sciences. https://doi.org/10.13140/2.1.2797.9526

Duan D, Fan K, Zhang D, Tan S, Liang M, Liu Y, Zhang J, Zhang P, Liu W, Qiu X, Kobinger GP (2015) Nanozyme-strip for rapid local diagnosis of Ebola. Biosens Bioelectron 74:134–141. https://doi.org/10.1016/j.bios.2015.05.025

Durán N, Marcato PD, Durán M, Yadav A, Gade A, Rai M (2011) Mechanistic aspects in the biogenic synthesis of extracellular metal nanoparticles by peptides, bacteria, fungi, and plants. Appl Microbiol Biotechnol 90(5):1609–1624. https://doi.org/10.1007/s00253-011-3249-8

Durán N, Cuevas R, Cordi L, Rubilar O, Diez MC (2014) Biogenic silver nanoparticles associated with silver chloride nanoparticles (Ag@ AgCl) produced by laccase from T*rametes versicolor*. Springerplus 3(1):645. https://doi.org/10.1186/2193-1801-3-645

Durán M, Silveira CP, Durán N (2015) Catalytic role of traditional enzymes for biosynthesis of biogenic metallic nanoparticles: a mini-review. IET Nanobiotechnol 9(5):314–323. https://doi.org/10.1049/iet-nbt.2014.0054

Dutta AK, Maji SK, Srivastava DN, Mondal A, Biswas P, Paul P, Adhikary B (2012) Synthesis of FeS and FeSe nanoparticles from a single source precursor: a study of their photocatalytic activity, peroxidase-like behavior, and electrochemical sensing of H_2O_2. ACS Appl Mater Interf 4(4):1919–1927. https://doi.org/10.1021/am300408r

Dwivedi S, AlKhedhairy AA, Ahamed M, Musarrat J (2013) Biomimetic synthesis of selenium nanospheres by bacterial strain JS-11 and its role as a biosensor for nanotoxicity assessment: a novel Se-bioassay. PloS One 8(3). https://doi.org/10.1371/journal.pone.0057404

El-Batal AI, ElKenawy NM, Yassin AS, Amin MA (2015) Laccase production by *Pleurotus ostreatus* and its application in synthesis of gold nanoparticles. Biotechnol Rep 5:31–39. https://doi.org/10.1016/j.btre.2014.11.001

Elegbede JA, Lateef A (2018) Valorization of corn-cob by fungal isolates for production of xylanase in submerged and solid state fermentation media and potential biotechnological applications. Waste Biomass Valor. 9(8):1273–1287. https://doi.org/10.1007/s12649-017-9932-y

Elegbede JA, Lateef A (2019) Green nanotechnology in Nigeria: the research landscape, challenges and prospects. Ann Sci Technol 4(2):6–38. https://doi.org/10.2478/ast-2019-0008

Elegbede JA, Lateef A (2019b) Green synthesis of silver (Ag), gold (Au) and silver-gold (Ag-Au) alloy nanoparticles: a review on recent advances, trends and biomedical applications. In: Verma DK, Goyal MR, Suleria HAR (eds) Nanotechnology and nanomaterial applications in food, health and biomedical sciences. Apple Academic Press Inc./CRC Press, Taylor and Francis Group, Oakville, Ontario, Canada, pp 3-89. ISBN 978-1-77188-764-9. https://doi.org/10.1201/9780429425660-1

Elegbede JA, Lateef A (2019c) Optimization of the production of xylanases in corncob-based media by *Aspergillus niger* and *Trichoderma longibrachiatum* using Taguchi approach. Acta Biol Szeged 63(1):51–58. https://doi.org/10.14232/abs.2019.1.51-58

Elegbede JA, Lateef A (2020) Nanotechnology in the built environment for sustainable development. IOP Conf Ser Mater Sci Eng 805:012044. https://doi.org/10.1088/1757-899X/805/1/012044

Elegbede JA, Lateef A, Azeez MA, Asafa TB, Yekeen TA, Oladipo IC, Adebayo EA, Beukes LS, Gueguim-Kana EB (2018) Fungal xylanases-mediated synthesis of silver nanoparticles for catalytic and biomedical applications. IET Nanobiotechnol 12(6):857–863. https://doi.org/10.1049/iet-nbt.2017.0299

Elegbede JA, Lateef A, Azeez MA, Asafa TB, Yekeen TA, Oladipo IC, Hakeem AS, Beukes LS, Gueguim-Kana EB (2019) Silver-gold alloy nanoparticles biofabricated by fungal xylanases exhibited potent biomedical and catalytic activities. Biotechnol Prog 35(5):e2829. https://doi.org/10.1002/btpr.2829

Elegbede JA, Lateef A, Azeez MA, Asafa TB, Yekeen TA, Oladipo IC, Aina DA, Beukes LS, Gueguim-Kana EB (2020) Biofabrication of gold nanoparticles using xylanases through valorization of corncob by *Aspergillus niger* and *Trichoderma longibrachiatum*: antimicrobial, antioxidant, anticoagulant and thrombolytic activities. Waste Biomass Valor 11(3):781–791. https://doi.org/10.1007/s12649-018-0540-2

Fang Y, Wang S, Liu Y, Xu Z, Zhang K, Guo Y (2018) Development of Cu nanoflowers modified the flexible needle-type microelectrode and its application in continuous monitoring glucose in vivo. Biosens Bioelectron 110:44–51. https://doi.org/10.1016/j.bios.2018.03.024

Faramarzi MA, Forootanfar H (2011) Biosynthesis and characterization of gold nanoparticles produced by laccase from *Paraconiothyrium variabile*. Colloids Surf B 87(1):23–27. https://doi.org/10.1016/j.colsurfb.2011.04.022

Ferrer M, Golyshina O, Beloqui A, Golyshin PN (2007) Mining enzymes from extreme environments. Curr Opin Microbiol 10(3):207–214. https://doi.org/10.1016/j.mib.2007.05.004

Gaidhani SV, Yeshvekar RK, Shedbalkar UU, Bellare JH, Chopade BA (2014) Bio-reduction of hexachloroplatinic acid to platinum nanoparticles employing *Acinetobacter calcoaceticus*. Process Biochem 49(12):2313–2319. https://doi.org/10.1016/j.procbio.2014.10.002

Gao L, Zhuang J, Nie L, Zhang J, Zhang Y, Gu N, Wang T, Feng J, Yang D, Perrett S, Yan X (2007) Intrinsic peroxidase-like activity of ferromagnetic nanoparticles. Nat Nanotechnol 2(9):577–583. https://doi.org/10.1038/nnano.2007.260

Gao Z, Xu M, Hou L, Chen G, Tang D (2013) Magnetic bead-based reverse colorimetric immunoassay strategy for sensing biomolecules. Anal Chem 85(14):6945–6952. https://doi.org/10.1021/ac401433p

Gawande MB, Branco PS, Varma RS (2013) Nano-magnetite (Fe_3O_4) as a support for recyclable catalysts in the development of sustainable methodologies. Chem Soc Rev 42(8):3371–3393. https://doi.org/10.1039/C3CS35480F

Gholami-Shabani M, Akbarzadeh A, Norouzian D, Amini A, Gholami-Shabani Z, Imani A, Chiani M, Riazi G, Shams-Ghahfarokhi M, Razzaghi-Abyaneh M (2014) Antimicrobial activity and physical characterization of silver nanoparticles green synthesized using nitrate reductase from *Fusarium oxysporum*. Appl Biochem Biotechnol 172(8):4084–4098. https://doi.org/10.1007/s12010-014-0809-2

Gholami-Shabani M, Shams-Ghahfarokhi M, Gholami-Shabani Z, Akbarzadeh A, Riazi G, Ajdari S, Amani A, Razzaghi-Abyaneh M (2015) Enzymatic synthesis of gold nanoparticles using sulfite reductase purified from Escherichia coli: a green eco-friendly approach. Process Biochem 50(7):1076–1085. https://doi.org/10.1016/j.procbio.2015.04.004

Guan Y, Li M, Dong K, Gao N, Ren J, Zheng Y, Qu X (2016) Ceria/POMs hybrid nanoparticles as a mimicking metallopeptidase for treatment of neurotoxicity of amyloid-β peptide. Biomaterials 98:92–102. https://doi.org/10.1016/j.biomaterials.2016.05.005

Guo S, Wang E (2011) Noble metal nanomaterials: controllable synthesis and application in fuel cells and analytical sensors. Nano Today 6(3):240–264. https://doi.org/10.1016/j.nantod.2011.04.007

Guo W, Hu Y, Wei H (2017) Enzymatically activated reduction-caged SERS reporters for versatile bioassays. Analyst 142(13):2322–2326. https://doi.org/10.1039/C7AN00552K

Guo L, Qian P, Yang M (2017) Determination of immunoglobulin G by a hemin–manganese (iv) oxide-labeled enzyme-linked immunosorbent assay. Anal Lett 50(11):1803–1811. https://doi.org/10.1080/00032719.2016.1251447

Gupta S, Singh SP, Singh R (2015) Synergistic effect of reductase and keratinase for facile synthesis of protein-coated gold nanoparticles. J Microbiol Biotechnol 25(5):612–619. https://doi.org/10.4014/jmb.1411.11022

Gurung N, Ray S, Bose S, Rai V (2013) A broader view: microbial enzymes and their relevance in industries, medicine, and beyond. BioMed Res Int. Article ID 329121. https://doi.org/10.1155/2013/329121

Han L, Zhang H, Chen D, Li F (2018) Protein-directed metal oxide nanoflakes with tandem enzyme-like characteristics: colorimetric glucose sensing based on one-pot enzyme-free cascade catalysis. Adv Funct Mater 28(17):1800018. https://doi.org/10.1002/adfm.201800018

Hayat A, Cunningham J, Bulbul G, Andreescu S (2015) Evaluation of the oxidase like activity of nanoceria and its application in colorimetric assays. Anal Chim Acta 885:140–147. https://doi.org/10.1016/j.aca.2015.04.052

He S, Guo Z, Zhang Y, Zhang S, Wang J, Gu N (2007) Biosynthesis of gold nanoparticles using the bacteria *Rhodopseudomonas capsulata*. Mater Lett 61(18):3984–3987. https://doi.org/10.1016/j.matlet.2007.01.018

He W, Liu Y, Yuan J, Yin JJ, Wu X, Hu X, Zhang K, Liu J, Chen C, Ji Y, Guo Y (2011) Au@Pt nanostructures as oxidase and peroxidase mimetics for use in immunoassays. Biomaterials 32(4):1139–1147. https://doi.org/10.1016/j.biomaterials.2010.09.040

He W, Zhou YT, Wamer WG, Hu X, Wu X, Zheng Z, Boudreau MD, Yin JJ (2013) Intrinsic catalytic activity of Au nanoparticles with respect to hydrogen peroxide decomposition and superoxide scavenging. Biomaterials 34(3):765–773. https://doi.org/10.1016/j.biomaterials.2012.10.010

He X, Zhang F, Liu J, Fang G, Wang S (2017) Homogenous graphene oxide-peptide nanofiber hybrid hydrogel as biomimetic polysaccharide hydrolase. Nanoscale 9(45):18066–18074. https://doi.org/10.1039/C7NR06525F

Hosseini M, Aghazadeh M, Ganjali MR (2017) A facile one-pot synthesis of cobalt-doped magnetite/graphene nanocomposite as peroxidase mimetics in dopamine detection. New J Chem 41(21):12678–12684. https://doi.org/10.1039/C7NJ02082A

Hu Y, Cheng H, Zhao X, Wu J, Muhammad F, Lin S, He J, Zhou L, Zhang C, Deng Y, Wang P (2017) Surface-enhanced Raman scattering active gold nanoparticles with enzyme-mimicking activities for measuring glucose and lactate in living tissues. ACS Nano 11(6):5558–5566. https://doi.org/10.1021/acsnano.7b00905

Hu M, Korschelt K, Daniel P, Landfester K, Tremel W, Bannwarth MB (2017) Fibrous nanozyme dressings with catalase-like activity for H_2O_2 reduction to promote wound healing. ACS Appl Mater Interf 9(43):38024–38031. https://doi.org/10.1021/acsami.7b12212

Hulkoti NI, Taranath TC (2014) Biosynthesis of nanoparticles using microbes—a review. Colloids Surf B 121:474–483. https://doi.org/10.1016/j.colsurfb.2014.05.027

Illanes A (2008) Enzyme biocatalysis. In: Principles and applications. Editorial Springer-Verlag New York Inc., United States, pp 57–106. https://doi.org/10.1007/978-1-4020-8361-7_2

Illanes A, Cauerhff A, Wilson L, Castro GR (2012) Recent trends in biocatalysis engineering. Biores Technol 115:48–57. https://doi.org/10.1016/j.biortech.2011.12.050

Ingle A, Gade A, Pierrat S, Sonnichsen C, Rai M (2008) Mycosynthesis of silver nanoparticles using the fungus *Fusarium acuminatum* and its activity against some human pathogenic bacteria. Curr Nanosci 4(2):141–144. https://doi.org/10.2174/157341308784340804

Jang EY, Son YJ, Park SY, Yoo JY, Cho YN, Jeong SY, Liu S, Son HJ (2018) Improved biosynthesis of silver nanoparticles using keratinase from *Stenotrophomonas maltophilia* R13: reaction optimization, structural characterization, and biomedical activity. Bioprocess Biosyst Eng 41(3):381–393. https://doi.org/10.1007/s00449-017-1873-0

Janoš P, Kuráň P, Pilařová V, Trögl J, Šťastný M, Pelant O, Henych J, Bakardjieva S, Životský O, Kormunda M, Mazanec K (2015) Magnetically separable reactive sorbent based on the $CeO_2/\gamma\text{-}Fe_2O_3$ composite and its utilization for rapid degradation of the organophosphate pesticide parathion methyl and certain nerve agents. Chem Eng J 262:747–755. https://doi.org/10.1016/j.cej.2014.10.016

Jiang T, Song Y, Wei T, Li H, Du D, Zhu MJ, Lin Y (2016) Sensitive detection of *Escherichia coli* O157: H7 using Pt–Au bimetal nanoparticles with peroxidase-like amplification. Biosens Bioelectron 77:687–694. https://doi.org/10.1016/j.bios.2015.10.017

Jin J, Zhu S, Song Y, Zhao H, Zhang Z, Guo Y, Li J, Song W, Yang B, Zhao B (2016) Precisely controllable core–shell Ag@ carbon dots nanoparticles: application to in situ super-sensitive monitoring of catalytic reactions. ACS Appl Mater Interf 8(41):27956–27965. https://doi.org/10.1021/acsami.6b07807

Jin L, Meng Z, Zhang Y, Cai S, Zhang Z, Li C, Shang L, Shen Y (2017) Ultrasmall Pt nanoclusters as robust peroxidase mimics for colorimetric detection of glucose in human serum. ACS Appl Mater Interf 9(11):10027–10033. https://doi.org/10.1021/acsami.7b01616

Johannes TW, Zhao H (2006) Directed evolution of enzymes and biosynthetic pathways. Curr Opin Microbiol 9(3):261–267. https://doi.org/10.1016/j.mib.2006.03.003

Kalimuthu K, Babu RS, Venkataraman D, Bilal M, Gurunathan S (2008) Biosynthesis of silver nanocrystals by *Bacillus licheniformis*. Colloids Sur B 65(1):150–153. https://doi.org/10.1016/j.colsurfb.2008.02.018

Kalishwaralal K, Banumathi E, Pandian SR, Deepak V, Muniyandi J, Eom SH, Gurunathan S (2009) Silver nanoparticles inhibit VEGF induced cell proliferation and migration in bovine retinal endothelial cells. Colloids Surf B 73(1):51–57. https://doi.org/10.1016/j.colsurfb.2009.04.025

Kalishwaralal K, Deepak V, Pandian SR, Kottaisamy M, BarathManiKanth S, Kartikeyan B, Gurunathan S (2010) Biosynthesis of silver and gold nanoparticles using *Brevibacterium casei*. Colloids Surf B 77(2):257–262. https://doi.org/10.1016/j.colsurfb.2010.02.007

Kamada K, Soh N (2015) Enzyme-mimetic activity of Ce-intercalated titanate nanosheets. J Phys Chem B 119(16):5309–5314. https://doi.org/10.1021/jp512038x

Karbasian M, Atyabi SM, Siadat SD, Momen SB, Norouzian D (2008) Optimizing nano-silver formation by *Fusarium oxysporum* PTCC 5115 employing response surface methodology. Am J Agric Biol Sci 3:433–437

Kim MC, Lee D, Jeong SH, Lee SY, Kang E (2016) Nanodiamond–gold nanocomposites with the peroxidase-like oxidative catalytic activity. ACS Appl Mater Interf 8(50):34317–34326. https://doi.org/10.1021/acsami.6b10471

Kitching M, Ramani M, Marsili E (2015) Fungal biosynthesis of gold nanoparticles: mechanism and scale up. Microb Biotechnol 8(6):904–917. https://doi.org/10.1111/1751-7915.12151

Konishi Y, Tsukiyama T, Tachimi T, Saitoh N, Nomura T, Nagamine S (2007) Microbial deposition of gold nanoparticles by the metal-reducing bacterium *Shewanella* algae. Electrochim Acta 53(1):186–192. https://doi.org/10.1016/j.electacta.2007.02.073

Kumar SA, Abyaneh MK, Gosavi SW, Kulkarni SK, Pasricha R, Ahmad A, Khan MI (2007) Nitrate reductase-mediated synthesis of silver nanoparticles from $AgNO_3$. Biotechnol Lett 29(3):439–445. https://doi.org/10.1007/s10529-006-9256-7

Lateef A, Adeeyo AO (2015) Green synthesis and antibacterial activities of silver nanoparticles using extracellular laccase of *Lentinus edodes*. Not Sci Biol 7(4):405–411. https://doi.org/10.15835/nsb749643

Lateef A, Adelere IA, Gueguim-Kana EB, Asafa TB, Beukes LS (2015) Green synthesis of silver nanoparticles using keratinase obtained from a strain of *Bacillus safensis* LAU 13. Int Nano Lett 5(1):29–35. https://doi.org/10.1007/s40089-014-0133-4

Lateef A, Azeez MA, Asafa TB, Yekeen TA, Akinboro A, Oladipo IC, Ajetomobi FE, Gueguim-Kana EB, Beukes LS (2015) *Cola nitida*-mediated biogenic synthesis of silver nanoparticles using seed and seed shell extracts and evaluation of antibacterial activities. Bionanoscience 5(4):196–205. https://doi.org/10.1007/s12668-015-0181-x

Lateef A, Ojo SA, Akinwale AS, Azeez L, Gueguim-Kana EB, Beukes LS (2015) Biogenic synthesis of silver nanoparticles using cell-free extract of *Bacillus safensis* LAU 13: antimicrobial, free radical scavenging and larvicidal activities. Biologia 70(10):1295–1306. https://doi.org/10.1515/biolog-2015-0164

Lateef A, Akande MA, Azeez MA, Ojo SA, Folarin BI, Gueguim-Kana EB, Beukes LS (2016) Phytosynthesis of silver nanoparticles (AgNPs) using miracle fruit plant (*Synsepalum dulcificum*) for antimicrobial, catalytic, anticoagulant, and thrombolytic applications. Nanotechnol Rev 5(6):507–520. https://doi.org/10.1515/ntrev-2016-0039

Lateef A, Akande MA, Ojo SA, Folarin BI, Gueguim-Kana EB, Beukes LS (2016b) Paper wasp nest-mediated biosynthesis of silver nanoparticles for antimicrobial, catalytic, anticoagulant, and thrombolytic applications. 3 Biotech 6(2):140. https://doi.org/10.1007/s13205-016-0459-x

Lateef A, Azeez MA, Asafa TB, Yekeen TA, Akinboro A, Oladipo IC, Azeez L, Ojo SA, Gueguim-Kana EB, Beukes LS (2016) Cocoa pod husk extract-mediated biosynthesis of silver nanoparticles: its antimicrobial, antioxidant and larvicidal activities. J Nanostruct Chem 6(2):159–169. https://doi.org/10.1007/s40097-016-0191-4

Lateef A, Azeez MA, Asafa TB, Yekeen TA, Akinboro A, Oladipo IC, Azeez L, Ajibade SE, Ojo SA, Gueguim-Kana EB, Beukes LS (2016) Biogenic synthesis of silver nanoparticles using a pod extract of *Cola nitida*: antibacterial and antioxidant activities and application as a paint additive. J Taibah Univ Sci 10(4):551–562. https://doi.org/10.1016/j.jtusci.2015.10.010

Lateef A, Ojo SA, Elegbede JA (2016) The emerging roles of arthropods and their metabolites in the green synthesis of metallic nanoparticles. Nanotechnol Rev 5(6):601–622. https://doi.org/10.1515/ntrev-2016-0049

Lateef A, Ojo SA, Folarin BI, Gueguim-Kana EB, Beukes LS (2016) Kolanut (*Cola nitida*) mediated synthesis of silver–gold alloy nanoparticles: antifungal, catalytic, larvicidal and thrombolytic applications. J Clust Sci 27(5):1561–1577. https://doi.org/10.1007/s10876-016-1019-6

Lateef A, Ojo SA, Azeez MA, Asafa TB, Yekeen TA, Akinboro A, Oladipo IC, Gueguim-Kana EB, Beukes LS (2016) Cobweb as novel biomaterial for the green and eco-friendly synthesis of silver nanoparticles. Appl Nanosci 6(6):863–874. https://doi.org/10.1007/s13204-015-0492-9

Lateef A, Ojo SA, Oladejo SM (2016) Anti-candida, anti-coagulant and thrombolytic activities of biosynthesized silver nanoparticles using cell-free extract of *Bacillus safensis* LAU 13. Process Biochem 51(10):1406–1412. https://doi.org/10.1016/j.procbio.2016.06.027
Lateef A, Ojo SA, Elegbede JA, Azeez MA, Yekeen TA, Akinboro A (2017) Evaluation of some biosynthesized silver nanoparticles for biomedical applications: hydrogen peroxide scavenging, anticoagulant and thrombolytic activities. J Clust Sci 28(3):1379–1392. https://doi.org/10.1007/s10876-016-1146-0
Lateef A, Ojo SA, Elegbede JA, Akinola PO, Akanni EO (2018) Nanomedical applications of nanoparticles for blood coagulation disorders. In: Environmental nanotechnology 2018. Springer, Cham, pp 243–277. https://doi.org/10.1007/978-3-319-76090-2_8
Lateef A, Elegbede JA, Akinola PO, Ajayi VA (2019) Biomedical applications of green synthesized-metallic nanoparticles: a review. Pan Afr J Life Sci 3:157–182
Lee N, Yoo D, Ling D, Cho MH, Hyeon T, Cheon J (2015) Iron oxide based nanoparticles for multimodal imaging and magnetoresponsive therapy. Chem Rev 115(19):10637–10689. https://doi.org/10.1021/acs.chemrev.5b00112
Li S, Yang X, Yang S, Zhu M, Wang X (2012) Technology prospecting on enzymes: application, marketing and engineering. Comput Struct Biotechnol J 2(3):e201209017. https://doi.org/10.5936/csbj.201209017
Li L, Zhang L, Carmona U, Knez M (2014) Semi-artificial and bioactive ferroxidase with nanoparticles as the active sites. Chem Comm 50(59):8021–8023. https://doi.org/10.1039/C4CC03477E
Li W, Zhang H, Zhang J, Fu Y (2015) Synthesis and sensing application of glutathione-capped platinum nanoparticles. Anal Methods 7(11):4464–4471. https://doi.org/10.1039/C5AY00365B
Li S, Li H, Chen F, Liu J, Zhang H, Yang Z, Wang B (2016) Strong coupled palladium nanoparticles decorated on magnetic graphene nanosheets as enhanced peroxidase mimetics for colorimetric detection of H_2O_2. Dyes Pigm 125:64–71. https://doi.org/10.1016/j.dyepig.2015.10.009
Li XJ, Wang YS, Yang SY, Tang X, Liu L, Zhou B, Wang XF, Zhu YF, Huang YQ, He SZ (2016) Determination of metallothioneins based on the enhanced peroxidase-like activity of mercury-coated gold nanoparticles aggregated by metallothioneins. Microchim Acta 183(7):2123–2129. https://doi.org/10.1007/s00604-016-1828-1
Lin Y, Ren J, Qu X (2014) Catalytically active nanomaterials: a promising candidate for artificial enzymes. Acc Chem Res 47(4):1097–1105. https://doi.org/10.1021/ar400250z
Lin S, Wu J, Yao J, Cao W, Muhammad F, Wei H (2018) Nanozymes for biomedical sensing applications: from in vitro to living systems. In: Biomedical applications of functionalized nanomaterials. Elsevier, Amsterdam, pp 171–209. https://doi.org/10.1016/B978-0-323-50878-0.00007-0
Liu X, Kokare C (2017) Microbial enzymes of use in industry. Biotechnology of microbial enzymes: production, biocatalysis and industrial applications. https://doi.org/10.1016/B978-0-12-803725-6.00011-X
Liu Y, Wu H, Chong Y, Wamer WG, Xia Q, Cai L, Nie Z, Fu PP, Yin JJ (2015) Platinum nanoparticles: efficient and stable catechol oxidase mimetics. ACS Appl Mater Interf 35:19709–19717. https://doi.org/10.1021/acsami.5b05180
Liu F, Tang J, Xu J, Shu Y, Xu Q, Wang H, Hu X (2016) Low potential detection of indole-3-acetic acid based on the peroxidase-like activity of hemin/reduced graphene oxide nanocomposite. Biosens Bioelectron 86:871–878. https://doi.org/10.1016/j.bios.2016.07.089
Liu Q, Yang Y, Lv X, Ding Y, Zhang Y, Jing J, Xu C (2017) One-step synthesis of uniform nanoparticles of porphyrin functionalized ceria with promising peroxidase mimetics for H_2O_2 and glucose colorimetric detection. Sens Actuators B Chem 240:726–734. https://doi.org/10.1016/j.snb.2016.09.049
Liu CP, Wu TH, Liu CY, Chen KC, Chen YX, Chen GS, Lin SY (2017) Self-supplying O_2 through the catalase-like activity of gold nanoclusters for photodynamic therapy against hypoxic cancer cells. Small 13(26):1700278. https://doi.org/10.1002/smll.201700278

Luo W, Zhu C, Su S, Li D, He Y, Huang Q, Fan C (2010) Self-catalyzed, self-limiting growth of glucose oxidase-mimicking gold nanoparticles. ACS Nano 4(12):7451–7458. https://doi.org/10.1021/nn102592h

Ma S, Cai Q, Lu K, Liao F, Shao M (2016) Bi-functional Au/FeS (Au/Co_3O_4) composite for in situ SERS monitoring and degradation of organic pollutants. J Nanopart Res 18(1):26. https://doi.org/10.1007/s11051-016-3335-z

Mishra A, Sardar M (2012) Alpha-amylase mediated synthesis of silver nanoparticles. Sci Adv Mater 4(1):143–146. https://doi.org/10.1166/sam.2012.1263

Mishra A, Sardar M (2014) Alpha amylase mediated synthesis of gold nanoparticles and their application in the reduction of nitroaromatic pollutants. Energy Environ 3(2):179–184. https://doi.org/10.1166/eef.2014.1077

Moglianetti M, De Luca E, Pedone D, Marotta R, Catelani T, Sartori B, Amenitsch H, Retta SF, Pompa PP (2016) Platinum nanozymes recover cellular ROS homeostasis in an oxidative stress-mediated disease model. Nanoscale 8(6):3739–3752. https://doi.org/10.1039/C5NR08358C

Moteshafi H, Mousavi SM, Shojaosadati SA (2012) The possible mechanisms involved in nanoparticles biosynthesis. J Ind Eng Chem 18(6):2046–2050. https://doi.org/10.1016/j.jiec.2012.05.025

Mukherjee P, Ahmad A, Mandal D, Senapati S, Sainkar SR, Khan MI, Parishcha R, Ajaykumar PV, Alam M, Kumar R, Sastry M (2001) Fungus-mediated synthesis of silver nanoparticles and their immobilization in the mycelial matrix: a novel biological approach to nanoparticle synthesis. Nano Lett 1(10):515–519. https://doi.org/10.1021/nl0155274

Narayanan KB, Sakthivel N (2010) Biological synthesis of metal nanoparticles by microbes. Adv Colloid Interf Sci 156(1–2):1–3. https://doi.org/10.1016/j.cis.2010.02.001

Niu X, He Y, Pan J, Li X, Qiu F, Yan Y, Shi L, Zhao H, Lan M (2016) Uncapped nanobranch-based CuS clews used as an efficient peroxidase mimic enable the visual detection of hydrogen peroxide and glucose with fast response. Anal Chim Acta 947:42–49. https://doi.org/10.1016/j.aca.2016.10.013

Ogi T, Saitoh N, Nomura T, Konishi Y (2010) Room-temperature synthesis of gold nanoparticles and nanoplates using *Shewanella* algae cell extract. J Nanopart Res 12(7):2531–2539. https://doi.org/10.1007/s11051-009-9822-8

Ojo SA, Lateef A, Azeez MA, Oladejo SM, Akinwale AS, Asafa TB, Yekeen TA, Akinboro A, Oladipo IC, Gueguim-Kana EB, Beukes LS (2016) Biomedical and catalytic applications of gold and silver-gold alloy nanoparticles biosynthesized using cell-free extract of *Bacillus safensis* LAU 13: antifungal, dye degradation, anti-coagulant and thrombolytic activities. IEEE Trans Nanobiosci 15(5):433–442. https://doi.org/10.1109/TNB.2016.2559161

Oladipo IC, Lateef A, Elegbede JA, Azeez MA, Asafa TB, Yekeen TA, Akinboro A, Gueguim-Kana EB, Beukes LS, Oluyide TO, Atanda OR (2017) *Enterococcus* species for the one-pot biofabrication of gold nanoparticles: characterization and nanobiotechnological applications. J Photochem Photobiol B 173:250–257. https://doi.org/10.1016/j.jphotobiol.2017.06.003

Oladipo IC, Lateef A, Azeez MA, Asafa TB, Yekeen TA, Akinboro A, Akinwale AS, Gueguim-Kana E.B, Beukes LS (2017b) Green synthesis and antimicrobial activities of silver nanoparticles using cell-free extracts of *Enterococcus* species. Not Sci Biol 9(2):196–203. https://doi.org/10.15835/nsb929938

Ovais M, Khalil AT, Islam NU, Ahmad I, Ayaz M, Saravanan M, Shinwari ZK, Mukherjee S (2018) Role of plant phytochemicals and microbial enzymes in biosynthesis of metallic nanoparticles. Appl Microbiol Biotechnol 102(16):6799–6814. https://doi.org/10.1007/s00253-018-9146-7

Ovais M, Khalil AT, Ayaz M, Ahmad I, Nethi SK, Mukherjee S (2018) Biosynthesis of metal nanoparticles via microbial enzymes: a mechanistic approach. Int J Mol Sci 19(12):4100. https://doi.org/10.3390/ijms19124100

Pandey PC, Singh R, Pandey Y (2015) Controlled synthesis of functional Ag, Ag–Au/Au–Ag nanoparticles and their Prussian blue nanocomposites for bioanalytical applications. RSC Adv 5(61):49671–49679. https://doi.org/10.1039/C5RA06251A

Pandian SR, Deepak V, Kalishwaralal K, Viswanathan P, Gurunathan S (2010) Mechanism of bactericidal activity of silver nitrate-a concentration dependent bi-functional molecule. Brazilian J Microbiol 41(3):805–809. https://doi.org/10.1590/S1517-83822010000300033
Pereira AR, Iost RM, Martins MV, Yokomizo CH, da Silva WC, Nantes IL, Crespilho FN (2011) Molecular interactions and structure of a supramolecular arrangement of glucose oxidase and palladium nanoparticles. Phys Chem Chem Phys 13(26):12155–12162. https://doi.org/10.1039/C1CP20432G
Poeta P, Dias AA, Igrejas G, Silva V, Bezerra R, Nunes CS (2018) Selection, engineering, and expression of microbial enzymes. In: Enzymes in human and animal nutrition. Academic Press, Cambridge, pp 1–29. https://doi.org/10.1016/B978-0-12-805419-2.00001-0
Ragg R, Natalio F, Tahir MN, Janssen H, Kashyap A, Strand D, Strand S, Tremel W (2014) Molybdenum trioxide nanoparticles with intrinsic sulfite oxidase activity. ACS Nano 8(5):5182–5189. https://doi.org/10.1021/nn501235j
Ragg R, Schilmann AM, Korschelt K, Wieseotte C, Kluenker M, Viel M, Völker L, Preiß S, Herzberger J, Frey H, Heinze K (2016) Intrinsic superoxide dismutase activity of MnO nanoparticles enhances the magnetic resonance imaging contrast. J Mater Chem B 4(46):7423–7428. https://doi.org/10.1039/C6TB02078J
Rai T, Panda D (2015) An extracellular enzyme synthesizes narrow-sized silver nanoparticles in both water and methanol. Chem Phys Lett 623:108–112. https://doi.org/10.1016/j.cplett.2015.02.003
Rai M, Yadav A, Gade A (2009) Silver nanoparticles as a new generation of antimicrobials. Biotechnol Adv 27(1):76–83. https://doi.org/10.1016/j.biotechadv.2008.09.002
Raynal M, Ballester P, Vidal-Ferran A, van Leeuwen PW (2014) Supramolecular catalysis. Part 2: artificial enzyme mimics. Chem Soc Rev 43:1734–1787. https://doi.org/10.1039/C3CS60037H
Ren X, Liu J, Ren J, Tang F, Meng X (2015) One-pot synthesis of active copper-containing carbon dots with laccase-like activities. Nanoscale 7(46):19641–19646. https://doi.org/10.1039/C5NR04685H
Ren C, Hu X, Zhou Q (2018) Graphene oxide quantum dots reduce oxidative stress and inhibit neurotoxicity in vitro and in vivo through catalase-like activity and metabolic regulation. Adv Sci 5(5):1700595. https://doi.org/10.1002/advs.201700595
Sabu, A (2003) Sources, properties and applications of microbial therapeutic enzymes. Ind J Biotechnol 2:334–341. https://hdl.handle.net/123456789/11329
Safaepour M, Shahverdi AR, Shahverdi HR, Khorramizadeh MR, Gohari AR (2009) Green synthesis of small silver nanoparticles using geraniol and its cytotoxicity against fibrosarcoma-wehi 164. Avicenna J Med Biotechnol 1(2):111
Salvio R, Cincotti A (2014) Guanidine based self-assembled monolayers on Au nanoparticles as artificial phosphodiesterases. RSC Adv 4(54):28678–28682. https://doi.org/10.1039/C4RA03150D
Samuel EL, Marcano DC, Berka V, Bitner BR, Wu G, Potter A, Fabian RH, Pautler RG, Kent TA, Tsai AL, Tour JM (2015) Highly efficient conversion of superoxide to oxygen using hydrophilic carbon clusters. Proc Natl Acad Sci 112(8):2343–2348. https://doi.org/10.1073/pnas.1417047112
Sanchez S, Demain A (2017) Useful microbial enzymes—an introduction. Biotechnology of microbial enzymes: production, biocatalysis and industrial applications. https://doi.org/10.1016/B978-0-12-803725-6.00001-7
Sanghi R, Verma P, Puri S (2011) Enzymatic formation of gold nanoparticles using *Phanerochaete chrysosporium*. Adv Chem Eng Sci 1(03):154. https://doi.org/10.4236/aces.2011.13023
Saranraj P, Naidu MA (2014) Microbial pectinases: a review. Global J Trad Med Syst 3(1):1–9
Schladt TD, Graf T, Tremel W (2009) Synthesis and characterization of monodisperse manganese oxide nanoparticles—evaluation of the nucleation and growth mechanism. Chem Mater 21(14):3183–3190. https://doi.org/10.1021/cm900663t
Senapati S, Ahmad A, Khan MI, Sastry M, Kumar R (2005) Extracellular biosynthesis of bimetallic Au–Ag alloy nanoparticles. Small 1(5):517–520. https://doi.org/10.1002/smll.200400053

Sennuga A, Van Marwijk J, Whiteley CG (2011) Ferroxidase activity of apoferritin is increased in the presence of platinum nanoparticles. Nanotechnology 23(3):035102. https://doi.org/10.1088/0957-4484/23/3/035102

Shaligram NS, Bule M, Bhambure R, Singhal RS, Singh SK, Szakacs G, Pandey A (2009) Biosynthesis of silver nanoparticles using aqueous extract from the compactin producing fungal strain. Process Biochem 44(8):939–943. https://doi.org/10.1016/j.procbio.2009.04.009

Shankar S, Jaiswal L, Aparna RS, Prasad RG (2014) Synthesis, characterization, in vitro biocompatibility, and antimicrobial activity of gold, silver and gold silver alloy nanoparticles prepared from *Lansium domesticum* fruit peel extract. Mater Lett 137:75–78. https://doi.org/10.1016/j.matlet.2014.08.122

Sharma H, Upadhyay SK (2020) Enzymes and their production strategies. In: Biomass, biofuels, biochemicals, pp 31–48. https://doi.org/10.1016/B978-0-12-819820-9.00003-X

Shedbalkar U, Singh R, Wadhwani S, Gaidhani S, Chopade BA (2014) Microbial synthesis of gold nanoparticles: current status and future prospects. Adv Colloid Interf Sci 209:40–48. https://doi.org/10.1016/j.cis.2013.12.011

Shin J, Lee S, Cha M (2017) Neuroprotective effect of single-wall carbon nanotubes with built-in peroxidase-like activity against β-amyloid-induced neurotoxicity. Med Chem Comm 8(3):625–632. https://doi.org/10.1039/C6MD00716C

Silva VB, Rodrigues TS, Camargo PH, Orth ES (2017) Detoxification of organophosphates using imidazole-coated Ag. Au and Ag-Au Nanoparticles. RSC Adv 7(65):40711–40719. https://doi.org/10.1039/C7RA07059D

Singh S (2019) Nanomaterials exhibiting enzyme-like properties (Nanozymes): current advances and future perspectives. Front Chem 7:46. https://doi.org/10.3389/fchem.2019.00046

Singh S, Dosani T, Karakoti AS, Kumar A, Seal S, Self WT (2011) A phosphate-dependent shift in redox state of cerium oxide nanoparticles and its effects on catalytic properties. Biomaterials 32(28):6745–6753. https://doi.org/10.1016/j.biomaterials.2011.05.073

Singh R, Kumar M, Mittal A, Mehta PK (2016) Microbial enzymes: industrial progress in 21st century. 3 Biotech 6(2):174. https://doi.org/10.1007/s13205-016-0485-8

Singh R, Shedbalkar UU, Nadhe SB, Wadhwani SA, Chopade BA (2017) Lignin peroxidase mediated silver nanoparticle synthesis in *Acinetobacter* sp. AMB Express 7(1):226. https://doi.org/10.1186/s13568-017-0528-5

Slocik JM, Naik RR, Stone MO, Wright DW (2005) Viral templates for gold nanoparticle synthesis. J Mater Chem 15(7):749–753. https://doi.org/10.1039/B413074J

Song W, Nie G, Ji W, Jiang Y, Lu X, Zhao B, Ozaki Y (2016) Synthesis of bifunctional reduced graphene oxide/CuS/Au composite nanosheets for in situ monitoring of a peroxidase-like catalytic reaction by surface-enhanced Raman spectroscopy. RSC Adv 6(59):54456–54462. https://doi.org/10.1039/C6RA09471F

Soozanipour A, Taheri-Kafrani A, Isfahani AL (2015) Covalent attachment of xylanase on functionalized magnetic nanoparticles and determination of its activity and stability. Chem Eng J 270:235–243. https://doi.org/10.1016/j.cej.2015.02.032

Tan LH, Xing H, Lu Y (2014) DNA as a powerful tool for morphology control, spatial positioning, and dynamic assembly of nanoparticles. Acc Chem Res 47(6):1881–1890. https://doi.org/10.1021/ar500081k

Thakkar KN, Mhatre SS, Parikh RY (2010) Biological synthesis of metallic nanoparticles. Nanomed Nanotechnol Biol Med 6(2):257–262. https://doi.org/10.1016/j.nano.2009.07.002

Tian L, Qi J, Qian K, Oderinde O, Liu Q, Yao C, Song W, Wang Y (2018) Copper (II) oxide nanozyme based electrochemical cytosensor for high sensitive detection of circulating tumor cells in breast cancer. J Electroanal Chem 812:1–9. https://doi.org/10.1016/j.jelechem.2017.12.012

Vernekar AA, Sinha D, Srivastava S, Paramasivam PU, D'Silva P, Mugesh G (2014) An antioxidant nanozyme that uncovers the cytoprotective potential of vanadia nanowires. Nat Comm 5(1):1–3. https://doi.org/10.1038/ncomms6301

Wadhwani SA, Shedbalkar UU, Singh R, Chopade BA (2018) Biosynthesis of gold and selenium nanoparticles by purified protein from *Acinetobacter* sp. SW 30. Enzyme Microb Technol 111:81–86. https://doi.org/10.1016/j.enzmictec.2017.10.007

Wang H, Jiang W, Wang Y, Liu X, Yao J, Yuan L, Wu Z, Li D, Song B, Chen H (2013) Catalase-like and peroxidase-like catalytic activities of silicon nanowire arrays. Langmuir 29(1):3–7. https://doi.org/10.1021/la304378w

Wang GL, Jin LY, Dong YM, Wu XM, Li ZJ (2015) Intrinsic enzyme mimicking activity of gold nanoclusters upon visible light triggering and its application for colorimetric trypsin detection. Biosens Bioelectron 64:523–529. https://doi.org/10.1016/j.bios.2014.09.071

Wang X, Hu Y, Wei H (2016) Nanozymes in bionanotechnology: from sensing to therapeutics and beyond. Inorg Chem Front 3(1):41–60. https://doi.org/10.1039/C5QI00240K

Wang X, Cao W, Qin L, Lin T, Chen W, Lin S, Yao J, Zhao X, Zhou M, Hang C, Wei H (2017) Boosting the peroxidase-like activity of nanostructured nickel by inducing its 3^+ oxidation state in $LaNiO_3$ perovskite and its application for biomedical assays. Theranostics 7(8):2277. https://doi.org/10.7150/thno.19257

Wang Y, He C, Li W, Zhang J, Fu Y (2017) Catalytic performance of oligonucleotide-templated Pt nanozyme evaluated by laccase substrates. Cat Lett 147(8):2144–2152. https://doi.org/10.1007/s10562-017-2106-5

Wei H, Wang E (2008) Fe_3O_4 magnetic nanoparticles as peroxidase mimetics and their applications in H_2O_2 and glucose detection. Anal Chem 80(6):2250–2254. https://doi.org/10.1021/ac702203f

Wei H, Wang E (2013) Nanomaterials with enzyme-like characteristics (nanozymes): next-generation artificial enzymes. Chem Soc Rev 42:6060–6093. https://doi.org/10.1039/c3cs35486e

Wei H, Wang Z, Zhang J, House S, Gao YG, Yang L, Robinson H, Tan LH, Xing H, Hou C, Robertson IM (2011) Time-dependent, protein-directed growth of gold nanoparticles within a single crystal of lysozyme. Nat Nanotechnol 6(2):93. https://doi.org/10.1038/nnano.2010.280

Wei J, Chen X, Shi S, Mo S, Zheng N (2015) An investigation of the mimetic enzyme activity of two-dimensional Pd-based nanostructures. Nanoscale 7(45):19018–19026. https://doi.org/10.1039/C5NR05675F

Wu J, Tan LH, Hwang K, Xing H, Wu P, Li W, Lu Y (2014) DNA sequence-dependent morphological evolution of silver nanoparticles and their optical and hybridization properties. J Am Chem Soc 136(43):15195–151202. https://doi.org/10.1021/ja506150s

Wu J, Li S, Wei H (2018) Multifunctional nanozymes: enzyme-like catalytic activity combined with magnetism and surface plasmon resonance. Nanoscale Horiz 3(4):367–382. https://doi.org/10.1039/c8nh00070k

Wu J, Wang X, Wang Q, Lou Z, Li S, Zhu Y, Qin L, Wei H (2019) Nanomaterials with enzyme-like characteristics (nanozymes): next-generation artificial enzymes (II). Chem Soc Rev 48(4):1004–1076. https://doi.org/10.1039/c8cs00457a

Xie J, Lee JY, Wang DI, Ting YP (2007a) Identification of active biomolecules in the high-yield synthesis of single-crystalline gold nanoplates in algal solutions. Small 3(4):672–682. https://doi.org/10.1002/smll.200600612

Xie J, Lee JY, Wang DI, Ting YP (2007b) Silver nanoplates: from biological to biomimetic synthesis. ACS Nano 1(5):429–439. https://doi.org/10.1021/nn7000883

Xie X, Xu W, Liu X (2012) Improving colorimetric assays through protein enzyme-assisted gold nanoparticle amplification. Acc Chem Res 45(9):1511–1520. https://doi.org/10.1021/ar300044j

Xu C, Liu Z, Wu L, Ren J, Qu X (2014) Nucleoside triphosphates as promoters to enhance nanoceria enzyme-like activity and for single-nucleotide polymorphism typing. Adv Funct Mater 24(11):1624–1630. https://doi.org/10.1002/adfm.201301649

Yao J, Cheng Y, Zhou M, Zhao S, Lin S, Wang X, Wu J, Li S, Wei H (2018) ROS scavenging Mn_3O_4 nanozymes for in vivo anti-inflammation. Chem Sci 9(11):2927–2933. https://doi.org/10.1039/C7SC05476A

Zarif F, Rauf S, Qureshi MZ, Shah NS, Hayat A, Muhammad N, Rahim A, Nawaz MH, Nasir M (2018) Ionic liquid coated iron nanoparticles are promising peroxidase mimics for optical determination of H_2O_2. Microchim Acta 185(6):302. https://doi.org/10.1007/s00604-018-2841-3
Zeng Y, Miao F, Zhao Z, Zhu Y, Liu T, Chen R, Liu S, Lv Z, Liang F (2017) Low-cost nanocarbon-based peroxidases from graphite and carbon fibers. Appl Sci 7(9):924. https://doi.org/10.3390/app7090924
Zhang L, Han L, Hu P, Wang L, Dong S (2013) TiO_2 nanotube arrays: intrinsic peroxidase mimetics. Chem Comm 49(89):10480–10482. https://doi.org/10.1039/C3CC46163G
Zheng AX, Cong ZX, Wang JR, Li J, Yang HH, Chen GN (2013) Highly-efficient peroxidase-like catalytic activity of graphene dots for biosensing. Biosens Bioelectron 49:519–524. https://doi.org/10.1016/j.bios.2013.05.038
Zhou H, Han T, Wei Q, Zhang S (2016) Efficient enhancement of electrochemiluminescence from cadmium sulfide quantum dots by glucose oxidase mimicking gold nanoparticles for highly sensitive assay of methyltransferase activity. Anal Chem 88(5):2976–2983. https://doi.org/10.1021/acs.analchem.6b00450
Zhou Y, Liu B, Yang R, Liu J (2017) Filling in the gaps between nanozymes and enzymes: challenges and opportunities. Bioconj Chem 28(12):2903–2909. https://doi.org/10.1021/acs.bioconjchem.7b00673

Microalgal Nanobiotechnology and Its Applications—A Brief Overview

I. A. Adelere and A. Lateef

1 Introduction

Nanotechnology is a branch of science that deals with the synthesis and application of nanosized materials. It is emerging as a fast growing field with many applications virtually in every aspect of life due to the production of nanomaterials. Nanoparticles are the building block of nanotechnology and are presently gaining importance as they possess very unique physicochemical properties. In nanotechnology, metallic and non-metallic nanoparticles are produced, but metallic nanoparticles are the most preferred due to their superior properties and functional versatility. Nanoparticles were previously produced only by physical and chemical methods like chemical reduction, electrochemical and photochemical reactions, attrition, and pyrolysis (Rodriguez-Sánchez et al. 2000; Balantrapu and Goia 2009; Chen et al. 2013). Basically, two approaches are adopted for the synthesis of nanoparticles, namely bottom-up and top-down approaches. The top-down approach involves removal of certain parts of bulk material using mechanical (physical) or chemical means, whereas in the bottom-up approach, the synthesis involves attachment of materials with a base substrate (Singh et al. 2010).

The common physical and chemical approaches used for nanoparticles synthesis have some major drawbacks which are the presence of toxic chemical contaminants from precursor chemicals, generation of harmful by-products, high production cost,

I. A. Adelere (✉)
Department of Microbiology, Federal University of Technology, Minna, Nigeria
e-mail: isiaka.ade@futminna.edu.ng

I. A. Adelere · A. Lateef
Laboratory of Industrial Microbiology and Biotechnology, Department of Pure and Applied Biology, Ladoke Akintola University of Technology, Ogbomoso, Nigeria

A. Lateef
Nanotechnology Research Group (NANO+), Ladoke Akintola University of Technology, Ogbomoso, Nigeria

A. Lateef et al. (eds.), *Microbial Nanobiotechnology*, Materials Horizons: From Nature to Nanomaterials, https://doi.org/10.1007/978-981-33-4777-9_8

and too much of energy consumption (Li et al. 1999). These challenges have drastically limited the areas of applications of nanotechnology and therefore necessitated the development of better alternative. However, the involvement of biological materials for nanoparticles synthesis has shown to be better alternative as it promotes the production of nanoparticles in a cost-effective manner. The procedure is simple, environmentally benign, relatively reproducible, and also suitable for the large-scale production. The integration of biological principles to nanotechnology has become an important area in nanoscience and had gained very high momentum in recent times (Adelere and Lateef 2016; Lateef et al. 2016a, 2018a, 2019; Elegbede and Lateef 2019a, b).

Nanobiotechnology is a newly developed area of nanoscience that combines biological principles with the existing nanotechnological procedures. It was developed to improve the previous strategies for nanoparticles synthesis and applications. It had received remarkable attentions as a result of its advancement over physical and chemical methods since it is simple, cost effective, sustainable, and ecological friendly. Biological principles have been adopted to synthesize metallic and nonmetallic nanoparticles of suitable physicochemical properties with potential applications in diverse areas of human endeavors. Researchers have exploited very considerable diversity of plants, arthropods, and microorganisms to fabricate a wide variety of functional nanomaterials (Lateef et al. 2016b, c, d, e, 2018b, 2020; Ojo et al. 2016; Oladipo et al. 2017; Elegbede et al. 2020). We have documented a comprehensive review on the utilization of agrowastes, enzymes, biological pigments, and metabolites of arthropods for the production of metallic nanoparticles (Adelere and Lateef 2016; Lateef et al. 2016a). The reviews revealed the emerging importance of biological resources to produce biocompatible and efficient nanoparticles. It has been suggested that biomolecules such as flavonoids, terpenoids, polyols, and organic acids (Shankar et al. 2004; Begum et al. 2009; Sharon et al. 2012) in the biological resources control the fabrication of nanoparticles (Ghashghaei and Emtiazi 2013).

The carbonyl functional group of aldehyde, ketone, and carboxylic acid present in biomolecules acts as reducing agent by donating electrons to metallic ions to form corresponding metallic nanoparticles (Shankar et al. 2004). Also, several protein molecules are recognized capable of reducing metallic ions to metallic nanoparticles. They bind with the metal through their free amine groups or carboxylate ions (Gopinath et al. 2012). They act as capping agents, thereby increasing their stability by preventing oxidation or agglomeration. Thus, enzymes being proteinous in nature have been used to produce metal nanoparticles (Lateef and Adeeyo 2015; Lateef et al. 2015; Adelere and Lateef 2016; Parashar et al. 2017; Singh et al. 2017; Ovais et al. 2018; Wadhwani et al. 2018; Elegbede et al. 2018, 2019, 2020).

Among the biological systems that have been used to synthesize nanoparticles, microalgae attract huge attention due to their ability to adsorb toxic metals and subsequently convert them to non-toxic form (Patel et al. 2015). Microalgae are capable of producing nanoparticles of various metals like silver, gold, palladium, and platinum (Ebrahiminezhad et al. 2016; Khalid et al. 2017; Priyadarshini et al. 2019; Sathishkumar et al. 2019). Among the algae, microalgae are recognized as the most suitable factories for nanoparticles synthesis, due to their high growth rate and high biomass productivity with a less cultivation time. They have high ability for heavy metal detoxification and also can produce varieties of important metabolites capable of nanoparticles fabrication (Ahmad et al. 2011). Algae generally do not require complex carbon and nitrogen sources for their growth. They utilize sunlight as source of energy, carbon dioxide as carbon source, and ammonium salts as nitrogen source. Hence, they are economic for nanotechnological processes in large scales (Ebrahiminezhad et al. 2014). The study of algae-mediated biosynthesis of nanomaterials is termed as phyconanotechnology (Sharma et al. 2016). The evolvement of nanobiotechnology had tremendously increase the popularity of nanotechnology as it attracted attentions of many researchers to the field which had led to the productions of varieties of biosynthesized nanomaterials with a wide range of applications. This article, therefore, reviewed various biosynthetic procedures involving the use of many microalgal species for nanoparticles production. The potential applications of nanomaterials fabricated through this process are also summarized.

2 Microalgae in Biotechnology

They are primitive microscopic plants that are present in aquatic and terrestrial ecosystems, representing a diverse variety of species. They can grow rapidly with a minimum growth requirement of sunlight, atmospheric CO_2, and a few common mineral salts to reproduce their biomass. Unlike higher plants, microalgae grow extremely rapidly. They double their mass on average ten times faster than higher plants. They have very promising applications for a large number of biotechnological areas, including cosmetics, pharmaceuticals, nutrition, food additives, aquaculture, and pollution control such as wastewater treatment (Mata et al. 2010; Ahmad et al. 2011; von Alvensleben and Heimann 2018; Rizwan et al. 2018; Xiong et al. 2018; Schulze et al. 2019). Microalgae produce many high value products including fatty acids, protein, carbohydrates, pigments, enzymes, and microalgal biomass, which are considered a healthy food supply (García et al. 2017; Molino et al. 2018; Galasso et al. 2019; Matos et al. 2020). Their lipids and carbohydrates can be further converted into biofuel. They are considered as an alternative renewable biofuel because of their very rapid growth rate and higher photosynthetic efficiency (Milano et al. 2016; Shuba and Kifle 2018; Menegazzo and Fonseca 2019). They are also an important for phycoremediation of toxic metals (Abinandan et al. 2018; Batool et al. 2019; Li et al. 2019; Naveed et al. 2019).

Two microalgae species, *Chlorella* and *Spirulina*, dominated the microalgal market food (Salati et al. 2017; Soni et al. 2017; Sathasivam et al. 2019), due to their high protein content and nutritive value. Their biomass is sold as tablets, capsules, and liquids. They can be used as a nutritional supplement or as a source of natural food colorants (Salim et al. 2019; Tang et al. 2020). Also, some microalgal species have good health-promoting effects, like protection against renal failure, alleviation of hyperlipidemia, suppression of hypertension, enhancement of intestinal *Lactobacillus* growth, suppression of elevated serum glucose level, promotion of mental health, and others (Caporgno and Mathys 2018; Charles et al. 2019; Koyande et al. 2019; Fields et al. 2020). Microalgae are more important as live feed in aquaculture (Walker and Berlinsky 2011). They are used to feed culture larvae, juvenile shell, finfish and also for raising the zooplankton required for feeding of juvenile animals (Taelman et al. 2013). The commonly used microalgae in aquaculture are species of *Chlorella*, *Tetraselmis*, *Isochrysis*, *Pavlova*, *Phaeodactylum*, *Chaetoceros*, *Nannochloropsis*, *Skeletonema*, and *Thalassiosira* (Roy and Pal 2015).

The use of biofertilizer is both economical and environment friendly. Members of cyanobacteria are rich sources of biologically active constituents that can be used as biofertilizers (Garcia-Gonzalez and Sommerfeld 2016). Cyanobacteria play an important role in fixing atmospheric nitrogen with the aid of their heterocysts. They produce complex organic carbon compounds that can bind to soil particles to improve its quality such as soil structure, soil permeability, and water-holding capacity of soil (Nisha et al. 2007). Phycoremediation is the process of using microalgae for the removal or biotransformation of organic pollutants from environments with concomitant biomass propagation (Renuka et al. 2015). The potential of microalgae in metal biosorption process has been studied extensively due to their ubiquitous occurrence in nature. Many algal genera are known to have abilities to accumulate heavy metals, thereby reducing their toxic effects in the environments (Jais et al. 2017). There are many features in algae that make them suitable for the bioremediation of heavy metals; these include high heavy metal tolerance, phototaxy, ability to grow both autotrophically and heterotrophically, large surface area/volume ratios, phytochelatin expression, and potential for genetic manipulation (Gaur and Rai 2001). The accumulation of heavy metals by algae provides an advantage for phytoremediation over other methods that are more costly and not environmental friendly (Khatoon and Pal 2015). Microalgae are effective in nanoparticles synthesis due to their rapid growth rate to produce large amount of biomass in a very short time and their ability to accumulate high quantity of metals followed by subsequent reduction to metallic nanoparticles. Their production costs are also negligible; therefore, the biogenic mode of nanoparticles synthesis is very important for economic prospects and applications. Moreover, nanoproducts, synthesized from algae, are expected to be biocompatible, therefore minimize environment and public health risks (Khatoon and Pal 2015).

3 Nanoparticles

Nanoparticles are particulate materials that have size ranging from 1 to 100 nm. They have large surface to volume ratio and many unique physicochemical properties which facilitate their applications in areas like imaging, catalysis, medical applications, energy-based research, and environmental applications (Khan et al. 2019). They are classified into different groups on the basis of their physicochemical properties. Those groups include metal NPs, carbon-based NPs, polymeric NPs, and ceramic NPs. Carbon nanotubes (CNTs) belong to group of carbon-based NPs. They are elongated, tubular structure, with average diameter of 1–2 nm (Ibrahim 2013). They can be predicted as metallic or semiconducting based on their diameter telicity (Aqel et al. 2012). They are widely produced by deposition of carbon precursors, especially the atomic carbons, vaporized from graphite by laser or by electric arc on to metal particles. Lately, they have also been produced through green approach (Wang et al. 2018). They have high electrical conductivity, strong strength, and they are very versatile (Astefanei et al. 2015) for different applications.

Metal NPs are basically made of metals precursors. Due to well-known localized surface plasmon resonance (LSPR) characteristics, they possess good optoelectrical properties. Metallic nanoparticles such as copper, silver, and gold have abroad absorption band in the visible zone of the electromagnetic solar spectrum. The facet, size, and shape controlled synthesis of metal NPs is important in cutting-edge materials (Dreaden et al. 2012). Metallic nanoparticles are widely recognized due to their broad range of applications such as catalysis, electronics, optics, and fabrication of nanodevices. Ceramics nanoparticles are inorganic non-metallic solids, synthesized through heating and successive cooling. They can be found in amorphous, polycrystalline, dense, porous, or hollow forms. They include oxides, carbonates, caribides, and phosphates of metals and metalloids. They are receiving great attention because of their application in many areas such as catalysis, photocatalysis, photodegradation of dyes, and imaging applications (Thomas et al. 2015).

Semiconductor nanoparticles have broad bandgaps and thereby showed significant alteration in their properties with bandgap tuning. They are very useful in photocatalysis, photo optics, and electronic devices (Koole et al. 2014). Varieties of semiconductor nanoparticles are very efficient in water splitting applications owing to their suitable bandgap and band edge positions (Hisatomi et al. 2015). Polymeric nanoparticles are mostly nanospheres or nanocapsular shaped. They are readily functionalized and thus find bundles of applications in many areas (Abouelmagd et al. 2016). Furthermore, lipid-based nanoparticles are the types that possess a solid core made of lipid with a matrix containing soluble lipophilic molecules. Lipid nanotechnology (Mashaghi et al. 2013) involves the fabrication of lipid-based nanoparticles for various applications such as drug carriers and delivery (Yingchoncharoen et al. 2016).

4 Algal Nanobiotechnology

The abundance and ease of availability of algae make them very essential for nanobiotechnological studies. The study of algae-mediated biosynthesis of nanomaterials had been taken to a newer branch named as phyconanotechnology (Sharma et al. 2015). Algae are relatively simple to grow, along with several other advantages such as synthesis of nanoparticles at low temperature with greater energy efficiency and less toxicity to the ecosystem. The synthesis of nanoparticles is dose dependent, and it is also related to the type of algae used. Microalgae have gained lot of interest in nanobiotechnology since they are very capable to bioremediate toxic metals and further convert them to non-toxic forms. They are effective in nanoparticles synthesis, as they grow rapidly in eco-friendly atmosphere producing large amount of biomass in a very short time and can accumulate high quantity of metals followed by subsequent reduction. Their production costs are also negligible. Quite a number of microalgal live cells and microalgal extracts has been utilized for the biosynthesis of nanomaterials as presented in Table 1.

The biosynthesis of nanoparticles using microalgae can be divided into four methods: the first method involves the exploitation of the biomolecules extracted from the disrupted cells of microalgae; the second method resides in the use of supernatants devoid of cells; this is obtained after the centrifugation or filtration of the culture media; the third method depends on harvested whole cells of microalgae, removed from their culture media and re-suspended in the corresponding salt solution to promote the biosynthesis of NPs; the last method relies on the use of living cells of microalgae maintained under their normal culturing conditions (Dahoumane et al. 2016). The design of photobioreactors for nanoparticles production is allowed by microalgae because the process often requires photosynthetic enzymatic machinery. Several microalgal species have been investigated for their potential to produce metallic nanoparticles. For instance, strains of *Leptolyngbya foveolarum, Anabaena flos-aquae*, and *Calothrix pulvinata* were successfully used for the biosynthesis of nanoparticles of Au, Ag, Pt, and Pd. After the synthesis within the cells, the NPs were released into the culture media, where they form stable colloids which ease their recovery (Brayner et al. 2007). Authors have reported the biogenic production of metallic nanoparticles using microalgal species belonging to Haptophyta, Ochrophyta, Chlorophyta, Bacillariophyta, Cyanobacteria, Euglenozoa, and Chlorophyta (Merin et al. 2010; Luangpipat et al. 2011; Mohseniazar et al. 2011; Dahoumane et al. 2012, 2014, 2016).

4.1 Phycosynthesis of Silver Nanoparticles (AgNPs)

The aqueous extract of *Chlorella pyrenoidosa* was used to synthesize AgNPs by Aziz et al. (2015). The synthesis was carried out by adding 10 ml of cell extract of *C. pyrenoidosa* to 90 ml of 1 mM $AgNO_3$ solution, followed by incubation at 28 ± 2 °C for 24 h. The biosynthesis was also performed under different pH to comprehensively study its effect on the process of synthesis. The biosynthesized AgNPs showed impressive antibacterial activities against some pathogens, possibly due to the intrinsic protein cap leading to an easier mode of entry into the bacterial cell. They had remarkable photocatalytic properties in comparison to chemically synthesized nanoparticles, due to their inherent porosity and therefore larger surface area. Khalifa et al. (2016) synthesized AgNPs through the aqueous extract of *Calothrix marchica, Anabaena oryzae,* and *Nostoc muscorum.* The biosynthesized AgNPs showed UV absorption peaks at wavelength varying from 400 to 450 nm. The cytotoxic activity of biosynthesized AgNPs was evaluated in vitro against EAC cell line. Results of

Table 1 Synthesis of nanoparticles by microalgae

Microalgal species	Reaction conditions	Mode of synthesis	Type of nanoparticles	Size (nm)	Shape	Application	References
Calothrix marchi	5 m of $AgNO_3$ (560 mg/l) + 5 ml cyanobacterial culture, at 25 °C for 28 days in the dark	Extracellular	AgNPs	–	–	Antitumor activity	Khalifa et al. (2016)
Spirulina platensis	1 g of culture pellet + 100 ml $AgNO_3$ (100 mg/l) at 28–30 °C for 24–72 h with repeated shaking	Extracellular	AgNPs	~6	Majorly spherical	–	Cepoi et al. (2015)
Nostoc linckia	0.3 g of culture pellet + 100 ml $AgNO_3$ (100 mg/l) at 28–30 °C for 24–72 h with repeated shaking	Extracellular	AgNPs	4–5	Majorly spherical	–	Cepoi et al. (2015)
Scenedesmus abundans	10 ml of cell-free extract + 90 ml of 1 mM $AgNO_3$ incubated at room temperature for 48 h	Extracellular	AgNPs	59–66	Spherical	Antibacterial activity	Aziz et al. (2014)
Spirulina platensis	2 ml of algal extract + 100 ml (1 mM) of $AgNO_3$ in drop wise at 60 °C for 10 min under constant stirring	Extracellular	AgNPs	30–50	Spherical	Antibacterial activity	Sharma et al. (2015)
Chlorella vulgaris	15 ml culture supernatant + 30 ml (5 mM) $AgNO_3$ incubated at 50 °Cfor 24 h	Extracellular	AgNPs	7	Spherical	Anticancer and antimicrobial	Ebrahiminezhad et al. (2016)

(continued)

Table 1 (continued)

Microalgal species	Reaction conditions	Mode of synthesis	Type of nanoparticles	Size (nm)	Shape	Application	References
Chlorococcum sp. MM11	0.4 g of the algal pellet + 5 mL of 0.1 M $FeCl_3$ and placed under dark conditions at 24 °C with shaking at 120 rpm for 48 h	Extracellular	FeNPs	20–50	Spherical	Metal remediation	Subramaniyam et al. (2015)
Dictyosphaerium spp HM1, HM2, and *Pectinodesmus* sp HM3	5 mM $AgNO_3$ + ethanolic extract of each microalgae in a ratio of 5:1 and incubated at 37 °C	Extracellular	AgNPs	22.5, 47.5, and 57.5	spherical and ovoid	Antimicrobial and anticancer activity	Khalid et al. (2017)
Coelastrella sp and *Phormidium* sp	500 mg of each algal culture + 100 ml of 1 mM $HAuCl_4$ incubated for 1 h in the dark conditions	Intracellular	AuNPs	25 and 30	Triangles and spherical	Antioxidant activity	MubarakAli et al. (2013)
Chlorella vulgaris	100 ml of algal broth + 10 ml of $HAuCl_4.3H_2O$ solution (0.2 g/l) incubated at 25 °C under illumination at 100 rpm for 48 h	Intracellular	AuNPs	40–60	Spherical and rod	–	Luangpipat et al. (2011)
Chlorella vulgaris	10 ml of crude extract of the alga + 50 ml of $PdCl_2$ solution (1 mM), mixture was held at 60 °C on magnetic heater stirrer under dark condition	Extracellular	PdNPs	50–90	Spherical	–	Arsiya et al. (2016)

(continued)

Table 1 (continued)

Microalgal species	Reaction conditions	Mode of synthesis	Type of nanoparticles	Size (nm)	Shape	Application	References
Nostoc commune	50 ml of EPS (10 mg/ml) of the culture + 30 mM $AgNO_3$ solution, the mixture was autoclaved at 15 psi and 121 °C for 5 min, centrifuged at 15,000 for 30 min at 20 °C	Extracellular	AgNPs	15–54	Spherical	Antibacterial and seed surface sterilizing activity	Morsy et al. (2014)
Acutodesmus dimorphus	10 ml of aqueous extract of the de-oiled biomass + 90 ml of 1 mM $AgNO_3$ stirred at room temp for 24 h	Extracellular	AgNPs	2–20	spherical	antioxidant	Chokshi et al. (2016)
Euglena gracilis	20% (v/v) of culture + 10^{-2} M each of Fe(II) and /Fe(III) ions at 20.0 ± 0.5 °C, 70–100 µmol m^{-2} s^{-1} PPF under ambient CO_2 conditions for 4 weeks	Intracellular	(Fh2L) nanoparticles	0.6–1.0	Spherical	Magnetic activity	Brayner et al. (2012)
Nitzschia diatoms	50 mL of culture suspension + 0.1 mM of $AuCl_3$ incubated for 24 h	Intracellular	AuNPs	43	Irregular	Antibacterial activity	Borase et al. (2017)
Scenedesmus sp	100 mM $AgNO_3$ + algal biomass to obtain a final concentration of 5 mM incubated at 28 °C for 72 h	Intracellular	AgNPs	15–20	Spherical	Antibacterial activity	Jena et al. (2014)

(continued)

Table 1 (continued)

Microalgal species	Reaction conditions	Mode of synthesis	Type of nanoparticles	Size (nm)	Shape	Application	References
Microcoleus sp	5 ml of cell filtrate + 95 mL of 1 mM $AgNO_3$ incubated at room temperature for 24 h	Extracellular	AgNPs	40–80	Spherical	Antibacterial activity	Sudha et al. (2013)
Amphora sp.	Aqueous extract of the diatom + $AgNO_3$ solution to a final concentration of 2×10^{-2} M $AgNO_3$ held at 35–40 °C under light	Extracellular	AgNPs	20–25	Spherical	Antibacterial activity	Jena et al. (2015)
Limnothrix sp	5 mg ml^{-1}C-phycocyanin + 10 ml of 1 mM $AgNO_3$ incubated at 25 °C, pH 7, under cool fluorescent light for 48 h	Extracellular	AgNPs	13–31	Elongated	–	Patel et al. (2015)
Cylindrospermumstagnale and *Microcheate* sp.	Aqueous extract of each species + 1 mM $AgNO_3$ held at 30 °C under light for 30–360 h	Extracellular	AgNPs	40 and 100	Spherical	–	Husain et al. (2015)
Spirulina platensis	1 g of algal biomass + 100 ml of 1 mM $HAuCl_4$ incubated at room temperature with shaking, held for 1–6 days	Extracellular	AuNPs	15–60	Spherical	–	Kalabegishvili et al. (2012)
Chlorella vulgaris and *Chaetoceros calcitrans*	10 mL of 10 mM $AgNO_3$ + 100 ml of algal extract heated at 60 °C for 1 h	Extracellular	AgNPs	50–70 and 30–35	Spherical	–	Karthikeyan et al. (2015)

trypan blue exclusion test revealed that EAC cells viability decreases as AgNPs concentration increases in all tested algae.

Aziz et al. (2014) synthesized AgNPs from cell-free extract of *Scenedesmus abundans*. The synthesis was carried out by mixing 10 ml of cell-free extract with 90 ml of 1 mM $AgNO_3$ and incubated at room temperature for 48 h. The particles exhibited maximum absorbance of 420 nm. They are well dispersed with the size ranging from 59 to 66 nm. These nanoparticles showed a broad spectrum of antimicrobial activity against both gram positive and gram negative bacteria. In a related study, Sharma et al. (2015) synthesized AgNPs using extract obtained from *Spirulina platensis* cell mass. The synthesis was done by adding 2 ml of pure microalgal extract dropwise into the 100 ml of 1 mM of $AgNO_3$ solution in 250 ml conical flask. The reaction mixture was held at 60 °C for 10 min under constant mechanical stirring. It was observed that reduction of Ag ions into AgNPs completed within 10 min. The color change was monitored visually, and the peak at 437 nm indicated the formation of AgNPs. The particles were majorly spherical in shape, monodispersed in nature with the size of 30–50 nm. They displayed potent antibacterial activities against some pathogenic bacterial isolates.

According to the study of Ebrahiminezhad et al. (2016), synthesis of AgNPs was done through the addition of 15 ml culture supernatant of *Chlorella vulgaris* to 30 ml $AgNO_3$ solution; the final $AgNO_3$ concentration was adjusted to 5 mM. The reaction was allowed to take place at 50 °C in a water bath for 24 h without stirring or shaking. UV–vis characterization of the AgNPs displayed SPR at 434 nm. Data from the FTIR analysis indicated that carbohydrates present in the supernatant functioned as the reducing and stabilizing agent for the produced AgNPs. The particles were fairly uniform and spherical shape with average mean size of 7 nm. They were stable over six months at room temperature with no observable flocculation or sedimentation traces. The biosynthesized AgNPs have promising biomedical application as they demonstrated very desirable anticancer and antimicrobial activities. Production of AgNPs by whole cell of microalga (in vivo) and fresh extract (in vitro) of *Chlorococcum humicola* was investigated by Jena et al. (2013). The in vitro phycoproduction of AgNPs was carried out by incubating 1 ml of 100 mM $AgNO_3$ with 19 ml of cell-free algal extract for 48 h at room temperature. The in vivo production was done by suspending the algal biomass in 47.5 ml of distilled water. Further, 2.5 ml of 100 mM $AgNO_3$ was added to get a 5 mM $AgNO_3$ concentration. The cultures were incubated at 28 °C for 48 h. After completion of the reaction, biomass was separated by centrifugation and stored at −20 °C till further characterization. The UV–visible absorption spectrum recorded for the particles peaked at 430 nm, while the FTIR analysis revealed the involvement of protein molecules in their formation. TEM analysis confirms that AgNPs synthesized were 16 nm in size. Also the particles showed a strong antibacterial effect against the pathogenic bacterial isolates.

The biosynthesis of AgNPs using *Spirulina platensis* aqueous extract was reported by Muthusamy et al. (2017). In the synthesis, 5 g of dried, finely powdered *Spirulina platensis* was taken in a 100-ml Erlenmeyer flask along with 50 ml of sterilized double-distilled water and then boiling the mixture for 10–15 min at 50 °C. The aqueous extract was treated with 1 mM $AgNO_3$ solution in an Erlenmeyer flask and incubated at room temperature. Optimization of the synthesis was held at pH values of 5, 6, 7, and 8 and various times of 0, 3, 6, 9, and 12 h. The UV–vis spectroscopy showed that the AgNPs synthesized have SPR around 450 nm. Particles were very dispersed with size range of 5–50 nm. They have face-centered cubic structure according to XRD result. The biosynthesized AgNPs have potential biomedical application as they demonstrated potent antibacterial activity against bacterial pathogens. The intracellular and extracellular biosynthesis of AgNPs by *Euglena intermedia* (EI) and *Euglena gracilis* (EG) strains were established by Li et al. (2014). The AgNPs produced have characteristic absorption peak of 420 nm. The TEM analyses of the in vitro and in vivo synthesized AgNPs indicated the sizes ranging from 15 to 60 nm and 6 to 24 nm in diameter, respectively. The AgNPs generally were polydispersed with spherical shape. The FTIR analysis identified amine as the predominant functional group responsible for the capping and stabilization of the particles.

Khalid et al. (2017) carried out a comparative study on three novel freshwater microalgae strains for AgNPs synthesis. In the study, ethanolic extract of *Dictyosphaerium* sp. strain HM1 (DHM1), *Dictyosphaerium* sp. strain HM2 (DHM2), and *Pectinodesmus* sp. strain HM3 (PHM3) was used to synthesize AgNPs using 5 mM solution of $AgNO_3$ with ethanolic extracts of microalgae in a ratio of 5:1 and incubated at 37 °C. SEM and TEM analyses showed homogeneous and uniformly distributed AgNPs with narrow size range for each species, but of different sizes and shapes. XRD results confirmed the crystalline structure, as well as formation of mono-centered and face-centered cubic structure of these particles. The biosynthesized AgNPs displayed significant activities against 14 pathogenic bacterial strains, *Candida albicans*, hepatocellular carcinoma (HepG2), breast cancer (MCF7) cell lines, and Newcastle disease virus (NDV). Mahdieh et al. (2012) treated $AgNO_3$ solution with *Spirulina platensis* for the synthesis of AgNPs. The synthesis was carried out by dispersing 5 g of a thoroughly washed *Spirulina platensis* biomass from an exponential growth phase in 100 ml of 1 m Maqueous $AgNO_3$ solution (pH 7) for 24 h. The entire process of metal ion reduction to nanoparticles was held at 25 °C. They particles were crystalline with an average size of 11.6 nm.

Biosynthesis of AgNPs was investigated using water soluble fraction of the extracellular polysaccharides (EPS) matrix of *Nostoc commune* (Morsy et al. 2014). In the synthesis, 50 ml of water containing EPS (10 mg/ml) of cyanobacterium *Nostoc commune* and 30 mM of $AgNO_3$ which was autoclaved at 15 psi and 121 °C for 5 min. Centrifugation was conducted at 15,000 × g for 30 min at 20 °C to precipitate the synthesized AgNPs. The particles displayed characteristic absorption peak of 415 nm, and the size was 15–54 nm. The AgNPs possessed strong antibacterial and surface sterilizing effect on seed crops against phytopathogenic fungi. Chokshi et al. (2016) reported the synthesis of AgNPs using de-oiled biomass of *Acutodesmus dimorphus*. In the synthesis, 10 ml aqueous extract of the de-oiled biomass was mixed with 90 ml of 1 mM $AgNO_3$ and stirred at room temperature for 24 h. The AgNPs were polydispersed, spherical, and 2–15 nm in size. The particles demonstrated remarkable free radicals scavenging activity. The green synthesis of AgNPs using the aqueous extract of microalgae, *Chlorella vulgaris* and the diatom, *Chaetoceros calcitrans* was investigated by Karthikeyan et al. (2015). The absorbance spectra of the biosynthesized AgNPs showed peaks at 436 and 420 nm by *C. vulgaris* and *C. calcitrans*, respectively. They are mainly spherical in shape with size ranging from 30 to 70 nm. FTIR analysis revealed that the biomolecules responsible for the reduction of Ag^+ to AgNPs are functional groups like amines, phenols and alcohols ethers, and aromatic rings.

Sudha et al. (2013) synthesized AgNPs by reacting the cellular metabolites of *Microcoleus* sp with $AgNO_3$ at room temperature for 24 h. The AgNPs are spherical in shape, well distributed with an average size of 40–80 nm. They showed remarkable antibacterial activity against some antibiotic resistance human bacterial pathogens. Most recently Adenigba et al. (2020) reported rapid biosynthesis of AgNPs through the mixture of 1 mM $AgNO_3$ with each of the cell-free aqueous extracts of *Nannochloropsis* sp and *Chlorella vulgaris.* The biosynthesized AgNPs displayed SPR at a wavelength of 426 nm for *C. vulgaris* (CV–AgNPs) and 430 nm for *Nannochloropsis* sp (NN–AgNPs). The particles were polydispersed and spherical in shape with size ranging from 27 to 90 nm and 41 to 88 nm for CV–AgNPs and NN–AgNPs, respectively. They demonstrated strong affinity for zinc and lead bioremediation from pharmaceutical effluent. The particles exhibited very potent antimicrobial activity against gram positive and gram negative bacteria. Similarly, Omomowo et al. (2020) synthesized AgNPs upon reacting the 10 ml aqueous extract of *Neodesmus pupukensis* (MG257914) with 90 ml of 1 mM $AgNO_3$ under ambient conditions. The biosynthesized AgNPs were spherical in shape and particle size ranged 52–179 nm. They exhibited strong antimicrobial and antioxidant activities which indicate their potential biotechnological application.

4.2 Phycosynthesis of Gold Nanoparticles (AuNPs)

Feurtet-Mazel et al. (2016) reported a successful intracellular biosynthesis of AuNPs by freshwater diatom *Eolimna minima* under a light intensity of 40 μmol photon m^{-2} s^{-1}, temperature of 20 °C, and a photoperiod of 12L/12D. The absorption spectrum of the biosynthesized AuNPs showed peak between 520 and 585 nm, while the particle sizes ranged from 5 to 100 nm. Dahoumane et al. (2014) gave an overview of biological pathways for the biosynthesis of gold colloids. The authors elucidated that when the living microalgal cultures were suspended in Au (III) solutions, the gold cations were taken up by the cells, migrated to the photosynthetic organelle, i.e., thylakoid, where they were reduced to AuNPs. They became stable gold colloids after being been released into the culture medium due to their interaction with the exopolysaccharides produced by the organisms. In a related study, Brayner et al. (2007) reported the intracellular synthesis of AuNPs by cyanobacterial strains of *Anabaena, Calothrix*, and *Leptolyngbya.* The intracellular nitrogenase enzyme produced by the strains played a very active role in the bioreduction of gold ion to AuNPs. The biosynthesized particles when released to the culture medium became stabilized by the algal polysaccharides. Furthermore, Dahoumane et al. (2012) investigated the cell-based bioreaction for the production of colloidal gold using microalgae, *Klebsormidium flaccidum*. In this study, green algae was grown 2 weeks in the incubator (20 °C, 16 h light/8 h dark) before the addition of $HAuCl_4$ aqueous solution. Strong color modification of the biomass and the culture medium from yellow to purple was observed within a few days suggesting gold metallic nanoparticles formation. The organism demonstrated strong adaptation to high concentration of metal ion as it showed no sign of serious physiological damage after nanoparticles recovery. The biosynthesized AuNPs were spherical with an average size of 8.6 ± 4.2 nm according to TEM analysis.

MubarakAli et al. (2013) conducted a comparative study on the synthesis and application of AuNPs by prokaryotic (*Phormidium* sp) and eukaryotic (*Coelastrella* sp) photosynthetic microorganisms. It was discovered that the bioreduction of gold ion to form AuNPs occurred intracellularly by protein molecules in the algal cells. The biosynthesized AuNPs by these prokaryote and eucakaryote were 25 nm sized nanotriangles and 30 nm sized spherical shape, respectively. They demonstrated potent potential biomedical applications. Also, Luangpipat et al. (2011) biosynthesized AuNPs by incubating the live cells of green microalgae *Chlorella vulgaris* with gold chloride solution. The reduction of gold ion by the biomolecules occurred predominantly in the cell cytoplasm. The biosynthesized AuNPs were mostly spheroidal or polyhedral in shape with diameter range of 40–60 nm. In a study conducted by Shakibaie et al. (2010), biogenic synthesis of AuNPs was reported using extract of disrupted cells of *Tetraselmis suecica*. The biosynthesized AuNPs displayed maximum SPR peak of 530 nm. They were crystalline in nature, polydispersed, with an average particle size of 79 nm. Moreover, the synthesis of bioactive AuNPs by locally isolated *Nitzschia* diatoms was reported by Borase et al. (2017). The particles displayed characteristic ruby red color with sharp absorbance peak at 529 nm. The biosynthesized AuNPs have irregular shape, average size of 43 nm with zeta potential of -16.8 mV. FTIR analysis revealed that fungal proteins and polysaccharides were actively involved in the gold ions reduction to AuNPs. The particles produced synergistic antibacterial effect when combined with conventional antibiotics against some clinical bacterial isolates.

The biosynthesis of AuNPs was also established by *Spirulina platensis* under ambient conditions (Kalabegishvili et al. 2012). Characterization of the AuNPs was done by UV–vis, TEM, SEM, EDAX, and XRD. Results of the complex analytical studies of the particles revealed that they were synthesized extracellularly through bioreduction of gold ions by proteins and enzymes produced by *Spirulina platensis* with UV maximum absorbance at 530 nm. The particles had crystalline property, spherical shape, and 30 nm average size. A large quantity of gold materials was synthesized through a simple, one-pot method (Xie et al. 2007). The synthesis was carried out by reacting chloroauric acid solution with the aqueous extract of *Chlorella vulgaris* at room temperature. It was observed that the protein molecules in the algal extract was predominantly involved in the reduction of gold ion to nanogold crystals. Precisely, a protein (gold shape-directing protein, GSP) with a molecular weight of approximately 28 kDa was identified to possess both reduction and shape-directing functionalities. The biosynthesized nanogold crystals were triangular and hexagonal in shape. The kinetics of the reduction reaction was manipulated by changing the

GSP concentration to produce nanoplates with lateral sizes ranging from nanometers to micrometers. Most recently, green synthesis of AuNPs was reported using the aqueous extract of microalgae, *Nannochloropsis* sp and *C. vulgaris* (Adenigba et al. 2020). The absorbance spectra of the biogenic AuNPs showed peak at 520 nm and 535 nm for *C. vulgaris* (CV–AuNPs) and *Nannochloropsis* sp (NN–AuNPs), respectively. They have spherical shape, polydispersed with size ranging from 7 to 20 nm (CV–AuNPs) and 7 to 26 nm (NN-AuNPs). The biosynthesized AuNPs displayed very considerable affinity to bioremediate lead and zinc from pharmaceutical effluent. Similarly, Omomowo et al. (2020) successfully biosynthesized AuNPs through the mixture of 90 ml of 1 mM gold chloride with 10 ml aqueous extract of *Neodesmus pupukensis* (MG257914) under ambient conditions. The biosynthesized AuNPs have circular shape with size ranging from 5 to 34 nm. They demonstrated considerable antimicrobial activity against some pathogenic microorganisms and strong antioxidant activity against DPPH radicals.

4.3 Others Nanoparticles

Biosynthesis of iron nanoparticles using soil microalgae, *Chlorococcum* sp. MM11, and their potential application in chromium remediation was investigated by Subramaniyam et al. (2015). The 0.1 M iron chloride solution was incubated with the exponentially growing culture of *Chlorococcum* sp. under shaking and dark conditions for a period of 48 h. Synthesis of nanoiron was confirmed by the transformation of reddish yellow of the iron chloride solution to yellowish brown colloid. The biosynthesized nanoiron displayed UV maximum absorbance at 291 nm. Characterization of the biosynthesized nanoiron by DLS and TEM confirmed the synthesis of spherical nanoiron with 20–50 nm size range. The FTIR analysis confirmed that biomolecules such as polysaccharides and glycoproteins contents of the algal cell wall function as the reducing and capping agent for the synthesized nanoiron. The phyco-synthesized nanoiron showed potential application for the bioremediation of chromium. Moreover, Brayner et al. (2007) reported the intracellular synthesis of platinum and palladium nanoparticles by strains of cyanobacteria. It was hypothesized that nitrogenase produced by the cyanobacteria bioreduced the platinum and palladium ion to their corresponding nanoparticles. The nanoparticles were discharged into the culture medium, and they were thereafter became stabilized by polysaccharides.

The biogenic production of cadmium sulfide (CdS) nanoparticles from cell-free algal extract was investigated by Rao and Pennathur (2017). The cell-free extract of *Chlamydomonas reinhardtii* was used to react with cadmium chloride and sodium sulfide salts at 65 °C for 20 min. The biosynthesized CdS nanoparticles were obtained by centrifugation at 8720 × g followed by characterization using UV–visible absorption spectroscopy, HRTEM, and FTIR. The biosynthesized CdS nanoparticles showed emission peaks at 430 nm and 470 nm. They have spherical morphology with 5 nm average size. They were stable in solution with a zeta potential of −30.7 mV. The CdS nanoparticles displayed a very remarkable photocatalytic activity as they degraded 90% of methylene blue within 90 min under UV light. Furthermore, Arsiya et al. (2016) established the biological production of palladium nanoparticles from the aqueous extract of *Chlorella vulgaris*. The biosynthesized palladium nanoparticles (PdNPs) were dark brown in color with an absorption peak between 410 and 420 nm. The particles have spherical shape and 15 nm average size. They were crystalline in nature according to XRD analysis while the FTIR investigation confirmed that amide and polyol groups present in *C. vulgaris* extract played a crucial function in the reduction of palladium ions for PdNPs formation. In another study, the intracellular production of superparamagnetic two-lines ferri-hydrite nanoparticles by microalgae *Euglena gracilis* was reported by Brayner et al. (2012). They are dark in color with spherical shape. The HRTEM analysis confirmed the diameter of the spherical nanoparticles as 0.6–1.0 nm. The nanoparticles have a well-controlled shape and size with a narrow size distribution.

5 Conclusion

Microalgae have shown to be a very good resource for the synthesis of various metallic nanoparticles with diverse areas of application. Microalgal-mediated nanoparticles synthesis allows better control of size, crystal, shape, and other physicochemical properties. Microalgae are considered as cell factories for nanoscale particle synthesis due to their high metal uptake capacity, high growth rate, and high biomass productivity with a less cultivation time. The green chemistry approach of microalgal nanoparticles synthesis could therefore be considered technically feasible owing to its economic viability, environmental friendliness, biocompatibility, and possibility of scaling up to produce particles of desirable physicochemical properties. Further investigations would involve exploring the potency of microalgae to synthesize other metallic nanoparticles that will be applicable in pharmaceutical, food, cosmetics, industry, and biomedical areas.

References

Abinandan S, Subashchandrabose SR, Venkateswarlu K, Megharaj M (2018) Microalgae–bacteria biofilms: a sustainable synergistic approach in remediation of acid mine drainage. Appl Microbiol Biotechnol 102:1131–1144. https://doi.org/10.1007/s00253-017-8693-7

Abouelmagd SA, Meng F, Kim BK, Hyun H, Yeo Y (2016) Tannic acid-mediated surface functionalization of polymeric nanoparticles. ACS Biomater Sci Eng 2:2294–2303. https://doi.org/10.1021/acsbiomaterials.6b00497

Adelere IA, Lateef A (2016) A novel approach to the green synthesis of metallic nanoparticles: the use of agro-wastes, enzymes, and pigments. Nanotechnol Rev 5:567–587. https://doi.org/10.1515/ntrev-2016-0024

Adenigba VO, Omomowo IO, Oloke JK, Fatukasi BA, Odeniyi MA, Adedayo AA (2020) Evaluation of microalgal-based nanoparticles in the adsorption of heavy metals from Wastewater. IOP Conf Series Mater Sci Eng C 805:012030. https://doi.org/10.1088/1757-899X/805/1/012030

Ahmad AL, Yasin NM, Derek CJC, Lim JK (2011) Optimization of microalgae coagulation process using chitosan. Chem Eng J 173:879–882. https://doi.org/10.1016/j.cej.2011.07.070

Aqel A, El-Nour KMA, Ammar RA, Al-Warthan A (2012) Carbon nanotubes, science and technology part (I) structure, synthesis and characterisation. Arab J Chem 5:1–23. https://doi.org/10.1016/j.arabjc.2010.08.022

Arsiya F, Sayadi MH, Sobhani S (2016) Green synthesis of palladium nanoparticles using *Chlorella vulgaris*. Mater Lett 186:113–115. https://doi.org/10.1016/j.matlet.2016.09.101

Astefanei A, Núñez O, Galceran MT (2015) Characterisation and determination of fullerenes: a critical review. Anal Chim Acta 882:1–21. https://doi.org/10.1016/j.aca.2015.03.025

Aziz N, Fatma T, Varma A, Prasad R (2014) Biogenic synthesis of silver nanoparticles using *Scenedesmus abundans* and evaluation of their antibacterial activity. Int J Nanoparticles. Article ID 689419. https://doi.org/10.1155/2014/689419

Aziz N, Faraz M, Pandey R, Shakir M, Fatma T, Varma A, Barman I, Prasad R (2015) Facile algae-derived route to biogenic silver nanoparticles: synthesis, antibacterial, and photocatalytic properties. Langmuir 31:11605–11612. https://doi.org/10.1021/acs.langmuir.5b03081

Balantrapu K, Goia DV (2009) Silver nanoparticles for printable electronics and biological applications. J Mater Res Technol 24:2828–2836. https://doi.org/10.1557/jmr.2009.0336

Batool S, Hussain A, Iqbal MA, Javid A, Ali W, Bukhari SM, Akmal M, Qazi JI (2019) Implication of highly metal-resistant microalgal-bacterial co-cultures for the treatment of simulated metal-loaded wastewaters. Int Microbiol 22:41–48. https://doi.org/10.1007/s10123-018-0025-y

Begum NA, Mondal S, Basu S, Laskar RA, Mandal D (2009) Biogenic synthesis of Au and Ag nanoparticles using aqueous solutions of black tea leaf extracts. Colloids Surf B Biointerf 71:113–118. https://doi.org/10.1016/j.colsurfb.2009.01.012

Borase HP, Patil CD, Suryawanshi RK, Koli SH, Mohite BV, Benelli G, Patil SV (2017) Mechanistic approach for fabrication of gold nanoparticles by *Nitzschia* diatom and their antibacterial activity. Bioprocess Biosyst Eng 40:1437–1446. https://doi.org/10.1007/s00449-017-1801-3

Brayner R, Barberousse H, Hemadi M, Djedjat C, Yéprémian C, Coradin T, Livage J, Fiévet F, Couté A (2007) Cyanobacteria as bioreactors for the synthesis of Au, Ag, Pd, and Pt nanoparticles via an enzyme-mediated route. J Nanosci Nanotechnol 7:2696–2708. https://doi.org/10.1166/jnn.2007.600

Brayner R, Coradin T, Beaunier P, Grenèche JM, Djediat C, Yéprémian C, Couté A, Fiévet F (2012) Intracellular biosynthesis of superparamagnetic 2-lines ferri-hydrite nanoparticles using *Euglena gracilis* microalgae. Colloids Surf B Biointerf 93:20–23. https://doi.org/10.1016/j.colsurfb.2011.10.014

Caporgno MP, Mathys A (2018) Trends in microalgae incorporation into innovative food products with potential health benefits. Front Nutr 5:58. https://doi.org/10.3389/fnut.2018.00058

Cepoi L, Rudi L, Chiriac T, Valuta A, Zinicovscaia I, Duca G, Kirkesali E, Frontasyeva M, Culicov O, Pavlov S, Bobrikov I (2015) Biochemical changes in cyanobacteria during the synthesis of silver nanoparticles. Can J Microbiol 61:13–21. https://doi.org/10.1139/cjm-2014-0450

Charles CN, Msagati T, Swai H, Chacha M (2019) Microalgae: an alternative natural source of bioavailable omega-3 DHA for promotion of mental health in East Africa. Sci Afr e00187. https://doi.org/10.1016/j.sciaf.2019.e00187

Chen H, Roco MC, Son J, Jiang S, Larson CA, Gao Q (2013) Global nanotechnology development from 1991 to 2012: patents, scientific publications, and effect of NSF funding. J Nanopart Res 15:1951. https://doi.org/10.1007/s11051-013-1951-4

Chokshi K, Pancha I, Ghosh T, Paliwal C, Maurya R, Ghosh A, Mishra S (2016) Green synthesis, characterization and antioxidant potential of silver nanoparticles biosynthesized from de-oiled biomass of thermotolerant oleaginous microalgae *Acutodesmus dimorphus*. RSC Adv 6:72269–72274. https://doi.org/10.1039/c6ra15322d

Dahoumane SA, Djediat C, Yéprémian C, Couté A, Fiévet F, Coradin T, Brayner R (2012) Recycling and adaptation of *Klebsormidium flaccidum* microalgae for the sustained production of gold nanoparticles. Biotechnol Bioeng 109:284–288. https://doi.org/10.1002/bit.23276

Dahoumane SA, Yéprémian C, Djédiat C, Couté A, Fiévet F, Coradin T, Brayner R (2014) A global approach of the mechanism involved in the biosynthesis of gold colloids using micro-algae. J Nanopart Res 16:2607. https://doi.org/10.1007/s11051-014-2607-8

Dahoumane SA, Mechouet M, Alvarez FJ, Agathos SN, Jeffryes C (2016) Microalgae: an outstanding tool in nanotechnology. Bionatura 1:196–201. https://doi.org/10.1002/bit.23276

Dreaden EC, Alkilany AM, Huang X, Murphy CJ, El-Sayed MA (2012) The golden age: gold nanoparticles for biomedicine. Chem Soc Rev 41:740–2779. https://doi.org/10.1039/c1cs15237h

Ebrahiminezhad A, Rasoul-Amini S, Ghoshoon MB, Ghasemi Y (2014) *Chlorella vulgaris*, a novel microalgal source for L-asparaginase production. Biocatal Agric Biotechnol 3:214–217. https://doi.org/10.1016/j.bcab.2013.10.005

Ebrahiminezhad A, Bagheri M, Taghizadeh SM, Berenjian A, Ghasemi Y (2016) Biomimetic synthesis of silver nanoparticles using microalgal secretory carbohydrates as a novel anticancer and antimicrobial. Adv Nat Sci Nanosci Nanotechnol Lett 015018. https://doi.org/10.1088/2043-6262/7/1/015018

Elegbede JA, Lateef A (2019a) Green nanotechnology in Nigeria: the research landscape, challenges and prospects. Ann Sci Technol 6-38. https://doi.org/10.2478/ast-2019-0008

Elegbede JA, Lateef A (2019b) Green synthesis of silver (Ag), gold (Au) and silver-gold (Ag-Au) alloy nanoparticles: a review on recent advances, trends and biomedical applications. In: Verma DK, Goyal MR, Suleria HAR (eds) Nanotechnology and nanomaterial applications in food, health and biomedical sciences. Apple Academic Press Inc./CRC Press, Taylor and Francis Group, Oakville, Ontario, Canada, pp 3–89. ISBN 978-1-77188-764-9. https://doi.org/10.1201/9780429425660-1

Elegbede JA, Lateef A, Azeez MA, Asafa TB, Yekeen TA, Oladipo IC, Adebayo EA, Beukes LS, Gueguim-Kana EB (2018) Fungal xylanases-mediated synthesis of silver nanoparticles for catalytic and biomedical applications. IET Nanobiotechnol 12:857–863. https://doi.org/10.1049/iet-nbt.2017.0299

Elegbede JA, Lateef A, Azeez MA, Asafa TB, Yekeen TA, Oladipo IC, Abbas SH, Beukes LS, Gueguim-Kana EB (2019) Silver-gold alloy nanoparticles biofabricated by fungal xylanases exhibited potent biomedical and catalytic activities. Biotechnol Prog 35:e2829. https://doi.org/10.1002/btpr.2829

Elegbede JA, Lateef A, Azeez MA, Asafa TB, Yekeen TA, Oladipo IC, Aina DA, Beukes LS, Gueguim-Kana EB (2020) Biofabrication of gold nanoparticles using xylanases through valorization of corncob by *Aspergillus niger* and *Trichoderma longibrachiatum*: antimicrobial, antioxidant, anticoagulant and thrombolytic activities. Waste Biomass Valor 11(3):781–791. https://doi.org/10.1007/s12649-018-0540-2

Feurtet-Mazel A, Mornet S, Charron L, Mesmer-Dudons N, Maury-Brachet R, Baudrimont M (2016) Biosynthesis of gold nanoparticles by the living freshwater diatom *Eolimna minima*, a species developed in river biofilms. Environ Sci Poll Res 23:4334–4339. https://doi.org/10.1007/s11356-015-4139-x

Fields FJ, Lejzerowicz F, Schroeder D, Ngoi SM, Tran M, McDonald D, Jiang L, Chang JT, Knight R, Mayfield S (2020) Effects of the microalgae *Chlamydomonas* on gastrointestinal health. J Funct Food 65:103738. https://doi.org/10.1016/j.jff.2019.103738

Galasso C, Gentile A, Orefice I, Ianora A, Bruno A, Noonan DM, Sansone C, Albini A, Brunet C (2019) Microalgal derivatives as potential nutraceutical and food supplements for human health: a focus on cancer prevention and interception. Nutrients 11:1226. https://doi.org/10.3390/nu11061226

García JL, de Vicente M, Galán B (2017) Microalgae, old sustainable food and fashion nutraceuticals. Microb Biotechnol 10:1017–1024. https://doi.org/10.1111/1751-7915.12800

Garcia-Gonzalez J, Sommerfeld M (2016) Biofertilizer and biostimulant properties of the microalga *Acutodesmus dimorphus*. J Appl Phycol 28:1051–1061. https://doi.org/10.1007/s10811-015-0625-2

Gaur JP, Rai LC (2001) Heavy metal tolerance in algae. In: Algal adaptation to environmental stresses. Springer, Berlin, Heidelberg, pp 363–388. https://doi.org/10.1007/978-3-642-59491-5_12

Ghashghaei S, Emtiazi G (2013) Production of calcite nanocrystal by a urease-positive strain of *Enterobacter ludwigii* and study of its structure by SEM. Curr Microbiol 67:406–413. https://doi.org/10.1007/s00284-013-0379-5

Gopinath V, MubarakAli D, Priyadarshini S, Priyadharsshini NM, Thajuddin N, Velusamy P (2012) Biosynthesis of silver nanoparticles from *Tribulus terrestris* and its antimicrobial activity: a novel biological approach. Colloids Surf B Biointerf 96:69–74. https://doi.org/10.1016/j.colsurfb.2012.03.023

Hisatomi T, Takanabe K, Domen K (2015) Photocatalytic water-splitting reaction from catalytic and kinetic perspectives. Catal Lett 145:95–108. https://doi.org/10.1007/s10562-014-1397-z

Husain S, Sardar M, Fatma T (2015) Screening of cyanobacterial extracts for synthesis of silver nanoparticles. World J Microbiol Biotechnol 31:1279e1283. https://doi.org/10.1007/s11274-015-1869-3

Ibrahim KS (2013) Carbon nanotubes-properties and applications: a review. Carbon Lett 14:131–144. https://doi.org/10.5714/cl.2013.14.3.131

Jais NM, Mohamed RMSR, Al-Gheethi AA, Hashim MA (2017) The dual roles of phycoremediation of wet market wastewater for nutrients and heavy metals removal and microalgae biomass production. Clean Technol Environ Policy 19:37–52. https://doi.org/10.1007/s10098-016-1235-7

Jena J, Pradhan N, Dash BP, Sukla LB, Panda PK (2013) Biosynthesis and characterization of silver nanoparticles using microalga *Chlorococcum humicola* and its antibacterial activity. Int J Nanomater Biostruct 3:1–8

Jena J, Pradhan N, Nayak RR, Dash BP, Sukla LB, Panda PK, Mishra BK (2014) Microalga *Scenedesmus* sp.: a potential low-cost green machine for silver nanoparticle synthesis. J Microbiol Biotechnol 24:522–533. https://doi.org/10.4014/jmb.1306.06014

Jena J, Pradhan N, Dash BP, Panda PK, Mishra BK (2015) Pigment mediated biogenic synthesis of silver nanoparticles using diatom *Amphora* sp. and its antimicrobial activity. J Saudi Chem Soc 19:661–666. https://doi.org/10.1016/j.jscs.2014.06.005

Kalabegishvili T, Kirkesali E, Rcheulishvili A (2012) Synthesis of gold nanoparticles by blue-green algae Spirulina platensis. Frank Lab. of Neutron Physics. 43. https://inis.iaea.org/search/search.aspx?orig_q=RN:43108872

Karthikeyan P, Mohan D, Abishek G, Priya R (2015) Synthesis of silver nanoparticles using Phytoplankton and its characteristics. Int J Fisher Aquat Stud 2:398–401

Khalid M, Khalid N, Ahmed I, Hanif R, Ismail M, Janjua HA (2017) Comparative studies of three novel freshwater microalgae strains for synthesis of silver nanoparticles: insights of characterization, antibacterial, cytotoxicity and antiviral activities. J Appl Phycol 29:1851–1863. https://doi.org/10.1007/s10811-017-1071-0

Khalifa KS, Hamouda RA, Hamza HA (2016) Antitumor activity of silver nanoparticles biosynthesized by micro algae. J Chem Pharm Res 8:1–6

Khan I, Saeed K, Khan I (2019) Nanoparticles: properties, applications and toxicities. Arab J Chem 12:908–931. https://doi.org/10.1016/j.arabjc.2017.05.011

Khatoon N, Pal R (2015) Microalgae in biotechnological application: a commercial approach. In: Plant biology and biotechnology. Springer, New Delhi, pp 27–47. https://doi.org/10.1007/978-81-322-2283-5_2

Koole R, Groeneveld E, Vanmaekelbergh D, Meijerink A, de Mello Donegá C (2014) Size effects on semiconductor nanoparticles. In: Nanoparticles. Springer, Berlin, Heidelberg, pp 13–51. https://doi.org/10.1007/978-3-662-44823-6_2

Koyande AK, Chew KW, Rambabu K, Tao Y, Chu DT, Show PL (2019) Microalgae: a potential alternative to health supplementation for humans. Food Sci Hum Wellbeing 8:16–24. https://doi.org/10.1016/j.fshw.2019.03.001

Lateef A, Adeeyo AO (2015) Green synthesis and antibacterial activities of silver nanoparticles using extracellular laccase of *Lentinus edodes*. Not Sci Biol 7:405–411. https://doi.org/10.15835/nsb749643

Lateef A, Adelere IA, Gueguim-Kana EB, Asafa TB, Beukes LS (2015) Green synthesis of silver nanoparticles using keratinase obtained from a strain of *Bacillus safensis* LAU 13. Int Nano Lett 5:29–35. https://doi.org/10.1007/s40089-014-0133-4

Lateef A, Ojo SA, Elegbede JA (2016a) The emerging roles of arthropods and their metabolites in the green synthesis of metallic nanoparticles. Nanotechnol Rev 5:601–622. https://doi.org/10.1515/ntrev-2016-0049

Lateef A, Ojo SA, Oladejo SM (2016b) Anti-candida, anti-coagulant and thrombolytic activities of biosynthesized silver nanoparticles using cell-free extract of *Bacillus safensis* LAU 13. Process Biochem 51:1406–1412. https://doi.org/10.1016/j.procbio.2016.06.027

Lateef A, Akande MA, Ojo SA, Folarin BI, Gueguim-Kana EB, Beukes LS (2016c) Paper wasp nest-mediated biosynthesis of silver nanoparticles for antimicrobial, catalytic, anti-coagulant and thrombolytic applications. 3 Biotech 6:140. https://doi.org/10.1007/s13205-016-0459-x

Lateef A, Ojo SA, Azeez MA, Asafa TB, Yekeen TA, Akinboro A, Oladipo IC, Gueguim-Kana EB, Beukes LS (2016d) Cobweb as novel biomaterial for the green and eco-friendly synthesis of silver nanoparticles. Appl Nanosci 6:863–874. https://doi.org/10.1007/s13204-015-0492-9

Lateef A, Ojo SA, Folarin BI, Gueguim-Kana EB, Beukes LS (2016e) Kolanut (*Cola nitida*) mediated synthesis of silver-gold alloy nanoparticles: antifungal, catalytic, larvicidal and thrombolytic applications. J Clust Sci 27(5):1561–1577. https://doi.org/10.1007/s10876-016-1019-6

Lateef A, Ojo SA, Elegbede JA, Akinola PO, Akanni EO (2018a) Nanomedical applications of nanoparticles for blood coagulation disorders. In: Dasgupta N, Ranjan S, Lichtfouse E (eds) Environmental nanotechnology, vol 1. Springer International Publishing AG, Cham, Switzerland, pp 243-277. ISBN 978-3-319-76089-6. https://doi.org/10.1007/978-3-319-76090-2_8

Lateef A, Folarin BI, Oladejo SM, Akinola PO, Beukes LS, Gueguim-Kana EB (2018b) Characterization, antimicrobial, antioxidant and anticoagulant activities of silver nanoparticles synthesized from *Petiveria alliacea* L. leaf extract. Prep Biochem Biotechnol 48:646–652. https://doi.org/10.1080/10826068.2018.1479864

Lateef A, Elegbede JA, Akinola PO, Ajayi VA (2019) Biomedical applications of green synthesized-metallic nanoparticles: a review. Pan Afr J Life Sci 3:157–182

Lateef A, Oladejo SM, Akinola PO, Aina DA, Beukes LS, Folarin BI, Gueguim-Kana EB (2020) Facile synthesis of silver nanoparticles using leaf extract of *Hyptis suaveolens* (L.) Poit for environmental and biomedical applications. IOP Conf Ser Mater Sci Eng 805:012042. https://doi.org/10.1088/1757-899X/805/1/012042

Li CZ, Bogozi A, Huang W, Tao NJ (1999) Fabrication of stable metallic nanowires with quantized conductance. Nat Nanotechnol 10:221. https://doi.org/10.1088/0957-4484/10/2/320

Li B, Zhang T, Yang Z (2019) Immobilizing unicellular microalga on pellet-forming filamentous fungus: Can this provide new insights into the remediation of arsenic from contaminated water? Bioresour Technol 284:231–239. https://doi.org/10.1016/j.biortech.2019.03.128

Li Y, Tang X, Song W, Zhu L, Liu X, Yan X, Jin C, Ren Q (2014) Biosynthesis of silver nanoparticles using *Euglena gracilis*, *Euglena intermedia* and their extract. IET Nanobiotechnol 9:19–26. https://doi.org/10.1049/iet-nbt.2013.0062

Luangpipat T, Beattie IR, Chisti Y, Haverkamp RG (2011) Gold nanoparticles produced in a microalga. J Nanopart Res 13:6439–6445. https://doi.org/10.1007/s11051-011-0397-9

Mahdieh M, Zolanvari A, Azimee AS (2012) Green biosynthesis of silver nanoparticles by *Spirulina platensis*. Sci Iran 19:926–929. https://doi.org/10.1016/j.scient.2012.01.010

Mashaghi S, Jadidi T, Koenderink G, Mashaghi A (2013) Lipid nanotechnology. Int J Mol Sci 14:4242–4282. https://doi.org/10.3390/ijms14024242

Mata TM, Martins AA, Caetano NS (2010) Microalgae for biodiesel production and other applications: a review. Renew Sustain Energ Rev 14:217–232. https://doi.org/10.1016/j.rser.2009.07.020

Matos J, Cardoso CL, Falé P, Afonso CM, Bandarra NM (2020) Investigation of nutraceutical potential of the microalgae *Chlorella vulgaris* and *Arthrospira platensis*. Int J Food Sci Technol 55:303–312. https://doi.org/10.1111/ijfs.14278

Menegazzo ML, Fonseca GG (2019) Biomass recovery and lipid extraction processes for microalgae biofuels production: a review. Renew Sustain Energ Rev 107:87–107. https://doi.org/10.1016/j.rser.2019.01.064

Merin DD, Prakash S, Bhimba BV (2010) Antibacterial screening of silver nanoparticles synthesized by marine micro algae. Asian Pac J Trop Med 3:797–799. https://doi.org/10.1016/s1995-7645(10)60191-5

Milano J, Ong HC, Masjuki HH, Chong WT, Lam MK, Loh PK, Vellayan V (2016) Microalgae biofuels as an alternative to fossil fuel for power generation. Renew Sustain Energ Rev 58:180–197. https://doi.org/10.1016/j.rser.2015.12.150

Mohseniazar M, Barin M, Zarredar H, Alizadeh S, Shanehbandi D (2011) Potential of microalgae and lactobacilli in biosynthesis of silver nanoparticles. Bioimpacts BI 1:149. https://doi.org/10.5681/bi.2011.020

Molino A, Iovine A, Casella P, Mehariya S, Chianese S, Cerbone A, Rimauro J, Musmarra D (2018) Microalgae characterization for consolidated and new application in human food, animal feed and nutraceuticals. Int J Environ Res Public Health 15:2436. https://doi.org/10.3390/ijerph15112436

Morsy FM, Nafady NA, Abd-Alla MH, Elhady DA (2014) Green synthesis of silver nanoparticles by water soluble fraction of the extracellular polysaccharides/matrix of the cyanobacterium *Nostoc commune* and its application as a potent fungal surface sterilizing agent of seed crops. Univ J Microbiol Res 2:36–43. https://doi.org/10.13189/ujmr.2014.020303

MubarakAli D, Arunkumar J, Nag KH, Sheik Syed Ishack KA, Baldev E, Pandiaraj D, Thajuddin N (2013) Gold nanoparticles from Pro and eukaryotic photosynthetic microorganisms—comparative studies on synthesis and its application on biolabelling. Colloids Surf B Biointerf 103:166–173. https://doi.org/10.1016/j.colsurfb.2012.10.014

Muthusamy G, Thangasamy S, Raja M, Chinnappan S, Kandasamy S (2017) Biosynthesis of silver nanoparticles from *Spirulina* microalgae and its antibacterial activity. Environ Sci Pollut Res 24:19459–19464. https://doi.org/10.1007/s11356-017-9772-0

Naveed S, Li C, Lu X, Chen S, Yin B, Zhang C, Ge Y (2019) Microalgal extracellular polymeric substances and their interactions with metal(loid)s: a review. Crit Rev Environ Sci Technol 49:1769–1802. https://doi.org/10.1080/10643389.2019.1583052

Nisha R, Kaushik A, Kaushik CP (2007) Effect of indigenous cyanobacterial application on structural stability and productivity of an organically poor semi-arid soil. Geoderma 138:49–56. https://doi.org/10.1016/j.geoderma.2006.10.007

Ojo SA, Lateef A, Azeez MA, Oladejo SM, Akinwale AS, Asafa TB, Yekeen TA, Akinboro A, Oladipo IC, Gueguim-Kana EB, Beukes LS (2016) Biomedical and catalytic applications of gold and silver-gold alloy nanoparticles biosynthesized using cell-free extract of *Bacillus safensis* LAU 13: antifungal, dye degradation, anti-coagulant and thrombolytic activities. IEEE Trans Nanobiosci 15:433–442. https://doi.org/10.1109/TNB.2016.2559161

Oladipo IC, Lateef A, Elegbede JA, Azeez MA, Asafa TB, Yekeen TA, Akinboro A, Gueguim-Kana EB, Beukes LS, Oluyide TO, Atanda OR (2017) *Enterococcus* species for the one-pot biofabrication of gold nanoparticles: characterization and nanobiotechnological applications. J Photochem Photobiol B Biol 173:250–257. https://doi.org/10.1016/j.jphotobiol.2017.06.003

Omomowo IO, Adenigba VO, Ogunsola SB, Adeyinka GC, Oluyide OO, Adedayo AA, Fatukasi BA (2020) Antimicrobial and antioxidant activities of algal-mediated silver and gold nanoparticles. IOP Conf Series Mater Sci Eng C805:012010. https://doi.org/10.1088/1757-899X/805/1/012010

Ovais M, Khalil AT, Islam NU, Ahmad I, Ayaz M, Saravanan M, Shinwari ZK, Mukherjee S (2018) Role of plant phytochemicals and microbial enzymes in biosynthesis of metallic nanoparticles. Appl Microbiol Biotechnol 102:6799–6814. https://doi.org/10.1007/s00253-018-9146-7

Parashar A, Kedare PS, Alex SA, Chandrasekaran N, Mukherjee A (2017) A novel enzyme-mediated gold nanoparticle synthesis and its application for in situ detection of horseradish peroxidase inhibitor phenylhydrazine. New J Chem 41:15079–15086. https://doi.org/10.1039/c7nj03783j

Patel V, Berthold D, Puranik P, Gantar M (2015) Screening of cyanobacteria and microalgae for their ability to synthesize silver nanoparticles with antibacterial activity. Biotechnol Rep 5:112–119. https://doi.org/10.1016/j.btre.2014.12.001

Priyadarshini E, Priyadarshini SS, Pradhan N (2019) Heavy metal resistance in algae and its application for metal nanoparticle synthesis. Appl Microbiol Biotechnol 103:3297–3316. https://doi.org/10.1007/s00253-019-09685-3

Rao MD, Pennathur G (2017) Green synthesis and characterization of cadmium sulphide nanoparticles from *Chlamydomonas reinhardtii* and their application as photocatalysts. Mater Res Bull 85:64–73. https://doi.org/10.1016/j.materresbull.2016.08.049

Renuka N, Sood A, Prasanna R, Ahluwalia AS (2015) Phycoremediation of wastewaters: a synergistic approach using microalgae for bioremediation and biomass generation. Int J Environl Sci Technol 12:1443–1460. https://doi.org/10.1007/s13762-014-0700-2

Rizwan M, Mujtaba G, Memon SA, Lee K, Rashid N (2018) Exploring the potential of microalgae for new biotechnology applications and beyond: a review. Renew Sustain Energ Rev 92:394–404. https://doi.org/10.1016/j.rser.2018.04.034

Rodriguez-Sánchez L, Blanco MC, Lopez-Quintela MA (2000) Electrochemical synthesis of silver nanoparticles. J Phys Chem B 104:9683–9688. https://doi.org/10.1021/jp001761r

Roy SS, Pal R (2015) Microalgae in aquaculture: a review with special references to nutritional value and fish dietetics. Proc Zool Soc 68:1–8. https://doi.org/10.1007/s12595-013-0089-9

Salati S, D'Imporzano G, Menin B, Veronesi D, Scaglia B, Abbruscato P, Mariani P, Adani F (2017) Mixotrophic cultivation of *Chlorella* for local protein production using agro-food by-products. Bioresour Technol 230:82–89. https://doi.org/10.1016/j.biortech.2017.01.030

Salim KP, Chandra RD, Heriyanto H, Susilaningsih D, Limantara L, Brotosudarmo TH (2019) Economically potential pigments from marine blue-green algae for the application in food and health. Ind J Nat Pig 1:37. https://doi.org/10.33479/ijnp.2019.01.2.37

Sathasivam R, Radhakrishnan R, Hashem A, AbdAllah EF (2019) Microalgae metabolites: a rich source for food and medicine. Saudi J Biol Sci 26:709–722. https://doi.org/10.1016/j.sjbs.2017.11.003

Sathishkumar RS, Sundaramanickam A, Srinath R, Ramesh T, Saranya K, Meena M, Surya P (2019) Green synthesis of silver nanoparticles by bloom forming marine microalgae *Trichodesmium erythraeum* and its applications in antioxidant, drug-resistant bacteria, and cytotoxicity activity. J Saudi Chem Soc 23:1180–1191. https://doi.org/10.1016/j.jscs.2019.07.008

Schulze PS, Hulatt CJ, Morales-Sánchez D, Wijffels RH, Kiron V (2019) Fatty acids and proteins from marine cold adapted microalgae for biotechnology. Algal Res 42:101604. https://doi.org/10.1016/j.algal.2019.101604

Shakibaie M, Forootanfar H, Mollazadeh-Moghaddam K, Bagherzadeh Z, Nafissi-Varcheh N, Shahverdi AR, Faramarzi MA (2010) Green synthesis of gold nanoparticles by the marine microalga *Tetraselmis suecica*. Biotechnol Appl Biochem 57:71–75. https://doi.org/10.1042/ba20100196

Shankar SS, Rai A, Ahmad A, Sastry M (2004) Rapid synthesis of Au, Ag, and bimetallic Au core–Ag shell nanoparticles using Neem (*Azadirachta indica*) leaf broth. J Colloid Interf Sci 275:496–502. https://doi.org/10.1016/j.jcis.2004.03.003

Sharma G, Jasuja ND, Kumar M, Ali MI (2015) Biological synthesis of silver nanoparticles by cell-fres extract of *Spirulina platensis*. J Nanotechnol Article ID 132675. https://doi.org/10.1155/2015/132675

Sharma A, Sharma S, Sharma K, Chetri SP, Vashishtha A, Singh P, Kumar R, Rathi B, Agrawal V (2016) Algae as crucial organisms in advancing nanotechnology: a systematic review. J Appl Phycol 28:1759–1774. https://doi.org/10.1007/s10811-015-0715-1

Sharon G, Stern R, Felner A, Sturtevant NR (2012) Meta-agent conflict-based search for optimal multi-agent path finding. *SoCS*1 39–40

Shuba ES, Kifle D (2018) Microalgae to biofuels: 'Promising' alternative and renewable energy, review. Renew Sustain Energ Rev 81:743–755. https://doi.org/10.1016/j.rser.2017.08.042

Singh V, Sengupta S, Solanki HS, Dhall R, Allain A, Dhara S, Pant P, Deshmukh MM (2010) Probing thermal expansion of graphene and modal dispersion at low-temperature using graphene nanoelectromechanical systems resonators. Nanotechnology 21:165204. https://doi.org/10.1088/0957-4484/21/16/165204

Singh R, Shedbalkar UU, Nadhe SB, Wadhwani SA, Chopade BA (2017) Lignin peroxidase mediated silver nanoparticle synthesis in Acinetobacter sp. AMB Express 7:226. https://doi.org/10.1186/s13568-017-0528-5

Soni RA, Sudhakar K, Rana RS (2017) *Spirulina*-From growth to nutritional product: a review. Trends Food Sci Technol 69:157–171. https://doi.org/10.1016/j.tifs.2017.09.010

Subramaniyam V, Subashchandrabose SR, Thavamani P, Megharaj M, Chen Z, Naidu R (2015) *Chlorococcum* sp. MM11—a novel phyco-nanofactory for the synthesis of iron nanoparticles. J Appl Phycol 27:1861–1869. https://doi.org/10.1007/s10811-014-0492-2

Sudha SS, Rajamanickam K, Rengaramanujam, J (2013) Microalgae mediated synthesis of silver nanoparticles and their antibacterial activity against pathogenic bacteria. Ind J Exp Biol 51:393–39. http://nopr.niscair.res.in/handle/123456789/17672

Taelman SE, De Meester S, Roef L, Michiels M, Dewulf J (2013) The environmental sustainability of microalgae as feed for aquaculture: a life cycle perspective. Bioresour Technol 150:513–522. https://doi.org/10.1016/j.biortech.2013.08.044

Tang DYY, Khoo KS, Chew KW, Tao Y, Ho SH, Show PL (2020) Potential utilization of bioproducts from microalgae for the quality enhancement of natural products. Bioresour Technol 304:122997. https://doi.org/10.1016/j.biortech.2020.122997

Thomas SC, Kumar MP, Talegaonkar S (2015) Ceramic nanoparticles: fabrication methods and applications in drug delivery. Curr Pharm Des 21:6165–6188. https://doi.org/10.2174/1381612821666151027153246

von Alvensleben N, Heimann K (2018) The potential of microalgae for biotechnology: a focus on carotenoids. Blue Biotechnol. Prod Mar Mol 1:117–142. https://doi.org/10.1002/9783527801718.ch4

Wadhwani SA, Shedbalkar UU, Singh R, Chopade BA (2018) Biosynthesis of gold and selenium nanoparticles by purified protein from *Acinetobacter* sp. SW 30. Enzyme Microb Technol 111:81–86. https://doi.org/10.1016/j.enzmictec.2017.10.007

Walker AB, Berlinsky DL (2011) Effects of partial replacement of fish meal protein by microalgae on growth, feed intake, and body composition of Atlantic cod. North Am J Aquacult 73:76–83. Retrieved from https://doi.org/10.1080/15222055.2010.549030

Wang JG, Liu H, Zhang X, Li X, Liu X, Kang F (2018) Green synthesis of hierarchically porous carbon nanotubes as advanced materials for high-efficient energy storage. Small 14:1703950. https://doi.org/10.1002/smll.201703950

Xie J, Lee JY, Wang DI, Ting YP (2007) Identification of active biomolecules in the high-yield synthesis of single-crystalline gold nanoplates in algal solutions. Small 3:672–682. https://doi.org/10.1002/smll.200600612

Xiong JQ, Kurade MB, Jeon BH (2018) Can microalgae remove pharmaceutical contaminants from water? Trends Biotechnol 36:30–44. https://doi.org/10.1016/j.tibtech.2017.09.003

Yingchoncharoen P, Kalinowski DS, Richardson DR (2016) Lipid-based drug delivery systems in cancer therapy: what is available and what is yet to come. Pharmacol Rev 68:701–787. https://doi.org/10.1124/pr.115.012070

Mushroom Nanobiotechnology: Concepts, Developments and Potentials

E. A. Adebayo, M. A. Azeez, M. B. Alao, M. A. Oke, and D. A. Aina

1 Mushrooms at a Glance

Historically, mushroom is a collective term for describing diverse species of macrofungi which the primitive Romans referred to as *food of the gods*, the Stone Aged Egyptians named it *the gifts from god of Osiris,* while to the earliest Chinese, it is *the elixir of life* (Smith et al. 2002). However, in the world today, mushrooms are described as fungi fruiting bodies of orders Basidiomycetes and Ascomycetes which play vital roles in the management of forest ecosystems as they have the potential of decomposing complex biomolecules and recycling nutrients within the ecosystem. Thus, they are found all over the world mainly, in soils rich in organic matter (humus), moist wood and animals waste. Mushrooms stand out in their habitats with the possession of umbrella like fruiting bodies, remarkable colours and sizes. Interestingly, mushrooms fruiting bodies disappear after a heavy rain or a sudden change in temperature, leaving behind the mycelia, and this has drawn the attention of mankind to the study of mushrooms. Recently in a study by Bal et al. (2017), approximately 10% of all mushroom species on Earth has been extensively studied. Similarly, about 2000 species of 31 genera are classified as edible mushrooms while a little over 10% of the extensively studied species are classified as poisonous or lethal (Teferi et al. 2013; Ruan-Soto et al. 2017).

E. A. Adebayo (✉) · M. A. Azeez
Department of Pure and Applied Biology, Ladoke Akintola University, Ogbomoso P.M.B 4000, Oyo State, Nigeria
e-mail: eaadebayo@lautech.edu.ng

M. B. Alao · M. A. Oke
Microbiology and Biotechnology Laboratory, Ladoke Akintola University of Technology, Ogbomoso P.M.B 4000, Oyo State, Nigeria

D. A. Aina
Department of Microbiology, Babcock University, Ilishan-Remo, Ogun State, Nigeria

A. Lateef et al. (eds.), *Microbial Nanobiotechnology*, Materials Horizons: From Nature to Nanomaterials, https://doi.org/10.1007/978-981-33-4777-9_9

From time immemorial, mushrooms are considered as just food and now due to availability of various nutraceuticals, mushrooms are recognized as healthy foods globally. They are of good taste and of high nutritional value (Adebayo et al. 2014a). They are abundant source of proteins, vitamins, minerals, polysaccharides and essential amino acids (Adebayo et al. 2014b; Adeeyo et al. 2016; Adebayo and Oloke 2017; Bamigboye et al. 2019) with low fat and calories (Valverde et al. 2015). *Pleurotus* species are high in proteins with about 5–49% of their dry weight being protein as well as rich in carbohydrates, vitamin, dietary fibres and important minerals like iron and potassium among others (Valverde et al. 2015; Adebayo and Oloke 2017). In general, edible mushroom species contain about 88–90% water, 4–9% carbohydrates, 0–3% fat, 1–2% ash (bromine, calcium, copper, manganese, iron, phosphorus, zinc, sodium and chlorine), vitamin A, vitamin B complexes of B5 (nicotinic acid), B1 (thiamine), B3 (pantetonic acid) and B2 (riboflavin) (Ndaw et al. 2000). Furthermore, mushrooms are rich in vitamins C and D. Due to the abundance of vitamins, mushrooms are reportedly good for man's nervous system.

Generally, mushrooms are important in alternative medicine and bioenhancement of synthetic drugs (Tabbouche et al. 2017; Yilmaz et al. 2016). Mushroom contains secondary metabolites which include polyketides, steroids, terpenes and phenolic compounds (Adebayo et al. 2012a). These secondary metabolites are distinguished by their medical properties and as well serve as antioxidant defence mechanisms of biological systems (Bal et al. 2017). Mushrooms are good natural sources of antioxidant compounds (Akgul et al. 2017; Adebayo et al. 2018). Previous studies had reported diverse species of mushrooms with medicinal values such as anti-carcinogenic (Vetchinkina et al. 2016), antimicrobial (Adebayo et al. 2014b), immunosuppressive (Oloke and Adebayo 2015), anti-inflammatory (Adebayo et al. 2012b), cytotoxic, cytostatic (Gurunathan et al. 2013), hypoglycemic (Mohanta et al. 2018) and antioxidant properties (Sevindik et al. 2017; Adebayo et al. 2018; Sevindik et al. 2018). Consequent upon these, mushrooms have been applied for different purposes by man for decades, and this has necessitated its exploration as natural resource in nanobiotechnology.

2 Nanotechnology and Nanobiotechnology

Nanotechnology entails designing, creating, developing and applying materials and devices on the scale of nanometre (nm) with 1 nm equivalent to a metre's one billionth or about the width of six atoms of carbon or ten molecules of water (Emerich and Thanos 2003; Sahoo and Labhasetwar 2003). Lately, nanomaterials are greatly applied in health care, electronics, cosmetics and other areas of research. On the other hand, nanobiotechnology is the interesting combination of nanotechnology and biotechnology which ensures that technology on nanoscale can be realistically amalgamated with the biology of molecules. Interestingly, machines of molecular or atomic concern are established through copying or fusing systems of diverse biology or simply by constructing minute tools for analysing or modifying different attributes of biological system on the backbone of molecules through this methodology. Biotechnology employs the techniques and knowledge of life science to modify or completely change processes in the cells through editing of genes and molecules culminating in service delivery and manufacture of products important in diverse fields. In biotechnology, several physiological procedures, most importantly the metabolism of the living subject is essential. Therefore, nanobiotechnology simplifies different approaches of biological science by consolidating the information technology applications and that of nanotechnology into the present issues of biology. Nanobiotechnology to some extent possesses capability of bridging preexisting boundaries between chemistry, biology and physics beyond the present understanding, imagination or idea (Elegbede and Lateef 2019). It is opinionated that bridging gaps of different science fields can stem multifunctional systems and devices of higher class and greater specificity, sensitivity and identification in the world of analysis and diagnosis (Niemeyer and Mirkin 2004).

The fundamental aspect of nanotechnology is the creation of nanoparticles which can be synthesized by either a top-down or a bottom-up approach. In the bottom-up technique, fundamental units are amalgamated into larger stable structures while the top-down approach utilizes larger initial structures by controlling the nanostructures (Agarwal et al. 2017). The methods of nanoparticles synthesis can be physical, chemical or biological otherwise called 'green chemistry'. The physical method requires high temperature and pressure, and it includes the laser ablation lithography (Zhang and Wang 2008) and high-energy irradiation (Mafune et al. 2001; Treguer et al. 1998) which are very expensive to achieve. The chemical method entails highly toxic compounds, and it includes electrochemistry, chemical reduction and photochemical reduction (Chen et al. 2001; Eustis et al. 2005; Frattini et al. 2005). On the other hand, green approach (biological) is an environmental and eco-friendly safe approach of nanoparticles synthesis which employs microorganisms or other living things like plants, animals and their metabolites (Adelere and Lateef 2016; Lateef et al. 2016a). This method remains the most researched area of nanoparticles synthesis which holds immense potential.

Biosynthesis of copper, platinum, gold, zinc, palladium, silver and selenium particles from bacteria, actinomycetes, fungi and yeasts had been globally reported (Shah et al. 2015). Green synthesis of nanoparticles is accomplished via adoption of biological materials as stabilizing and reducing agents that are eco-friendly and non-toxic (Jegadeeswaran et al. 2012). The inorganic metal ions can be converted into metal nanoparticles via the reductive capacities of these biological systems rich in metabolites and other important proteins (Sanghi and Verma 2009a; Makarov et al. 2014). Interestingly, the abundant availability of several biological resources has expanded the scope of biological synthesis technique of nanomaterials. Some of the biofactories that have been previously reported are fungi (Verma et al. 2010), mushrooms (Devika et al. 2012; Yehia and Al-Sheikh 2014; Owaid et al. 2017a), bacteria (Lateef et al. 2016b; Ojo et al. 2016; Oladipo et al. 2017a, b), plants (Lateef et al. 2016c, d, 2017, 2018; Azeez et al. 2017; Adebayo et al. 2019a, b; Aina et al. 2019), enzymes (Faramarzi and Forootanfar 2011; Durán et al. 2014, 2015; Khan et al. 2015; Lateef et al. 2015; Lateef and Adeeyo 2015; Elegbede et al. 2018, 2019, 2020) and metabolites obtained from arthropods (Xu et al. 2013; Singh et al. 2014; Lateef et al. 2016e, f).

Generally, nanoparticles are characterized based on some advanced techniques of spectroscopy and microscopy. The UV–visible (UV-vis) spectrum of mushroom-mediated silver nanoparticles revealed surface plasmon resonance bands ranged between 400 and 460 nm (Bhat et al. 2011; Devika et al. 2012; Anthony et al. 2013; Arun et al. 2014; Chowdhury et al. 2014; Ekar et al. 2015; Elamawi et al. 2018; Aygün et al. 2020) and 500–550 nm for gold nanoparticles (Philip 2009; Bhat et al. 2013; Owaid et al. 2017a; Bawadekji et al. 2018; Owaid et al. 2019). FTIR spectra revealed polysaccharides, oligosaccharides, proteins and enzymes as reducing and capping/stabilization agents in mushroom-mediated synthesis of nanoparticles (Bhat et al. 2011; Anthony et al. 2013; Eskandari-Nojedehi et al. 2018), heterocyclic compounds (Arun et al. 2014), phenols (Owaid et al. 2017a), polyphenols, amines and alkaloids (Bawadekji et al. 2018), alcohols and ethers (Karthikeyan et al. 2019). The structural features such as particle size and morphology obtained using scanning electron microscopy (SEM), transmission electron microscopy (TEM) and dynamic light scattering (DLS) are as follows: spherical, particle size (9–21 nm) with agglomeration (Aygün et al. 2019), spherical and size ranging between 51 and 99 nm (Arun et al. 2014), monodispersed spherical with size range 10–15 (Anthony et al. 2013), spherical in shape with particle size 10.3–38.7 nm (Bawadekji et al. 2018) and polydispersed spherical with size range of 20–25 nm (Bhat et al. 2011). Except with some exceptional cases, the shape of most mushroom-mediated nanoparticles is spherical which may be monodispersed or polydispersed with varying sizes. The X-ray diffraction (XRD) pattern of the mushroom-synthesized nanoparticles usually have numbers of Bragg reflections of nanocrystals corresponding to the (1 1 1), (2 0 0), (2 2 0) and (3 1 1) sets of lattice planes (Devika et al. 2012; Bhat et al. 2013; Chowdhury et al. 2014; Eskandari-Nojedehi et al. 2018; Karthikeyan et al. 2019). The implication is that most mushroom-mediated nanoparticles are highly crystalline.

Biosynthesized nanoparticles have been investigated as catalytic, antimicrobial and biosensing agents in fields such as agriculture, health care, consumer products, environment, foods and energy industries among others. In ancient Indian Ayurvedic medicine book *Charak Samhita,* the therapeutic properties of different metals are in details revealing the use of silver as earlier as 300 BC (Galib et al. 2011). Silver nanoparticles (AgNPs) had been widely used out of the different nanoparticles due to its applicability in biomedical, drug delivery, water treatment, agricultural, electronic devices, adhesives and in several industrial processes (Srikar et al. 2015; Velusamy et al. 2016). Biosynthesized AgNPs are applied in biomedical sciences to improve topical antimicrobial gel formulation, blood-contacting implants, wound healing, medical catheters, orthopaedic applications, coating of contact lenses, dental instruments and endodontic filling materials (Sathishkumar et al. 2018). Other metallic nanoparticles such as palladium, platinum and gold are applied in the cosmetic industry as well as in medical and pharmaceutical industries. Gold nanoparticles are important in separation science and clinical disease diagnosis and treatment with several applications in the biomedicals/pharmaceuticals. Platinum nanoparticles had been universally applied in their pure or alloyed state, mostly in biomedical field (Shah et al. 2015). It is also useful in catalysis, optoelectronics, sensory and antimicrobials. Non-noble metallic nanoparticles of selenium, iron, zinc oxide and copper are also useful in agricultural, industrial and biomedical fields with great importance in antimicrobials and cosmetics (Shah et al. 2015).

3 Mushrooms in Nanobiotechnology

Mushrooms, like other fungi, are important in nanobiotechnology due to the abundant production of important biomolecules vital to the synthesis of nanoparticles. More so, the biomolecules are produced extracellularly saving the cost of downstream process. In addition, mushrooms have bioaccumulation and high wall-binding ability with high stability reported in most mushroom-mediated nanoparticles, and this has been linked to protein capping and amide linkages (Anthony et al. 2013). Mushroom biomolecules of importance in nanobiotechnology have been basically implicated in the process of stabilizing and reducing metal ions to nanoparticles such as the reductase enzyme. Furthermore, mushroom mycelia surfaces are sticky and negatively charged which thus serve as adhesive and electrostatic force during the formation of nanoparticles (Afshar and Sedaghat 2016). Most mushrooms of importance in nanobiotechnology are edible and highly medicinal, thus serving as biofactories for the production of safe and medically important nanoparticles.

The earliest research on mushroom-mediated nanoparticles was in 2004 when fibres of nanosizes were produced from *Schizophyllum commune* such as Schizophyllan, which is a purified β-1, 3-glucan used to synthesis nanoparticles (Numata et al. 2004). The utilization of mushroom extracts from *Volvoriella volvaceae, Lentinula edodes, Ganoderma lucidum, Agaricus bisporus, Agrocybe aegerita, Pleurotus sajor caju, P. florida* and others had been reported with their potential in reducing metal ions into metal nanoparticles (Owaid et al. 2015; Owaid and Ibraheem 2017; Owaid 2019). Mushrooms are capable of large biomass productions which are useful in nanoparticles synthesis (Owaid et al. 2017b). Oyster mushrooms remain the most explored mushroom in nanobiotechnology. In addition, of all oyster mushrooms utilized in nanobiotechnology, *Pleurotus ostreatus* is the most explored with 28%. This is closely followed by *Pleurotus sajor caju* with 21%, *Pleurotus florida* with 17% and *P. cornucopiae* with 7%. Other oyster mushroom species such as *Pleurotus giganteus, P. salmoneostramineus, P. citrinopileatus, P. eous, P. djamor, P. pulmonarius, P. platypus* and *P. djamor* var roseus are of 3% each. Compound secreted by mushrooms (edible and medicinal) which are used for nanoparticles formation produce nanoparticles that are mostly monodispersed and stable. Furthermore, mushroom provides exceedingly interesting ability to modulate and create different nanoparticles sizes and shapes in addition to lasting stability due to their slower kinetics (Zhao et al. 2018).

4 Biosynthesis of Novel Metal Nanoparticles by Mushrooms

Mushroom mycelia and fruiting bodies contain various essential amino acids, proteins and polysaccharides that can be utilized for producing nanoparticles of iron, silver, selenium, gold and others either intracellularly or extracellularly (Owaid et al. 2017c). Nanoparticles of silver are produced from silver nitrate ($AgNO_3$) as reported in several studies (Paul et al. 2015). Of all oyster mushroom-mediated nanoparticles, 66% are reportedly silver nanoparticles (AgNPs), while 25% are of gold nanoparticles (AuNPs) and zinc oxide (ZnONPs). Only 9% are of iron nanoparticles (FeNPs), cadmium sulphide nanoparticles (CdSNPs) and selenium nanoparticles (SeNPs) (Paul et al. 2015). The remaining 9% approximately are made up of non-metal nanoparticles (Paul et al. 2015). The biological synthesis of mushroom nanoparticles is stable, environmental-friendly and not toxic. Interestingly, the knowledge of synthesizing nanoparticles from mushroom is increasing daily (Shivashankar et al. 2013; Moghaddam et al. 2015).

4.1 Mushroom-Mediated Silver Nanoparticles (AgNPs) and Their Applications

Extracts of edible mushrooms have been found useful in the production of silver nanoparticles (AgNPs). Generally, the mechanism of synthesizing AgNPs is not fully understood; however, some studies have provided some explanations in this front as presented in Fig. 1. Extracts of mushrooms are usually added to aqueous solution of 10^{-3} M $AgNO_3$ to form a reaction mixture that is immediately accompanied by shades of colour changes culminating in reddish brown colour to produce AgNPs. The process of nanoparticles formation may take place at ambient temperature or aided by other conditions such as sonication, photocatalysis, placement in the dark cupboard among others. Different sizes and shapes of mushroom-mediated AgNPs have been reported with major applications in biomedical fields. Spherical AgNPs from *Pleurotus florida* were reportedly effective against bacteria such as *Proteus mirabilis, Salmonella typhii, Staphylococcus aureus* and *P. alcalifaciens* (Bhat et al. 2011). Irregularly shaped AgNPs of *A. bisporus, Calocybe indica, P. florida* and *P. platypus* were effective against *Klebsiella pneumoniae, E. coli, S. aureus, Enterobacter aerogenes* and *Pseudomonas putida* (Sujatha et al. 2013). Synthesized AgNPs of *Pleurotus giganteus* were spherical crystals of 2–20 nm size with minimum inhibitory concentration of 10–15 μg/ml against *S. aureus, E. coli, B. subtilis* and *P. aeruginosa.* More so, a particular study revealed the *P. giganteus*-mediated silver nanoparticles exhibiting a concentration-dependent α-Amylase inhibition (Debnath et al. 2019).

Ganoderma has been extensively studied with great applications as anticancer, antibacterial, hepatoprotective, antioxidant, antitumour, hypocholesterolemic, anti-inflammatory, antidiabetic, anti-HIV and antiproliferative among several other activities. According to Mohanta et al. (2018), common species like *G. capense,*

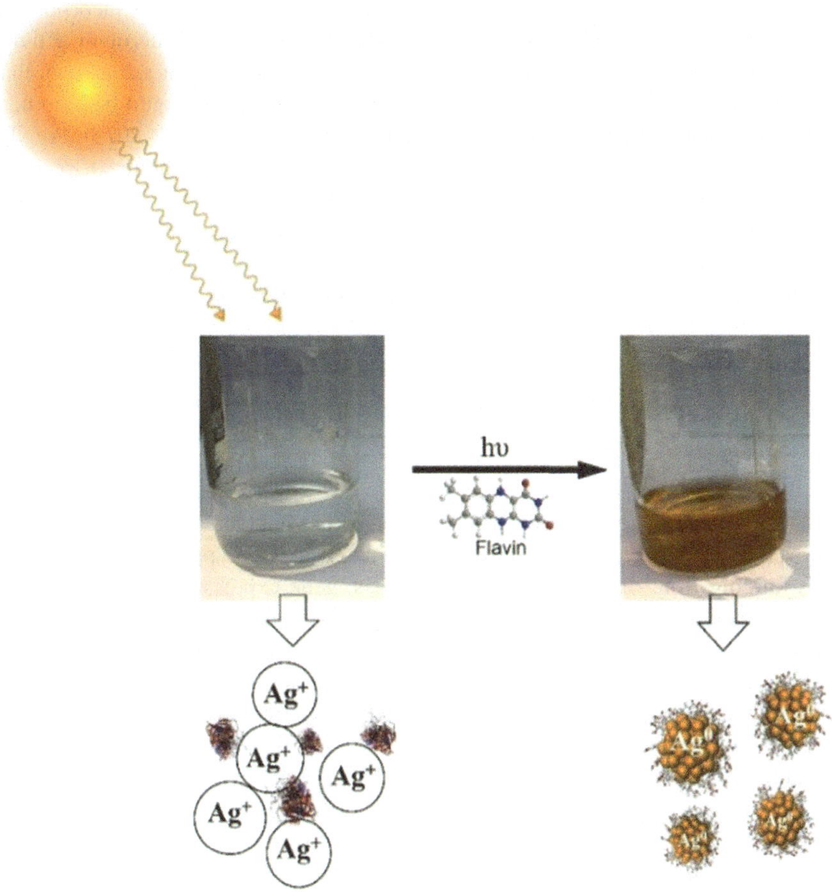

Fig. 1 Biosynthesis of mushroom silver nanoparticles (Bhat et al. 2011)

G. lucidum, G. applanatum and *G. tsugae* have found diverse applications in biomedicine. *Ganoderma applanatum*-mediated AgNPs have been shown to manifest great antimicrobial activity against pathogenic bacteria; *Listeria innocua, B. subtilis, Micrococcus luteus, E. coli* and *Streptococcus faecalis*. In a similar study, Aygün et al. (2020) reported 9–21 nm, spherically shaped AgNPs biosynthesized by *G. lucidum* with antimicrobial activity against *C. albicans, P. aeruginosa, S. aureus, E. coli, E. hirae, L. pneumophila* and *B. cereus*. Biosynthesis of 2 nm spherical AgNPs by *Ganoderma* mushroom extract has been reported (Ekar et al. 2015). AgNPs produced by *Ganoderma lucidum* have demonstrated interesting antimicrobial activity by enhancing the effectiveness of tetracycline when used in combination and singly against *E. coli* and *S. aureus* as reported by Karwa et al. (2011). In a related development, a 133 nm, spherical *G. applanatum*-mediated AgNPs demonstrated antibacterial activity against *S. aureus, E. coli, S. epidermidis, Shigella flexneri, Vibrio cholerae* and *B. subtilis* as documented by Mohanta et al. (2016). Cotton fabrics incorporated with 75 nm cubic AgNPs of *G. lucidum* showed promising antimicrobial activity against *Pseudomonas, Streptococcus* and *Proteus* species. Thus, it was shown that the dressing material incorporated with AgNPs was sterile and could be commercially used for the treatment of wounds and infections (Paul et al. 2015). AgNPs of *Ganoderma neo-japonicum* Imazeki extract generated little amount of reactive oxygen species with great promises in apoptosis due to their cytotoxic effects, and this can be built upon in the race to discover potent anticancer drugs (Gurunathan et al. 2013). Furthermore, Anthony et al. (2013) demonstrated the antibacterial activity of *Tricholoma matsutake*-mediated AgNPs against *B. subtilis* and *E. coli.*

4.2 Mushroom-Mediated Gold Nanoparticles (AuNPs) and Their Applications

Mushroom-mediated gold nanoparticles (AuNP) are another interesting members of novel metal nanoparticles discovered to have promising applications in several fields ranging from disease diagnosis, catalysis, gene expression, nanoelectronics and nonlinear optics (Bhat et al. 2013). AuNPs biosynthesis with mushroom extracts had been revealed to have different shapes such as spherical, hexagonal and triangular as well as 5-150 nm sizes (Sen et al. 2013). Some spherical AuNPs of sizes 16–91 nm from dried mushroom and 23–100 nm for AuNPs from fresh mushroom of *Pleurotus cornucopiae* var. *citrinopileatus* have been reported (Owaid et al. 2017a), while Owaid (2019) reported the presence of triangular gold nanoparticles from *Lentinula edodes* aqueous extract. Antiproliferative effect of irregularly shaped AuNPs from *P. florida* against human cancer cell lines has been reported (Bhat et al. 2011). AuNPs of *Pleurotus florida* were irregular in shape with sizes 12–15 nm and demonstrated anticancer potential against common cancer cell lines of HeLa (Human cervix), K-562 (Human chronic myelogenous leukaemia bone marrow), MDA-MB (Human adenocarcinoma mammary gland) and A-549 (Human lung carcinoma) as well as no lethal activity recorded in African green monkey kidney normal cell (Bhat et al. 2013). Spherical AuNPs of sizes 10.3–38.7 nm synthesized from *Pleurotus ostreatus* showed antimicrobial activity against *Enterococcus faecalis, E. coli, Klebsiella pneumoniae, P. aeruginosa* and *Staphylococcus aureus,* while antifungal activity was reported against *C. albicans* (Bawadekji et al. 2018). Fabricated AuNPs of 10–50 nm size and spherically shaped obtained using *Agaricus bisporus* exhibited high antifungal activity against *Aspergillus flavus* (Eskandari-Nojedehi et al. 2018).

4.3 Mushroom-Mediated Selenium Nanoparticles (SeNPs) and Their Applications

Mushrooms have been used to facilitate selenium reduction in sodium selenite, which is inorganic or 1,5-diphenyl-3-selenopentanedione-1,5 (DAPS-25), which is organic. In this development, spherical SeNPs were reportedly synthesized by *Lentinula edodes* by Vetchinkina et al. (2013). More so, different shapes and sizes of selenium nanoparticles have been produced with *Grifila frondosa, P. ostreatus* and *G. lucidum* cultivated on germanium (GeO_2) and selenium (Na_2SeO_3)-rich media (Vetchinkina et al. 2016). In addition, mushroom metabolites such as proteins and polysaccharides have been widely reported to be important in synthesizing SeNPs with great antioxidant potential (Xiao et al. 2017). Chen et al. (2015) reported promising antitumor activity of SeNPs when used in combination with polysaccharides from *Pleurotus tuber-regium.* In a related study, Wang et al. (2014) reported improved immune modulatory potential of SeNPs synthesized from polysaccharides of *Ganoderma lucidum* origin. Generally, according to Wu et al. (2013), polysaccharides and protein complexes from diverse mushrooms are of immense importance to SeNPs biosynthesis which is of great stability and controllable sizes.

4.4 Mushroom-Mediated Cadmium Sulphide Nanoparticles (CdSNPs) and Their Applications

Research on the biological synthesis of cadmium sulphide nanoparticles has intensified due to its application in quantum dotting, biosensory, photocatalysis, solar cells and diode, drug delivery and treatment. Mushroom protein thiol groups have been heavily linked to the supply of sulphur important to the change of poisonous cadmium to safe cadmium sulphide; thus, leading to the secretion of CdSNPs with autocap and high stability (Sanghi and Verma 2009b). Under room temperature, *Coriolus versicolour*-mediated CdSNPs have been produced extracellularly, while 4–5 nm spherical quantum dots CdSNPs were produced using *P. ostreatus* (Borovaya et al. 2015). Antibacterial and antifungal activities of *Ganoderma lucidum*-mediated CdSNPs have also been reported (Raziya et al. 2006).

4.5 Mushroom-Mediated ZnNPs, FeNPs and Other Metal Nanoparticles with Their Applications

The biosynthesis of ZnNPs with various mushrooms extracts as bioreductant has been explored by a number of researchers with zinc nanoparticles having size range of 3–201 nm (Kowalczyk et al. 2011). In a study by Senapati and Sarkar (2014), cubic *Pleurotus ostreatus*-mediated zinc sulphide nanoparticles demonstrated linkage between size and density of dislocation. Also, in the XRD analysis, strain increased with increase in *P. ostreatus* extract concentration. Furthermore, *P. ostreatus*-mediated zinc sulphide nanoparticles have found applications in several industrial processes mainly in coating surfaces. Iron nanoparticles have been produced using mushrooms, most especially diverse species of *Pleurotus* with the secretion of the nanoparticles detected at wavelength of 226–276 nm where siderophore was important in the transportation of iron within the mycelium (Mazumdar and Haloi 2011). Other fungi have been employed in iron nanoparticles synthesis; however, there is no recent information on mushroom-mediated iron nanoparticles; hence, this is an emerging field of mushroom nanobiotechnology requiring more research. Faraji et al. (2010) reported the potentials of synthesizing more metallic nanoparticles from mushrooms and in the world today. The synthesis of copper oxide, aluminium oxide, magnesium oxide nanoparticles among others is gradually gaining attentions in mushroom nanobiotechnology (Table 1).

5 Mechanism of Mushroom-Mediated Nanoparticles Synthesis

Light sensitivity and water solubility are the two important criteria in the green synthesis of nanoparticles, and interestingly, mushrooms contain compounds such as riboflavin that possess these characteristics. Mushroom riboflavin acts as enzyme in redox reactions when in the form of flavin adenine dinucleotide (FAD) and as bound coenzyme when in form of flavin mononucleotide (FMN). Furthermore, the common result of degradation of riboflavin and its derivatives in FMN and FAD in the presence of light is lumiflavin. Thus, in mushroom nanobiotechnology, ions of metals are reduced to nanoparticles due to the presence of flavo proteins, also known as flavin in the mushroom extracts. In the process of nanoparticles synthesis with mushroom, photons of energy are absorbed in the presence of sunlight which consequently excite the flavin which is abundant in several mushroom species to donate electrons or act as oxidizing agent important in nanoparticles production (Bhat et al. 2011) (Fig. 2). Flavonoids have been demonstrated for the green synthesis of nanoparticles (Sathishkumar et al. 2018).

Generally, several studies have strongly linked the mechanism of mushroom nanoparticles biosynthesis to the presence of important metal-reducing enzymes in mushroom species that catalyse the production of nanometric elemental metal from metal ions. According to Ahmad et al. (2003), biosynthesis of mushroom-mediated nanoparticles is marked with clear colour change and optical properties alteration as determined by resonance bands of surface plasmon in UV–vis spectroscopy. In addition, several factors have been known to influence mushroom nanoparticles biosynthesis and some of these factors that have been extensively discussed in literature include the species of the mushroom and certain characteristics of the mushroom such as capping, temperature, medium of dispersion and pH (Khandel and Shahi

Table 1 Mushroom-mediated nanoparticles and their applications

Mushroom	Localization	Nanoparticles	Size (nm)	Shape	Application	References
Agaricus bisporus	Intracellular	Ag	15–20	Spherical	Antimicrobial activity	Sudhakar et al. (2014)
	Extracellular	Ag	30	Spherical	Antimicrobial activity	Haq et al. (2015a, b)
	Extracellular	ZnS	3–201	Cubic	Selenite uptake rate, Nano-tunned devixe	Senapati et al. (2015)
Calocybe indica	Intracellular	Ag	100	Spherical	Antibacterial activity	Mirunalini et al. (2012)
Coriolus versicolor	Intracellular	Ag	10	Spherical	Antimicrobial activity	Sanghi and Verma (2009a)
	Extracellular	CdS	100–200	Spherical	Semiconductor quantum dots	Chowdhury et al. (2014)
Lentinula edodes	Extracellular	Ag		–	Antibacterial activity	Sujatha et al. (2013)
	Intracellular	Au	5–15	Spherical	Anticancer activity	Vetchinkina et al. (2016)
Helvella lacunose	Extracellular	Ag	8–50	–	Antimicrobial activity	Sudhakar et al. (2014)
Ganoderma appalanatum	Extracellular	Ag	3–15	–	Antimicrobial activity	Philip (2009)
Ganoderma lucidium	Extracellular	Ag	10–70	Spherical	Antimicrobial activity	Karwa et al. (2011)
	Intracellular	Au	5–15	Spherical	Anticancer activity	Vetchinkina et al. (2016)
Ganoderma neo-japonicum	Extracellular	Ag	5–8	Spherical	Anticancer activity	Gurunathan et al. (2013)
Grifola frondosa	Intracellular	Au	5–15	Spherical	Anticancer activity	Vetchinkina et al. (2016)
Fomes fomenterieus	Extracellular	Ag	8–50	Spherical	Antibacterial activity	Deepak et al. (2015)

(continued)

Table 1 (continued)

Mushroom	Localization	Nanoparticles	Size (nm)	Shape	Application	References
Pleurotus ostreatus	Extracellular	Ag	5–40	Spherical cubic	Dye decolorization	Karthikeyan et al. (2019)
	Intracellular	Au	5–15	Spherical	Anticancer activity	Vetchinkina et al. (2016)
	Extracellular	ZnS	2–5	Spherical with crystalline	Selenite uptake rate, Nano-tunned device	Senapati and Sarkar (2014)
Pleurotus florida	Extracellular	Ag	20–50		Antibacterial activity	Sujatha et al. (2013)
	Polysaccharides/glucan	Au	12–15	Crystalline	–	Bhat et al. (2013)
	Extracellular	Au	10–50	Triangular, spherical	–	Bhat et al. (2013)
Pleurotus tuber-regium	Polysaccharides/protein	Se	–		Cellular uptake, antiproliferative activity, anticancer activity	Wong (2020)
Pleurotus rhinoceros	Polysaccharides/protein	Se	–	Spherical	Cellular uptake, antiproliferative activity, anticancer activity	Wu et al. (2013)
Pleurotus sajor caju	Extracellular	Ag	5–50	–	Antimicrobial activity	Sudhakar et al. (2014)
	Extracellular	Ag	35	Spherical	Dye decolorization	Nithya and Ragunathan (2009)
Pleurotus citrinopileatus	Extracellular	Ag	6–10	Core shell	Antibacterial activity	Maurya et al. (2016)
P. comucopiae var. citrinopileatus	Intracellular	Ag	10–20	Spherical	Antifungal activity	Owaid et al. (2015)
Pleurotus platypus	Intracellular	Ag	5–50	–	Antibacterial activity	Sujatha et al. (2013)

(continued)

Table 1 (continued)

Mushroom	Localization	Nanoparticles	Size (nm)	Shape	Application	References
Agrocybe aegerita	Extracellular	Ag	–	–	–	Muthu and Shanmugasundaram (2015)
Volvariella volvacea	Extracellular	Ag	5	–	Antibacterial activity	Bhat et al. (2011)
	Extracellular	Au	20–150	Triangular, spherical, prism, hexagonal	Anticancer activity	Philip (2009)
Tricholoma matsutake	Extracellular	Ag	10–20	–	Antimicrobial activity	Anthony et al. (2013)
Schizophyllum commune	Schizophyllan	Ag	–	–	Biomedical application	Arun et al. (2014)

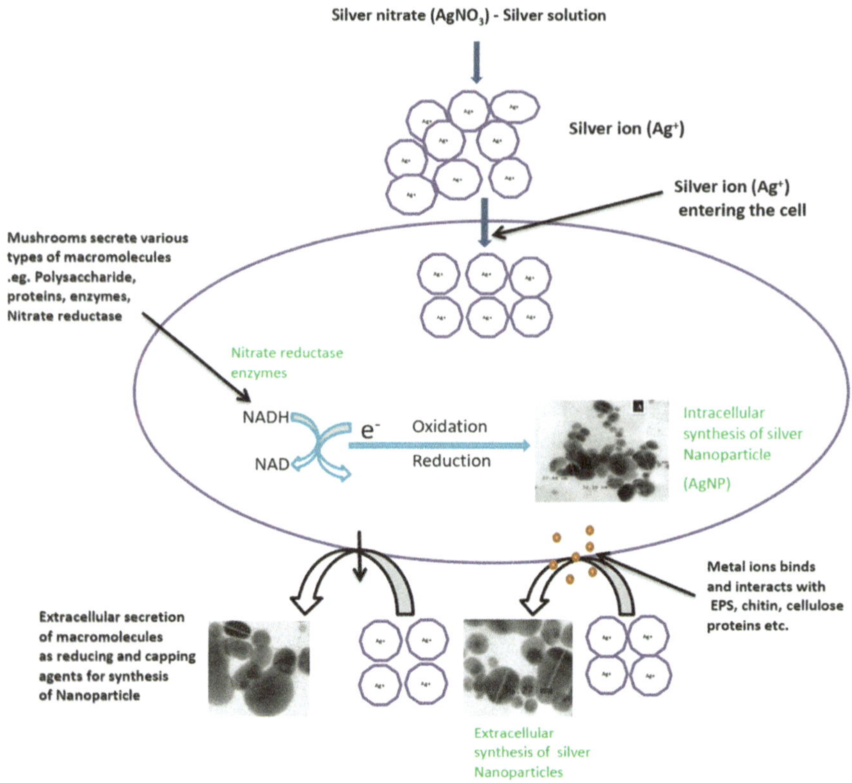

Fig. 2 Mechanism of mushroom biosynthesis of silver nanoparticles (AgNPs)

2018; Lee and Jun 2019). More so, the higher the intensity of the dispersion colour, the higher the resonance of surface plasmon which depends on the absorbance and sizes of the nanoparticles (Adeeyo and Odiyo 2018; Bhangale et al. 2019; Lee and Jun 2019).

According to Gudikandula et al. (2017), several biomolecules in mushrooms can reduce and stabilize metal ions to nanoparticles, most especially the biomolecules involved in the transfer of electrons in complex pathways of NADPH/NADH to NAD^+/NAD^+, while Baymiller et al. (2017) added to this claim by postulating that reductase enzymes of nicotinamide adenine dinucleotide (NADH) and NADH dependent nitrate are integral to mushroom nanoparticles biosynthesis. A study by Hietzschold et al. (2019) demonstrated the importance of NADPH over other nitrate reductases in nanoparticles biosynthesis and thus, shedding more light on the biological synthesis of nanoparticles without emphasis on reductase. However, Durán et al. (2005) previously showed that anthraquinones and nitrate reductase are vital in fungal nanoparticles biosynthesis, while Kumar et al. (2007) stressed the importance of quinones and extracellular NAFPH-dependent nitrate reductase on nanoparticles synthesis.

6 Factors Affecting Biosynthesis of Nanoparticles

Several factors that could affect mushroom extract-mediated biosynthesis of nanoparticles have been reported. Some of these factors include temperature, concentration of the raw materials, pressure, the alkalinity or acidity of the solution, time, magnitude of the particle, size of the pore, environment and accumulation of the mushroom extracts among others (Baker et al. 2013). Though biosynthesis of nanoparticles using mushroom extracts is a biological method of synthesis considered to be safer, biocompatible, ecological-friendly, cost-effective and widely acceptable, the conditions of synthesis such as exposure to light intensity, duration of exposure, pH of the reactants and concentration of the precursor may largely influence the shape, size and activity of the synthesized nanoparticles as found in other studies (Mudunkotuwa et al. 2013; Narayanan et al. 2015).

6.1 Influence of PH

The pH of the reacting medium is a very important parameter to be considered during biosynthesis of nanoparticles. The pH value of the medium influences the size and the consistency of the synthesized nanoparticles (Gardea-Torresdey et al. 1999; Armendariz et al. 2004). Soni and Prakash (2011) opined that varying the pH value of the solution would affect the size and shape of the specific compounded nanoparticles. However, various extracts from different type of mushrooms have different pH values; hence, full consideration should be given to pH values during nanoparticles fabrication. Several researchers had revealed that bigger nanoparticles sizes are built at lower pH value in contrast to higher pH value (Dubey et al. 2010).

6.2 Effect of Temperature

Temperature remains one of the key factors that affect nanoparticles synthesis using biological methods. The biological methods of nanoparticles synthesis require less temperatures below 100 °C or ambient temperature when contrasted with the physical and chemical methods. However, sequel to the utmost temperature (>350 °C) required by the physical method and the fewer than 350 °C temperature required by the chemical methods, nanoparticles setting will be unbeatable by temperature of the medium of reaction (Rai et al. 2006). Andreescu et al. (2007) showed the formation of silver nanoparticles at higher temperatures. An increase in temperature in mushroom extracts can lead to an increase in sharpness of absorption peaks of nanoparticles. The increase in temperature may trigger the rate of reaction, which enhances the synthesis of nanoparticles. The sharpness of the absorbance peaks relies on the magnitude of the compounded nanoparticles with superior temperature (Fayaz et al. 2009; Shaligram et al. 2009).

6.3 *Effect of Pressure*

Pressure is one of the key factors that affect the synthesis of nanoparticles using biological extracts. According to Abhilash and Pandey (2012), the applications of pressure to the reaction medium altered the magnitude and shape of the biosynthesized nanoparticles. The metal ions reduction rate by using biological extracts had been reported to be very swift at atmospheric pressure conditions (Tran et al. 2013).

6.4 *Effect of Incubation Time*

Incubation time refers to the period needed for every reaction process to take place. However, several reports have shown that incubation time affects the synthesis of nanoparticles using extracts (Tran et al. 2013; Darroudi et al. 2011). However, the characteristics of the biosynthesized nanoparticles are altered by the period of maturation which largely depends on the synthesis steps, light exposure and how it is stored (Akbari et al. 2011; Mudunkotuwa et al. 2013). It is reported that the sharpness of absorption peaks in both silver and gold nanoparticles is affected by an increase in time contact (Dubey et al. 2010). The shelf life and potency of biosynthesized nanoparticles would be affected with long time storage (Patra and Baek 2014).

6.5 *Other Factors Affecting Mushroom-Mediated Biosynthesis of Nanoparticles*

The permeability of biosynthesized nanoparticles determines the nanoparticles quality and application (Park et al. 2011). Mushrooms are rich in subordinate metabolites which serve as diminishing and balancing agents for synthesizing nanoparticles. However, mushroom species and its extraction methodology influence the composition of these metabolites. Furthermore, different mushroom species are embedded with unique intracellular and extracellular enzymes which may affect mushroom biosynthesis of nanoparticles (Mohanpuria et al. 2008).

7 Challenges of Mushroom Nanobiotechnology

Despite the wide range benefits of mushroom-mediated nanoparticles, there are still several shortcomings and issues to defeat before it can be employed extensively. Mushroom nanobiotechnology as a newly developing field has some challenges, most especially in the aspect of commercialization. Such problems include profit-making, effectiveness of innovation, funding, raw materials and other needed resources and market availability. In addition to these specific problems, almost all other challenges of nanotechnology abound in mushroom nanobiotechnology. It is believed that there is no single technology without challenges. Nanotechnology has challenges, and no one can give vivid solutions to the problems that nanotechnology causes. It is therefore necessary to monitor human and animal population for any adverse effect of nanomaterial exposure in the environment. The challenges become increasingly strenuous in exceedingly difficult grid like food requiring high standard and regulations. Another challenge is the development of applicable methodology to find and ascertain how poisonous the manufactured nanomaterials could be. However, suggesting models to forecast the impact of these nanomaterials on the well-being of human and its surroundings would be an unavoidable challenge. Furthermore, the problem of reverse system developments to appraise detailed influence of the manufactured nanomaterials on well-being and the surroundings over the entire span of life with respect to the life cycle problem abound. Another major challenge is the development of the tools that could be used to appropriately evaluate the possible danger to the health of humans and to the environment.

The use of nanoparticles in different fields will keep developing; nevertheless, there is a reason to assess the toxicity aftermath and concentration in the environment and the outcome on the people and animal health. This calls for more research in the optimization of various reaction states to achieve better grip over magnitude, structure and monodiversity of the nanoparticles. Also, nanoparticles stability is also a very crucial variable to be reviewed. It is imperative to carry out and overcome all these restrictions. More so, the correct process of nanoparticles formation with the use of mushroom (fungi) has not been distinctly and thoroughly comprehended yet. Therefore, more study and experimental trials are needed in this aspect to detail the exact mechanism in other to identify the responsible biomolecules (enzymes and proteins) involved in reduction and stabilization of nanoparticles (Khandel and Shahi 2018). The potentials of nanobiotechnology are massive in terms of improving existing products and producing entirely new products. Although, a lot of questions need to be answered in this new field majorly requiring new rules. Governments all over the world should appraise likely dangers and a suitable law-binding answer to the wide application of this advanced technology.

8 Mushroom Nanobiotechnology: Current and Future Prospects

Nanotechnology is an international business enterprise influencinguniversities, industries and regulatory agencies. Though still at its primitive stage of expansion and in all aspects, nanobiotechnology's evolution is multifaceted and rapid. Nanobiotechnology will provide chances for creating new materials in a way that will strengthen human's capability to grow faster with enablement to evolving more dependable and thoughtful problem-solving structures. It is a multidisciplinary field that has great potential to bring the science of indecipherable minute gadget to proximity of real life. At some point, the impact of these evolutions will be so enormous that it will most likely impact nearly all areas of science and technology. In medicine, a broad range of uses is presented by nanotechnology especially in the area of drug delivery and tissue biomimetics. There is no doubt that this technology has brought the hope of giving birth to new things and materials that may culminate in effective management and finding lasting solutions to life threatening clinical conditions. There are many arguments about the implication of nanobiotechnology in the days ahead which could give birth to possible enactment on several new products and gadgets, which might be of great use in the medical, biomaterials, energy and electrical field. Buzea et al. (2007) reported that the approach might aggravate several issues related to the emergence of a new technology like the impact of nanomaterials on the environment, its toxicity and the its likely aftermath on the world's economy, likewise as hypothesizing on several future destructions that might likely occur. The above agitations brought about an argument amidst concerned citizens and the government on if exceptional enactment of certain status is needed. Despite these arguments, nanotechnology still provides monumental expectation for the days ahead. Sahoo and Labhasetwar (2003) said this might result into various new inventions by acting a leading part in different applicable fields like the use of biology and physiology to clinical medicine including the delivery of drug and healing quality of gene to molecular imaging, biomarkers and biosensors. The use of one of the above mentioned being the major objective of the immediate period of time will be aimed at a specific drug healing power and ways for early detection and cure for illnesses.

Biosynthesis of nanoparticles from mushroom cannot be overemphasized in recent times due to available reports on its applications in areas like agriculture, electronics, food, optical, medicine, cosmetics and textile. The utilization of mushroom for nanoparticles synthesis has an edge of simple operation and downstream procedure. The applications of nanotechnology required ample assessment of its pros and cons. A lot of researchers who are in opposition of the use of this technology also concur that the growth in nanotechnology should proceed since this area guarantees massive advantages nevertheless investigations have to be done so as to ascertain its hazardous effect on the environment, other living organisms and well-being of people. In a nutshell, if all goes well, nanobiotechnology will become an unavoidable part of human's daily living, economic hub, promising technology in the industries and securing many lives.

9 Safety Evaluation of Nanoparticles

Although mushroom-mediated nanoparticles are believed to be safe, the safety of nanoparticles in general is a debate in the today's world. Nanoparticles have possible threats akin to atmospheric aerosol particles following its very nanosized feature which gives a special edge (Li et al. 2007). These particles have the potential to give rise to various medical science of the ventilatory, circulatory and digestive system

(Nijhara and Balakrishnan 2006). In mice, the introduction of carbon nanotube directly into their trachea showed that carbon nanotubes have the ability to bring about diverse disease of the lung, e.g., necrosis of lung, pulmonary fibrosis, peribranchial inflammation and epithelioid granuloma (Oberdörster et al. 2015). According to Lam et al. (2004), carbon nanotube produces more toxicity than thermal black. Various studies have shown that there are many ports in the human body that can serve as passage for nanomaterials into the human body (Hoet et al. 2004). Some biosynthesized nanoparticles have also shown cytotoxicity and genotoxicity in *Allium cepa* assay (Yekeen et al. 2017a, b).

Unexpected exposure while manufacturing or use will possibly take place through respiratory system, from which a quick transition is feasible to other important parts of the body through the circulatory system. At the level of cells, nanoparticles have been reported to be a likely carrier of a genetic code (Williams 2004). Nanoparticles can break into the nervous system which maybe exactly through the nerve cells of the nose or through the blood stream via the olfactory lobe (Oberdorster 2004). Elder et al. (2006) observed accumulation of nanoparticles that is composed of carbon and manganese in the olfactory lobe in monkeys and rats through the olfactory pathway. However, this shows that delivery caused by nanoparticles provides a medium of proxy course bypassing the blood brain obstacles in the times ahead. Moreover, this can lead to an inflamed outcome/response in the brain as suggested, but more studies are required to ascertain these effects. Radomski et al. (2005) reported the pro-accumulation influence of nanotubes on blood platelets *in the glass studies* quickening of vascular thrombosis in rat. It was also reported that fullerenes do not have the nature of inducing platelet accumulation. Therefore, fullerenes might be a secure way in contrast to nanotubes in creating a delivery system based on nanoparticles (Medina et al. 2007). Nanoparticles toxicity can also infer to circulatory system, leading to inflammatory disease of the bowel. Also, the poison level of the nanoparticles may be with respect to its potential to instigate the discharge of pro-inflammatory mediators proceeding to inflammatory feedback and damage of organs of the body. Nanoparticles can potentially enter the blood stream, break into various parts and structure of the human body and most likely cause toxicity if taken in. Above all, the application of nanoparticles in humans necessitates more studies and adequate cautions.

10 Conclusion

Mushrooms have made great improvements in nanobiotechnology with prominent applications in biomedical and industries that clearly promised further advancements in future. Mushrooms nanobiotechnology-based treatment has offered very effective, efficient, durable and eco-friendly approaches. The safe nature, ease of production, high stability, long shelf life, active of metabolites and enzymes with ease of extraction are factors that favoured mushrooms in nanobiotechnology. These methods are more cost-effective, less time and energy consuming with very less waste generations than conventional bulk materials-based methods. The impact of mushroom-mediated nanoparticles in the advancement of methods, techniques and applications in drug delivery, biomedical and other paramedical industries will be more promising in the years to come.

References

Abhilash, Pandey BD (2012) Synthesis of zinc-based nanomaterials: a biological perspective. IET Nanobiotechnol 6:144–148. https://doi.org/10.1049/iet-nbt.2011.0051

Adebayo EA, Oloke JK (2017) Oyster mushroom (*Pleurotus* species); a natural functional food. J Microbiol Biotechnol Food Sci 7(3):254–264. https://doi.org/10.15414/jmbfs.2017/18.7.3.254-264

Adebayo EA, Oloke JK, Ayandele AA, Adegunlola CO (2012a) Phytochemical, antioxidant and antimicrobial assay of mushroom metabolite from *Pleurotus pulmonarius*–LAU 09 (JF736658). J Microbiol Biotechnol Res 2(2):366–374

Adebayo EA, Oloke JK, Majolagbe ON, Ajani RA, Bora TC (2012b) Antimicrobial and anti-inflammatory potential of polysaccharide from *Pleurotus pulmonarius* LAU 09. Afr J Microbiol Res 6(13):3315–3323

Adebayo EA, Oloke JK, Azeez MA, Omomowo IO, Bora TC (2014a) Assessment of the genetic diversity among ten genotypes of *Pleurotus* (Oyster mushroom) using nutrient and mineral compositions. Scientia Horticul 166:59–67. https://doi.org/10.1016/j.scienta.2013.12.010

Adebayo EA, Oloke JK, Aina DA, Bora TC (2014b) Antioxidant and nutritional importance of some *Pleurotus* species. J Microbiol Biotechnol Food Sci 3(4):289–294. https://www.researchgate.net/publication/260020139

Adebayo EA, Martınez-Carrera D, Morales P, Sobal M, Escudero H, Meneses ME, Avila-Nava A, Castillo I, Bonilla M (2018) Comparative study of antioxidant and antibacterial properties of the edible mushrooms *Pleurotus levis, P. ostreatus, P. pulmonarius* and *P. tuber-regium*. Int J Food Sci Technol 53:1316–1330. https://doi.org/10.1111/ijfs.13712

Adebayo EA, Oke MA, Lateef A, Oyatokun AA, Abisoye OD, Adiji IP, Fagbenro DO, Amusan TV, Badmus JA, Asafa TB, Beukes LS, Gueguim-Kana EB, Abbas SH (2019a) Biosynthesis of silver, gold and silver–gold alloy nanoparticles using *Persea americana* fruit peel aqueous extract for their biomedical properties. Nanotechnol Environ Eng 4:13. https://doi.org/10.1007/s41204-019-0060-8

Adebayo EA, Ibikunle JB, Oke AM, Lateef A, Azeez MA, Adeboye OO, Ajala VA, Olowoporoku TB, Okunlola OC, Ogundele OA, Badmus JA, Asafa TB, Beukes LS, Gueguim-Kana EB, Abbas SH (2019b) Antimicrobial and antioxidant activity of silver, gold and silver-gold alloy nanoparticles phytosynthesized using extract of *Opuntia ficus-indica*. Rev Adv Mater Sci 58:313–326. https://doi.org/10.1515/rams-2019-0039

Adeeyo AO, Lateef A, Gueguim-Kana EB (2016) Optimization of the production of extracellular polysaccharide from the Shiitake medicinal mushroom *Lentinus edodes* (Agaricomycetes) using mutation and a genetic algorithm-coupled artificial neural network (GA-ANN). Int J Med Mushrooms 18(7):571–581. https://doi.org/10.1615/IntJMedMushrooms.v18.i7.20

Adeeyo AO, Odiyo JO (2018) Biogenic synthesis of silver nanoparticle from mushroom exopolysaccharides and its potentials in water purification. Open Chem 5:64–75. https://doi.org/10.2174/1874842201805010064

Adelere IA, Lateef A (2016) A novel approach to the green synthesis of metallic nanoparticles: the use of agro-wastes, enzymes and pigments. Nanotechnol Rev 5(6):567–587. https://doi.org/10.1515/ntrev-2016-0024

Afshar P, Sedaghat S (2016) Bio-synthesis of silver nanoparticles using water extracts of *Satureja hortensis* L and evaluation of the antibacterial properties. Curr Nanosci 12(1):90–93. https://doi.org/10.2174/1573413711666150529202238

Agarwal HS, Kumar VS, Rajeshkumar S (2017) A review on green synthesis of zinc oxide nanoparticles—An eco-friendly approach. Resou Eff Technol 3:406–413. https://doi.org/10.1016/j.reffit.2017.03.002

Ahmad A, Mukherjee P, Senapati S, Mandal D, Kahn MI, Kumar R (2003) Extracellular biosynthesis of silver nanoparticles using the fungus *Fusarium oxysporum*. Colloids Surf B Biointerf 28:313–318. https://doi.org/10.1016/S0927-7765(02)00174-1

Aina DA, Owolo O, Lateef A, Aina FO, Abbas SH, Adeoye-Isijola M, Okon V, Asafa TB, Elegbede JA, Olukanni OD, Adediji I (2019) Biomedical applications of *Chasmanthera dependens* stem extract mediated silver nanoparticles as antimicrobial, antioxidant, anticoagulant, thrombolytic, and larvicidal agents. Karbala Int J Modern Sci 5(2):71–80. https://doi.org/10.33640/2405-609X.1018

Akbari B, Tavandashti MP, Zandrahimi M (2011) Particle size characterization of nanoparticles—a practical approach. Iran J Mat Sci Eng 8:48–56

Akgul H, Sevindik M, Coban C, Alli H, Selamoglu Z (2017) New Approaches in traditional and complementary alternative medicine practices: *Auricularia auricula* and *Trametes versicolor*. J Tradit Med Clin Nat 6:239. https://doi.org/10.4172/2573-4555.1000239

Andreescu D, Eastman C, Balantrapu K, Goia DV (2007) A simple route for manufacturing highly dispersed silver nanoparticles. J Mater Res 22:2488–2496. https://doi.org/10.1557/jmr.2007.0308

Anthony KJ, Murugan M, Jeyaraj M, Rathinam NK, Sangiliyandi G (2013) Synthesis of silver nanoparticles using pine mushroom extract: a potential antimicrobial agent against *E. coli* & *B. subtilis*. J Ind Eng Chem 5:1–7. https://doi.org/10.1016/j.jiec.2013.10.008

Armendariz V, Herrera I, Peralta-Videa JR (2004) Size controlled gold nanoparticle formation by *Avena sativa* biomass: use of plants in nanobiotechnology. J Nanopart Res 6:377–382. https://doi.org/10.1007/s11051-004-0741-4

Arun G, Eyini M, Gunasekaran P (2014) Green synthesis of silver nanoparticles using the mushroom fungus *Schizophyllum commune* and its biomedical applications. Biotechnol Bioprocess Eng 19:1083–1090. https://doi.org/10.1007/s12257-014-0071-z

Aygün A, Özdemir S, Gülcan M, Cellat K, Şen F (2020) Synthesis and characterization of Reishi mushroom-mediated green synthesis of silver nanoparticles for the biochemical applications. J Pharma Biomed Anal 178:112970. https://doi.org/10.1016/j.jpba.2019.112970

Azeez MA, Lateef A, Asafa TB, Yekeen TA, Akinboro A, Oladipo IC, Gueguim-Kana EB, Beukes LS (2017) Biomedical applications of cocoa bean extract-mediated silver nanoparticles as antimicrobial, larvicidal and anticoagulant agents. J Clust Sci 28:149–164. https://doi.org/10.1007/s10876-016-1055-2

Baker S, Rakshith D, Kavitha KS, Santosh P, Kavitha HU, Rao Y, Satish S (2013) Plants: emerging as nanofactories towards facile route in the synthesis of nanoparticles. Bio Impacts 3:111–117. https://doi.org/10.5681/bi.2013.012

Bal C, Akgul H, Sevindik M (2017) Determination of the anti-oxidative activities of six mushrooms. Fresenius Environ Bull 26:6246–6252

Bamigboye CO, Oloke JK, Dames JF, Burton M, Lateef A (2019) Optimization of the process for producing biomass and exopolysaccharide from the King tuber oyster mushroom, *Pleurotus tuber-regium* (Agaricomycetes), for biotechnological applications. Int J Med Mushrooms 21(4):311–322. https://doi.org/10.1615/IntJMedMushrooms.2019030357

Bawadekji A, Oueslati MH, Al Ali, Basha J (2018) Biosynthesis of gold nanoparticles using *Pleurotus ostreatus* (Jacq. ex. Fr.) Kummer extract and their antibacterial and antifungal activities. J Appl Environ Biol Sci 8(3):142–147

Baymiller M, Huang F, Rogelj S (2017) Rapid one-step synthesis of gold nanoparticles using the ubiquitous coenzyme NADH. Matters 3(7):c201705000007. https://doi.org/10.19185/matters.201705000007

Bhangale HG, Bachhav SG, Nerkar DM, Sarode KM, Patil DR (2019) Study on optical properties of green synthesized silver nanoparticles for surface plasmon resonance. J Nanosci Tech 5:658–661. https://doi.org/10.30799/jnst.230.19050203

Bhat R, Deshpande R, Ganachari SV, Huh DS, Venkataraman A (2011) Photo-irradiated biosynthesis of silver nanoparticles using edible mushroom *Pleurotus florida* and their antibacterial activity studies. Bioinorg Chem Appl Article ID 650979. https://doi.org/10.1155/2011/650979
Bhat R, Deshpande R, Ganachari SV, SungHuh D, Venkataraman A (2013) Photo-bio-synthesis of irregular shaped functionalized gold nanoparticles using edible mushroom *Pleurotus florida* and its anticancer evaluation. J Photochem Photobiol B Biol 125:63–69. https://doi.org/10.1016/j.jphotobiol.2013.05.002
Borovaya M, Pirko Y, Krupodorova T, Naumenko A, Blume Y, Yemets A (2015) Biosynthesis of cadmium sulphide quantum dots by using *Pleurotus ostreatus* (Jacq.) P. Kumm. Biotechnol Biotechnol Equip 29(6):1156–1163. https://doi.org/10.1080/13102818.2015.1064264
Buzea C, Pacheco I, Robbie K (2007) Nanomaterials and nanoparticles: sources and toxicity. Biointerphases 2: MR17. https://doi.org/10.1116/1.2815690
Chen W, Cai W, Zhang L, Wang G, Zhang L (2001) Sonochemical processes and formation of gold nanoparticles within pores of mesoporous silica. J Colloid Interf Sci 238:291–295. https://doi.org/10.1006/jcis.2001.7525
Chen T, Wong KH, Wu HL, Zheng W, Wong MK, Man WYC (2015). *Pleurotus tuber-regium* polysaccharide functionalized nano-selenium hydrosol with anti-tumor activity and preparation method thereof. U.S. Patent No. 9,072,669. U.S. Patent and Trademark Office, Washington, DC
Chowdhury S, Basu A, Kundu S (2014) Green synthesis of protein capped silver nanoparticles from phytopathogenic fungus *Macrophomina phaseolina* (Tassi) gold with antimicrobial properties against multidrug-resistant bacteria. Nanoscale Res Lett 9:2–11. https://doi.org/10.1186/1556-276X-9-365
Darroudi M, Ahmad MB, Zamiri R, Zak AK, Abdullah AH, Ibrahim NA (2011) Time-dependent effect in green synthesis of silver nanoparticles. Int J Nanomed 6:677–681. https://doi.org/10.2147/IJN.S17669
Debnath G, Das P, Krishna S (2019) Green synthesis of silver nanoparticles using mushroom extract of *Pleurotus giganteus*: characterization, antimicrobial, and α-Amylase inhibitory activities. BioNanoScience 9:611–619. https://doi.org/10.1007/s12668-019-00650-y
Deepak K, Rahi, Barwal M (2015) Biosynthesis of silver nanoparticles by their antibacterial and antibiotic activity enhancing potential. World J Pharm Pharmaceut Sci 4:1234–1247
Devika R, Elumalai S, Manikandan E, Eswaramoorthy D (2012) Biosynthesis of silver nanoparticles using the fungus *Pleurotus ostreatus* and their antibacterial activity. Sci Rep 1:10–14. https://doi.org/10.4172/scientificreports.557
Dubey SP, Lahtinen M, Sillanpaa M (2010) Tansy fruit mediated greener synthesis of silver and gold nanoparticles. Process Biochem 45:1065–1071. https://doi.org/10.1016/j.procbio.2010.03.024
Durán N, Marcato PD, Alves OL, De Souza GIH, Esposito E (2005) Mechanistic aspects of biosynthesis of silver nanoparticles by several *Fusarium oxysporum* strains. J Nanobiotechnol 3:8. https://doi.org/10.1186/1477-3155-3-8
Durán N, Cuevas R, Cordi L, Rubilar O, Diez MC (2014) Biogenic silver nanoparticles associated with silver chloride nanoparticles (Ag@ AgCl) produced by laccase from *Trametes versicolor*. SpringerPlus 3(1):645. https://doi.org/10.1186/2193-1801-3-645
Durán M, Silveira CP, Durán N (2015) Catalytic role of traditional enzymes for biosynthesis of biogenic metallic nanoparticles: a mini-review. IET Nanobiotechnol 9:314–323. https://doi.org/10.1049/iet-nbt.2014.0054
Ekar SU, Khollam YB, Koinkar PM, Mirji SA, Mane RS, Naushad M, Jadhav SS (2015) Biosynthesis of silver nanoparticles by using *Ganoderma* mushroom extract. Modern Phys Lett B 29(6–7):1540047. https://doi.org/10.1142/S0217984915400473
Elamawi RM, Al-Harbi RE, Hendi AA (2018) Biosynthesis and characterization of silver nanoparticles using *Trichoderma longibrachiatum* and their effect on phytopathogenic fungi. Egypt J Biol Pest Control 28:28. https://doi.org/10.1186/s41938-018-0028-1
Elder A, Gelein R, Silva V, Feikert T, Opanashuk L, Carter J, Potter R, Maynard A, Ito Y, Finkelstein J, Oberdörster G (2006) Translocation of inhaled ultrafine manganese oxide particles to the central nervous system. Environ Health Perspect 114:1172–1178. https://doi.org/10.1289/ehp.9030

Elegbede JA, Lateef A (2019) Green nanotechnology in Nigeria: the research landscape, challenges and prospects. Ann Sci Technol 4(2):6–38. https://doi.org/10.2478/ast-2019-0008

Elegbede JA, Lateef A, Azeez MA, Asafa TB, Yekeen TA, Oladipo IC, Adebayo EA, Beukes LS, Gueguim-Kana EB (2018) Fungal Xylanase-mediated synthesis of silver nanoparticles for catalytic and biomedical applications. IET Nanobiotechnol 12(6):857–863. https://doi.org/10.1049/iet-nbt.2017.0299

Elegbede JA, Lateef A, Azeez MA, Asafa TB, Yekeen TA, Oladipo IC, Abbas SH, Beukes LS, Gueguim-Kana EB (2019) Silver-gold alloy nanoparticles biofabricated by fungal xylanases exhibited potent biomedical and catalytic activities. Biotechnol Progr 35:e2829. https://doi.org/10.1002/btpr.2829

Elegbede JA, Lateef A, Azeez MA, Asafa TB, Yekeen TA, Oladipo IC, Aina DA, Beukes LS, Gueguim-Kana EB (2020) Biofabrication of gold nanoparticles using xylanases through valorization of corncob by *Aspergillus niger* and *Trichoderma longibrachiatum*: antimicrobial, antioxidant, anticoagulant and thrombolytic activities. Waste Biomass Valor 11(3):781–791. https://doi.org/10.1007/s12649-018-0540-2

Emerich DF, Thanos CG (2003) Nanotechnology and medicine. Expert Opin Biol Ther 3:655–663. https://doi.org/10.1517/14712598.3.4.655

Eskandari-Nojedehi M, Jafarizadeh-Malmiri H, Rahbar-Shahrouzi J (2018) Hydrothermal green synthesis of gold nanoparticles using mushroom (*Agaricus bisporus*) extract: physico-chemical characteristics and antifungal activity studies. Green Process Synth 7:38–47. https://doi.org/10.1515/gps-2017-0004

Eustis S, Hsu HY, El-Sayed MA (2005) Gold nanoparticle formation from photochemical reduction of Au3+ by continuous excitation in colloidal solutions: a proposed molecular mechanism. J Phys Chem B 109:4811–4815. https://doi.org/10.1021/jp0441588

Faraji M, Yamini Y, Rezaee M (2010) Magnetic nanoparticles: synthesis, stabilization, functionalization, characterization and applications. J Iran Chem Soc 7:1–37. https://doi.org/10.1007/BF03245856

Faramarzi MA, Forootanfar H (2011) Biosynthesis and characterization of gold nanoparticles produced by laccase from *Paraconiothyrium variabile*. Colloids Surf B Biointerf 87(1):23–27. https://doi.org/10.1016/j.colsurfb.2011.04.022

Fayaz AM, Balaji K, Kalaichelvan PT, Venkatesan R (2009) Fungal based synthesis of silver nanoparticles-an effect of temperature on the size of particles. Colloids Surf B Biointerf 74:123–126. https://doi.org/10.1016/j.colsurfb.2009.07.002

Frattini A, Pellegri N, Nicastro D, deSanctis O (2005) Effect of amine groups in the synthesis of Ag nanoparticles using aminosilanes. Mater Chem Phys 949:148–152. https://doi.org/10.1016/j.matchemphys.2005.04.023

Galib, Barve M, Mashru M, Jagtap C, Patgiri BJ, Prajapati PK (2011) Therapeutic potentials of metals in ancient India: a review through Charaka Samhita. J Ayurveda Integr Med 2(2):55–63. https://doi.org/10.4103/0975-9476.82523

Gardea-Torresdey JL, Tiemann KJ, Gamez G, Dokken K, Icano-Aguilera I, Furenlid LR, Renner MW (1999) Recovery of gold (III) by *Alfalfa* biomass and binding characterization using X-ray Microfluorescence. Adv Environ Res 3:83–93. https://doi.org/10.1021/es991325m

Gudikandula K, Vadapally P, Charya MAS (2017) Biogenic synthesis of silver nanoparticles from white rot fungi: their characterization and antibacterial studies. Open Nano 2:64–78. https://doi.org/10.1016/j.onano.2017.07.002

Gurunathan S, Raman J, Malek SN, Ajohn P, Vikineswary S (2013) Green synthesis of silver nanoparticles using *Ganoderma neo-japonicum* Imazeki: a potential cytotoxic agent against breast cancer cells. Int J Nanomed 8:4399. https://doi.org/10.2147/IJN.S51881

Haq MU, Rathod V, Shivraj N, Singh D, Yasin M, Singh AK (2015a) Silver nanoparticles from mushroom *Agricus bisporus* and their activity against multi-drug resistant strains of *Klebsiella* sp., *Pseudomonas* sp. & *Acinetobacter* sp. Int J Nat Prod Res 5:20–26

Haq MU, Rathod V, Singh D, Singh AK, Ninganagouda S, Hiremath J (2015b) Dried mushroom *Agaricus bisporus* mediated synthesis of silver nanoparticles from Bandipora District (Jammu and Kashmir) and their efficacy against methicillin-resistant *Staphylococcus aureus* (MRSA) Strains. Int J Nanosci 5:1–8

Hietzschold S, Walter A, Davis C, Taylor AA, Sepunaru L (2019) Does nitrate reductase play a role in silver nanoparticle synthesis? Evidence for NADPH as the sole reducing agent. ACS Sustain Chem Eng 7(9):8070–8076. https://doi.org/10.1021/acssuschemeng.9b00506

Hoet PHM, Brüske-Hohlfeld I, Salata OV (2004) Nanoparticles—known and unknown health risks. J Nanobiotechnol 2:12. https://doi.org/10.1186/1477-3155-2-12

Jegadeeswaran P, Shivaraj R, Venckatesh R (2012) Green synthesis of silver nanoparticles from extract of *Padina tetrastromatica* leaf. Digest J Nanomater Biostruct 7(3):991–998

Karthikeyan V, Ragunathan R, Jesteena J, Kabesh K (2019) Green synthesis of silver nanoparticles and application in dye decolorization by *Pleurotus ostreatus* (MH591763). Global J Bio-Sci Biotechnol 8:80–86

Karwa A, Gaikwar S, Rai M (2011) Mycosynthesis of silver nanoparticles using Lingzhi or Reishi medicinal mushroom, *Ganoderma lucidum* (W. Curt.:Fr.) P. Karst. and their role as antimicrobials and antibiotic activity enhancers. Int J Med Mushroom 13:483–491. https://doi.org/10.1615/intjmedmushr.v13.i5.80

Khan S, Rizvi SMD, Avaish M, Arshad M, Bagga P, Khan MS (2015) A novel process for size controlled biosynthesis of gold nanoparticles using bromelain. Mater Lett 159:373–376. https://doi.org/10.1016/j.matlet.2015.06.118

Khandel O, Shahi SK (2018) Mycogenic nanoparticles and their bioprospective applications: current status and future challenges. J Nanostruct Chem 8:369–391. https://doi.org/10.1007/s40097-018-0285-2

Kowalczyk B, Lagzi I, Grzybowski BA (2011) Nanoseparations: strategies for size and shape selective purification of nanoparticles. Curr Opin Colloids Interf Sci 16:135–148. https://doi.org/10.1016/j.cocis.2011.01.004

Kumar SA, Abyaneh MK, Gosavi SW, Kulkarni SK, Pasricha N, Ahmad A (2007) Nitrate reductase-mediated synthesis of silver nanoparticles from $AgNO_3$. Biotechnol Lett 29:439–445. https://doi.org/10.1007/s10529-006-9256-7

Lam CW, James JT, McCluskey R, Hunter RL (2004) Pulmonary toxicity of single-wall carbon nanotubes in mice 7 and 90 days after intratracheal instillation. Toxicol Sci 77:126–134. https://doi.org/10.1093/toxsci/kfg243

Lateef A, Adeeyo AO (2015) Green synthesis and antibacterial activities of silver nanoparticles using extracellular laccase of *Lentinus edodes*. Not Sci Biol 7(4):405–411. https://doi.org/10.15835/nsb749643

Lateef A, Adelere IA, Gueguim-Kana EB, Asafa TB, Beukes LS (2015) Green synthesis of silver nanoparticles using keratinase obtained from a strain of *Bacillus safensis* LAU 13. Int Nano Lett 5:29–35. https://doi.org/10.1007/s40089-014-0133-4

Lateef A, Ojo SA, Elegbede JA (2016a) The emerging roles of arthropods and their metabolites in the green synthesis of metallic nanoparticles. Nanotechnol Rev 5(6):601–622. https://doi.org/10.1515/ntrev-2016-0049

Lateef A, Ojo SA, Oladejo SM (2016b) Anti-candida, anti-coagulant and thrombolytic activities of biosynthesized silver nanoparticles using cell-free extract of *Bacillus safensis* LAU 13. Process Biochem 51(10):1406–1412. https://doi.org/10.1016/j.procbio.2016.06.027

Lateef A, Ojo SA, Folarin BI, Gueguim-Kana EB, Beukes LS (2016c) Kolanut (*Cola nitida*) mediated synthesis of silver-gold alloy nanoparticles: antifungal, catalytic, larvicidal and thrombolytic applications. J Clust Sci 27(5):1561–1577. https://doi.org/10.1007/s10876-016-1019-6

Lateef A, Akande MA, Azeez MA, Ojo SA, Folarin BI, Gueguim-Kana EB, Beukes LS (2016d) Phytosynthesis of silver nanoparticles (AgNPs) using miracle fruit plant (*Synsepalum dulcificum*) for antimicrobial, catalytic, anti-coagulant and thrombolytic applications. Nanotechnol Rev 5(6):507–520. https://doi.org/10.1515/ntrev-2016-0039

Lateef A, Akande MA, Ojo SA, Folarin BI, Gueguim-Kana EB, Beukes LS (2016e) Paper wasp nest-mediated biosynthesis of silver nanoparticles for antimicrobial, catalytic, anti-coagulant and thrombolytic applications. 3 Biotech 6:140 http://dx.doi.org/10.1007/s13205-016-0459-x

Lateef A, Ojo SA, Azeez MA, Asafa TB, Yekeen TA, Akinboro A, Oladipo IC, Gueguim-Kana EB, Beukes LS (2016f) Cobweb as novel biomaterial for the green and eco-friendly synthesis of silver nanoparticles. Appl Nanosci 6(6):863–874. http://dx.doi.org/10.1007/s13204-015-0492-9

Lateef A, Ojo SA, Elegbede JA, Azeez MA, Yekeen TA, Akinboro A (2017) Evaluation of some biosynthesized silver nanoparticles for biomedical applications: hydrogen peroxide scavenging, anticoagulant and thrombolytic activities. J Clust Sci 28(3):1379–1392. https://doi.org/10.1007/s10876-016-1146-0

Lateef A, Folarin BI, Oladejo SM, Akinola PO, Beukes LS, Gueguim-Kana EB (2018) Characterization, antimicrobial, antioxidant and anticoagulant activities of silver nanoparticles synthesized from *Petiveria alliacea* L. leaf extract. Prep Biochem Biotechnol 48(7):646–652. https://doi.org/10.1080/10826068.2018.1479864

Lee SH, Jun BH (2019) Silver nanoparticles: synthesis and application for nanomedicine. Int J Mol Sci 20:E865. https://doi.org/10.3390/ijms20040865

Li Z, Hulderman T, Salmen R, Chapman R, Leonard SS, Young SH, Shvedova A, Luster MI, Simeonova PP (2007) Cardiovascular effects of pulmonary exposure to single-wall carbon nanotubes. Environ Health Perspect 115:377–382. https://doi.org/10.1289/ehp.9688

Mafune F, Jun-ya K, Yoshihiro T, Tamotsu K (2001) Dissociation and aggregation of gold nanoparticles under laser irradiation. J Phys Chem B 105:9050–9056. https://doi.org/10.1021/jp0111620

Makarov VV, Love AJ, Sinitsyna OV, Makarova SS, Yaminsky IV, Taliansky ME, Kalinina NO (2014) Green nanotechnologies: synthesis of metal nanoparticles using plants. Acta Nat 6:35–44

Maurya S, Bhardwaj AK, Gupta KK, Agarwal S, Kushwaha A, Chturvedi VK, Pathak RK, Gopal R, Uttam KN, Singh AK, Verma V, Singh MP (2016) Green synthesis of silver nanoparticles using *Pleurotus* and its bactericidal activity. J Cell Mol Biol 62:1–3. https://doi.org/10.4172/1165-158X.1000131

Mazumdar H, Haloi N (2011) A study on biosynthesis of iron nanoparticles by *Pleurotus* sp. J Microbiol Biotechnol Res 1(3):39–49

Medina C, Santos-Martinez MJ, Radomski A, Corrigan OI, Radomski MW (2007) Nanoparticles: pharmacological and toxicological significance. Br J Pharmacol 150:552–558. https://doi.org/10.1038/sj.bjp.0707130

Mirunalini S, Arulmozhi V, Deepalakshmi K, Krishnaveni M (2012) Intracellular biosynthesis and antibacterial activity of silver nanoparticles using edible mushrooms. Not Sci Biol 4:55–61. https://doi.org/10.15835/nsb448051

Moghaddam AB, Namvar F, Moniri M, Tahir P, Azizi S, Mohamad R (2015) Nanoparticles biosynthesized by Fungi and yeast: a review of their preparation, properties, and medical applications. Molecules 16540. https://doi.org/10.3390/molecules200916540

Mohanpuria P, Rana NK, Yadav SK (2008) Biosynthesis of nanoparticles: technological concepts and future applications. J Nanopart Res 10:507–517. https://doi.org/10.1007/s11051-007-9275-x

Mohanta YK, Singdevsachan SK, Parida UK, Panda SK, Mohanta TK, Bae H (2016) Green synthesis and antimicrobial activity of silver nanoparticles using wild medicinal mushroom *Ganoderma applanatum* (Pers.) Pat. from Similipal Biosphere Reserve, Odisha, India. IET Nanobiotechnol 10:184–189. https://doi.org/10.1049/iet-nbt.2015.0059

Mohanta YK, Nayak D, Biswas K, Singdevsachan SK, Abd_Allah EF, Hashem A, Alqarawi AA, Yadav D, Mohanta TK (2018) Silver nanoparticles synthesized using wild mushroom show potential antimicrobial activities against food borne pathogens. Molecules 23:655. https://doi.org/10.3390/molecules2303065

Mudunkotuwa A, Pettibone JM, Grassian VH (2013) Environmental implications of nanoparticle aging in the processing and fate of copper-based nanomaterials. Environ Sci Technol 46:7001–7010. https://doi.org/10.1021/es203851d

Muthu N, Shanmugasundaram K (2015) Comparative study of phytochemicals in aqueous and silver nanoparticles extracts of *Acrocybe aegerita*, (V.BRIG.) singer black poplar mushroom. Int J Pharm Biol Sci 6:190–197

Narayanan KB, Park HH, Han SS (2015) Synthesis and characterization of biomatrixed-gold nanoparticles by the mushroom *Flammulina velutipes* and its heterogeneous catalytic potential. Chemosphere 141:169–175. https://doi.org/10.1016/j.chemosphere.2015.06.101

Ndaw S, Bergaentzle M, Aoude-Werner D, Hasselmann C (2000) Extraction procedures for the liquid chromatographic determination of thiamin, riboflavin and vitamin B6 in foodstuffs. Food Chem 71:129–138. https://doi.org/10.1016/S0308-8146(00)00135-7

Niemeyer CM, Mirkin CA (2004) Nanobiotechnology: concepts, applications and perspectives. Wiley, New York, p 491. ISBN: 978-3-527-30658-9

Nijhara R, Balakrishnan K (2006) Bringing nanomedicines to market: regulatory challenges, opportunities, and uncertainties. Nanomedicine 2:127–136. https://doi.org/10.1016/j.nano.2006.04.005

Nithya R, Ragunathan R (2009) Synthesis of silver nanoparticle using Pleurotus sajor caju and its antimicrobial study. Digest J Nanomater Biostruct 4(4):623–629

Numata M, Hasegawa T, Fujisawa T, Sakurai K, Shinkai S (2004) Β-1,3-glucan (Schizophyllan) can act as a one-dimensional host for creation of novel poly (aniline) nanofiber structures. Org Lett 6:4447–4450. https://doi.org/10.1021/ol0483448

Oberdorster E (2004) Manufactured nanomaterials (fullerenes, C60) induce oxidative stress in the brain of juvenile largemouth bass. Environ Health Perspect 112:1058–1062. https://doi.org/10.1289/ehp.7021

Oberdörster G, Castranova V, Asgharian B, Sayre P (2015) Inhalation exposure to carbon nanotubes (CNT) and carbon nanofibers (CNF): methodology and dosimetry. J Toxicol Environ Health B Crit Rev 18(3–4):121–212. https://doi.org/10.1080/10937404.2015.1051611

Ojo SA, Lateef A, Azeez MA, Oladejo SM, Akinwale AS, Asafa TB, Yekeen TA, Akinboro A, Oladipo IC, Gueguim-Kana EB, Beukes LS (2016) Biomedical and catalytic applications of gold and silver-gold alloy nanoparticles biosynthesized using cell-free extract of *Bacillus safensis* LAU 13: antifungal, dye degradation, anti-coagulant and thrombolytic activities. IEEE Transact NanoBiosci 15(5):433–442. https://doi.org/10.1109/TNB.2016.2559161

Oladipo IC, Lateef A, Azeez MA, Asafa TB, Yekeen TA, Akinboro A, Akinwale AS, Gueguim-Kana EB, Beukes LS (2017a) Green synthesis and antimicrobial activities of silver nanoparticles using cell-free extracts of *Enterococcus* species. Not Sci Biol 9(2):196–203. http://dx.doi.org/10.15835/nsb929938

Oladipo IC, Lateef A, Elegbede JA, Azeez MA, Asafa TB, Yekeen TA, Akinboro A, Gueguim-Kana EB, Beukes LS, Oluyide TO, Atanda OR (2017b) *Enterococcus* species for the one-pot biofabrication of gold nanoparticles: characterization and nanobiotechnological applications. J Photochem Photobiol B Biol 173:250–257. https://doi.org/10.1016/j.jphotobiol.2017.06.003

Oloke JK, Adebayo EA (2015) Effectiveness of immunotherapies from oyster mushroom (*Pleurotus* species) in the management of immunocompromised patients. Int J Immunol 3(2–1):8–20

Owaid MN (2019) Green synthesis of silver nanoparticles by *Pleurotus* (oyster mushroom) and their bioactivity: review. Environ Nanotechnol Monit Manage 12:100256. https://doi.org/10.1016/j.enmm.2019.100256

Owaid MN, Ibraheem IJ (2017) Mycosynthesis of nanoparticles using edible and medicinal mushrooms. Eur J Nanomed 9(1):5–23. https://doi.org/10.1515/ejnm-2016-0016

Owaid MN, Raman J, Lakshmanan H, Al-Saeedi SS, Sabaratnam V, Ali Abe I (2015) Mycosynthesis of silver nanoparticles from *Pleurotus corncopiaevar. citrinopileatus* and its inhibitory effect against *Candida* sp. Adv Mater Lett 153:186–190. https://doi.org/10.1016/j.matlet.2015.04.023

Owaid MN, Al-Saeedic SS, Ali Abed I (2017a) Biosynthesis of gold nanoparticles using yellow oyster mushroom *Pleurotus cornucopiae* var. *citrinopileatus*. Environ Nanotechnol Monit Manag 8:157–162. https://doi.org/10.1016/j.enmm.2017.07.004

Owaid MN, Abed IA, Al-Saeedi SSS (2017b) Applicable properties of the bio-fertilizer spent mushroom substrate in organic systems as a byproduct from the cultivation of *Pleurotus* spp. Inf Process Agric 4:78–82. https://doi.org/10.1016/j.inpa.2017.01.001

Owaid MN, Barish A, Shariati MA (2017c) Cultivation of *Agaricus bisporus* (button mushroom) and its usages in the biosynthesis of nanoparticles. Open Agric 2(1):537–543. https://doi.org/10.1515/opag-2017-0056

Owaid MN, Rabeea MA, Aziz AA, Jameel MS, Dheyab MA (2019) Mushroom-assisted synthesis of triangle gold nanoparticles using the aqueous extract of fresh *Lentinula edodes* (shiitake). Omphalotaceae. Environ Nanotechnol Monit Manag 12:100270. https://doi.org/10.1016/j.enmm.2019.100270

Park Y, Hong YN, Weyers A, Kim YS, Linhatdt RJ (2011) Polysaccharides and phytochemicals: a natural reservoir for the green synthesis of gold and silver nanoparticles. IET Nanobiotechnol 5:69–78. https://doi.org/10.1049/iet-nbt.2010.0033

Patra JK, Baek KH (2014) Green nanobiotechnology: factors affecting synthesis and characterization techniques. J Nanomater 20:1–12. https://doi.org/10.1155/2014/417305

Paul S, Sasikumari CS, Singh AR (2015) Fabrication of silver nanoparticles synthesized from *Ganoderma lucidum* into the cotton fabric and its antimicrobial property. Int J Pharm Pharmaceut Sci 7(8):53–56

Philip D (2009) Biosynthesis of Au, Ag and Au–Ag nanoparticles using edible mushroom extract. Spectrochim Acta Part A Mole Biomol Spectr 73(2):374–381. https://doi.org/10.1016/j.saa.2009.02.037

Radomski A, Jurasz P, Onso-Escolano D, Drews M, Morandi M, Malinski T (2005) Nanoparticle-induced platelet aggregation and vascular thrombosis. Br J Pharmacol 146:882–893. https://doi.org/10.1038/sj.bjp.0706386

Rai A, Singh A, Ahmad A, Sastry M (2006) Role of halide ions and temperature on the morphology of biologically synthesized gold nanoparticles. Langmuir 22:736–741. https://doi.org/10.1021/la052055q

Raziya S, Durga B, Rajamahanthe SG, Govindh B, Annapurna N (2006) Synthesis and characterization of CdS nanoparticles using Reishi mushroom. Int J Adv Technol Eng Sci 4:220–227

Ruan-Soto F, Ordaz-Velázquez M, García-Santiago W (2017) Traditional processing and preservation of wild edible mushrooms in Mexico. Ann Food Process Preserv 2(1):1013–1018

Sahoo KS, Labhasetwar V (2003) Nanotech approaches to drug delivery and imaging. Drug Discov Today 8(24):1112–1120. https://doi.org/10.1016/s1359-6446(03)02903-9

Sanghi R, Verma P (2009a) Biomimetic synthesis and characterization of protein capped silver nanoparticles. Bioresour Technol 100:501–504. https://doi.org/10.1016/j.biortech.2008.05.048

Sanghi R, Verma P (2009b) A facile green extracellular biosynthesis of CdS nanoparticles by immobilized fungus. Chem Eng J 155:886–891. https://doi.org/10.1016/j.cej.2009.08.006

Sathishkumar P, Gu FL, Zhan Q, Palvannan T, Mohd Yusoff AR (2018) Flavonoids mediated "Green" nanomaterials: a novel nanomedicine system to treat various diseases—current trends and future perspective. Mater Lett 210:26–30. https://doi.org/10.1016/j.matlet.2017.08.078

Sen I, Maity K, Islam SS (2013) Green synthesis of gold nanoparticles using a glucan of an edible mushroom and study of catalytic activity. Carbohydr Polym 91:518–528. https://doi.org/10.1016/j.carbpol.2012.08.058

Senapati US, Sarkar D (2014) Characterization of biosynthesized zinc sulphide nanoparticles using edible mushroom *P. ostreatus*. Indian J Phys 88:557–562. https://doi.org/10.1007/s12648-014-0456-z

Senapati US, Jha DK, Sarkar D, Res J (2015) Structural, optical, thermal and electrical properties of fungus guided biosynthesized zinc sulphide nanoparticles. Res J Chem Sci 5:33–40

Sevindik M, Akgul H, Bal C (2017) Determination of oxidative stress status of *Ompholatus olearius* gathered from Adana and Antalya provinces in Turkey. Sakarya Univ J Sci 21(3):324–327. https://doi.org/10.16984/saufenbilder.09547

Sevindik M, Akgul H, Dogan M, Akata I, Selamoglu Z (2018) Determination of antioxidant, antimicrobial, DNA protective activity and heavy metals content of *Laetiporus sulphureus*. Fresenius Environ Bull 27(3):1946–1952

Shah M, Fawcett D, Sharma S, Tripathy SK, Poinern GEJ (2015) Green synthesis of metallic nanoparticles via biological entities. Materials 8:7278–7308. https://doi.org/10.3390/ma8115377

Shaligram NS, Bule M, Bhambure R, Singhal RS, Singh SK, Szakacs G, Pandey A (2009) Biosynthesis of silver nanoparticles using an aqueous extract from the compacting producing fungal strain. Process Biochem 44:939–943. https://doi.org/10.1016/j.procbio.2009.04.009

Shivashankar M, Premkumari B, Chandan N (2013) Biosynthesis, partial characterization and antimicrobial activities of silver nanoparticles from *Pleurotus* species. Int J Integr Sci Innov Technol 2:13–23

Singh N, Chaudhary A, Abraham J (2014) Susceptibility testing of methicillin resistant *Staphylococcus aureus* (MRSA) and biological role of silver nanoparticles of honey against MRSA. J Biologically Active Prod Nat 4(5–6):332–342. https://doi.org/10.1080/22311866.2014.957097

Smith JE, Rowan NJ, Sullivan R (2002) Medicinal mushrooms: a rapidly developing area of biotechnology for cancer therapy and other bioactivities. Biotechnol Lett 24(22):1839–1845. https://link.springer.com/article/10.1023/A:1020994628109

Soni N, Prakash S (2011) Factors affecting the geometry of silver nanoparticles synthesis in *C. tropicum* and *F. oxusporum*. Current Res Nanotechnol 2:112–121. https://doi.org/10.3844/ajnsp.2011.112.121

Srikar SK, Giri DD, Pal DB, Mishra PK, Upadhyay SN (2015) Green synthesis of silver nanoparticles: a review. Green Sustain Chem 6:34–56. https://doi.org/10.4236/gsc.2016.61004

Sudhakar T, Nanda A, Babu SG, Janani S, Evans MD, Markose TK (2014) Synthesis of silver nanoparticles from edible mushroom and its antimicrobial activity against human pathogen. Int J PharmTech Res 6:1718–1723

Sujatha S, Tamilselvi Subha K, Subha K, Panneerselvam A (2013) Studies on the biosynthesis of silver nanoparticles using mushroom and its antibacterial activities. Int J Curr Microbiol Appl Sci 2:605–614

Tabbouche S, Gürgen A, Yildiz S (2017) Antimicrobial and anti-quorum sensing activity of some wild mushrooms collected from Turkey. MSU J Sci 5(2):453–457. https://doi.org/10.18586/msufbd.347692

Teferi Y, Muleta M, Woyessa D (2013) Mushroom consumption habits of Wacha Kebele residents, southwestern Ethiopia. Global Res J Agri Biol Sci 4(1):6–16

Tran QH, Nguyen VQ, Le AT (2013) Silver nanoparticles: synthesis, properties, toxicology, applications and perspectives. Adv Nat Sci Nanosci Nanotechnol 4:1–20. https://doi.org/10.1088/2043-6262/4/3/033001

Treguer M, de Cointet C, Remita H, Khatouri J, Mostafavi M, Amblard J, Belloni J, de Keyzer R (1998) Dose rate effect on radiolytic synthesis of gold-silver bimetallic clusters in solution. J Phys Chem B 102:4310–4321. https://doi.org/10.1021/jp981467n

Valverde ME, Hernández-Pérez T, Paredes-López O (2015) Edible mushrooms: improving human health and promoting quality life. Int J Microbiol Article ID 376387. https://doi.org/10.1155/2015/376387

Velusamy P, Venkat Kumar G, Jeyanthi V, Das J, Pachaiappan R (2016) Bio-inspired green nanoparticles: synthesis, mechanism, and antibacterial application. Toxicol Res 32:95–102. https://doi.org/10.5487/TR.2016.32.2.095

Verma VC, Kharwar RN, Gange AC (2010) Biosynthesis of antimicrobial silver nanoparticles by the endophytic fungus *Aspergillus clavatus*. Nanomed 5(1):33–40. https://doi.org/10.2217/nnm.09.77

Vetchinkina E, Loshchinina E, Kursky V, Nikitina V (2013) Reduction of organic and inorganic selenium compounds by the edible medicinal basidiomycete *Lentinula edodes* and the accumulation of elemental selenium nanoparticles in its mycelium. J Microbiol 51(6):829–835. https://doi.org/10.1007/s12275-013-2689-5

Vetchinkina EP, Loshchinina EA, Vodolazov IR, Kursky VF, Dykman LA, Nikitina VE (2016) Biosynthesis of nanoparticles of metals and metalloids by Basidiomycetes. Preparation of gold nanoparticles by using purified fungal phenol oxidases. Appl Microbiol Biotechnol 5:76–80. https://doi.org/10.1007/s00253-016-7893-x

Wang J, Zhang Y, Yuan Y, Yue T (2014) Immunomodulatory of selenium nano-particles decorated by sulfated *Ganoderma lucidum* polysaccharides. Food Chem Toxicol 68:183–189. https://doi.org/10.1016/j.fct.2014.03.003

Williams D (2004) The risks of nanotechnology. Med Device Technol 15:9–10

Wong KH (2020) Preparation of highly stable selenium nanoparticles with strong anti-tumor activity using tiger milk mushroom. https://www.polyu.edu.hk/itdo/cntfiles/initiatives/initiative_file_2_137_TC.pdf. Accessed on 28 Jun 2020

Wu H, Zhu H, Li X, Liu Z, Zheng W, Chen T (2013) Induction of apoptosis and cell cycle arrest in A549 human lung adenocarcinoma cells by surface-capping selenium nanoparticles: an effect enhance by a polysaccharide-protein complex from *Polyporus rhinocerus*. J Agric Food Chem 61:1–30. https://doi.org/10.1021/jf403564s

Xiao Y, Huang Q, Zheng Z, Guan H, Liu S (2017) Construction of a *Cordyceps sinensis* exopolysaccharide-conjugated selenium nanoparticles and enhancement of their antioxidant activities. Int J Biol Macromol 99:483–491. https://doi.org/10.1016/j.ijbiomac.2017.03.016

Xu S, Yong L, Wu P (2013) One-pot, green, rapid synthesis of flowerlike gold nanoparticles/reduced graphene oxide composite with regenerated silk fibroin as efficient oxygen reduction electrocatalysts. ACS Appl Mater Interf 5(3):654–662. https://doi.org/10.1021/am302076x

Yehia RS, Al-Sheikh H (2014) Biosynthesis and characterization of silver nanoparticles produced by *Pleurotus ostreatus* and their anticandidal and anticancer activities. World J Microbiol Biotechnol 30:2797–2803. https://doi.org/10.1007/s11274-014-1703-3

Yekeen TA, Azeez MA, Lateef A, Asafa TB, Oladipo IC, Badmus JA, Adejumo SA, Ajibola AA (2017a) Cytogenotoxicity potentials of Cocoa pod and bean-mediated green synthesized silver nanoparticles on *Allium cepa* cells. Caryologia Int J Cytol Cytosystem Cytogenet 70(4):366–377. https://doi.org/10.1080/00087114.2017.1370260

Yekeen TA, Azeez MA, Akinboro A, Lateef A, Asafa TB, Oladipo IC, Oladokun SO, Ajibola AA (2017b) Safety evaluation of green synthesized C*ola nitida* pod, seed and seed shell extracts-mediated silver nanoparticles (AgNPs) using *Allium cepa* assay. J Taibah Univ Sci 11(6):895–909. https://doi.org/10.1016/j.jtusci.2017.06.005

Yilmaz A, Yildiz S, Çelik A (2016) Determination of heavy metal and radioactivity in *Agaricus campestris* mushroom collected from kahramanmaraş and erzurum proviences. Turk J Agri Food Sci Technol 4(3):208–215. https://doi.org/10.24925/turjaf.v4i3.208-215.596

Zhang G, Wang DJ (2008) Fabrication of heterogeneous binary arrays of nanoparticles via colloidal lithography. J Am Chem Soc 130:5616–5617. https://doi.org/10.1021/ja710771j

Zhao X, Zhou L, Riaz Rajoka MS, Yan L, Jiang C, Shao D, Zhu J, Shi J, Huang Q, Yang H, Jin M (2018) Fungal silver nanoparticles: synthesis, application and challenges. Crit Rev Biotechnol 38(6):817–835. https://doi.org/10.1080/07388551.2017.1414141

Microbial-Mediated Nanoparticles for Sustainable Environment: Antimicrobial and Photocatalytic Applications

S. B. Jaffri and K. S. Ahmad

1 Introduction

Nanotechnology and nanoscience have been associated with novel developments and advancements aimed at human welfare since many decades. Among different nanoscale materials synthesized using varied nano-routes, nanoparticles (NPs) have gained an immensely significant position due to their facile fabrication methods, cost effectiveness and eco-friendly attributes. The syntheses of NPs have been usually emphasized upon more than other nanomaterials by the virtue of their favorable aspects, i.e., catalytic, electrical, optical, magnetic, chemical and photoelectrochemical properties. Such unique physicochemical aspects of NPs can be attributed to a myriad of factors, especially their minimal size besides the quantum effect and facile tunability of NPs electronic energy bands by means of size effect. Another significant cause of such extraordinary characteristics can be an augmentation in the surface area to the mass ratio of NPs leading to an ascendency in the performance of the surface atoms in comparison to the atoms existing in the internal region of particles.

Nano-biotechnology is a progressing field nanotechnology related with the utilization of biotic components for nanoscale developments and is impacting human and his ecospheric region in an influential way (Zhao et al. 2020; Lestrell et al. 2020). NPs, which usually are referred to those particles which possess a size of 100 nm or less in at least one of their dimensional features, differ greatly from the bulk portion of their counterparts. Therefore, NPs are proven to be highly reactive in nanoscale toward other molecules than their same bulk fragment due to an enhancement in the surface to volume ratio. Prodigious deals of industries have been making use of NPs, especially biomedicine (Lateef et al. 2018; Elegbede and Lateef 2019). For

S. B. Jaffri · K. S. Ahmad (✉)
Department of Environmental Sciences, Fatima Jinnah Women University, The Mall, 46000 Rawalpindi, Pakistan
e-mail: chemist.phd33@yahoo.com; dr.k.s.ahmad@fjwu.edu.pk

A. Lateef et al. (eds.), *Microbial Nanobiotechnology*, Materials Horizons: From Nature to Nanomaterials, https://doi.org/10.1007/978-981-33-4777-9_10

instance, polymeric micelles NPs are used as an active delivery agent of drugs in case of tumor treatment. Iron oxide NPs coated with polymer can also be employed of for division of bacterial clusters, leading to the greater potential of these NPs in curing chronic diseases instigated due to bacteria. Furthermore, NPs have also been filled with various proteinaceous substances equipped with an inherent ability of stimulation of the immune response in humans and other organisms (Vigneshwaran et al. 2006).

Over the years, novel compounds are fabricated (Mannam et al. 2017; Navalakhe et al. 2009) aimed at different applications with an advancement in applications of NPs. There has been a greater employment of various synthetic routes for NPs based on the physical, chemical and biological procedures (Adebayo et al. 2019; Elegbede et al. 2020). Variety of techniques, e.g., precipitative, thermal, pyrolytic, sol–gel, and hydrothermal have been used for NPs synthesis. However, physicochemical modes, though very effective and applied ones, have been rigorously challenged for their elevated capital and operational costs, complexity in operation requiring extraordinarily alleviated temperature and pressure ranges, and above all release of environmentally toxic substance having profound persistence in form of by-products (Lateef et al. 2016a; Jaffri and Ahmad 2017; Azeez et al. 2020). Often during such chemical and physical synthesis, environmental concerns are highly neglected. This can be comprehended from the example of chemically processed synthesis of quantum dots of cadmium selenide marked by imposition of the significant pollutant atmospheric, hydrospheric, and lithospheric load in form of various organic and inorganic substances. Furthermore, other dominant pollutants in the similar way can be different types of soils, iron, chloride ions, sulfur, and sodium that are released into the immediate environment. Additionally, utilization of different organophosphorus solvents makes the cost of overall chemical routes reaching up to 90% of the total NPs generation cost (Sirinakis et al. 2003). In order to gain the benefits related with the nanoscale particles in different fields in a sustainable way, green chemists have proposed and tested the solution to these limitations associated with physicochemical routes. Greener synthesis of NPs is an effective alternative to physical and chemical routes. It is a novel, efficacious, toxicity-free and economically viable procedure. Greener routes employ the utilization of natural biotic resources, particularly renewable ones that have good replenishment and production spans such as plant parts, microbes, human cells, and metabolites of arthropods among others (Adelere and Lateef 2016; Lateef et al. 2016a, b, c). These bio-entities are completely toxicity-free, abundantly available and cost effective in provisioning natural reducing cum stabilizing agents as a replacement for chemically obnoxious counterparts.

Among biological reducing agents, microbial species have been highly cherished as nano-biofactories for reduction and stabilization of different NPs having profound ability for alterations in tuning the shapes and sizes of NPs in addition to optical characteristics (Gutiérrez et al. 2012). Nano-biofactories are usually based on bacterial, fungal, yeast, actinomycetes, algal, cyanobacterial, and viral genesis which are employed for NPs production such as magnetite, silver, platinum, gold, magnesium, copper, palladium, and zinc. Bacteria have been designated as one of these nano-biotechnological tools among all bio-entities in case of microbial resources for the green synthesis of NPs. This is due to the augmented mean generation rate and genetic transformation potentials of the bacterial cells (Majumdar et al. 2020). The reduction procedure of metallic ions by bacterial cells is usually a result of their effort of defeating metallic toxicity to their cells. Secondary metabolites are also produced in the secondary phase of the bacterial maturation phase inside the culturing media, which act as vigorous stabilizers for NPs synthesis. The concept of stable NPs production by bacterial cells also arises from bacterial cells' potential in active prevention of the particle agglomeration. Microbial and specifically bacterial-driven NPs synthesis can be technically referred to as a by-product of the resistive mechanism established by microbial cells in response toward metals invading cell. This signifies a direct consequence of microbial species evolution in the higher concentration ecospheric zones. In a more specialized manner, green microbial NPs are fabricated when the microbial organisms are grabbing target ions provided by the immediate environment and causing reduction of the metallic ions by means of enzymatic substances produced by different cellular activities. Through this process, the toxic metallic ions are converted to their non-toxic forms by means of the reduction to elemental forms. In another alternative pathway of microbial NPs synthesis, there is an attachment of the metallic ions onto the microbial cells, resulting in production of the stabilized NP complexes, nullifying the need for extra energy requirements. In case of microbial synthesis of NPs, the synthetic site possesses a significant position. For intracellular procedure, there exists a transportation of ions inside microbial cell from NPs in the presence of the enzymatic substances. But extracellular procedures are based on the metallic ions ensnaring over microbial cells surficial level leading to the reduction aided by microbial enzymes (Li et al. 2011; Zhang et al. 2011).

Microbial potential of metallic ions reduction and stabilization by acting as a bio-template have been utilized for a wide range of elements for development of NPs having well-defined structural, compositional, morphological, optical, and electrochemical aspects. Different researchers have investigated the auspiciousness of microbial green NPs (Hulkoti and Taranath 2014; Siddiqi and Husen 2016; Prasad et al. 2016; Gahlawat and Choudhury 2019). However, to the best of our knowledge, no review has been published signifying the role of microbial green NPs synthesized via nano-biotechnological route in strengthening environmental sustainability by highlighting the role of green NPs in solving environmental challenges in a sustainable manner.

Therefore, the current chapter has been compiled to signify the profound involvement of the microbially synthesized green NPs in terms of their candidacy for becoming sustainable environmental detoxifiers. This chapter comprises of the detailed account of microbial NPs in terms of environmental sustainability in addition to the environmental remediation carried out with these nano-biotechnological tools. The discussion over environmental role, i.e, antimicrobial and photocatalytic activities, played by green NPs synthesized via microbial route has been limited to silver (AgNPs) considering the excellent behavior of AgNPs in exceeding other NPs in terms of catalytic and antimicrobial performance. In the final portion of the chapter, there is a brief summarization of the closing comments in terms of future perspectives and challenges regarding microbial NPs.

2 Methods

2.1 Searching Approach

The current chapter is based on the meticulous analysis of the significant roles of microbial entities in stabilizing and reducing metallic ions consolidating environmental sustainability by means of data assortment from varied electronic databases that included Google Scholar, Web of Science, PubMed, Cochrane, Science Direct, ProQuest, and Embase covering published research articles, reviews, perspectives, and chapters from Springer, Springer Nature, Elsevier, Taylor and Francis, Jstore, Royal Society of Chemistry, American Chemical Society, and The Institute of Electrical and Electronics Engineers. The investigative pattern has been based on content analysis. Selected studies emphasize the synthesis of microbial NPs from different sources, their applications specifically antimicrobial and photo-catalysis. Publications spanning over a range of 2000–2020 were selected for analysis but the ones from 2015 to 2020 were majorly focused. The searching strategy was varied by feeding terms as: ["microbial synthesis", "green synthesis", "microbial nanoparticles", "microbial nanoparticles applications", "microbial nanoparticles photo-catalysis", "environmental sustainability," and "cleaner production"].

3 Green Chemistry and Ecospheric Sustainability

Current era has been marked by a greater number of environmental challenges in comparison to the last decades in terms of ecological toxicity, hydrological extremes, climatic disruption, scarcity of resources, and pressure over exhaustible sources. Scientific community has been eagerly contributing to solve these issues of environment challenges through the development of efficient contraptions for solving energy crises (Jaffri and Ahmad 2020a), smart agricultural practices (Iftikhar et al. 2019), and environmental cleanup strategies (Iram et al. 2019, 2020); however, for an overall environment to gain balance and human issues to be solved, a consolidated sustainability approach is required. By considering the eco-friendly procedures of green chemistry, this objective can be fulfilled. Microbes being nano-biotechnological tools are in complete conformity with the roles of green chemistry, and thus, they have a significant role in strengthening environmental sustainability in terms of both cleaner production and environmental remediators. Generally, sustainability is marked by establishment of the association between dynamically altering anthropogenic economic systems and transforming ecospheric systems in a slower pace. Additionally, sustainability also points toward the life continuation of human, his flourishing, culture development equally taking into account the integrity and protection of ecological life. Or in a simpler way, sustainability is the achievement of that economic phase by different countries and nations in which the resources are being driven from the natural environment by human for fulfilling his needs for survival without inducing any alleviation in the potential of ecospheric zone for providing resources to future generations. Sustainable development is another extension of this concept which has been adopted by United Nations delineated in the Rio Declaration and Agenda 21 to be achieved by international community.

Different nations working toward sustainability achievement in terms of food, energy and finance have now also been afflicted with the need for considering their environmental issues from a critical point of view. Environment is the ultimate recipient of all the pollutants and contaminants released from any preparative and industrial procedure. In order to not making it a dumping bin for future generations, different industries have now been pressurized to adopt sustainable approaches. In this regard, three approaches are quite significant, i.e., either to completely prevent the chemical toxins release or to minimize them or more specifically to switch to environmentally cleaner production strategies. The later approach is marked by fulfilling all rules of green chemistry (Esposito and Antonietti 2015). Microbial synthesis of NPs is one of such approaches aimed at environmental integrity maintenance showing cleaner production having an association with the maneuvers, ecological sustainability and enlargement of leftover decline, reprocessing, and re-claiming at the innovativeness level. The scope of such process is both microeconomic and ecological (Khalili et al. 2015). The profound role of nano-biotechnology can be comprehended from the fact that it directly contributes to the environmental management in a sustainable mode by adopting cleaner production procedures and ecospheric managerial tools aimed at gaining an efficiency in the generation process, using renewable inputs and sustainable waste production. The rational utilization of nature's resources and reduced waste production can prove an effective tool toward environmental sustainability. Interestingly, the microbial synthetic processes can also be further extended to sustainability by means of analysis of the life cycle of synthesized product and eco-designs. This trend has been increasingly adopted in various fields since it gives hint regarding development of the novel products and services besides rational utilization of exhaustible resources and thus indirectly minimizing the ecological impacts (Casamayor and Su 2013; Chou 2014; Kuo et al. 2016).

There have been many decades to the microbes mediated green NPs synthesis in a cost-effective manner; however, it is entirely impossible to outline a general framework of mechanistic rules expressing the underlying procedures associated with the biogenesis. It is partly due to the fact that microbial community is an immensely diverse niche, where metallic ion stabilizers can be bacterial, fungal, algal, yeast, or viral agents. Yet in a more generalized manner to comprehend green NPs synthesis via microbial route, it can be said that microbial strains are specialized in detoxification of the metallic ions by means of different reactions leading to the reduction or precipitation of the metallic ions in soluble form to their eco-friendly and insoluble forms having sizes in nano-range.

4 Microbes Mediated Green NPs

Considering nature's tools as raw materials for NPs synthesis via sophisticated and resourceful approaches, chemists and biotechnologists have been considering microbial species for sustainable nanoscale material synthesis. This green approach is inclusive of both the conventional green methods adopted for NPs synthesis and also other size control materials (Kang et al. 2008) using microbial nano-factories as bio-templates (Singh et al. 2016a, c) (Fig. 1). Great deals of microorganisms have been used as nano-biotechnological tools having different sizes and morphologies aimed at different applications (Karthik et al. 2020) as shown in Table 1. Among all the microbial agents used for NPs synthesis, bacterial and fungal species have been greatly focused by the virtue of their bio-transformative and metal bio-accumulative potential. The auspiciousness of fungal and bacterial strains in making green NPs is expressed by the ability of fungi to be scaled up by means of using a fermentation method known as thin solid substrate. Fungi are the major secretors of extracellular enzymatic substances which can be used for the comprehensive manufacturing of enzymes. Another good feature associated with fungi is the facile handling of fungal biomass. Similarly, the adoption of bacterial cells for microbial green synthesis is marked by its advantageousness in terms of easier colony forming aspects. NPs synthesis with bacterial precursors is also marked by benefits of the focused evolution or manipulations at genetic level for achieving an overexpression in the targeted

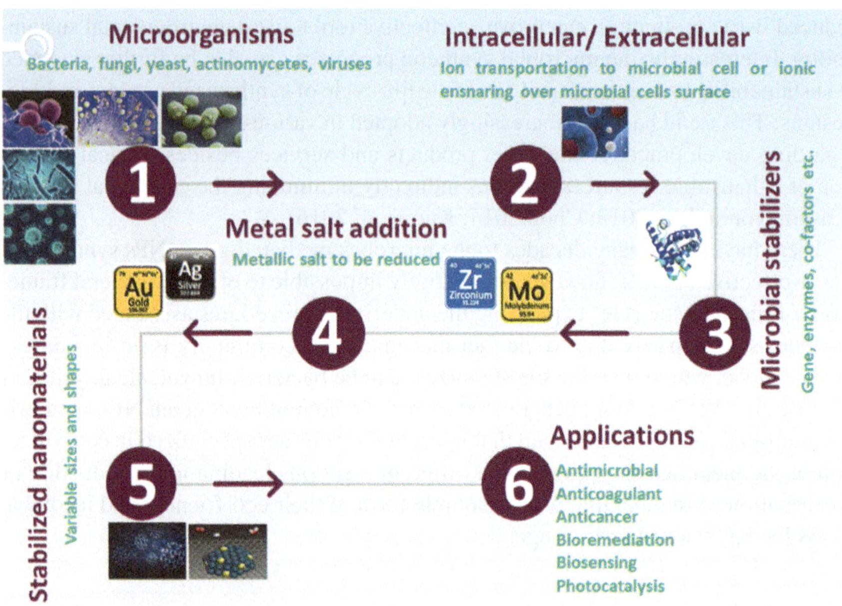

Fig. 1 General synthetic route elucidation expressing the utilization of different bio-nano-factories used for the fabrication of green nanoparticles via facile pathway

Table 1 Different types of microorganisms used for biomimetic synthesis of metallic and metallic oxide nanoparticles in a cost-effective and sustainable manner

NP	Microbial agent	Strain source	Microbial agent type	Size (nm)	Morphology	Crystal structure	Application	Reference
Ag	*Hormoconis resinae*	Soil samples, India	Fungal	20–80	Spherical	Face-centered cubic	–	Varshney et al. (2009)
Au, Pd	*Escherichia coli*	University of Birmingham, UK	Bacterial	1.5–2.5	–	–	Catalytic	Deplanche et al. (2012)
Ag	*Rhodococcus* sp.	–	Bacterial	10	Spherical	Face-centered cubic	–	Otari et al. (2012)
ZnO	*Aeromonas hydrophila*	Institute of Microbial Technology (IMTECH), Chandigarh	Bacterial	57.72	Spherical and oval	Wurtzite	Antimicrobial	Jayaseelan et al. (2012)
Ag	*Bacillus amyloliquefaciens*	Microbiology laboratory, Odisha	Bacterial	10–100	Spherical	–	Antimicrobial	Behera et al. (2013)
Au	*Penicillium crustosum*	Soil samples, Iran	Fungal	<100	Spherical	–	–	Barabadi et al. (2014)
Ag	*Actinomycetes* sp.	Soil samples, Iran	Bacterial	10–20	Spherical	–	Therapeutic	Abdeen et al. (2014)
Ag	*Gluconacetobacter xylinus*	Institute of Microbiology, Chinese Academy of Science	Bacterial	40–60	Spherical	Face-centered cubic	Antimicrobial	Li et al. (2015)
Ag	*Leuconostoc mesenteroides* T3	Water kefir grain	Bacterial	40	–	Face-centered cubic	Antimicrobial	Davidović et al. (2015)
Au	*Streptomyces clavuligerus*	Bhitarkanika Mangrove Forest	Bacterial	8.2	Spherical	Face-centered cubic	Anticancer	Kumar et al. (2015)

(continued)

Table 1 (continued)

NP	Microbial agent	Strain source	Microbial agent type	Size (nm)	Morphology	Crystal structure	Application	Reference
Ag	*Listeria monocytogenes* *Bacillus subtilis* *Streptomyces anulatus*	Agriculture Research Services (ARS), Chandigarh	Bacterial	>11.53	Rod, spherical, hexagonal, triangular	–	Antimicrobial and mosquitocidal	Soni and Prakash (2015)
Ag	*Bacillus safensis* LAU 13	Feather waste dump	Bacterial	5–95	Spherical	Face-centered cubic	Antimicrobial, free radical scavenging larvicidal, Anticandida, anticoagulant and thrombolytic	Lateef et al. (2015, 2016d)
Se, Te	*Stenotrophomonas maltophilia* SeITE02 *Ochrobactrum* sp. MPV1	Environmental isolates	Bacterial	Variable	Spherical	–	Antimicrobial	Zonaro et al. (2015)
Se	*Ralstonia eutropha*	IMTECH, Chandigarh	Bacterial	40–120	Spherical	Hexagonal	Antimicrobial	Srivastava and Mukhopadhyay(2015)
Au	*Bhargavaea indica* DC1	*Panax ginseng* rhizospheric soil, Korea	Bacterial	30–100	Flower	–	–	Singh et al. (2016)
Ag	*Weissella oryzae* DC6	Mountain ginseng, Korea	Bacterial	10–30	Spherical	Face-centered cubic	Antimicrobial	Singh et al. (2016b)
Ag	*Bacillus methylotrophicus*	Ginseng field soils, Korea	Bacterial	10–30	Spherical	–	Antimicrobial	Wang et al. (2015)

(continued)

Table 1 (continued)

NP	Microbial agent	Strain source	Microbial agent type	Size (nm)	Morphology	Crystal structure	Application	Reference
Ag	*Bacillus* sp.	Hot spring in Iran	Bacterial	~9 to 32	Spherical	–	Antimicrobial	Deljou and Goudarzi (2016)
Ag	*Escherichia coli*	School of Biosciences culture library	Bacterial	5–25	Spherical		Antimicrobial	Divya et al. (2016)
Ag	*Ochrobactrum rhizosphaerae*	–	Bacterial	5–25	Spherical	–	Antimicrobial	Gahlawat et al. (2016)
Ag	*Bacillus amyloliquefaciens* *Bacillus subtilis*	Institute of Biotechnology, China	Bacterial	15.9–80	Spherical	Face-centered cubic	Antimicrobial and mosquitocidal	Fouad et al. (2017)
Ag	*Aeromonas* sp. THG-FG1.2	Soil samples, Korea	Bacterial	8–16	Spherical	Face-centered cubic	Antimicrobial	Singh et al. (2017)
Ag	*Novosphingobium* sp. THG-C3	Soil samples, Korea	Bacterial	8–25	Spherical	Face-centered cubic	Antimicrobial	Du et al. (2017)
Se	*Vibrio natriegens*	–	Bacterial	100–400	Spherical	–	Bioremediation	Fernández-Llamosas et al. (2017)
Ag	*Bacillus brevis* NCIM 2533	National Collection of Industrial Microorganisms, Pune	Bacterial	41–68	Spherical	–	Antibacterial	Saravanan et al. (2018b)

(continued)

Table 1 (continued)

NP	Microbial agent	Strain source	Microbial agent type	Size (nm)	Morphology	Crystal structure	Application	Reference
Ag	*Phenerochaete chrysosporium* MTCC-787	Institute of Microbial Technology (IMTECH), Chandigarh	Fungal	34–90	Spherical	–	Antibacterial	Saravanan et al. (2018a)
Ag	*Bacillus endophyticus* SCU-L	saline soil, China	Bacterial	5.1	Spherical	Cubic face-centers	Antibacterial	Gan et al. (2018)
Ag	*Bacillus* sp. SBT8	Soil samples, Turkey	Bacterial	1–20	Spherical and pseudo-spherical	–	–	Yurtluk et al. (2018)
Ag	*Bacillus mojavensis* BTCB15	Soil samples, Pakistan	Bacterial	2.3	Irregular	Face-centered cubic	Antimicrobial	Iqtedar et al. (2019)
ZrO2	*Pseudomonas aeruginosa*	Department of Bioengineering, NIT-Agartala	Bacterial	15	Spherical	Monoclinic and tetragonal	Tetracycline adsorption	Debnath et al. (2020)
CdO	*Penicillium oxalicum*	Quaid-i-Azam University, Islamabad, Pakistan	Fungal	40–60	Spherical	Octahedral	Antimicrobial	Azam et al. (2020)
Ag	*Trichoderma atroviride*	Plants at Oman	Fungal	10–15	Spherical	–	Antibacterial	Abdel-Azeem et al. (2020)

enzymes. Such approach has been recently adopted by using various bacterial strains, e.g., *Bacillus* sp. and *Pseudomonas* sp., for conversion of metal ions to their elemental forms in nanoscale as elucidated in Table 1.

Microbial green NPs synthesized with bacterial strains, especially lactic acid bacteria (LAB), are highly cherished by the dint of their safe use in handling and processing of the food grade materials referred to as "generally recognized as safe" when it comes to the production and preservation of food. Additionally, some strains of LAB are also associated with the exhibition of the probiotic characteristics for humans and other animal species in case of ingestion and contribute to the general promotion of health (George Kerry et al. 2018). LAB has been used for the fabrication of selenium (Moreno-Martin et al. 2017), gold (Nair and Pradeep 2002), and Ag NPs (Garmasheva et al. 2016). In a generalized manner, LAB acts as an efficient nano-biofactory for NPs synthesis. LAB is a gram-positive bacterial strain, and thus, it has a thicker cellular wall comprising of chemical constituents such as peptidoglycan, proteinaceous substances, polysaccharides, and lipo-teichoic acid. In terms of NPs, the different layers in LAB cell are the functional attachment sites for metallic ions to be biosorbed and bioreduced as a result of comparatively negative electro-kinetic characteristics leading to the attraction of metallic cations for initiation of the NPs greener synthesis. Furthermore, there is also considerable evidence suggesting the green synthesis of metallic or metallic oxide NPs to be highly impacted by reducing microbial species' potential in tolerating the metallic ions. Augmented metallic stress will ultimately contribute to the microbial activity by initiating an interaction between metallic ions and bacterial cells giving rise to particles have nano-range size.

Biosorption and bioaccumulation have been identified as eminent bacterial strategies being adopted for combating metal toxicity. Biosorption activates passive mechanisms and are primarily non-metabolically aided procedures associated with the attachment, ionic exchange, chelation, and precipitative mechanisms completely dependent upon the composition of bacterial cellular wall in terms of different functional groups presence (Sintubin et al. 2009). In the consequent manner, there is bioaccumulation based on the entrance of the metallic ions inside cell body and interaction between metal ion and bacterial cell's intracellular structures is initiated. An exclusive example of this phenomenon are some of the microbes showing coping mechanisms in response to heavy metal triggered damage by means of direct deliverance of metals, e.g., Hg, Pb, and As from periplasmic to cytoplasm regions for the metal binding proteinaceous substances to start working. Thus, the heavy metals detoxification potential of microbes is explained by this phenomenon. This process is highly facilitated by the existence of reductase enzymes reducing metallic salts to nanoscale materials and leading to green NPs having narrower size distribution and lesser extent of polydispersity (Hulkoti and Taranath 2014). The involvement of various membrane transporters has been known for the transportation of the transition metallic cations, however, the exact underlying mechanism needs to be explored yet. *Synechocystis* and *Synechococcus* sp. of cyanobacteria have an ability to assemble the proteins and cations in form of metallic clusters, e.g., Fe, Mn, and Cu, inside their cellular regions and later consuming them for different metabolic processes, e.g., photosynthetic and respiratory processes.

Green NPs have been synthesized using different microbes following intra- and extracellular approaches in which metallic ions are converted to their nano-range counterparts by means of different microbial substances, e.g., genes, proteins, co-factors, enzymes, and peptides having the ability to reduce metal ions subsequently ending up in the formation of stabilized nanoparticles (Chellamuthu et al. 2019). In case of the adoption of microbes' mediated green NPs synthesis at commercial scale, extracellular synthesis has been cherished more than intracellular one because it is a downstream procedure even with minute quantity of biomass handling. Furthermore, an augmented supply of enzymatic substances extracellularly secreted proteins finally results in an improvement in yield in addition to imparting a great deal of stability extent to the corresponding NPs.

In the order of preference for selection of different microbes for metallic NPs synthesis, bacteria have been ranked first followed by fungi, yeast, actinomycetes, and viruses. Bacterial cells are capable of synthesizing green NPs by both extra-cellular and intracellular operating mechanisms. The first ever microbial green NPs synthesized were gold NPs reported by Beveridge and Murray, which were deposited on the cell wall of *Bacillus subtilis* in an extracellular manner (Beveridge and Murray 1980). In additional study, AgNPs were synthesized by using *Pseudomonas stutzeri* AG259 having Ag resistance, which expressed an accumulation of the Ag NPs in an intracellular fashion spanning over a range of few—200 nm. In this report, reductase enzyme dependent on NADH was involved supplying electronic species and itself undergoing oxidation to NAD^+ (Klaus-Joerger et al. 2001). The electronic transfer led to the reduction of the ionic silver to nanoscale NPs. Recent investigations on bacteria such as *Aeromonas hydrophila, Bacillus amyloliquefaciens, Bacillus mojavensis* BTCB15, *Bacillus endophyticus* SCU-L, *Vibrio natriegens, Novosphingobium* sp. THG-C3, *Ochrobactrum rhizosphaerae, Weissellaoryzae* DC6, and *Lactobacillus plantarum* sampled from different resources (Table 1) showing bacterial-mediated NPs synthesis is defensive of the sole utilization of microbes for green synthesis without utilization of any external reducing agents.

Varied kinds of fungal strains, e.g., *Penicillium crustosum, Hormoconis resinae, Phenerochaete chrysosporium* MTCC-787, *Penicillium oxalicum, Trichoderma atro-viride* among others have also been explored for the synthesis of metallic NPs since fungal strains are characterized by comparatively higher productivity rates when it comes to NPs generation in addition to an elevated tolerance indices toward different metals in terms of higher capacity for attachment with the fungal cell walls of the biomass. Another aspect making fungi as a preferred nano-biotechnological tool is the downstream dispensation and treatments of the biomass that are comparatively easier in case of fungi than bacterial strains or viruses. Additionally, the fungus-driven green synthesis is marked by economically viable results for NPs generation by the virtue of an augmented potential for bioaccumulation in response to the metallic ions. Nevertheless, the influence of different operational parameters on NPs green synthesis cannot be overlooked.

In addition to bacterial and fungal strains, actinomycetes have also been utilized for the manufacturing of different extracellular enzymatic species and different types of secondary metabolites. Due to this potential, they have also been used for the greener synthesis of NPs. Their capacity is unmatched in case of generation of different types of biologically active substances containing higher proteinaceous content. Although these nano-biofactories are capable of production of NPs in both extracellular and intracellular modes, however, extracellular reduction has been the most frequently adopted pathway possessing a multitude of applications in varied commercial fields. Actinobacterium *Rhodococcus* NCIM 2891 has been used by Otari et al. (2012) for the synthesis of AgNPs having spherical shape possessing 10 nm size. The results were indicative of the bio-mineralization of Ag^{+} in an intra-cellular mode resulting from cell wall enzymatic substances leading to formation of Ag nuclei. In another report, *Streptomyces* sp. LK-3, having a marine origin was used for green synthesis of AgNPs (Karthik et al. 2014).

Yeast cells have also been used for the biogenic synthesis of metallic NPs by the virtue of the inherent potential of absorption and accumulation in higher quantities of toxic metallic ions from their immediate ambience (Shah et al. 2015). They are also known for adaptation toward metal toxicity circumstances by utilization of varied detoxifying mechanistic routes, e.g., bio-precipitative mechanisms, chelation, and sequestration in intracellular mode. Such an extraordinary combating potential of yeasts has been used by different scientific groups. For instance, *Yarrowia lipolytica* having a marine genesis and is ascomycetous yeast strain has been used for the biogenic synthesis of Ag NPs in a cell concomitant way (Apte et al. 2013). Results for *Yarrowia lipolytica*-based AgNPs were suggestive of the melanin to be an influential factor in causing bio-mineralization of the metallic ions. Furthermore, the melanin pigment-based AgNPs expressed a remarkable antibiofilm activity toward *Salmonella paratyphi*.

Interestingly, the borderline between living and non-living organisms, i.e., viruses, has also been employed in the sustainable synthesis of biomimetic NPs in a sustainable and cost-effective manner as per their thicker external surficial coating of the capsid proteinaceous substances associated with the providence of the remarkably favorable media for reaction between viruses and metallic ions (Kobayashi et al. 2012). By means of incorporating techniques of genetic engineering, the protein cages of viruses can be made into monodispersed entities having considerable strength and ability to be transformed in any manner. Thus, this type of modification has made them a template for biogenic synthesis. More specifically, they can be used for the deposition of different materials or can be precisely engineered for the creation of three-dimensional receptacles aimed at drug delivery (Zeng et al. 2013). In the same manner, they can be effectively made use of for the development of different types of conjugates and composites in nano-regime with metallic NPs having a profound significance in the drug delivery and therapy of cancer. In this regard, Mao et al. (2003) has reported the utilization of M13 bacteriophage focused at the semiconductor nanocrystals undergoing the process of nucleation and positioning.

5 Characterization of Biomimetic NPs

For the validation of microbes mediated green NPs, number of analytical techniques are employed (Fig. 2). These techniques range from physicochemical to bio-analytical ones. Different types of isolations methods have been used and microbial strains are derived from various sources, e.g., pedospheric samples, hot springs, mountain samples, environmental isolates, and mangrove forest samples to mention a few. They have been cultured by means of different assays on nutrient broths in the biotechnological labs for utilization as bio-precursors for metallic NPs synthesis after incubation at certain conditions. The microbial strains are also identified via DNA analysis after isolation followed by PCR amplification and sequencing and used accordingly. Usually, biogenic NPs are analyzed for optical parameters via ultraviolet-visible spectrophotometry. The direct and indirect bandgaps (Eqs. 1 and 2, respectively) and other optical responses can also be calculated by using Tauc's plot from this technique.

$$\alpha = \frac{A(hv - E_g)^2}{hv} \tag{1}$$

$$\alpha = \frac{A(hv - E_g)^{1/2}}{hv} \tag{2}$$

Another widely used technique for analysis of vibrational parameters and presence of organic functional groups is done via Fourier transform infrared spectroscopy in a range of 400–4000 cm^{-1}. Another similar technique is the utilization of Raman

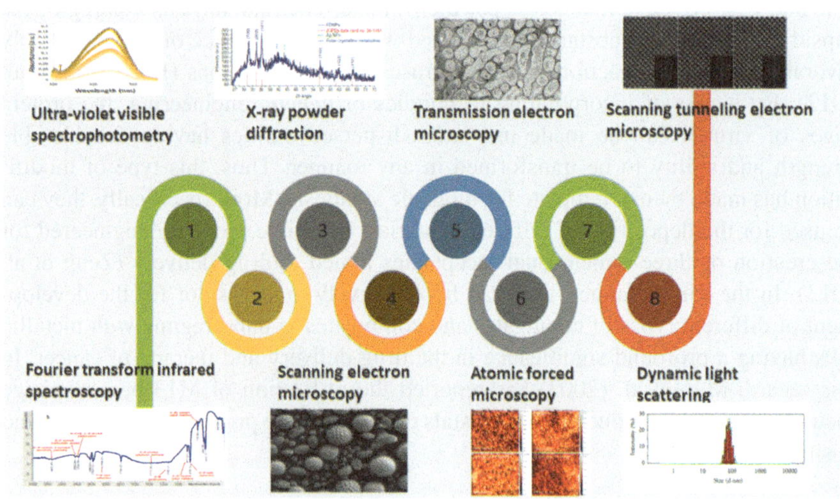

Fig. 2 Different types of analytical techniques used for validation of microbial-mediated green nanoparticles

spectrophotometry used for analysis of vibrational and rotational aspects. Usually, this technique makes use of the potassium bromide palletization method for analysis. Many researches in the present era are focused on analysis of the geometrical shapes and crystal features of the nanoparticles. This is usually done via X-ray powder diffraction. This is an important technique since it gives the average crystallite size of NPs by means of Debye Scherrer's formula given by Eq. 3. Furthermore, this technique also reveals the crystal shapes shown in Table 1.

$$D = \left[\frac{\kappa\lambda}{\beta\cos\theta}\right] x\text{Å} \tag{3}$$

In addition to the optical, crystalline, and vibrational analysis, the morphological features exploration is highly incumbent for the comprehension and prediction of NPs' efficacy in different types of applications. For instance, the role played by spherical NPs will be different in comparison to triangular nanoparticles. This type of analysis is done by variety of techniques but predominantly, scanning electron microscopy, transmission electron microscopy, atomic force microscopy, or scanning tunneling electron microscopy. The microscopic images are used for calculation of the NPs size and used in varied kinds of applications. NPs sizes in different researches are also measured via dynamic light scattering which also provides valuable information on the polydispersity extent of NPs. In a number of research institutes, scanning electron microscopy is usually coupled with energy-dispersive X-ray spectroscopy which discloses the elemental composition of NPs tested. In addition to NPs, morphology, many researches have included the utilization biological transmission electron microscopy for correct identification of the bio-entities being used. Apart from these basic techniques, great deal of other analysis have been done in different research institutes for analysis of biogenic NPs; e.g., thermogravimetric analysis is done for examining the thermal features of synthesized NPs and their response over elevated temperatures. In addition to NPs investigations, high performance liquid chromatography (HPLC) and gas chromatography–mass spectrometry (GC-MS) are sophisticated techniques that can be used for the identification of the enzymatic products, co-factors, and other microbial secretions for their use in the synthesis of NPs. Additionally, microbial capping agents responsible for reducing metallic ions can also be studied via sodium dodecylsulfate polyacrylamide gel electrophoresis (SDS-PAGE) and matrix-assisted laser desorption/ionization mass spectrometry (MALDI-TOF-MS).

6 Silver: A Nano-biotechnological Tool for Eco-detoxification

6.1 Nano-photocatalytic Potential

Hydrospheric contamination arising from the indiscriminate dumping of waste materials into water bodies has been following an increasing trend. In this regard, different pollutants in an untreated form are being dumped into waterbodies partly by the dint of non-existence of treatment plants at industrial and partly due to the lack of resources of industries to bear the capital and operational costs of treatment plants. In any case, waterbodies are being harmed. Organic dyes have been one of such pollutant groups that are used in a multitude of industries and are being dumped regularly. Such coloring agents are being accumulated into the hydrospheric zones leading to eutrophication, alleviation in the re-oxygenation potential and imparting severe harm toward the aquatic biota by becoming an obstacle in hindering the solar light infiltration in waterbody (Jaffri and Ahmad 2017, 2018; Dasgupta et al. 2020). Such pollutants are usually persistent organic pollutants (POPs) and are not easier targets for microbial degradation in addition to showing poor percentages of removal by costly physical and chemical methods such as flocculation, adsorptive, ultra-filtration, redox treatments, electrochemical, and electro-coagulation. In this regard, the photo-catalysis-based environmental detoxification with NPs has been proven to be an effective method signifying a facile and cost-effective mode. Nano-photocatalytic degradation has been marked by a rapid oxidative rate, no toxic compounds releases and driven by sunlight without requiring complex machinery and other operational requirements (Noman et al. 2020).

AgNPs synthesized by microbial route is an effective nano-biotechnological tool that has been employed for photo-catalysis of various dyes shown in Table 2 achieving impressive aromatization percentages. The underlying mechanism responsible for the effective nano-photocatalytic degradation percentages with AgNPs is due to the adsorptive interaction established between the organic dyes and AgNPs acting as an adsorbent with positive ions for the negatively charged dye molecules to be attached on it (Jiang et al. 2005; Jaffri and Ahmad 2020b). In a recent study, AgNPs were synthesized using microbial precursors—*Bacillus subtilis* (SJ 15) having been isolated from the pedospheric samples for the photo-catalysis of thiazine dye known as methylene blue which is known for causing a number of health issues (Roshmi et al. 2017). Impressive detoxification rates obtained for AgNPs based assay can be ascribed to the adsorption as well the decomposition phenomenon. The incidence of solar light on reaction mixture of methylene blue dye and AgNPs triggers the photocatalytic process due to the excitation of the surface plasmon resonance, i.e., vacillations of charge density having an ability of the propagation at the interfacial region of metallic and dielectric intermediate (García 2011). AgNPs synthesized from *Bacillus subtilis* (SJ 15) displayed an effective removal of methylene blue in 4 h with the yield reaching up to $18 \pm 0.3\%$ for AgNPs taken, i.e., 20 mg L^{-1}. In a

Table 2 Nano-photocatalytic role of microbes mediated metallic and metallic oxides nanoparticles against target pollutants in photocatalytic assay

NP	Microbial agent	Target pollutants	Detoxification extent (%)	Reference
Cu	*Escherichia* sp.	Congo red Malachite green Direct blue-1 Deactive black-5	97.07 90.55 88.42 83.61	Noman et al. (2020)
Ag	*Bacillus amyloliquefaciens*	4-nitrophenol	98	Samuel et al. (2020)
Ag	*Bacillus licheniformis*	Methylene blue	94	
Au	*Bacillus marisflavi*	Congo red	98	Nadaf and Kanase (2019)
Fe	*Bacillus* sp.	Methyl violet Methylene blue Rhodamine-B	85.4 71.5	Walujkar et al. (2019)
Au	*Trichoderma* sp.	2-nitrophenol 3-nitrophenol 4-nitrophenol 2-nitroaniline 3-nitroaniline	82.2–97.5	Qu et al. (2018)
Au	*Staphylococcus warneri*	2-nitroaniline 4-nitroaniline 2-nitrophenol 4-nitrophenol	>80	Nag et al. (2018)
Au	*Acinetobacter* sp. SW30	Direct black 22 Reactive yellow 186	93 83	Wadhwani et al. (2018)
Au, Ag-Au	*Bacillus safensis* LAU 13	Malachite green	>90	Ojo et al. (2016)
TiO_2 Doped TiO_2	*Bacillus amyloliquefaciens*	Reactive Red 31	75.83 (undoped) 90.98 (doped)	Khan and Fulekar (2016)

very recent study, excellent degradation of methylene blue was achieved by microbial AgNPs synthesized using *Bacillus licheniformis* M09 in less than 3 h (Momin et al. 2019).

Upon exposure of AgNPs to the visible solar light, there is a consequent excitation of the electron by photonic species having an energy (hυ) which is either equivalent to or is greater than the bandgap. In this way, there is electronic excitation between valence and conduction band leading to the leaving of hole in the valence band. This leftover hole is known for reacting with the water molecules and adsorption on the AgNPs surface for the production of a highly powerful oxidizing agent hydroxyl-free radical. Furthermore, there is also another reaction in which the excited electron reacts with O_2 for the generation of the reducing agent, i.e., superoxide ions

O_2^-. Such a combination of oxidizing and reducing agents contributed in the nano-photocatalytic reaction destroys the dye molecules which were previously adsorbed on the biomimetic AgNPs surface (Ameta et al. 2013). In addition to methylene blue, other types of dyes have also been efficiently degraded with biogenically synthesized AgNPs. In a recent study, 4-nitrophenol (4-NP) was effectively degraded by *Bacillus amyloliquefaciens* MSR5 AgNPs achieving up to 98% of removal efficiency within 15 min (Samuel et al. 2020). *Aeromonas* sp.based AgNPs were used for photocatalytic removal of brilliant green. The results for this dye were supportive of the achievement of 92.62% degradation in 120 min (Gurme et al. 2019). AgNPs synthesized from *Microchaete* NCCU-342, a cyanobacterium, has excelled in degrading methyl red in 120 min reaching up to 84.60% removal (Husain et al. 2019).

6.2 Antimicrobial Potential

Harmful microbial agents have been responsible for a number of diseases in human, animals, and plants. These microbes have been dealt with by good antibiotics developed over time, yet the expression of multidrug resistance by bacterial and other microbial agents is an emerging issue. Microbes driven green NPs can be the best solution to treat such microbes proven by variety of investigations (Elegbede et al. 2019). In this regard, AgNPs nano-formulations have been proven to be an effective against a multitude of microbes. In a recent report, multi-drug-resistant bacterial strains of *Escherichia coli, Klebsiella pneumoniae, Salmonella typhimurium, Staphylococcus aureus,* and *Bacillus subtilis* were successfully inhibited by *Bacillus subtilis* (SJ 15)-mediated AgNPs showing impressive zones of inhibition and the antibacterial activity expressed an elevation with an increase in the AgNPs dose (Roshmi et al. 2017). In another recent work, AgNPs developed by using *Crella-cyathophora,* which is an endosymbiotic actinomycetes having marine background, was used for antibiofilm activity against *Pseudomonas aeruginosa, Bacillus subtilis,* and *Staphylococcus aureus* expressing remarkable inhibition activity besides to testing of these NPs for anticancer activity (Hamed et al. 2020). *Streptacidiphilus durhamensis* HGG16n, having its origin as acidophilic actinobacterium, has been used for synthesis of AgNPs and proven effective against *Pseudomonas aeruginosa, Staphylococcus aureus, Proteus mirabilis, Escherichia coli, Klebsiella pneumoniae,* and *Bacillus subtilis* (Buszewski et al. 2018). *Bacillus subtilis* ATCC 6653, *Staphylococcus aureus* ATCC 25923, and *Escherichia coli* ATCC 10536 have also been effectively inhibited by cyanobacteria-mediated AgNPs (Keskin et al. 2016). *Bacillus licheniformis* M09-derived optimized AgNPs expressed a remarkable dose-dependent inhibition of food borne pathogens—*Bacillus subtilis* subsp. spizizenii ATCC 6633, *Staphylococcus aureus* ATCC 6538, *Escherichia coli* ATCC 25922, and *Pseudomonas aeruginosa* ATCC 10145 via well diffusion assay and broth dilution procedure (Momin et al. 2019). Rayaman et al. (2018) reported the use of *Aspergillus niger* fungal strains for the biogenic synthesis of AgNPs and found an effective antimicrobial activity against *Escherichia coli*, methicillin-resistant *Staphylococcus aureus*-LY, *Pseudomonas aeruginosa*, and *Candida albicans*.

Commendable potential of AgNPs can be ascribed to its ability of disrupting the bacterial membrane' integrity through development of adsorptive interactions, bacterial cellular division shutting down accomplished via complete prevention of DNA replication since AgNPs interact with the enzymatic thiol groups and bases having phosphorus. AgNPs are also effective antimicrobial agents because they can form free radicals leading to induction of oxidative stress (Soenen et al. 2011). In addition to the AgNPs' inherent potential in prevention of different microbes, the morphological aspects can also play an influential role, and thus, variable extents of antimicrobial activities are being reported for differently shaped AgNPs (Pal et al. 2007; Paredes et al. 2014). Furthermore, this activity also results from electrostatic attraction developed between the bacterial cells having negative charge and NPs being positively charged. Albeit, there is a great deal of dissimilarity between gram-positive and gram-negative bacteria in terms of their cell membrane, but majority of them express a negative charge by the dint of carboxyl, phosphatic and amino groups. Gram-negative bacteria's cell wall is composed of a layer of lipopolysaccharide that has lipids and polysaccharides having been bonded to one another through covalent linkage with a negative charge. Even then, the positively charged AgNPs are provided with weaker penetrability barricade. However, in case of the gram-positive bacteria, the cell wall is made up of thick layer having peptidoglycans in which there is a considerable cross-linking of the polysaccharide chains arranged in a linear manner through shorter peptides leading to the formation of a rigid structure having three dimensions. Such rigidity in the cell wall layer hinders AgNPs attachment to the cell wall, and penetration is affected (Tamboli and Lee 2013).

7 Conclusions and Future Prospects

The tremendous work done in the last decades in the field of environmental nano-biotechnology for fabrication of metallic NPs using microbial nano-factories as bio-templates for stabilizing and reducing purposes has been commendable. A great deal of metallic, metal oxide, metal chalcogenide, and semiconductor materials has been fabricated using number of microbial species, i.e., bacteria, fungi, yeast, actinomycetes, and viruses. Reduction of NPs done via microbial route signifies not only a facile and eco-friendly mode but is also economically valid in comparison to costly and complicated physicochemical reactions marked by inducing profound toxicity toward environment. Through microbes mediated biomimetic NPs synthesis, NPs of various shapes, sizes, and polydispersity extent can be efficiently developed and used for different applications, especially for detoxification of the ecospheric zones. Though great deal of work has been done and different microbes have been proven to be adopted as effective nano-biotechnological tools for NPs synthesis, yet keeping in mind the huge diversity of microbial world, many species needs to be explored in future researches for hunting the best of agents for NPs make up to be adopted at commercial scale. In the similar manner, there is also an urgent need to organize studies to better comprehend different mechanisms, e.g., cellular, biological, chemical, and molecular that involved in the facilitation of microbes mediated synthesis.

Microbial agents are nature's resource that can be replenished by different means and are precursors for biogenic synthesis. Since they produce NPs and induce no toxicity to the ecological niches, therefore this synthetic mode can be referred to as green chemistry since it is in conformity with all principles of green chemistry. Furthermore, this synthetic mode is also strengthening the concept of sustainability and sustainable development in a direct manner by means of cleaner production and no harmful waste generation. Microbes mediated NPs have been employed for a variety of applications including environmental detoxification, biomedical, and biosensor employments owing to their outstanding properties, antimicrobial potential, and optical characteristics. Number of persistent organic pollutants, especially organic dyes, have been effectively removed by microbes mediated NPs in nano-photocatalytic assays. Furthermore, the antimicrobial roles of these NPs against number of plant and human pathogens have also been shown by different researches. Such results for microbes driven NPs are supportive of them to be used as an auspicious candidate for environmental remediation of toxic contaminants and as antimicrobial agents.

With the great potential to be adopted as nano-biotechnological tools in future, some other challenges to be solved by biotechnologists are the efforts to comprehend and outline a set of mechanistic rules operational in this type of synthesis, which are yet missing. Furthermore, there is another limitation of these NPs in adopting them at commercial scale due to their large scale production that needs to be addressed. Improvement in these aspects will not only convert the research-based results into practically beneficial work, but it will also pave the way for further better investigations in this field.

References

Abdeen S, Geo S, Praseetha PK, Dhanya RP (2014) Biosynthesis of silver nanoparticles from Actinomycetes for therapeutic applications. J Int Nano Dimen 2:155–162

Abdel-Azeem A, Nada AA, O'Donovan A, Thakur VK, Elkelish A (2020) Mycogenic silver nanoparticles from endophytic *Trichoderma atroviride* with antimicrobial activity. J Renew Mat 8:171–185. https://doi.org/10.32604/jrm.2020.08960

Adebayo AE, Oke AM, Lateef A, Oyatokun AA, Abisoye OD, Adiji IP, Fagbenro DO, Amusan TV, Badmus JA, Asafa TB, Beukes LS (2019) Biosynthesis of silver, gold and silver–gold alloy nanoparticles using *Persea americana* fruit peel aqueous extract for their biomedical properties. Nanotech Envir Eng 4:13. https://doi.org/10.1007/s41204-019-0060-8

Adelere IA, Lateef A (2016) A novel approach to the green synthesis of metallic nanoparticles: the use of agro-wastes, enzymes and pigments. Nanotechnol Rev 5(6):567–587. https://doi.org/10.1515/ntrev-2016-0024

Ameta A, Ameta R, Ahuja M (2013) Photocatalytic degradation of methylene blue over ferric tungstate. Sci Rev Chem Commun 3:172–180

Apte M, Sambre D, Gaikawad S, Joshi S, Bankar A, Kumar AR, Zinjarde S (2013) Psychrotrophic yeast *Yarrowialipolytica* NCYC 789 mediates the synthesis of antimicrobial silver nanoparticles via cell-associated melanin. AMB Exp 3:32. https://doi.org/10.1186/2191-0855-3-32

Azam Z, Ayaz A, Younas M, Qureshi Z, Arshad B, Zaman W, Ullah F, Nasar MQ, Bahadur S, Irfan MM, Hussain S (2020) Microbial synthesized cadmium oxide nanoparticles induce oxidative stress and protein leakage in bacterial cells. Microb Pathogen 104188. https://doi.org/10.1016/j.micpath.2020.104188

Azeez L, Lateef A, Adejumo AL, Adeleke JT, Adetoro RO, Mustapha Z (2020) Adsorption behaviour of rhodamine B on hen feather and corn starch functionalized with green synthesized silver nanoparticles (AgNPs) mediated with cocoa pods extracts. Chem Africa 3:237–250. https://doi.org/10.1007/s42250-019-00113-7

Barabadi H, Honary S, Ebrahimi P, Mohammadi MA, Alizadeh A, Naghibi F (2014) Microbial mediated preparation, characterization and optimization of gold nanoparticles. J Brazil Microbiol 45:1493–1501. https://doi.org/10.1590/S1517-83822014000400046

Behera SS, Jha S, Arakha M, Panigrahi TK (2013) Synthesis of Silver Nanoparticles from microbial source-a green synthesis approach, and evaluation of its antimicrobial activity against *Escherichia coli*. Int J Eng Res Appl 3:58–62

Beveridge TJ, Murray RG (1980) Sites of metal deposition in the cell wall of *Bacillus subtilis*. J Bacteriol 141:876–887

Buszewski B, Railean-Plugaru V, Pomastowski P, Rafińska K, Szultka-Mlynska M, Golinska P, Wypij M, Laskowski D, Dahm H (2018) Antimicrobial activity of biosilver nanoparticles produced by a novel *Streptacidiphilus durhamensis* strain. J MicrobiolImmunol Infect 51:45–54. https://doi.org/10.1016/j.jmii.2016.03.002

Casamayor JL, Su D (2013) Integration of eco-design tools into the development of eco-lighting products. J Clean Prod 47:32–42. https://doi.org/10.1016/j.jclepro.2013.02.011

Chellamuthu P, Tran F, Silva KP, Chavez MS, El-Naggar MY, Boedicker JQ (2019) Engineering bacteria for biogenic synthesis of chalcogenide nanomaterials. Micro Biotechnol 12:161–172. https://doi.org/10.1111/1751-7915.13320

Chou JR (2014) An ARIZ-based life cycle engineering model for eco-design. J Clean Prod 66:210–223. https://doi.org/10.1016/j.jclepro.2013.11.037

Dasgupta N, Ranjan S, Lichtfouse E (2020) Environmental nanotechnology, vol 3. Springer. ISBN 978-3-030-26672-1

Davidović S, Miljković M, Lazić V, Jović D, Jokić B, Dimitrijević S, Radetić M (2015) Impregnation of cotton fabric with silver nanoparticles synthesized by dextran isolated from bacterial species *Leuconostoc mesenteroides* T3. Carbohydr Polym 131:331–336. https://doi.org/10.1016/j.carbpol.2015.06.024

Debnath B, Majumdar M, Bhowmik M, Bhowmik KL, Debnath A, Roy DN (2020) The effective adsorption of tetracycline onto zirconia nanoparticles synthesized by novel microbial green technology. J Environ Manage 261:110235. https://doi.org/10.1016/j.jenvman.2020.110235

Deljou A, Goudarzi S (2016) Green extracellular synthesis of the silver nanoparticles using thermophilic *Bacillus* sp. AZ1 and its antimicrobial activity against several human pathogenetic bacteria. J Iran Biotechnol 14:25. https://doi.org/10.15171/ijb.1259

Deplanche K, Merroun ML, Casadesus M, Tran DT, Mikheenko IP, Bennett JA, Zhu J, Jones IP, Attard GA, Wood J, Selenska-Pobell S (2012) Microbial synthesis of core/shell gold/palladium nanoparticles for applications in green chemistry. J Royal Soc Int 9:1705–1712. https://doi.org/10.1098/rsif.2012.0003

Divya K, Kurian LC, Vijayan S, Manakulam Shaikmoideen J (2016) Green synthesis of silver nanoparticles by *Escherichia coli*: analysis of antibacterial activity. J Water Environ Nanotechnol 1:63–74. https://doi.org/10.7508/JWENT.2016.01.008

Du J, Singh H, Yi TH (2017) Biosynthesis of silver nanoparticles by *Novosphingobium* sp. THG-C3 and their antimicrobial potential. Artif Cells Nanomed Biotechnol 45:211–217. https://doi.org/10.1080/21691401.2016.1178135

Elegbede JA, Lateef A (2019) Green synthesis of silver (Ag), gold (Au) and silver-gold (Ag-Au) alloy nanoparticles: a review on recent advances, trends and biomedical applications. In: Verma DK, Goyal MR, Suleria HAR (eds) Nanotechnology and nanomaterial applications in food, health and biomedical sciences. https://doi.org/10.1201/9780429425660-1. Apple Academic Press Inc./CRC Press, Taylor and Francis Group, Oakville, Ontario, Canada, pp 3–89. ISBN 978-1-77188-764-9

Elegbede JA, Lateef A, Azeez MA, Asafa TB, Yekeen TA, Oladipo IC, Hakeem AS, Beukes LS, Gueguim-Kana EB (2019) Silver-gold alloy nanoparticles biofabricated by fungal xylanases exhibited potent biomedical and catalytic activities. Biotechnol Progr 35:e2829. https://doi.org/10.1002/btpr.2829

Elegbede JA, Lateef A, Azeez MA, Asafa TB, Yekeen TA, Oladipo IC, Aina DA, Beukes LS, Gueguim-Kana EB (2020) Biofabrication of gold nanoparticles using xylanases through valorization of corncob by *Aspergillus niger* and *Trichoderma longibrachiatum*: antimicrobial, antioxidant, anticoagulant and thrombolytic activities. Waste Biomass Valor 11:781–791. https://doi.org/10.1007/s12649-018-0540-2

Esposito D, Antonietti M (2015) Redefining biorefinery: the search for unconventional building blocks for materials. Chem Soc Rev 44:5821–5835. https://doi.org/10.1039/C4CS00368C

Fernández-Llamosas H, Castro L, Blázquez ML, Díaz E, Carmona M (2017) Speeding up bioproduction of selenium nanoparticles by using *Vibrio natriegens* as microbial factory. Sci Rep 7:1–9. https://doi.org/10.1038/s41598-017-16252-1

Fouad H, Hongjie L, Yanmei D, Baoting Y, El-Shakh A, Abbas G, Jianchu M (2017) Synthesis and characterization of silver nanoparticles using *Bacillus amyloliquefaciens* and *Bacillus subtilis* to control filarial vector *Culexpipiens*pallens and its antimicrobial activity. Artif Cells Nanomed Biotechnol 45:1369–1378. https://doi.org/10.1080/21691401.2016.1241793

Gahlawat G, Choudhury AR (2019) A review on the biosynthesis of metal and metal salt nanoparticles by microbes. RSC Adv 9:12944–12967. https://doi.org/10.1039/C8RA10483B

Gahlawat G, Shikha S, Chaddha BS, Chaudhuri SR, Mayilraj S, Choudhury AR (2016) Microbial glycolipoprotein-capped silver nanoparticles as emerging antibacterial agents against cholera. Microb Cell Fact 15:25. https://doi.org/10.1186/s12934-016-0422-x

Gan L, Zhang S, Zhang Y, He S, Tian Y (2018) Biosynthesis, characterization and antimicrobial activity of silver nanoparticles by a halotolerant *Bacillus endophyticus* SCU-L. Prep Biochem Biotechnol 48:582–588. https://doi.org/10.1080/10826068.2018.1476880

García MA (2011) Surface plasmons in metallic nanoparticles: fundamentals and applications. J Phys D 44:283001. https://doi.org/10.1088/0022-3727/44/28/283001

Garmasheva I, Kovalenko N, Voychuk S, Ostapchuk A, Livins'ka O, Oleschenko L (2016) *Lactobacillus* species mediated synthesis of silver nanoparticles and their antibacterial activity against opportunistic pathogens in vitro. BioImpacts 6:219. https://doi.org/10.15171/bi.2016.29

George Kerry R, Patra JK, Gouda S, Park Y, Shin HS, Das G (2018) Benefaction of probiotics for human health: a review. J Food Drug Anal 26:927–939. https://doi.org/10.1016/j.jfda.2018.01.002

Gurme ST, Aware CB, Surwase SN, Chavan CS, Jadhav JP (2019) Synthesis of melanin mediated silver nanoparticles from *Aeromonas* sp. SNS using response surface methodology: characterization with the biomedical applications and photocatalytic degradation of brilliant green. J Poly Environ 27:2428–2438. https://doi.org/10.1007/s10924-019-01529-5

Gutiérrez CW, Velázquez JJS, Yacamán J (2012) Procedures for the synthesis and capping of metal nanoparticles. Methods Mol Biol 906:3–19. https://doi.org/10.1007/978-1-61779-953-2_1

Hamed AA, Kabary H, Khedr M, Emam AN (2020) Antibiofilm, antimicrobial and cytotoxic activity of extracellular green-synthesized silver nanoparticles by two marine-derived actinomycete. RSC Adv 10(17):10361–10367. https://doi.org/10.1039/C9RA11021F

Hulkoti NI, Taranath T (2014) Biosynthesis of nanoparticles using microbes—a review. Colloid Surf B 121:474–783. https://doi.org/10.1016/j.colsurfb.2014.05.027

Husain S, Afreen S, Yasin D, Afzal B, Fatma T (2019) Cyanobacteria as a bioreactor for synthesis of silver nanoparticles-an effect of different reaction conditions on the size of nanoparticles and their dye decolorization ability. J Microbiol Methods 162:77–82. https://doi.org/10.1016/j.mimet.2019.05.011

Iftikhar S, Saleem M, Ahmad KS, Jaffri SB (2019) Synergistic mycoflora–natural farming mediated biofertilization and heavy metals decontamination of lithospheric compartment in a sustainable mode via *Helianthus annuus*. J IntEnvirSci Tech 1–18. https://doi.org/10.1007/s13762-018-02180-8

Iqtedar M, Aslam M, Akhyar M, Shehzaad A, Abdullah R, Kaleem A (2019) Extracellular biosynthesis, characterization, optimization of silver nanoparticles (AgNPs) using *Bacillus mojavensis* BTCB15 and its antimicrobial activity against multidrug resistant pathogens. Prep Biochem Biotechnol 49:136–142. https://doi.org/10.1080/10826068.2018.1550654

Iram S, Basri R, Ahmad KS, Jaffri SB (2019) Mycological assisted phytoremediation enhancement of bioenergy crops Zea mays and Helianthus annuus in heavy metal contaminated lithospheric zone. Soil Seed Contam 28:411–430. https://doi.org/10.1080/15320383.2019.1597011

Iram S, Iqbal A, Ahmad KS, Jaffri SB (2020) Congruously designed eco-curative integrated farming model designing and employment for sustainable encompassments. Environ Sci Poll Res 1–18. https://doi.org/10.1007/s11356-020-08499-5

Jaffri SB, Ahmad KS (2017) Augmented photocatalytic, antibacterial and antifungal activity of prunosynthetic silver nanoparticles. Artif Cells Nanomed Biotechnol 46:127–137. https://doi.org/10.1080/21691401.2017.1414826

Jaffri SB, Ahmad KS (2018) *Prunus cerasifera* Ehrh. Fabricated ZnO nano falcates and its photocatalytic and dose dependent *in vitro* bio-activity. Open Chem 16:141–154. https://doi.org/10.1515/chem-2018-0022

Jaffri SB, Ahmad KS (2020a) Interfacial engineering revolutionizers: perovskite nanocrystals and quantum dots accentuated performance enhancement in perovskite solar cells. Critic Rev Solid State Mater Sci 1–30. https://doi.org/10.1080/10408436.2020.1758627

Jaffri SB, Ahmad KS (2020b) Biomimetic detoxifier *Prunus cerasifera* Ehrh. silver nanoparticles: innate green bullets for morbific pathogens and persistent pollutants. Environ Sci Poll Res. https://doi.org/10.1007/s11356-020-07626-6

Jayaseelan C, Rahuman AA, Kirthi AV, Marimuthu S, Santhoshkumar T, Bagavan A, Gaurav K, Karthik L, Rao KB (2012) Novel microbial route to synthesize ZnO nanoparticles using *Aeromonas hydrophila* and their activity against pathogenic bacteria and fungi. SpectrochimActa A 90:78–84. https://doi.org/10.1016/j.saa.2012.01.006

Jiang ZJ, Liu CY, Sun LW (2005) Catalytic properties of silver nanoparticles supported on silica spheres. J Phy Chem 109:1730–1735. https://doi.org/10.1021/jp046032g

Kang SH, Bozhilov KN, Myung NV, Mulchandani A, Chen W (2008) Microbial synthesis of CdS nanocrystals in genetically engineered *E. coli*. Angewandte Chem Int Ed 47:5186–5189. https://doi.org/10.1002/anie.200705806

Karthik L, Kirthi AV, Ranjan S, Srinivasan VM (2020) Biological synthesis of nanoparticles and their applications. CRC Press, 2 Jan 2020

Karthik L, Kumar G, Kirthi AV, Rahuman AA, Rao KB (2014) *Streptomyces* sp. LK3 mediated synthesis of silver nanoparticles and its biomedical application. Bioproc Biosys Eng 37:261–267. https://doi.org/10.1007/s00449-013-0994-3

Keskin S, Oya N, KoçberberKılıç N, Dönmez G, Tekinay T (2016) Green synthesis of silver nanoparticles using cyanobacteria and evaluation of their photocatalytic and antimicrobial activity. J Nano Res 40:120–127. https://doi.org/10.4028/www.scientific.net/JNanoR.40.120

Khalili NR, Duecker S, Ashton W, Chavez F (2015) From cleaner production to sustainable development: the role of academia. J Clean Prod 96:30–43. https://doi.org/10.1016/j.jclepro.2014.01.099

Khan R, Fulekar MH (2016) Biosynthesis of titanium dioxide nanoparticles using *Bacillus amyloliquefaciens* culture and enhancement of its photocatalytic activity for the degradation of a sulfonated textile dye Reactive Red 31. J Colloid Inter Sci 475:184–191. https://doi.org/10.1016/j.jcis.2016.05.001

Klaus-Joerger T, Joerger R, Olsson E, Granqvist CG (2001) Bacteria as workers in the living factory: metal-accumulating bacteria and their potential for materials science. Trends Biotechnol 19:15–20. https://doi.org/10.1016/S0167-7799(00)01514-6

Kobayashi M, Tomita S, Sawada K, Shiba K, Yanagi H, Yamashita I, Uraoka Y (2012) Chiral meta-molecules consisting of gold nanoparticles and genetically engineered tobacco mosaic virus. Optics Exp 20:24856–24863. https://doi.org/10.1364/OE.20.024856

Kumar CG, Poornachandra Y, Chandrasekhar C (2015) Green synthesis of bacterial mediated anti-proliferative gold nanoparticles: inducing mitotic arrest (G2/M phase) and apoptosis (intrinsic pathway). Nanoscale 7:18738–18750. https://doi.org/10.1039/C5NR04577K

Kuo TC, Smith S, Smith GC, Huang SH (2016) A predictive product attribute driven eco-design process using depth-first search. J Clean Prod 112:3201–3210. https://doi.org/10.1016/j.jclepro.2015.10.005

Lateef A, Akande MA, Ojo SA, Folarin BI, Gueguim-Kana EB, Beukes LS (2016a) Paper wasp nest-mediated biosynthesis of silver nanoparticles for antimicrobial, catalytic, anticoagulant, and thrombolytic applications. 3 Biotech 6:140. https://doi.org/10.1007/s13205-016-0459-x

Lateef A, Ojo SA, Akinwale AS, Azeez L, Gueguim-Kana EB, Beukes LS (2015) Biogenic synthesis of silver nanoparticles using cell-free extract of *Bacillus safensis* LAU 13: antimicrobial, free radical scavenging and larvicidal activities. Biologia 70:1295–1306. https://doi.org/10.1515/biolog-2015-0164

Lateef A, Ojo SA, Azeez MA, Asafa TB, Yekeen TA, Akinboro A, Oladipo IC, Gueguim-Kana EB, Beukes LS (2016b) Cobweb as novel biomaterial for the green and eco-friendly synthesis of silver nanoparticles. Appl Nanosci 6(6):863–874. https://doi.org/10.1007/s13204-015-0492-9

Lateef A, Ojo SA, Elegbede JA (2016c) The emerging roles of arthropods and their metabolites in the green synthesis of metallic nanoparticles. Nanotechnol Rev 5(6):601–622. https://doi.org/10.1515/ntrev-2016-0049

Lateef A, Ojo SA, Elegbede JA, Akinola PO, Akanni EO (2018) Nanomedical applications of nanoparticles for blood coagulation disorders. In: Dasgupta N, Ranjan S, Lichtfouse E (eds) Environmental nanotechnology. Environmental chemistry for a sustainable world, vol 14. Springer International Publishing AG, Cham, Switzerland, pp 243–277. https://doi.org/10.1007/978-3-319-76090-2_8. ISBN 978-3-319-76089-6

Lateef A, Ojo SA, Oladejo SM (2016d) Anti-candida, anti-coagulant and thrombolytic activities of biosynthesized silver nanoparticles using cell-free extract of *Bacillus safensis* LAU 13. Process Biochem 51(10):1406–1412. http://dx.doi.org/10.1016/j.procbio.2016.06.027

Lestrell E, Patolsky F, Voelcker NH, Elnathan R (2020) Engineered nano-bio interfaces for intracellular delivery and sampling: applications, agency and artefacts. Mater Today 33:87–104. https://doi.org/10.1016/j.mattod.2019.08.012

Li X, Xu H, Chen Z-S, Chen G (2011) Biosynthesis of nanoparticles by microorganisms and their applications. J Nanomater 1–16. https://doi.org/10.1155/2011/270974

Li Z, Wang L, Chen S, Feng C, Chen S, Yin N, Yang J, Wang H, Xu Y (2015) Facilely green synthesis of silver nanoparticles into bacterial cellulose. Cellulose 22:373–383. https://doi.org/10.1007/s10570-014-0487-9

Majumdar M, Shivalkar S, Pal A, Verma ML, Sahoo AK, Roy DN (2020) Nanotechnology for enhanced bioactivity of bioactive compounds. In: Biotechnological production of bioactive compounds. Elsevier, pp 433–466, 1 Jan 2020. https://doi.org/10.1016/B978-0-444-64323-0.00015-1

Mannam R, Kumar ES, DasGupta N, Rao MR (2017) Reversible p-type conductivity in H passivated nitrogen and phosphorous codoped ZnO thin films using rapid thermal annealing. Appl Surf Sci 400:312–317. https://doi.org/10.1016/j.apsusc.2016.12.146

Mao C, Flynn CE, Hayhurst A, Sweeney R, Qi J, Georgiou G, Iverson B, Belcher AM (2003) Viral assembly of oriented quantum dot nanowires. Proceed Nat Acad Sci 100:6946–6951. https://doi.org/10.1073/pnas.0832310100

Momin B, Rahman S, Jha N, Annapure US (2019) Valorization of mutant *Bacillus licheniformis* M09 supernatant for green synthesis of silver nanoparticles: photocatalytic dye degradation, antibacterial activity, and cytotoxicity. Bioprocess Biosyst Eng 42:541–553. https://doi.org/10.1007/s00449-018-2057-2

Moreno-Martin G, Pescuma M, Pérez-Corona T, Mozzi F, Madrid Y (2017) Determination of size and mass-and number-based concentration of biogenic SeNPs synthesized by lactic acid bacteria by using a multimethod approach. Anal Chim Acta 992:34–41. https://doi.org/10.1016/j.aca.2017.09.033

Nadaf NY, Kanase SS (2019) Biosynthesis of gold nanoparticles by *Bacillus marisflavi* and its potential in catalytic dye degradation. J Arab Chem 12:4806–4814. https://doi.org/10.1016/j.arabjc.2016.09.020

Nag S, Pramanik A, Chattopadhyay D, Bhattacharyya M (2018) Green-fabrication of gold nanomaterials using *Staphylococcus warneri* from Sundarbans estuary: an effective recyclable nanocatalyst for degrading nitro aromatic pollutants. Environ Sci Poll Res 25:2331–2349. https://doi.org/10.1007/s11356-017-0617-7

Nair B, Pradeep T (2002) Coalescence of nanoclusters and formation of submicron crystallites assisted by Lactobacillus strains. Cryst Growth Des 2:293–298. https://doi.org/10.1021/cg0255164

Navalakhe RK, DasGupta N, Das BK (2009) Fabrication and characterization of straight and compact S-bend optical waveguides on a silicon-on-insulator platform. Appl Optic 48:G125–G130. https://doi.org/10.1364/AO.48.00G125

Noman M, Shahid M, Ahmed T, Niazi MB, Hussain S, Song F, Manzoor I (2020) Use of biogenic copper nanoparticles synthesized from a native *Escherichia* sp. as photocatalysts for azo dye degradation and treatment of textile effluents. Environ Poll 257:113514. https://doi.org/10.1016/j.envpol.2019.113514

Ojo SA, Lateef A, Azeez MA, Oladejo SM, Akinwale AS, Asafa TB, Yekeen TA, Akinboro A, Oladipo IC, Gueguim-Kana EB, Beukes LS (2016) Biomedical and catalytic applications of gold and silver-gold alloy nanoparticles biosynthesized using cell-free extract of *Bacillus safensis* LAU 13: antifungal, dye degradation, anti-coagulant and thrombolytic activities. IEEE Trans Nanobiosci 15:433–442. https://doi.org/10.1109/TNB.2016.2559161

Otari SV, Patil RM, Nadaf NH, Ghosh SJ, Pawar SH (2012) Green biosynthesis of silver nanoparticles from an actinobacteria *Rhodococcus* sp. Mater Lett 72:92–94. https://doi.org/10.1016/j.matlet.2011.12.109

Pal S, Tak YK, Song JM (2007) Does the antibacterial activity of silver nanoparticles depend on the shape of the nanoparticle? A study of the gram-negative bacterium *Escherichia coli*. Appl Environ Microbiol 73:1712–1720. https://doi.org/10.1128/AEM.02218-06

Paredes D, Ortiz C, Torres R (2014) Synthesis, characterization, and evaluation of antibacterial effect of Ag nanoparticles against *Escherichia coli* O157: H7 and methicillin-resistant *Staphylococcus aureus* (MRSA). J Int Nanomed 9:1717. https://doi.org/10.2147/IJN.S57156

Prasad R, Pandey R, Barman I (2016) Engineering tailored nanoparticles with microbes: quo vadis? Wiley Interdis Rev 8:316–330. https://doi.org/10.1002/wnan.1363

Qu Y, Li X, Lian S, Dai C, Jv Z, Zhao B, Zhou H (2018) Biosynthesis of gold nanoparticles using fungus *Trichoderma* sp. WL-Go and their catalysis in degradation of aromatic pollutants. IET Nanobiotechnol 13:12–17. https://doi.org/10.1049/iet-nbt.2018.5177

Rayaman P, Ocsoy I, Gurer US (2018) Green synthesis and characterization of silver nanoparticles using the fungus *A. niger* and bioactive potential against microorganisms and cancer cells. Latin Am J Pharm 37:979–986

Roshmi T, Jishma P, Radhakrishnan EK (2017) Photocatalytic and antibacterial effects of silver nanoparticles fabricated by *Bacillus subtilis* SJ 15. Inorg Nano-Metal Chem 47:901–908. https://doi.org/10.1080/15533174.2016.1228668

Samuel MS, Jose S, Selvarajan E, Mathimani T, Pugazhendhi A (2020) Biosynthesized silver nanoparticles using *Bacillus amyloliquefaciens*; Application for cytotoxicity effect on A549 cell line and photocatalytic degradation of p-nitrophenol. J Photochem Photobiol B 202:111642. https://doi.org/10.1016/j.jphotobiol.2019.111642

Saravanan M, Arokiyaraj S, Lakshmi T, Pugazhendhi A (2018a) Synthesis of silver nanoparticles from *Phenerochaete chrysosporium* (MTCC-787) and their antibacterial activity against human pathogenic bacteria. Microb Pathogen 117:68–72. https://doi.org/10.1016/j.micpath.2018.02.008

Saravanan M, Barik SK, MubarakAli D, Prakash P, Pugazhendhi A (2018b) Synthesis of silver nanoparticles from *Bacillus brevis* (NCIM 2533) and their antibacterial activity against pathogenic bacteria. Microb Pathogen 116:221–226. https://doi.org/10.1016/j.micpath.2018.01.038

Shah M, Fawcett D, Sharma S, Tripathy SK, Poinern GE (2015) Green synthesis of metallic nanoparticles via biological entities. Materials 8:7278–7308. https://doi.org/10.3390/ma8115377

Siddiqi KS, Husen A (2016) Fabrication of metal nanoparticles from fungi and metal salts: scope and application. Nanoscale Res Lett 11:98. https://doi.org/10.1186/s11671-016-1311-2

Singh H, Du J, Yi TH (2017) Biosynthesis of silver nanoparticles using *Aeromonas* sp. THG-FG1. 2 and its antibacterial activity against pathogenic microbes. Artif Cells Nanomed Biotechnol 45:584–590. https://doi.org/10.3109/21691401.2016.1163715

Singh P, Kim YJ, Wang C, Mathiyalagan R, Yang DC (2016a) Microbial synthesis of flower-shaped gold nanoparticles. Artif Cells Nanomed Biotechnol 44:1469–1474. https://doi.org/10.3109/21691401.2015.1041640

Singh P, Kim YJ, Wang C, Mathiyalagan R, Yang DC (2016b) *Weissella oryzae* DC6-facilitated green synthesis of silver nanoparticles and their antimicrobial potential. Artif Cells Nanomed Biotechnol 44:1569–1575. https://doi.org/10.3109/21691401.2015.1064937

Singh P, Singh H, Kim YJ, Mathiyalagan R, Wang C, Yang DC (2016c) Extracellular synthesis of silver and gold nanoparticles by *Sporosarcina koreensis* DC4 and their biological applications. Enzym Microb Technol 86:75–83. https://doi.org/10.1016/j.enzmictec.2016.02.005

Sintubin L, De Windt W, Dick J, Mast J, Van Der Ha D, Verstraete W, Boon N (2009) Lactic acid bacteria as reducing and capping agent for the fast and efficient production of silver nanoparticles. Appl Microbiol Biotechnol 84:741–749. https://doi.org/10.1007/s00253-009-2032-6

Sirinakis G, Zhao ZY, Sevryugina Y, Tayi A, Carpenter M (2003) Tailored nanomaterials: selective & sensitive chemical sensors for hydrocarbon analysis. Doctoral thesis. Washington Ave, Albany, New York, United States: School of NanoSciences and Nano Engineering, University of Albany, SUNY

Soenen SJ, Rivera-Gil P, Montenegro JM, Parak WJ, De Smedt SC, Braeckmans K (2011) Cellular toxicity of inorganic nanoparticles: common aspects and guidelines for improved nanotoxicity evaluation. Nano Today 6:446–465. https://doi.org/10.1016/j.nantod.2011.08.001

Soni N, Prakash S (2015) Antimicrobial and mosquitocidal activity of microbial synthesized silver nanoparticles. Parasitol Res 114:1023–1030. https://doi.org/10.1007/s00436-014-4268-z

Srivastava N, Mukhopadhyay M (2015) Green synthesis and structural characterization of selenium nanoparticles and assessment of their antimicrobial property. Bioproc Biosys Eng 38:1723–1730. https://doi.org/10.1007/s00449-015-1413-8

Tamboli DP, Lee DS (2013) Mechanistic antimicrobial approach of extracellularly synthesized silver nanoparticles against gram positive and gram negative bacteria. J Hazard Mat 260:878–884. https://doi.org/10.1016/j.jhazmat.2013.06.003

Varshney R, Mishra AN, Bhadauria S, Gaura MS (2009) Novel microbial route to synthesize silver nanoparticles using fungus *Hormoconisresinae*. J Digest Nanomat Biostruct 4:349–355

Vigneshwaran N, Kathe AA, Varadarajan PV, Nachane RP, Balasubramanya RH (2006) Biomimetics of silver nanoparticles by white rot fungus, Phanerochaete chrysosporium. Colloid Surf B 53:55–59. https://doi.org/10.1016/j.colsurfb.2006.07.014

Wadhwani SA, Shedbalkar UU, Nadhe S, Singh R, Chopade BA (2018) Decolorization of textile dyes by combination of gold nanocatalysts obtained from *Acinetobacter* sp. SW30 and $NaBH_4$. Environ Tech Innov 9:186–197. https://doi.org/10.1016/j.eti.2017.12.001

Walujkar SA, Jadhav SP, Patil SS, Patil SC, Sharma AS, Pawar KD (2019) Utilizing the iron tolerance potential of *Bacillus* species for biogenic synthesis of magnetite with visible light active catalytic activity. Colloid Surf B 177:470–478. https://doi.org/10.1016/j.colsurfb.2019.02.033

Wang C, Kim YJ, Singh P, Mathiyalagan R, Jin Y, Yang DC (2015) Green synthesis of silver nanoparticles by *Bacillus methylotrophicus*, and their antimicrobial activity. Artif Cells Nanomed Biotechnol 44:1127–1132. https://doi.org/10.3109/21691401.2015.1011805

Yurtluk T, Akçay FA, Avcı A (2018) Biosynthesis of silver nanoparticles using novel *Bacillus* sp. SBT8. Prep Biochem Biotechnol 7;48(2):151–159. https://doi.org/10.1080/10826068.2017.1421963

Zeng Q, Wen H, Wen Q, Chen X, Wang Y, Xuan W, Liang J, Wan S (2013) Cucumber mosaic virus as drug delivery vehicle for doxorubicin. Biomaterials 34:4632–4642. https://doi.org/10.1016/j.biomaterials.2013.03.017

Zhang X, Yan S, Tyagi RD, Surampalli RY (2011) Synthesis of nanoparticles by microorganisms and their application in enhancing microbiological reaction rates. Chemosphere 82:489–494. https://doi.org/10.1016/j.chemosphere.2010.10.023

Zhao L, Lu L, Wang A, Zhang H, Huang M, Wu H, Xing B, Wang Z, Ji R (2020) Nano-biotechnology in agriculture: use of nanomaterials to promote plant growth and stress tolerance. J Agric Food Chem 68:1935–1947. https://doi.org/10.1021/acs.jafc.9b06615

Zonaro E, Lampis S, Turner RJ, Qazi SJ, Vallini G (2015) Biogenic selenium and tellurium nanoparticles synthesized by environmental microbial isolates efficaciously inhibit bacterial planktonic cultures and biofilms. Front Microbiol 6:584. https://doi.org/10.3389/fmicb.2015.00584

Beneficial Microbes as Novel Microbial Cell Factories in Nanobiotechnology: Potentials in Nanomedicine

E. A. Adebayo, I. C. Oladipo, J. A. Badmus, and A. Lateef

1 Introduction

The uniqueness of microorganisms and their unpredictable nature and biosynthetic capabilities with unusual amenability in a given environment and cultural conditions have made them quality agents for solving particularly difficult life problems such as health challenges. Generally, probiotics are microbes known to play valuable roles in health care delivery and nutrition. Probiotics as defined by FAO/WHO are live microbes which confer adequate protection against illness when administered in right doses (FAO/WHO 2006; Ozen, and Dinleyici 2015). It can also otherwise be defined as live microbial feed supplements which improve the colon health of animals and human by modulating the intestinal microbiota (Nagpal et al. 2012) and fortify mucosal defense system against pathogens (Yazdi et al. 2017). Probiotics are important for sustaining gastrointestinal tract stability, averting colon infection

E. A. Adebayo · I. C. Oladipo · J. A. Badmus · A. Lateef
Nanotechnology Research Group (NANO +), Ladoke Akintola University of Technology, PMB 4000, Ogbomoso, Nigeria

E. A. Adebayo · A. Lateef
Department of Pure and Applied Biology, Ladoke Akintola University of Technology, PMB 4000, Ogbomoso, Nigeria

I. C. Oladipo
Department of Science Laboratory Technology, Ladoke Akintola University of Technology, PMB 4000, Ogbomoso, Nigeria

J. A. Badmus
Department of Biochemistry, Ladoke Akintola University of Technology, PMB 4000, Ogbomoso, Nigeria

A. Lateef (✉)
Laboratory of Industrial Microbiology and Nanobiotechnology, Ladoke Akintola University of Technology, PMB 4000, Ogbomoso, Nigeria
e-mail: agbaje72@yahoo.com; alateef@lautech.edu.ng

A. Lateef et al. (eds.), *Microbial Nanobiotechnology*, Materials Horizons: From Nature to Nanomaterials, https://doi.org/10.1007/978-981-33-4777-9_11

and eventually serve as vital agents for the intestinal well-being (Gu et al. 2008). Probiotics function by limiting gut pathogen aggregation and lowering symptoms specific for diseases that are induced by immune response dysregulations (Basavaraju and Jamil, 2014). Other essential health benefit includes sound immune system response, vitamin production, reduction of serum cholesterol, anticarcinogenic and antimicrobial effects (Yazdi et al. 2017; Miller et al. 2016; West 2016).

In the recent years, beneficial microbes have played vital roles in nanotechnology, especially in biosynthesis of nanoparticles. Nanomaterials with sizes range between 1 and 100 nm are prominent nanoproducts for nanoscience and nanotechnological research. Nanomaterials, in particular metal nanoparticles, have attracted special attention in different fields of applied sciences, ranging from material science to biotechnology (Huang et al. 2007; Nath and Banerjee 2013). Nanotechnology is the use of engineering and technology techniques in a nanoscale to produce novel materials and devices. This literally translates to any technology performed on a nanoscale with useful applications (Govindasamy et al. 2013). Nanoparticles synthesis in recent times has gained unprecedented attention in different spheres of research such as materials science, chemistry, physics, life sciences, medicines and engineering (Ojo et al. 2016). Different methods have been used for the synthesis of mono- and bimetallic nanoparticles, and these include laser ablation (Liz-Marzan and Philipse 1995), laser irradiation (Chen and Yeh 2001), sono-chemical (Anandan et al. 2008; Wani and Ahmad 2013), sputter deposition (Okazaki et al. 2008), solvothermal (Ahmad et al. 2013) and biological or green method (Ahmad et al. 2003).

However, green synthesis remains an important alternative route that has drawn tremendous interests because of reliability, stability, lesser use of harmful chemicals, cost effectiveness and eco-friendliness (Salem et al. 2014). Generally, biological resources that include bacteria (Elbeshehy et al. 2015; Kushwaha et al. 2015; Singh et al. 2015a, b; Raj et al. 2016), bacteriophage (Ahiwale et al. 2017), fungi (Mishra et al. 2014; Priyadarshini et al. 2014; Dhanasekar et al. 2015; Gopinath et al. 2015; Ramalingmam et al. 2015; Eugenio et al. 2016; Yin et al. 2016; Hamedi et al. 2018; Vijayanandan and Balakrishnan 2018; Spagnoletti et al. 2019), enzymes (Talekar et al. 2014; Khan et al. 2015; Lateef and Adeeyo 2015; Lateef et al. 2015a; Rai and Panda 2015; Elegbede et al. 2018, 2019, 2020), plants (Salem et al. 2014; Singh et al. 2015a, b; Anand et al. 2015; Bogireddy et al. 2015; Khatami et al. 2015; Lateef et al. 2015b, 2016a, b, c, d, 2018a; Shanmugam et al. 2016; Adelere et al. 2017; Azeez et al. 2017, 2019a, b), algae (Aziz et al. 2015; da Silva Ferreira et al. 2017; Ramakrishna et al. 2016; Sonker et al. 2017) and exudates from arthropods (Sreelakshmi et al. 2011; El-Desouky and Ammar 2016; Lateef et al. 2016e, f, g) are involved in the processes of green synthesis of metallic nanoparticles. Synthesis of nanoparticles using extracts from green sources has been shown to be a veritable alternative to the more intricate chemical mode of synthesis. This is because extracts from the natural sources act simultaneously as reducing and capping agents in a one-pot synthesis using the availability of wide ranges of phytochemicals in plants and enzymes, protein, amino acids and polysaccharides in bacteria and fungi (Adelere and Lateef 2016). The green synthesis has been proved to be benign to the environment unlike chemical and physical methods (Sreelakshmy et al. 2016).

Application of these particles is due to their nanoscale size, increased surface area and various shapes. Green synthesized nanoparticles have useful wide spectrum of applications in biomedicine, sciences, agriculture, engineering and industry but not limited, to drug delivery, cancer treatment, gene therapy and DNA analysis, antibacterial agents, anticoagulants, thrombolytic agents, biosensors, optics, catalyses, separation science and magnetic resonance imaging (MRI) (Haverkamp et al. 2007; Zheng et al. 2010; Kim et al. 2013; Shankar et al. 2014; Kumari et al. 2015; Ojo et al. 2016; Fariq et al. 2017; Lateef et al. 2017; Olajire et al. 2017; Azeez et al. 2018; Lateef et al. 2018b, 2019, 2020). Howbeit, nanoparticles from beneficial microbes are expected to have prominent influences on human health considering the health benefits of the organisms themselves, such as improvement of intestinal well-being through the regulatory actions of microbiota, immune system stimulation and development, lactose intolerance reduction, nutrient availability, synthesis and enhancement and risk reduction of many diseases (Kumar et al. 2010; Nagpal et al. 2010, 2012; Nagpal and Kaur 2011; Gungor et al. 2015; Patten and Laws 2015; Aoyagi et al. 2017; Mokkala et al. 2017). The present review summarizes the prosperity of nanoparticles biosynthesized using beneficial microbes and their health benefits.

2 Roles of Beneficial Microbes in Food Materials

Beneficial microbes have been reported to improve the quality and nutritional value of food materials. Recent reports available indicated that probiotic bacteria are incorporated into dairy products such as cheese, yogurt, ice cream and other dairy desserts. The probiotics inclusion in food enhanced prolonged storage of perishables, good taste and higher demand of the foods (Ranadheera et al. 2010). Higher levels of viable microorganisms present in probiotics improved therapeutic level of food materials, exhibited better health benefits, increased energy and nutrients value to serve as functional foods (Lahteinen et al. 2010; Granato et al. 2010). Probiotics such as *Lactobacillus acidophilus, L. plantarum, L. curvatus, L. sake, L. pentosaceus, L. rhamnosus, L. fermentum* and *Bifidobacterium animalis* subsp. *lactis* were reported to produce polydextose, oligofructose and other valuable products which enhanced the growth, viability of the organisms and fatty acid production in fermented milk (Oliveira et al. 2009; Ji et al. 2015). Fermented meat products such as sausage, cheese and other fermented dairy products are good sources of probiotic foods (Ferrando et al. 2015; Han et al. 2017). Different groups or species of beneficial microbes with unique beneficial effects have been previously reported as summarized in Table 1.

Table 1 Some health benefits of beneficial microbes

Species	Bioactivities	References
L. rhamnosus, L. reuteri, L. delbrueckii subsp. *lactis*	Inhibit the growth of *Klebsiella oxytoca, Proteus mirabilis, E. coli, Citrobacter freundii, Enterobacter cloacae, Enterococcus faecalis, Enterococcus faecium*	Jacobsen et al. (1999)
L. paracasei	Prevention of gastrointestinal diseases	Gardiner et al. (2002)
L. acidophilus, L. reuteri, L. casei, B. animalis	Anti-candidiasis, immune modulating responses, inhibit oral bacteria growth	Meurman (2005)
L. plantarum	Antibacterial activities against multi-drug resistant pathogens	Thiruneelakandan et al. (2013)
Saccharomyces boulardii, L. casei, Bacillus clausii, Enterococcus faecium	Acute diarrhea treatment	Canani et al. (2007)
L. delbrueckii var *bulgaricus, L. acidophilus, S. thermophilus, B. bifidum*	Acute diarrhea treatment	Canani et al. (2007)
S. boulardii	Acute diarrhea treatment in children	Szajewska et al. (2007)
B. longum, B. breve, B. infantis, L. casei, L. plantarum, L. acidophilus, L. delbrueckii subsp. *bulgaricus, S. salivarius*	Prevention of hepatic damage and bacterial sepsis treatment	Ewaschuk et al. (2007)
L. acidophilus, B. infantis	Reduction of incidence and severity of necrotizing enterocolitis	Lin et al. (2005)
L. rhamnosus, L. reuteri	Bacterial vaginosis (BV) treatment	Martinez et al. (2009) Recine et al. (2016)
L. paracasei, L. plantarum, L. acidophilus, L. delbrueckii subsp. *bulgaricus, B. longum, B. breve, B. infantis*	Reduction in active inflammation and ulcerative colitis (UC) treatment	Miele et al. (2009)
L. rhamnosus, L. acidophilus, S. boulardii, L. casei, L. plantarum	Prevention of *Clostridium difficile*-associated diarrhea	Johnston et al. (2012)
L. salivarius, L. acidophilus, L. rhamnosus, L. casei,L. johnsonii, L. gasseri, L. plantarum, Lactococcus lactis	Treatment of *Helicobacter pylori* infection	Patel et al. (2014) O'Connor et al. (2013)
B. animalis subsp. *Lactis*	Bacterial translocation during diabetes	Amar et al. (2011)

(continued)

Table 1 (continued)

Species	Bioactivities	References
L. paracasei, L. acidophilus, B. lactis, S. faecium	Treatment of irritable bowel syndrome (IBS) and chronic idiopathic constipation (CIC)	Ford et al. (2014)
B. bifidum, L. casei, L. acidophilus,	Treatment of rheumatoid arthritis	Zamani et al. (2016, 2017)
B. lactis, L. rhamnosus, L. casei	Treatment of upper respiratory tract infections	Esposito et al. (2014)
L. rhamnosus	Treatment of respiratory illness in children	Kumpu et al. (2012)

3 Synergy Between Microbes and Metals

Inorganic materials and biological bodies are indispensable to each other since the beginning of this world. Their interactions have led to the existence of well-organized deposit of minerals on earth (Li et al. 2011). Recent scientific innovations have led credence to the profound interactions between inorganic molecules and biological entities, of which beneficial microbes are not left out (Akhtar and Pathak 2017). Many bacteria, fungi and plants are capable of producing inorganic nanoparticles intracellularly or extracellularly (Suresh et al. 2004). Bacteria, fungi and plants have been known to play a crucial role in biosorption, oxidation and reduction of metal ions, precipitation, impaired transport system and efflux system even in the presence of high concentration and toxic metal ions (Ojoawo et al. 2017). Organisms adapt by evolving mechanisms to cope within the environments containing high level of metals in order to survive and replicate. These adaptive mechanisms have positively allowed the organisms to alter the chemical nature of the metal to non-toxic and also the formation of the metal nanoparticles (Durán et al. 2007; Husseiny et al. 2007).

Bhattacharya and Mukherjee (2008) reported that nanoparticles synthesized by a biogenic enzymatic process which are far more superior in diverse ways to the ones synthesized by chemical methods. Although the chemical methods produce large quantities of nanoparticles with distinctive features of well-defined size and shape within a short time, but these particles are not only complex but are costly and some of their products are toxic to the environment and are implicated in some human health challenges. Unlike nanoparticles synthesized by chemical methods, enzymatic process does not involve the use of expensive chemicals, and the enzymatic route is not as as energy intensive as the chemical method. Biosynthesized nanoparticles are produced when microorganisms attract and attach to target ions and reduce it to elemental metal through enzyme induced cell activities. The site of actions dictates the classification of either it is intracellular or extracellular synthesis of nanoparticles. The intracellular method involves transportation of ions into the microbial cell and nanoparticles are formed through the catalytic activities of enzymes, while in the extracellular process, metal ions are trapped on the cell surface and enzymatically reduced to nanoparticles (Zhang et al. 2011).

4 Bio-inspired Synthesis of Nanoparticles Using Beneficial Microbes

The simplicity in procedures, the stability of nanoparticles and the applications in drug delivery, biological imaging, gene silencing, and antimicrobial procedures have made the synthesis of nanoparticles using natural organisms a major threshold in the field of nanotechnology (Wei and Qian 2008). Bacteria possess detoxification mechanisms (Mergeay et al. 2003), and the pivotal roles of microorganisms generally in the synthesis of nanoparticles have been observed (Klaus et al. 1999). Table 2 summarized the importance of different nanoparticles biosynthesized by beneficial microbes. Different types of beneficial microbes that include species of *Lactobacillus, Bacillus, Enterococcus* and *Rhodobacter* have been used to synthesize silver, gold, silver-gold alloy, titanium oxide, tellurium, selenium, zinc oxide, iron oxide and magnesium oxide nanoparticles. However, there is dearth of information on the use of *Bifidobacterium* for the synthesis of nanoparticles. These particles have been evaluated as antimicrobial, catalytic, anticoagulant, thrombolytic, cholesterol-reducing, antitumor, nanopreservative, larvicidal, antioxidant and nutrient supplementation agents. These applications traverse biomedical, environmental and nutritional fields.

Although, the degree of metal nanoparticles synthesized by beneficial microbes is far less compared with other types of microbes, nevertheless, their involvement in the biosynthesis of nanoparticles could be exploited to deliver novel nanoplatforms and products that can extend the frontier of applications of both beneficial microbes and the nanoparticles in a synergistic manner. Similarly, some nanoparticles have also been used to improve the performance of probiotics by enhancing their stability and viability (Table 3). Therefore, beneficial microbes can be easily deployed in the fields of nanobiotechnology and nanomedicine with limited concerns for safety. In the ensuing section, some applications of nanotechnology in the treatment of few devastating diseases are considered, with the view of projecting the possibility of relevance of beneficial microbes in nanomedical field.

Table 2 Importance of nanoparticles biosynthesized by some beneficial microbes

Microbes*	Type of nanoparticles	Benefits or importance	References
Lactobacillus species	Gold and Silver–Gold alloy	Antimicrobial activity against multi-drug resistant hospital isolates	Pratik et al. (2012)
Bacillus safensis LAU 13	Gold and Silver–Gold alloy	Antifungal, dye degradation, anticoagulant and thrombolytic activities	Ojo et al. (2016)
Lactobacillus rhamnosus MT14	Titanium	Antibacterial activities	Aldujaili et al. (2015)
Lactobacillus sp	Titanium	None	Azhar et al. (2011)
Lactobacillus acidophilus NCDC 15	Selenium	Livestock supplementation	Visha et al. (2015)
Lactobacillus plantarum PTCC 1058	Tellurium	Reduction in cholesterol and triglycerides activities	Mirjani et al. (2015)
Lactobacillus plantarum	Silver	Antimicrobial activities	Govindasamy et al. (2013)
Lactobacillus sp VIZYLAC	Silver	Antibacterial and antifungal	Chaudhari et al. (2012)
Lactobacillus casei subsp. *casei* DSM 20 011	Silver	None	Korbekandi et al. (2012)
Lactobacillus sp VRS-2	Silver	None	Ranganath et al. (2012)
Bacillus safensis LAU 13	Silver	Antimicrobial, free radical scavenging and larvicidal activities	Lateef et al. (2015c, d)
Bacillus safensis LAU 13	Silver	Anticandida, anticoagulant, thrombolytic activities	Lateef et al. (2016h)
Lactobacillus sp Sporelac	Silver and Titanium dioxide	Antimicrobial activities	Jha and Prasad (2010)
Lactobacillus sp	Titanium	None	HkeemIbrahem et al. (2014)

(continued)

Table 2 (continued)

Microbes*	Type of nanoparticles	Benefits or importance	References
Lactobacillus casei; Streptococcus thermophilus; Bifidobacterium BB-12; *Lactobacillus acidophilus* (LA-5); *Lactobacillus helveticus* (LH-B02)	Selenium	Production of selenium-enriched yoghurt	Eszenyi et al. (2011)
Lactobacillus fermentum ATCC 9338; PTCC 1638	Silver	Antimicrobial activities	Omidi et al. (2014)
Bacillus licheniformis Dahb1	Silver	Antibiofilm activities	Shanthi et al. (2016)
Enterococcus gallinarum, E. faecium, E. casseliflavus	Silver	Antibacterial activities, Nanopreservative in paints	Oladipo et al. (2017a)
E. gallinarum, E. faecium, E. casseliflavus	Gold	Larvicidal, dye degradation, antioxidant and thrombolytic activities	Oladipo et al. (2017b)
Yeast MKY3	Silver	None	Kowshik et al. (2002)
Rhodobacter sphaeroides	Silver	None	Bai et al. (2011)
Lactobacillus delbrueckii	Silver	Antibacterial activities	Saravanan et al. (2011)
Lactobacillus sporogens	Zinc oxide	Antibacterial activities	Mishra et al. (2013)
L. plantarum ATCC 8014, *L. johnsonii*	Selenium	Anticandida activities	Kheradmand et al. (2014)
Lactobacillus kimchicus DCY51^T	Gold	Free radical scavenging activities	Markus et al. (2016)

(continued)

Table 2 (continued)

Microbes*	Type of nanoparticles	Benefits or importance	References
L. acidophilus 58p UCM B-2637, CCM 4833^T, G1/1(3) UCM B-2691 *L. fermentum* 32/4(16) UCM B-2661, 215 UCM B-2675, CCM 7192*T*, *L. plantarum* G3/3(13) UCM B-2705, 92T UCM B-2629, *L. casei* CCM 7088^T, *L. rhamnosus* CCM 1825^T	Silver	Antimicrobial activities	Garmasheva et al. (2016)
Brevibacterium linens NCIM 2149	Silver	Antibacterial	Nithya and Ragunathan (2012)
Lactobacillus sporogenes	Selenium	Antibiofilm activities	Kaur et al. (2018)
Lactobacillus mindensis	Silver oxide	None	Dhoondia and Chakraborty (2012)
Lactobacillus casei	Selenium	Antitumor and antibacterial activities	Xu et al. (2018a)
Lactobacillus casei	Selenium	Antioxidant activities	Xu et al. (2018b)
Lactobacillus fermentum	Iron oxide	None	Ghandehari et al. (2018)
Lipase of *Lactobacillus plantarum*	Gold	Catalysis	Khan et al. (2019)
Lactobacillus sp	Magnesium oxide	Anticancer activity	Mohanasrinivasan et al. (2018)
Lactobacillus sp	Gold, Silver and Silver–Gold	None	Nair and Pradeep (2002)
Lactobacillus sp	Silver	Suppression of oxidative stress	Dakhil (2017)
Glycolipids of *Lactobacillus casei*	Gold	None	Kato et al. (2018)
Lactobacillus delbrueckii subsp. *bulgaricus*	Silver	None	Sani et al. (2017)

(continued)

Table 2 (continued)

Microbes*	Type of nanoparticles	Benefits or importance	References
Lactobacillus casei	Iron oxide	None	Torabian et al. (2018)
Lactobacillus acidophilus, L. delbrueckii subsp. *bulgaricus, L. reroute*	Selenium	None	Moreno-Martin et al. (2017)
Lactobacillus plantarum, Lactobacillus ruminis	Silver	None	de Lourdes et al. (2017)
Lactobacillus acidophilus	Selenium nanostructures	None	Kaprellyanz et al. (2018); Yusof et al. (2020)
Lactobacillus plantarum	Zinc	None	Bolla et al. (2020)
Lactobacillus kefiri	Platinum	None	Bolla et al. (2020)
Lactobacills casei	Selenium	Prevention of oxidative damage	Qiao et al. (2020)
Lactobacillus brevis	Silver	Antibacterial, antioxidant	Rajoka et al. (2020)
Lactobacills casei	Copper	Antibacterial and anticancer	Kouhkan et al. (2020)

*The isolates have been reported for beneficial roles by the authors

Table 3 Impact of nanoparticles on the potentials of beneficial microbes

Probiotics	Methods of protection	Type of nanoparticles	Impact	References
Pediococcus pentosaceus	Encapsulation	Microgels doped with magnesium oxide nanoparticles	MgO NPs neutralized the hydrogen ions in the gastric fluids, thereby reducing acid-induced degradation of the probiotics	Yao et al. (2018)
Lactobacillus acidophilus	Encapsulation	Chitosan Nanoparticles	Enhances the viability and survival of them against gastrointestinal environmental conditions	Ebrahimnejad et al. (2017)
Lactobacillus acidophilus	Encapsulation	Nanostructured polyelectrolyte layers of Chitosan and carboxymethyl cellulose	Losses occurring between freezing and freeze-drying were found to be lower for the coated cells	Priya et al. (2011)
Lactobacillus rhamnosus	Encapsulation	Alginate and carrageenan microcapsules	Thermotolerance and freeze-drying resistance	Cheow and Hadinoto (2013)
Lactobacillus plantarum	Encapsulation	Polyethylene oxide nanofibres	Drug delivery system for bacteriocins and the encapsulation of probiotic lactic acid bacteria	Heunis et al. (2010)

5 Medical Importance of Nanotechnology in the Treatment of Some Debilitating Diseases: Potential Contributions from Beneficial Microbes

5.1 *Alzheimer's Disease*

Alzheimer's disease (AD) is a proficient memory and cognitive malfunction due to an irreversible degeneration of neurons that culminates in total loss of autonomy and eventually death. In the USA, AD is the sixth leading cause of death and the fifth leading cause of death in Americans age 65 years. Between 2000 and 2013, deaths resulting from heart disease, stroke and prostate cancer reduced by 14, 23 and

11%, respectively, whereas deaths from AD increased by 71% (Alzeimer's Association 2015). There was no evident cure for Alzheimer's disease over the years until strategies inspired by nanotechnology were used (Keating 2005). Owing to the multifunctional potential of nanoparticles, synthesized PEGylated nanoparticles are capable of mimicking the physiological transport mechanism across the blood-brain barrier (BBB). The barrier is an essential physical confine made of cells protecting the brain from the hazardous foreign body in the blood flow. The production of nanoparticles ensures conveyance of conventional pharmaceutics and genes, siRNA and antibodies to the brain in vitro and in vivo. Through this method, the threat of AD has being reduced drastically (Carradori et al. 2018). In relation to probiotics, it has been envisaged that modulation of gut microbes that may influence adaptive immune response could play important role in the management of Alzheimer (Pérez Martínez et al. 2014).

5.2 *Parkinson's Disease*

Parkinson's disease is a long-term degenerating disorder of the central nervous system that mainly affects the gross and fine motor coordinating system. Symptoms involve shaking, rigidity and slow movement and walking with difficulty or movement of any kind. Thinking and behavioral troubles may as well develop over time (Sveinbjornsdottir 2016). Parkinson's disease has been treated with several drugs, but not without side effects. L-Dopa is one of the drugs used and by no means an ideal drug because it has to be administered on the patient in large dose to make sure that a small fraction of the drug bypass the blood-brain barrier (BBB) and as a result trigger unwanted side effects on other parts of the body like writhing, jerking or free flowing movements and nodding (dyskinesia) (Lockman et al. 2002). Nanotechnology has provided more efficient nanocarrier drug that could be used devoid of any side effect and even with better therapeutic efficiency. Nanocarriers are polymers, amphiphilic lipids or solid colloidal particles which enclose targeted drugs. Nanoparticles such as gold nanoparticles (NPs), liposomes, ceramic NPs, solid lipid NPs, polymer-drug conjugates and carbon nanotubes are being used as nanocarriers (Lockman et al. 2002).

Drugs like dopamine could be enclosed inside nanoparticles sphere or linked to the surface, and once the drug is encapsulated with the nanoparticles, it is protected until it reaches the target site such as the substantia nigra in the brain and degradation occurs. Protection is highly needed to prevent substance in the blood causing premature dopamine breakdown and more importantly to check unnecessary high concentrations of dopamine in the bloodstream (Bhaskar et al. 2010). Through nanotechnology, nanostructured drugs delivery system targeted to specific sites in the body, bio-compatible replacements for body parts and fluids, anticarcinogenic emissions, self-diagnostics biosensors, labs-on-a-chip and material for bone and tissue regeneration has been accomplished (Shrivastava and Dash 2009; Patra et al. 2018; Herlem et al. 2019).

Probiotics have also been proposed to be involved in the mechanisms that lead to development of Parkinson's disease. In a mouse model, Perez-Pardo et al. (2018) induce Parkinson's disease using pesticide rotenone with the associated consequences of reduced incidence of *Bifidobacterium* and imbalance in the gut microbiota. In another study, administration of different species of *Bifidobacterium* along with resistance training significantly decreased depression-anxiety scores and body mass index in healthy elderly subjects (Inoue et al. 2018). Thus, nanobiotechnology infused with probiotics could be a novel way to prevent or reduce the incidence of Parkinson's disease. For instance, a healthy bifidobacterial population can be achieved through encapsulation with nanoparticles.

5.3 Cancer

Cancer is a disease at the molecular level, which is second to cardiovascular-related ailment with regards to the number of mortality and morbidity. A lot of drugs have been formulated, synthesized and/or isolated from the natural sources to treat the disease. Toxicity to non-cancerous cells has been a major concern and a reason to continue mining for the effective drugs with less toxicity on the non-cancerous cells. The major cause of toxicity is the lowest effective drug concentration at the cancer site, which calls for an effective drug delivery system at the site for improving the utility (Fink et al. 1996). Nano drug delivery system is a critical and alternative way to carry chemotherapeutic drug at the right time, dose and the controlled measure and release to the required site to attain optimum therapeutic effect devoid of or at lower toxicity. The small size of nanoparticles and its preparation using biodegradable materials are fundamental essential properties of nanoparticles. Its smaller size allows easy transference through small capillaries into the target cells, while biodegradable materials allow controlled drug release in the target site (Shinde 2012). Drug carriers in nanoscale are safer because of reduce side effects of the active drugs occasioned by enhanced permeability, retentive effect, site-specific drug release and accumulation (Greish 2010). In addition, nanoparticles size variability is an added advantage in anticancer therapy because of increased intratumoral distribution and accumulation for better therapeutic response (Hwang et al. 2017). Efficacy of drug infused nanoparticles has been demonstrated which has led to proliferations of a number drug-loaded nanoparticles such as Caelyx (R) (Doxorubicin encapsulated within nanoliposomes), Oncospar (R) and Copaxone (R) (PEGylated protein conjugates) amongst others (Shankarappa et al. 2014).

Hwang and co-workers used Fucoidan, a naturally occurring sulfated polysaccharide to fabricate nanoparticles with cisplatin a chemotherapeutic agent known to exhibit indiscrimination between cancerous and non-cancerous cells. The fabrication improves the immunity of RAW 264.7 cells against cisplatin and increase in manifold cytotoxicity of cisplatin when compared with cisplatin treated HCT-8 cells alone. This implies that the synthesized nanoparticles lower the toxicity of the chemotherapeutic drugs against non-cancerous cells and also up-regulate the drug's efficacy against cancer cells (Hwang et al. 2017). Bacterial megnetosomes serve as a delivery vehicle for doxorubicin, a cardiotoxic chemotherapeutic drug in a study conducted by Sun et al. (2007) The study revealed that the loaded doxorubicin was more effective in H22 tumor-bearing mice than doxorubicin alone with significantly lower cardiac toxicity (Sun et al. 2007). Antitumoral nanoparticles drugs Abraxane (R) containing human albumin and paclitaxel have been shown to decrease the drug-induced toxicity and hypersensitive reactions, in addition to increased antitumor action of the drug (Gradishar 2006).

Mitoxantrone intraperitoneal application has been employed for the treatment of peritoneal carcinomatosis from colorectal cancers (Cotte et al. 2009), ovarian (Muggia 2006) and other surface malignancies (Van der Speeten et al. 2009), but therapeutic efficiency is fraught with austere systemic toxicity with high mortality which hamper its overall usefulness. In the experiment conducted by Chang et al. (2010), cationic mitoxantrone (MTO) was loaded on the Cardiolipin, an anionic liposome by electrostatic surface complexation. The deliverance of this drug from the encapsulated Cardiolipin liposomes on the target site showed prolonged retention, effective anticancer activity and negligible cardiotoxicity (Chang et al. 2010). In addition, to evade multi-drug resistance that lowers MTO efficiency, the drug was packaged in D-α-Tocopherolpolethylene 1000 succinate (TPGS) via disulfide linkage and assembled into micelle by Qiao et al. (2018). The investigators reported significant cytotoxicity against resistant tumor cells (MDA-MB-231/MDR) in vitro and in vivo antitumor activity in xenograft nude mice when compared with either free MTO or disulfide free prodrug micelle (Qiao et al. 2018).

The gold nanoparticles-labeled DNA was found to enhance optical detection when the atomic force microscope and optical measurements were applied. Csáki et al. (2003) study of gold nanoparticles functionalized with thiol modified oligonucleotide shows association between the surface density of the nanoparticles and the signal in optical transmission, which allows for quantification of the nanoparticles coverage using only the optical signal. The report from the study concludes that contrast generation for optical readout as a novel detection method for DNA chips and for single molecule techniques (Csáki et al. 2003). In another similar work, Shenoy et al. (2006) conjugated coumarin with poly ethylene glycol (PEG) spacer and gold nanoparticles. It was shown that using intracellular trafficking Keck3-D fusion microscope, the functionalized gold nanoparticles were rapidly internalized in the cells and localized in the perinuclear region. The work shows that using PEG spacer, the gold nanoparticles can be encapsulated with different naturally important ligands like fluorescent dyes and antibodies so as to target and induce the stimuli at the target site (Shenoy et al. 2006). Furthermore, in another study, anti-HERS and anti-CXCR4 monoclonal antibodies were conjugated separately to the short and long gold nanorods. The two conjugated short and long nanorods are essential for selective imaging of two different sites in a sample when laser irradiation wavelength is on the longitudinal plasmon resonance. The two rods had been to be useful as optoacoustic imaging for the prompt and accurate detection of cancer cells (Li et al. 2007).

In cancer therapy, a novel treatment regime could be availed through the combination of activities of beneficial microbes and nanotechnology. Several beneficial microbes have been shown to have anticancer properties (Wu et al. 2016; Saxami et al. 2017; Chondrou et al. 2018), whose efficacies can be improved through combination with nanotechnology. The nano-probiotic platforms among other things would achieve improved diagnosis through imaging when coupled with nanoparticles such as gold and enhanced therapy through the combined anticancer activities of nanoparticles and beneficial microbes or their products. The sensitivity and specificity of the treatment would undoubtedly increase, while side effects can be reduced to the barest minimum.

6 Conclusions and Future Prospects

In this short review, we considered the involvement of beneficial microbes, particularly bacteria in the green and ecofriendly synthesis of metallic nanoparticles. Although, fewer numbers of these organisms have been reported so far for the biofabrication of nanoparticles, their active roles in nanobiotechnology are indisputably a welcome development that would further enhance the safety of biosynthesized nanoparticles for biomedical and food applications. Thus, the entrance of beneficial microbes into the field of nanobiotechnology is expected to spur new lines of research activities that would expand the roles of beneficial microbes at delivering novel products and services for mankind. It would be of interest for researches to evolve in the areas of evaluation of biosafety and biocompatibility of beneficial microbes-mediated nanoparticles, development of nano-based sensors, enhancement of antimicrobial efficacy of metabolites such as nisin and bacteriocin in food applications, promotion of viability and stability of probiotics in the gut, improved nutrient and mineral supplementation using nano-based platforms, drug delivery, anticancer activities and more efficacious fermentative abilities of probiotics. Nanobiotechnology and beneficial microbes can be a launch pad of synergy to afford mankind the opportunity to combat debilitating disorders that include Alzheimer's and Parkinson's diseases and cancer.

Acknowledgements The authors are grateful to the authorities of LAUTECH, Ogbomoso, Nigeria, for the provision of some of the facilities used in our works that are cited in this review.

References

Adelere IA, Lateef A (2016) A novel approach to the green synthesis of metallic nanoparticles: the use of agro-wastes, enzymes and pigments. Nanotechnol Rev 5:567–587. https://doi.org/10.1515/ntrev-2016-0024

Adelere IA, Lateef A, Aboyeji DO, Abdulsalam R, Adabara NU, Bala JD (2017) Biosynthesis of silver nanoparticles using aqueous extract of *Buchholzia coriacea* (wonderful kola) seeds and their antimicrobial activities. Ann Food Sci Technol 18:671–679

Ahiwale SS, Bankar AV, Tagunde S, Kapadnis BP (2017) A bacteriophage mediated gold nanoparticles synthesis and their anti-biofilm activity. Indian J Microbiol 57:188–194. https://doi.org/10.1007/s12088-017-0640-x

Ahmad S, Senapati MI, Khan KR, Sastry M (2003) Extracellular biosynthesis of monodisperse gold nanoparticles by a novel extremophilic actinomycete, *Thermomonospora* sp. Langmuir 19:3550–3553. https://doi.org/10.1021/la026

Ahmad T, Wani IA, Lone IH, Ganguly A, Manzoor N, Ahmad A, Ahmed J, Al-Shihri AS (2013) Antifungal activity of gold nanoparticles prepared by solvothermal method. Mater Res Bull 48:12–20. https://doi.org/10.1016/j.materresbull.2012.09.069

Akhtar N, Pathak K (2017) Probiotics as a tool to biosynthesize metallic nanoparticles: research reports and patents survey. Recent Pat Drug Delivery Formulation 11:5–18. https://doi.org/10.2174/1872211311666170313124335

Aldujaili NH, Abdullah NY, Khaqani RL, Al-tfaly SA, Al-Shammary AH (2015) Biosynthesis and antibacterial activity of titanium nanoparticles using *Lactobacillus*. Int J Recent Sci Res 6:7741–7751

Alzheimer's, Association (2015) 2015 Alzheimer's disease facts and figures. Alzheimer's Dementia: J Alzheimer's Assoc 11:332–384. https://doi.org/10.1016/j.jalz.2015.02.003

Amar J, Chabo C, Waget A, Klopp P, Vachoux C, Bermúdez-Humarán LG, Smirnova N, Bergé M, Sulpice T, Lahtinen T, Ouwehand A, Langella P, Rautonen N, Sansonetti PJ, Burcelin J (2011) Intestinal mucosal adherence and translocation of commensal bacteria at the early onset of type 2 diabetes: molecular mechanisms and probiotic treatment. EMBO Mol Med 3:559–572. https://doi.org/10.1002/emmm.201100159

Anand K, Gengan RM, Phulukdaree A, Chuturgoon A (2015) Agroforestry waste *Moringa oleifera* petals mediated green synthesis of gold nanoparticles and their anti-cancer and catalytic activity. J Ind Eng Chem 21:1105–1111. https://doi.org/10.1016/j.jiec.2014.05.021

Anandan S, Grieser F, Ashokkumar M (2008) Sonochemical synthesis of Au-Ag core-shell bimetallic nanoparticles. J Phys Chem C 112:15102–15109. https://doi.org/10.1021/jp806960r

Aoyagi Y, Park S, Matsubara S, Honda Y, Amamoto R, Kushiro A, Miyazaki K, Shephard RJ (2017) Habitual intake of fermented milk products containing *Lactobacillus casei* strain Shirota and a reduced risk of hypertension in older people. Benef Microb 8:23–29. https://doi.org/10.3920/BM2016.0135

Azeez MA, Lateef A, Asafa TB, Yekeen TA, Akinboro A, Oladipo IC, Gueguim-Kana EB, Beukes LS (2017) Biomedical applications of cocoa bean extract-mediated silver nanoparticles as antimicrobial, larvicidal and anticoagulant agents. J Clust Sci 28:149–164. https://doi.org/10.1007/s10876-016-1055-2

Azeez L, Lateef A, Adebisi SA, Oyedeji AO (2018) Novel biosynthesized silver nanoparticles from cobweb as adsorbent for Rhodamine B: equilibrium isotherm, kinetic and thermodynamic studies. Appl Water Sci 8:32. https://doi.org/10.1007/s13201-018-0676-z

Azeez L, Adejumo AL, Lateef A, Adebisi SA, Adetoro RO, Adewuyi S, Tijani KO, Olaoye S (2019a) Zero-valent silver nanoparticles attenuate Cd and Pb toxicities on *Moringa oleifera* via immobilization and induction of phytochemicals. Plant Physiol Biochem 139:283–292. https://doi.org/10.1016/j.plaphy.2019.03.030

Azeez L, Lateef A, Wahab AA, Rufai MA, Salau AK, Ajayi IO, Ajayi EM, Maryam AK, Adebisi B (2019b) Phytomodulatory effects of silver nanoparticles on *Corchorus olitorius*: its antiphytopathogenic and hepatoprotective potentials. Plant Physiol Biochem 136:109–117. https://doi.org/10.1016/j.plaphy.2018.12.006

Azhar AM, Ladan SB, Ebrahimi MT, Heydari M (2011) *Lactobacillus*-mediated biosynthesis of titanium nanoparticles in MRS-broth medium. Brno Czech Rep EU 9:21–23

Aziz N, Faraz M, Pandey R, Shakir M, Fatma T, Varma A, Barman I, Prasad R (2015) Facile algae-derived route to biogenic silver nanoparticles: synthesis, antibacterial, and photocatalytic properties. Langmuir 31:11605–11612. https://doi.org/10.1021/acs.langmuir.5b03081

Bai HJ, Yang BS, Chai CJ, Yang GE, Jia WL, Yi ZB (2011) Green synthesis of silver nanoparticles using *Rhodobacter sphaeroides*. World J Microbiol Biotechnol 27:2723. https://doi.org/10.1007/s11274-011-0747-x

Basavaraju B, Jamil K (2014) Identification and characterization of probiotics from new sources. Int J Sci Res 3:837–841

Bhaskar S, Tian F, Stoeger T (2010) Multifunctional nanocarriers for diagnostics, drug delivery and targeted treatment across blood-brain barrier: perspectives on tracking and neuroimaging. Part Fibre Toxicol 7:3. https://doi.org/10.1186/1743-8977-7-3

Bhattacharya R, Mukherjee P (2008) Biological properties of "naked" metal nanoparticles. Adv Drug Deliv Rev 60:1289–1306. https://doi.org/10.1016/j.addr.2008.03.013

Bogireddy NKR, Anand KKH, Mandal BK (2015) Gold nanoparticles-synthesis by *Sterculia acuminata* extract and its catalytic efficiency in alleviating different organic dyes. J Mol Liq 211:868–875. https://doi.org/10.1016/j.molliq.2015.07.027

Bolla PA, Sanz A, Huggias S, Ruggera JF, Serradell MA, Casella ML (2020) Regular arrangement of Pt nanoparticles on S-layer proteins isolated from *Lactobacillus kefiri*: synthesis and catalytic application. Mol Catal 481:110262. https://doi.org/10.1016/j.mcat.2018.12.011

Canani RB, Cirillo P, Terrin G, Cesarano L, Spagnuolo MI, De Vincenzo A, Albano F, Passariello A, De Marco G, Manguso F, Guarino A (2007) Probiotics for treatment of acute diarrhea in children: randomised clinical trial of five different preparations. BMJ 335:340. https://doi.org/10.1136/bmj.39272.581736.55

Carradori D, Balducci C, Re F, Brambilla D, Le Droumaguet B, Flores O, Gaudin A, Mura S, Forloni G, Ordoñez-Gutierrez L, Wandosell F, Masserini M, Couvreur P, Nicolas J, Andrieux K (2018) Antibody-functionalized polymer nanoparticle leading to memory recovery in Alzheimer's disease-like transgenic mouse model. Nanomed Nanotechnol Biol Med 14:609–618. https://doi.org/10.1016/j.nano.2017.12.006

Chang RS, Kim J, Lee HY, Han SE, Na J, Kim K (2010) Reduced dose-limiting toxicity of intraperitoneal mitoxantrone chemotherapy using cardiolipin-based anionic liposomes. Nanomed Nanotechnol Biol Med 6:769–776. https://doi.org/10.1016/j.nano.2010.05.003

Chaudhari PR, Masurkar SA, Shidore VB, Kamble SP (2012) Antimicrobial activity of extracellularly synthesized silver nanoparticles using *Lactobacillus* species obtained from VIZYLAC capsule. J Appl Pharmaceut Sci 2:25–29

Chen YH, Yeh CS (2001) New approach for the formation of alloy nanoparticles: laser synthesis of gold-silver alloy from gold-silver. colloidal mixtures. Chem Comm 4:371–372. https://doi.org/10.1039/B009854J

Cheow WS, Hadinoto K (2013) Biofilm-like *Lactobacillus rhamnosus* probiotics encapsulated in alginate and carrageenan microcapsules exhibiting enhanced thermotolerance and freeze-drying resistance. Biomacromolecules 14:3214–3222. https://doi.org/10.1021/bm400853d

Chondrou P, Karapetsas A, Kiousi DE, Tsela D, Tiptiri-Kourpeti A, Anestopoulos I, Kotsianidis I, Bezirtzoglou E, Pappa A, Galanis A (2018) *Lactobacillus paracasei* K5 displays adhesion, anti-proliferative activity and apoptotic effects in human colon cancer cells. Benef Microb 9:975–983. https://doi.org/10.3920/BM2017.0183

Cotte E, Passot G, Mohamed F, Vaudoyer D, Gilly FN, Glehen O (2009) Management of peritoneal carcinomatosis from colorectal cancer: current state of practice. Cancer J 15:243–248. https://doi.org/10.1097/PPO.0b013e3181a58d67

Csáki A, Kaplanek P, Möller R, Fritzsche W (2003) The optical detection of individual DNA-conjugated gold nanoparticle labels after metal enhancement. Nanotechnology 14:1262. https://doi.org/10.1088/0957-4484/14/12/006

da Silva Ferreira V, ConzFerreira ME, Lima LMT, Frasés S, de Souza W, Sant'Anna C (2017) Green production of microalgae-based silver chloride nanoparticles with antimicrobial activity against pathogenic bacteria. Enzyme Microb Technol 97:114–121. https://doi.org/10.1016/j.enzmictec.2016.10.018

Dakhil AS (2017) Biosynthesis of silver nanoparticle (AgNPs) using *Lactobacillus* and their effects on oxidative stress biomarkers in rats. J King Saud Univ Sci 29:462–467. https://doi.org/10.1016/j.jksus.2017.05.013

de Lourdes Reyes-Escogido M, Meneses-Rodríguez D, Guardado-Mendoza R (2017) Carbohydrate source affects the synthesis of silver nanoparticles by *Lactobacillus plantarum* 1449 and *Lactobacillus ruminis* 1313. IET Nanobiotechnol 11:1035–1039. https://doi.org/10.1049/iet-nbt.2017.0107

Dhanasekar NN, Rahul GR, Narayanan KB, Raman G, Sakthivel N (2015) Green chemistry approach for the synthesis of gold nanoparticles using the fungus *Alternaria* sp. J Microbiol Biotechnol 25:1129–1135. https://doi.org/10.4014/jmb.1410.10036

Dhoondia ZH, Chakraborty H (2012) *Lactobacillus* mediated synthesis of silver oxide nanoparticles. Nanomater Nanotechnol 2:2–15. https://doi.org/10.5772/55741

Durán N, Marcato PD, De Souza D, Alves OL, Esposito E (2007) Antibacterial effect of silver nanoparticles produced by fungal process on textile fabrics and their effluent treatment. J Biomed Nanotechnol 3:203–208. https://doi.org/10.1166/jbn.2007.022

Ebrahimnejad P, Khavarpour M, Khalili S (2017) Survival of *Lactobacillus acidophilus* as probiotic bacteria using chitosan nanoparticles. Int J Eng Transact A: Basics 30:456–463

Elbeshehy EK, Elazzazy AM, Aggelis G (2015) Silver nanoparticles synthesis mediated by new isolates of *Bacillus* spp., nanoparticle characterization and their activity against Bean Yellow Mosaic Virus and human pathogens. Front Microbiol 6: 453. https://doi.org/10.3389/fmicb.2015.00453

El-Desouky TA, Ammar HA (2016) Honey mediated silver nanoparticles and their inhibitory effect on aflatoxins and ochratoxin A. J Appl Pharmaceut Sci 6:083–090. https://doi.org/10.7324/JAPS.2016.60615

Elegbede JA, Lateef A, Azeez MA, Asafa TB, Yekeen TA, Oladipo IC, Adebayo EA, Beukes LS, Gueguim-Kana EB (2018) Fungal xylanases-mediated synthesis of silver nanoparticles for catalytic and biomedical applications. IET Nanobiotechnol 12:857–863. https://doi.org/10.1049/iet-nbt.2017.0299

Elegbede JA, Lateef A, Azeez MA, Asafa TB, Yekeen TA, Oladipo IC, Abbas SH, Beukes LS, Gueguim-Kana EB (2019) Silver-gold alloy nanoparticles biofabricated by fungal xylanases exhibited potent biomedical and catalytic activities. Biotechnol Progr 35:e2829. https://doi.org/10.1002/btpr.2829

Elegbede JA, Lateef A, Azeez MA, Asafa TB, Yekeen TA, Oladipo IC, Aina DA, Beukes LS, Gueguim-Kana EB (2020) Biofabrication of gold nanoparticles using xylanases through valorization of corncob by *Aspergillus niger* and *Trichoderma longibrachiatum*: antimicrobial, antioxidant, anticoagulant and thrombolytic activities. Waste Biomass Valor 11(3):781–791. https://doi.org/10.1007/s12649-018-0540-2

Esposito E, Rigante D, Principi N (2014) Do children's upper respiratory tract infections benefit from probiotics? BMC Infect Dis 14:194. https://doi.org/10.1186/1471-2334-14-194

Eszenyi P, Sztrik A, Babka B, Prokisch J (2011) Elemental, nano-sized (100–500 nm) selenium production by probiotic lactic acid bacteria. Int J Biosci Biochem Bioinformat 1:148–152

Eugenio M, Müller N, Frasés S, Almeida-Paes R, Lima LMT, Lemgruber L, Farina M, de Souza W, Sant'Anna C (2016) Yeast-derived biosynthesis of silver/silver chloride nanoparticles and their antiproliferative activity against bacteria. RSC Adv 6:9893–9904. https://doi.org/10.1039/C5RA22727E

Ewaschuk J, Endersby R, Thiel D, Diaz H, Backer J, Ma M, Churchill T, Madsen K (2007) Probiotic bacteria prevent hepatic damage and maintain colonic barrier function in a mouse model of sepsis. Hepatology 46:841–850. https://doi.org/10.1002/hep.21750

FAO/WHO (2006) Probiotics in food, health and nutritional properties and guidelines for evaluation. In: FAO food and nutritional paper. no. 85. WHO/FAO, Rome

Fariq A, Khan T, Yasmin A (2017) Microbial synthesis of nanoparticles and their potential applications in biomedicine. J Appl Biomed 15:241–248. https://doi.org/10.1016/j.jab.2017.03.004

Ferrando V, Quiberoni A, Reinhemer J, Suárez V (2015) Resistance of functional *Lactobacillus plantarum* strains against food stress conditions. Food Microbiol 48:63–71. https://doi.org/10.1016/j.fm.2014.12.005

Fink D, Nebel S, Aebi S, Zheng H, Cenni B, Nehme A, Christen RD, Howell SB (1996) The role of DNA mismatch repair in platinum drug resistance. Cancer Res 56:4881–4886

Ford AC, Quigley EM, Lacy BE, Lembo AJ, Saito YA, Schiller LR, Soffer EE, Spiegel BM, Moayyedi P (2014) Efficacy of prebiotics, probiotics, and synbiotics in irritable bowel syndrome and chronic idiopathic constipation: systematic review and meta-analysis. Am J Gastroenterol 109:1547. https://doi.org/10.1038/ajg.2014.202

Gardiner GE, Bouchier P, O'Sullivan E, Kelly J, Collins JK, Fitzgerald G, Ross RP, Stanton C (2002) A spray-dried culture for probiotic Cheddar cheese manufacture. Int Dairy J 12:749–756. https://doi.org/10.1016/S0958-6946(02)00072-9

Garmasheva I, Kovalenko N, Voychuk S, Ostapchuk A, Livins'ka O, Oleschenko L (2016) *Lactobacillus* species mediated synthesis of silver nanoparticles and their antibacterial activity against opportunistic pathogens *in vitro*. BioImpacts: BI 6:219–223. https://doi.org/10.15171/bi.2016.29

Ghandehari F, Fani M, Rezaee M (2018) Biosynthesis of iron oxide nanoparticles by cytoplasmic extract of bacteria *Lactobacillus fermentum*. J Med Chem Sci 1:28–30. https://doi.org/10.26655/JMCHEMSCI.2018.9.2

Gopinath K, Karthika V, Sundaravadivelan C, Gowri S, Arumugam 2015. Mycogenesis of cerium oxide nanoparticles using *Aspergillus niger* culture filtrate and their applications for antibacterial and larvicidal activities. J Nanostruct Chem 5:295–303. https://doi.org/10.1007/s40097-015-0161-2

Govindasamy T, Vidya S, Vinola JS, Jayasudha B, Selvasundhari L, Sivakami R, Anthoni SA (2013) Antimicrobial activity of silver nanoparticles synthesized by marine *Lactobacillus* sp against multiple drug resistance pathogens. Sci Technol Arts Res J 2:05–09

Gradishar WJ (2006) Albumin-bound paclitaxel: a next-generation taxane. Expert Opin Pharmacother 7:1041–1053. https://doi.org/10.1517/14656566.7.8.1041

Granato D, Branco GF, Cruz AG, Faria JDAF, Shah NP (2010) Probiotic dairy products as functional foods. Compr Rev Food Sci Food Saf 9:455–470. https://doi.org/10.1111/j.1541-4337.2010.00120.x

Greish K (2010) Enhanced Permeability and Retention (EPR) effect for anticancer nanomedicine drug targeting. In: Grobmyer S, Moudgil B (eds) Cancer nanotechnology. Methods in molecular biology (Methods and Protocols), vol 624, pp 25–37. Humana Press. https://doi.org/10.1007/978-1-60761-609-2_3

Gu RX, Zhen-Quan Yang ZQ, Li ZH, Chen SL, Luo ZL (2008) Probiotic properties of lactic acid bacteria isolated from stool samples of longevous people in regions of Hotan, Xinjiang and Bama, Guangxi, China. Anaerobe 14:313–317. https://doi.org/10.1016/j.anaerobe.2008.06.001

Gungor OE, Kirzioglu Z, Kivanc M (2015) Probiotics: can they be used to improve oral health? Benef Microb 6:647–656. https://doi.org/10.3920/BM2014.0167

Hamedi S, Shojaosadati SA, Shokrollahzadeh S, Hashemi-Najaf Abadi S (2018) Controlled green synthesis of silver nanoparticles using culture supernatant of filamentous fungus. Iranian J Chem Chem Eng (IJCCE) 36:33–42

Han Q, Kong B, Chen O, Sun F, Zhang H (2017) In vitro comparison of probiotic properties of lactic acid bacteria isolated from Harbin dry sausages and selected probiotics. J Funct Foods 32:391–400. https://doi.org/10.1016/j.jff.2017.03.020

Haverkamp RG, Marshall AT, van Agterveld D (2007) Pick your carats: nanoparticles of gold-silver-copper alloy produced *in vivo*. J Nanopart Res 9:697–700. https://doi.org/10.1007/s11051-006-9198-y

Herlem G, Picaud F, Girardet C, Micheau O (2019) Carbon nanotubes: synthesis, characterization, and applications in drug-delivery systems. In: Nanocarriers for drug delivery, pp 469–529. Elsevier. https://doi.org/10.1016/B978-0-12-814033-8.00016-3

Heunis TDJ, Botes M, Dicks LMT (2010) Encapsulation of *Lactobacillus plantarum* 423 and its bacteriocin in nanofibers. Probiotics Antimicrob Proteins 2:46–51. https://doi.org/10.1007/s12602-009-9024-9

HkeemIbrahem K, Salman JAS, Ali FA (2014) Biosynthesis of titanium nanoparticles using locally *Lactobacillus* spp. isolated from different sources. Am J Pharm Tech Res 4:397–407

Huang CC, Yang Z, Lee KH, Chang HT (2007) Synthesis of highly fluorescent gold nanoparticles for sensing mercury (II). Angew Chemie 119:6948–6952. https://doi.org/10.1002/ange.200700803

Husseiny M, El-Aziz M, Badr Y, Mahmoud MA (2007) Biosynthesis of gold nanoparticles using *Pseudomonas aeruginosa*. Spectrochim Acta A: Mol Biomol Spectr 67:1003–1006. https://doi.org/10.1016/j.saa.2006.09.028

Hwang P, Lin X-Z, Kuo KL, Hsu FY (2017) Fabrication and cytotoxicity of fucoidan-cisplatin nanoparticles for macrophage and tumor cells. Materials 10:291. https://doi.org/10.3390/ma10030291

Inoue T, Kobayashi Y, Mori N, Sakagawa M, Xiao JZ, Moritani T, Sakane N, Nagai N (2018) Effect of combined bifidobacteria supplementation and resistance training on cognitive function, body composition and bowel habits of healthy elderly subjects. Benef Microb 9:843–853. https://doi.org/10.3920/BM2017.0193

Jacobsen CN, Nielsen VR, Hayford AE, Møller PL, Michaelsen KF, Pærregaard A, Sandstrom B, Tvede M, Jakobsen M (1999) Screening of probiotic activities of forty-seven strains of *Lactobacillus* spp. by *in vitro* techniques and evaluation of the colonization ability of five selected strains in humans. Appl Environ Microbiol 65:4949–4956

Jha AK, Prasad K (2010) Biosynthesis of metal and oxide nanoparticles using *Lactobacilli* from yoghurt and probiotic spore tablets. Biotechnol J 5:285–291

Ji K, Jang NY, Kim YT (2015) Isolation of lactic acid bacteria showing antioxidative and probiotic activities from kimchi and infant feces. J Microbiol Biotechnol 25:1568–1577. https://doi.org/10.1128/AEM.65.11.4949-4956

Johnston BC, Stephanie SY, Joshua ZG, Kristian T, Per OV, Mark L, Gordon HG (2012) Probiotics for the prevention of *Clostridium difficile*–associated diarrhea-a systematic review and meta-analysis. Ann Intern Med 157:878–888. https://doi.org/10.7326/0003-4819-157-12-201212180-00563

Kaprellyanz L, Zykova N, Petrosyants A, Zykov A (2018) Development of biotechnology of getting selenium nanostructures with *Lactobacillus acidophilus* culture. EUREKA: Life Sci 1:54–60

Kato Y, Kikuchi F, Imura Y, Yoshimura E, Suzuki M (2018) Various shapes of gold nanoparticles synthesized by glycolipids extracted from *Lactobacillus casei*. In: Biomineralization, pp 259–265. Springer, Singapore. https://doi.org/10.1007/978-981-13-1002-7_27

Kaur R, Kaudal T, Sharma A (2018) Probiotic mediated synthesis of selenium nanoparticles: characterization and biofilm scavenging analysis. Res J Life Sci Bioinform Pharmaceut Chem Sci 4:291–304. https://doi.org/10.26479/2018.0403.26

Keating CD (2005) Nanoscience enables ultrasensitive detection of Alzheimer's biomarker. Proc Natl Acad Sci (USA) 102:2263–2264. https://doi.org/10.1073/pnas.0500024102

Khan S, Rizvi SMD, Avaish M, Arshad M, Bagga P, Khan MS (2015) A novel process for size controlled biosynthesis of gold nanoparticles using bromelain. Mater Lett 159:373–376. https://doi.org/10.1016/j.matlet.2015.06.118

Khan I, Nagarjuna R, Dutta JR, Ganesan R (2019) Towards single crystalline, highly monodisperse and catalytically active gold nanoparticles capped with probiotic *Lactobacillus plantarum* derived lipase. Appl Nanosci 9:1101–1109. https://doi.org/10.1007/s13204-018-0735-7

Khatami M, Pourseyedi S, Khatami M, Hamidi H, Zaeifi M, Soltani L (2015) Synthesis of silver nanoparticles using seed exudates of *Sinapis arvensis* as a novel bioresource, and evaluation of their antifungal activity. Bioresour Bioprocess 2:19. https://doi.org/10.1186/s40643-015-0043-y

Kheradmand E, Rafii F, Yazdi MH, Sepahi AA, Shahverdi AS, Oveisi MR (2014) The antimicrobial effects of selenium nanoparticle-enriched probiotics and their fermented broth against *Candida albicans*. DARU J Pharmaceut Sci 22:48. https://doi.org/10.1186/2008-2231-22-48

Kim HK, Choi MJ, Cha SH, Koo YK, Jun SH, Cho S, Park Y (2013) Earthworm extracts utilized in the green synthesis of gold nanoparticles capable of reinforcing the anticoagulant activities of heparin. Nanoscale Res Lett 8:542. https://doi.org/10.1186/1556-276X-8-542

Klaus T, Joerger R, Olsson E, Granqvist CG (1999) Silver-based crystalline nanoparticles, microbially fabricated. Proc Natl Acad Sci (USA) 96:13611–13614. https://doi.org/10.1073/pnas.96.24.13611

Korbekandi H, Iravani S, Abbasi S (2012) Optimization of biological synthesis of silver nanoparticles using *Lactobacillus casei* subsp. *casei*. J Chem Technol Biotechnol 87:932–937. https://doi.org/10.1002/jctb.3702

Kouhkan M, Ahangar P, Babaganjeh LA, Allahyari-Devin M (2020) Biosynthesis of copper oxide nanoparticles using *Lactobacillus casei* subsp. *casei* and its anticancer and antibacterial activities. Current Nanosci 6(1):101–111. https://doi.org/10.2174/1573413715666190318155801

Kowshik M, Ashtaputre S, Kharrazi S, Vogel W, Urban J, Kulkarni SK, Paknikar KM (2002) Extracellular synthesis of silver nanoparticles by a silver-tolerant yeast strain MKY3. Nanotechnology 14:95. https://doi.org/10.1088/0957-4484/14/1/321

Kumar M, Kumar A, Nagpal R, Mohania D, Behare P, Verma V, Kumar P, Poddar D, Aggarwal PK, Henry CJK, Jain S (2010) Cancer-preventing attributes of probiotics: an update. Int J Food Sci Nutr 61:473–496. https://doi.org/10.3109/09637480903455971

Kumari MM, Jacob J, Philip D (2015) Green synthesis and applications of Au-Ag bimetallic nanoparticles. Spectrochim Acta A: Mol Biomol Spectr. 137:185–192. https://doi.org/10.1016/j.saa.2014.08.079

Kumpu M, Kekkonen RA, Kautiainen H, Järvenpää S, Kristo A, Huovinen P, Pitkäranta A, Korpela R, Hatakka K (2012) Milk containing probiotic *Lactobacillus rhamnosus* GG and respiratory illness in children: a randomized, double-blind, placebo-controlled trial. Eur J Clin Nutr 66:1020–1023. https://doi.org/10.1038/ejcn.2012.62

Kushwaha A, Singh VK, Bhartariya J, Singh P, Yasmeen K (2015) Isolation and identification of *E. coli* bacteria for the synthesis of silver nanoparticles: characterization of the particles and study of antibacterial activity. Eur J Exp Biol 5:65–70

Lahteinen T, Malinen E, Koort JMK Mertaniemi-Hannus U, Hankimo T, Karikoski N, Pakkanen S, Laine H, Sillanpaa H, Soderholm H, Palva A (2010) Probiotic properties of *Lactobacillus* isolates originating from porcine intestine and feces. Anaerobe 16:293–300. https://doi.org/10.1016/j.anaerobe.2009.08.002

Lateef A, Adeeyo AO (2015) Green synthesis and antibacterial activities of silver nanoparticles using extracellular laccase of *Lentinus edodes* Not. Sci Biol 7:405–411. https://doi.org/10.15835/nsb.7.4.9643

Lateef A, Adelere IA, Gueguim-Kana EB (2015a) The biology and potential biotechnological applications of *Bacillus safensis*. Biology 70:411–419. https://doi.org/10.1515/biolog-2015-0062

Lateef A, Adelere IA, Gueguim-Kana EB, Asafa TB, Beukes LS (2015b) Green synthesis of silver nanoparticles using keratinase obtained from a strain of *Bacillus safensis* LAU 13. Int Nano Lett 5:29–35. https://doi.org/10.1007/s40089-014-0133-4

Lateef A, Azeez MA, Asafa TB, Yekeen TA, Akinboro A, Oladipo IC, Ajetomobi FE, Gueguim-Kana EB, Beukes LS (2015c) *Cola nitida*-mediated biogenic synthesis of silver nanoparticles using seed and seed shell extracts and evaluation of antibacterial activities. BioNanoSci 5:196–205. https://doi.org/10.1007/s12668-015-0181-x

Lateef A, Ojo SA, Akinwale AS, Azeez L, Gueguim-Kana EB, Beukes LS (2015d) Biogenic synthesis of silver nanoparticles using cell-free extract of *Bacillus safensis* LAU 13: antimicrobial, free radical scavenging and larvicidal activities. Biology 70:1295–1306. https://doi.org/10.1515/biolog-2015-0164

Lateef A, Akande MA, Azeez MA, Ojo SA, Folarin BI, Gueguim-Kana EB, Beukes LS (2016a) Phytosynthesis of silver nanoparticles (AgNPs) using miracle fruit plant (*Synsepalum dulcificum*) for antimicrobial, catalytic, anti-coagulant and thrombolytic applications. Nanotechnol Rev 5:507–520. https://doi.org/10.1515/ntrev-2016-0039

Lateef A, Azeez MA, Asafa TB, Yekeen TA, Akinboro A, Oladipo IC, Azeez L, Ajibade SE, Ojo SA, Gueguim-Kana EB, Beukes LS (2016b) Biogenic synthesis of silver nanoparticles using a pod extract of *Cola nitida*: Antibacterial, antioxidant activities and application as a paint additive. J Taibah Univ Sci 10:551–562. https://doi.org/10.1016/j.jtusci.2015.10.010

Lateef A, Azeez MA, Asafa TB, Yekeen TA, Akinboro A, Oladipo IC, Azeez L, Ojo SA, Gueguim-Kana EB, Beukes LS (2016c) Cocoa pod extract-mediated biosynthesis of silver nanoparticles: its antimicrobial, antioxidant and larvicidal activities. J Nanostruct Chem 6:159–169. https://doi.org/10.1007/s40097-016-0191-4

Lateef A, Ojo SA, Elegbede JA (2016d) The emerging roles of arthropods and their metabolites in the green synthesis of metallic nanoparticles. Nanotechnol Rev 5:601–622. https://doi.org/10.1515/ntrev-2016-0049

Lateef A, Ojo SA, Azeez MA, Asafa TB, Yekeen TA, Akinboro A, Oladipo IC, Gueguim-Kana EB, Beukes LS (2016e) Cobweb as novel biomaterial for the green and eco-friendly synthesis of silver nanoparticles. Appl Nanosci 6:863–874. https://doi.org/10.1007/s13204-015-0492-9

Lateef A, Ojo SA, Folarin BI, Gueguim-Kana EB, Beukes LS (2016f) Kolanut (*Cola nitida*) mediated synthesis of silver-gold alloy nanoparticles: antifungal, catalytic, larvicidal and thrombolytic applications. J Clust Sci 27:1561–1577. https://doi.org/10.1007/s10876-016-1019-6

Lateef A, Akande MA, Ojo SA, Folarin BI, Gueguim-Kana EB, Beukes LS (2016g) Paper wasp nest-mediated biosynthesis of silver nanoparticles for antimicrobial, catalytic, anti-coagulant and thrombolytic applications. 3Biotech 6:140. https://doi.org/10.1007/s13205-016-0459-x

Lateef A, Ojo SA, Oladejo SM (2016h) Anti-candida, anti-coagulant and thrombolytic activities of biosynthesized silver nanoparticles using cell-free extract of *Bacillus safensis* LAU 13. Process Biochem 51:1406–1412. https://doi.org/10.1016/j.procbio.2016.06.027

Lateef A, Ojo SA, Elegbede JA, Azeez MA, Yekeen TA, Akinboro A (2017) Evaluation of some biosynthesized silver nanoparticles for biomedical applications: hydrogen peroxide scavenging, anticoagulant and thrombolytic activities. J Clust Sci 28:1379–1392. https://doi.org/10.1007/s10876-016-1146-0

Lateef A, Folarin BI, Oladejo SM, Akinola PO, Beukes LS and Gueguim-Kana EB (2018a) Characterization, antimicrobial, antioxidant and anticoagulant activities of silver nanoparticles synthesized from *Petiveria alliacea* L. leaf extract. Prep Biochem Biotechnol 48(7):646–652. https://doi.org/10.1080/10826068.2018.1479864

Lateef A, Ojo SA, Elegbede JA, Akinola PO, and Akanni EO (2018b) Nanomedical applications of nanoparticles for blood coagulation disorders. In: Dasgupta N, Ranjan S. Lichtfouse E (eds) Environmental nanotechnology, vol 1. Springer International Publishing AG, Cham, Switzerland. https://doi.org/10.1007/978-3-319-76090-2_8. ISBN 978-3-319-76089-6. Pp. 243-277

Lateef A, Elegbede JA, Akinola PO, Ajayi VA (2019) Biomedical applications of green synthesized-metallic nanoparticles: a review. Pan Afr J Life Sci 3: 157–182. https://doi.org/10.36108/pajols/9102/30(0170)

Lateef A, Oladejo SM, Akinola PO, Aina DA, Beukes LS, Folarin BI, Gueguim-Kana EB (2020) Facile synthesis of silver nanoparticles using leaf extract of Hyptis suaveolens (L.) Poit for environmental and biomedical applications. IOP Conf Ser.: Mater Sci Eng 805:012042. https://doi.org/10.1088/1757-899X/805/1/012042

Li PC, Wei CW, Liao CK, Chen CD, Pao KC, Wang CRC, Wu YN, Shieh DB (2007) Photoacoustic imaging of multiple targets using gold nanorods. IEEE Trans Ultrason Ferroelectr Freq Control 54:1642–1647

Li X, Xu H, Chen ZS, Chen G (2011) Biosynthesis of nanoparticles by microorganisms and their applications. J Nanomater. Article ID 270974. https://doi.org/10.1155/2011/270974

Lin HC, Su BH, Chen AC, Lin TW, Tsai CH, Yeh TF, Oh W (2005) Oral probiotics reduce the incidence and severity of necrotizing enterocolitis in very low birth weight infants. Pediatrics 115:1–4. https://doi.org/10.1542/peds.2004-1463

Liz-Marzan LM, Philipse AP (1995) Stable hydrosols of metallic and bimetallic nanoparticles immobilized on imogolite fibres. J Phys Chem 99:15120–15128. https://doi.org/10.1021/j100041a031

Lockman PR, Mumper RJ, Khan MA, Allen DD (2002) Nanoparticle technology for drug delivery across the blood-brain barrier. Drug Dev Ind Pharm 28:1–13. https://doi.org/10.1081/DDC-120001481

Markus J, Mathiyalagan R, Kim YJ, Abbai R, Singh P, Ahn S, Perez ZEJ, Hurh J, Yang DC (2016) Intracellular synthesis of gold nanoparticles with antioxidant activity by probiotic *Lactobacillus kimchicus* DCY51T isolated from Korean kimchi. Enzyme Microb Technol 95:85–93. https://doi.org/10.1016/j.enzmictec.2016.08.018

Martinez RCR, Franceschini SA, Patta MC, Quintana SM, Candido RC, Ferreira JC, De Martinis ECP, Reid G (2009). Improved treatment of vulvovaginal candidiasis with fluconazole plus probiotic *Lactobacillus rhamnosus* GR-1 and *Lactobacillus reuteri* RC-14. Lett Appl Microbiol 48:269–274. https://doi.org/10.1111/j.1472-765X.2008.02477.x

Mergeay M, Monchy S, Vallaeys T, Auquier V, Benotmane A, Bertin P, Taghavi S, Dunn J, Van Der Lelie D, Wattiez R (2003) *Ralstonia metallidurans*, a bacterium specifically adapted to toxic metals: towards a catalogue of metal-responsive genes. FEMS Microbiol Rev 27:385–410. https://doi.org/10.1016/S0168-6445(03)00045-7

Meurman JH (2005) Probiotics: do they have a role in oral medicine and dentistry? Eur J Oral Sci 113:188–196. https://doi.org/10.1111/j.1600-0722.2005.00191.x

Miele E, Pascarella F, Giannetti E, Quaglietta L, Baldassano RN, Staiano A (2009) Effect of a probiotic preparation (VSL#3) on induction and maintenance of remission in children with ulcerative colitis. Am J Gastroenterol 104:437–443. https://doi.org/10.1038/ajg.2008.118

Miller H, Ferris R, Phelps BR (2016) The effect of probiotics on CD4 counts among people living with HIV: a systematic review. Benef Microb 7:345–351. https://doi.org/10.3920/BM2015.0163

Mirjani R, Mohammad AF, Mohammad S, Neda S, Mohammad RK, Ahmad RS (2015) Biosynthesis of tellurium nanoparticles by *Lactobacillus plantarum* and the effect of nanoparticle-enriched probiotics on the lipid profiles of mice. IET Nanobiotechnol 9:300–305. https://doi.org/10.1049/iet-nbt.2014.0057

Mishra M, Paliwal JS, Singh SK, Selvarajan E, Subathradevi C, Mohanasrinivasan V (2013) Studies on the inhibitory activity of biologically synthesized and characterized zinc oxide nanoparticles using *Lactobacillus sporogens* against *Staphylococcus aureus*. J Pure Appl Microbiol 7:1263–1268

Mishra A, Kumari M, Pandey S, Chaudhry V, Gupta KC, Nautiyal CS (2014) Biocatalytic and antimicrobial activities of gold nanoparticles synthesized by *Trichoderma* sp. Bioresour Technol 166:235–242. https://doi.org/10.1016/j.biortech.2014.04.085

Mohanasrinivasan V, Devi CS, Mehra A, Prakash S, Agarwal A, Selvarajan E, Naine SJ (2018) Biosynthesis of MgO nanoparticles using *Lactobacillus* sp. and its activity against human leukemia cell lines HL-60. BioNanoScience 8:249–253. https://doi.org/10.1007/s12668-017-0480-5

Mokkala K, Röytiö H, Ekblad U, Laitinen K (2017) Opportunities for probiotics and polyunsaturated fatty acids to improve metabolic health of overweight pregnant women. Benef Microb 8:3–15. https://doi.org/10.3920/BM2016.0068

Moreno-Martin G, Pescuma M, Pérez-Corona T, Mozzi F, Madrid Y (2017) Determination of size and mass-and number-based concentration of biogenic SeNPs synthesized by lactic acid bacteria by using a multimethod approach. Analy Chim Acta 992:34–41. https://doi.org/10.1016/j.aca.2017.09.033

Muggia FM (2006) New and emerging intraperitoneal (IP) drugs for ovarian cancer treatment. Sem Oncol 33:S18–S24. https://doi.org/10.1053/j.seminoncol.2006.11.006

Nagpal R, Kaur A (2011) Synbiotic effect of various prebiotics on *in-vitro* activities of probiotic lactobacilli. Ecol Food Nutr 50:63–68. https://doi.org/10.1080/03670244.2011.539161

Nagpal R, Kumar A, Arora S (2010) *In-vitro* probiotic potential of lactobacilli from indigenous fermented milk products. Int J Probiotics Prebiotics 5:105–112

Nagpal R, Kumar A, Kumar M, Behare PV, Jain S, Yadav H (2012) Probiotics, their health benefits and applications for developing healthier foods: a review. FEMS Microbiol Lett 334:1–15. https://doi.org/10.1111/j.1574-6968.2012.02593.x

Nair B, Pradeep T (2002) Coalescence of nanoclusters and formation of submicron crystallites assisted by *Lactobacillus* strains. Cryst Growth Design 2:293–298. https://doi.org/10.1021/cg0255164

Nath D, Banerjee P (2013) Green nanotechnology-a new hope for medical biology. Environ Toxicol Pharmacol 36:997–1014. https://doi.org/10.1016/j.etap.2013.09.002

Nithya R, Ragunathan R (2012) Synthesis of silver nanoparticles using a probiotic microbe and its antibacterial effect against multidrug resistant bacteria. Afr J Biotechnol 11:11013–11021. https://doi.org/10.5897/AJB12.439

O'Connor A, Molina-Infante J, Gisbert JP, O'Morain C (2013) Treatment of *Helicobacter pylori* Infection 2013. Helicobacter 18(Suppl. 1):58–65. https://doi.org/10.1111/hel.12075

Ojo SA, Lateef A, Azeez MA, Oladejo SM, Akinwale AS, Asafa TB, Yekeen TA, Akinboro A, Oladipo IC, Gueguim-Kana EB, Beukes LS (2016) Biomedical and catalytic applications of gold and silver-gold alloy nanoparticles biosynthesized using cell-free extract of *Bacillus Safensis* LAU 13: antifungal, dye degradation, anti-coagulant and thrombolytic activities. IEEE Trans Nanobiosci 15:1536–1541. https://doi.org/10.1109/TNB.2016.2559161

Ojoawo SO, Lateef A, Oyeniran FA, Kupoluyi OT, Opatola OS, Daramola JO (2017) Bioaccumulation of heavy metals in steel processing industrial effluents using *Bacillus safensis* LAU 13. J Environ Biotechnol Res 6:58–63

Okazaki KI, Kiyama T, Hirahara K, Tanaka N, Kuwabata S, Torimoto T (2008) Single-step synthesis of gold-silver alloy nanoparticles in ionic liquids by sputter deposition technique. Chem Comm 6:691–693. https://doi.org/10.1039/B714761A

Oladipo IC, Lateef A, Azeez MA, Asafa TB, Yekeen TA, Akinboro A, Akinwale AS, Gueguim-Kana EB, Beukes LS (2017a) Green synthesis and antimicrobial activities of silver nanoparticles using cell free-extracts of *Enterococcus* species. Not. Sci Biol 9:196–203. https://doi.org/10.15835/nsb929938

Oladipo IC, Lateef A, Elegbede JA, Azeez MA, Asafa TB, Yekeen TA, Akinboro A, Gueguim-Kana EB, Beukes LS, Oluyide TO, Atanda OR (2017b) *Enterococcus* species for the one-pot biofabrication of gold nanoparticles: characterization and nanobiotechnological applications. J Photochem Photobiol B: Biol 173:250–257. https://doi.org/10.1016/j.jphotobiol.2017.06.003

Olajire AA, Abidemi JJ, Lateef A, Benson NU (2017) Adsorptive desulphurization of model oil by Ag nanoparticles-modified activated carbon prepared from brewer's spent grains. J Environ Chem Eng 5:147–159. https://doi.org/10.1016/j.jece.2016.11.033

Oliveira RPS, Florence ACR, Silva RC, Perego P, Converti A, Gioielli LA (2009) Effects of different prebiotics on the fermentation kinetics, probiotic survival and fatty acids profiles in nonfat symbiotic fermented milk. Int J Food Microbiol 128:467–472. https://doi.org/10.1016/j.ijfoodmicro.2008.10.012

Omidi B, Hashemi S, Mansour B, Kambiz L (2014) Biosynthesis of silver nanoparticles by *Lactobacillus fermentum*. Bull Environ Pharmacol Life Sci 3:186–192

Ozen M, Dinleyici EC (2015) The history of probiotics: the untold story. Benef Microb 6:159–165. https://doi.org/10.3920/BM2014.0103

Patel A, Shah N, Prajapati JB (2014) Clinical application of probiotics in the treatment of *Helicobacter pylori* infection-a brief review. J Microbiol Immunol Infect 47:429–437. https://doi.org/10.1016/j.jmii.2013.03.010

Patra JK, Das G, Fraceto LF, Campos EVR, del Pilar Rodriguez-Torres M, Acosta-Torres LS, Diaz-Torres LA, Grillo R, Swamy MK, Sharma S, Habtemariam S, Shin HS (2018) Nano based drug delivery systems: recent developments and future prospects. J Nanobiotechnol 16:71. https://doi.org/10.1186/s12951-018-0392-8

Patten DA, Laws AP (2015) *Lactobacillus*-produced exopolysaccharides and their potential health benefits: a review. Benef Microb 6:457–471. https://doi.org/10.3920/BM2014.0117

Pérez Martínez G, Bäuerl C, Collado MC (2014) Understanding gut microbiota in elderly's health will enable intervention through probiotics. Benef Microb 5:235–246. https://doi.org/10.3920/BM2013.0079

Perez-Pardo P, Dodiya HB, Engen PA, Naqib A, Forsyth CB, Green SJ, Garssen J, Keshavarzian A, Kraneveld AD (2018) Gut bacterial composition in a mouse model of Parkinson's disease. Benef Microb 9:799–814. https://doi.org/10.3920/BM2017.0202

Priya AJ, Vijayalakshmi SP, Raichur AM (2011) Enhanced survival of probiotic *Lactobacillus acidophilus* by encapsulation with nanostructured polyelectrolyte layers through layer-by-layer approach. J Agri Food Chem 59:11838–11845. https://doi.org/10.1021/jf203378s

Priyadarshini E, Pradhan N, Sukla LB, Panda PK, Mishra BK (2014) Biogenic synthesis of floral-shaped gold nanoparticles using a novel strain *Talaromyces flavus*. Ann Microbiol 64:1055–1063. https://doi.org/10.1007/s13213-013-0744-4

Qiao H, Zhu Z, Fang D, Sun Y, Kang C, Di L, Zhang L, Gao Y (2018) Redox-triggered mitoxantrone prodrug micelles for overcoming multidrug-resistant breast cancer. J Drug Targeting 26:75–85. https://doi.org/10.1080/1061186X.2017.1339195

Qiao L, Dou X, Yan S, Zhang B, Xu C (2020) Biogenic selenium nanoparticles synthesized by *Lactobacillus casei* ATCC 393 alleviate diquat-induced intestinal barrier dysfunction in C57BL/6 mice through their antioxidant activity. Food Funct 11(4):3020–3031. https://doi.org/10.1039/D0FO00132E

Rai T, Panda D (2015) An extracellular enzyme synthesizes narrow-sized silver nanoparticles in both water and methanol. Chem Phys Lett 623:108–112. https://doi.org/10.1016/j.cplett.2015.02.003

Raj R, Dalei K, Chakraborty J, Das S (2016) Extracellular polymeric substances of a marine bacterium mediated synthesis of CdS nanoparticles for removal of cadmium from aqueous solution. J Colloid Interf Sci 462:166–175. https://doi.org/10.1016/j.jcis.2015.10.004

Rajoka MS, Mehwish HM, Zhang H, Ashraf M, Fang H, Zeng X, Wu Y, Khurshid M, Zhao L, He Z (2020) Antibacterial and antioxidant activity of exopolysaccharide mediated silver nanoparticle synthesized by *Lactobacillus brevis* isolated from Chinese koumiss. Colloids Surf B: Biointerf 186:110734. https://doi.org/10.1016/j.colsurfb.2019.110734

Ramakrishna M, Babu DR, Gengan RM, Chandra S, Rao GN (2016) Green synthesis of gold nanoparticles using marine algae and evaluation of their catalytic activity. J Nanostruct Chem 6:1–13. https://doi.org/10.1007/s40097-015-0173-y

Ramalingmam P, Muthukrishnan S, Thangaraj P (2015) Biosynthesis of silver nanoparticles using an endophytic fungus, *Curvularia lunata* and its antimicrobial potential. J Nanosci Nanoeng 1:241–247

Ranadheera RDCS, Baines SK, Adams MC (2010) Importance of food in probiotic efficacy. Food Res Int 43:1–7. https://doi.org/10.1016/j.foodres.2009.09.009

Ranganath E, Rathod V, Banu A (2012) Screening of *Lactobacillus* spp, for mediating the biosynthesis of silver nanoparticles from silver nitrate. IOSR J Pharm 2:237–241

Recine N, Palma E, Domenici L, Giorgini M, Imperiale L, Sassu C, Musella A, Marchetti C, Muzii L, Panici PB (2016) Restoring vaginal microbiota: biological control of bacterial vaginosis. A prospective case–control study using *Lactobacillus rhamnosus* BMX 54 as adjuvant treatment against bacterial vaginosis. Arch Gynecol Obstetr 293:101–107. https://doi.org/10.1007/s00404-015-3810-2

Salem WM, Haridy M, Sayed WF, Hassan NH (2014) Antibacterial activity of silver nanoparticles synthesized from latex and leaf extract of *Ficus sycomorus*. Ind Crops Prod 62:228–234. https://doi.org/10.1016/j.indcrop.2014.08.030

Sani NJ, Aminu BM, Mukhtar MD (2017) Eco-friendly synthesis of silver nanoparticles using *Lactobacillus delbrueckii* subsp. *bulgaricus* isolated from kindrimo (locally fermented milk) in Kano State, Nigeria. Bayero J Pure Appl Sci 10:481–488

Saravanan M, Nanda A, Kingsley SJ (2011) *Lactobacillus delbrueckii* mediated synthesis of silver nanoparticles and their evaluation of antibacterial efficacy against MDR clinical pathogens. In: Nanoscience, Engineering and Technology (ICONSET), 2011 international conference on, pp 386–390. IEEE

Saxami G, Karapetsas A, Chondrou P, Vasiliadis S, Lamprianidou E, Kotsianidis I, Ypsilantis P, Botaitis S, Simopoulos C, Galanis A (2017) Potentially probiotic *Lactobacillus* strains with anti-proliferative activity induce cytokine/chemokine production and neutrophil recruitment in mice. Benef Microb 8:615–623. https://doi.org/10.3920/BM2016.0202

Shankar S, Jaiswal L, Aparna RSL, Prasad RGSV (2014) Synthesis, characterization, *in vitro* biocompatibility, and antimicrobial activity of gold, silver and gold silver alloy nanoparticles prepared from *Lansium domesticum* fruit peel extract. Mater Lett 137:75–78. https://doi.org/10.1016/j.matlet.2014.08.122

Shankarappa SA, Koyakutty M, Nair SV (2014) Efficacy versus toxicity—the Ying and Yang in translating nanomedicines. Nanomater Nanotechnol 4:4–23. https://doi.org/10.5772/59127

Shanmugam C, Sivasubramanian G, Parthasarathi B, Baskaran K, Balachander R, Parameswaran VR (2016) Antimicrobial, free radical scavenging activities and catalytic oxidation of benzyl alcohol by nano-silver synthesized from the leaf extract of *Aristolochia indica* L.: a promenade towards sustainability. Appl Nanosci 6:711–723. https://doi.org/10.1007/s13204-015-0477-8

Shanthi S, Jayaseelan BD, Velusamy P, Vijayakumar S, Chih CT, Vaseeharan B (2016) Biosynthesis of silver nanoparticles using a probiotic *Bacillus licheniformis* Dahb1 and their antibiofilm activity and toxicity effects in *Ceriodaphnia cornuta*. Microb Pathog 93:70–77. https://doi.org/10.1016/j.micpath.2016.01.014

Shenoy D, Fu W, Li J, Crasto C, Jones G, DiMarzio C, Sridhar S, Amiji M (2006) Surface functionalization of gold nanoparticles using hetero-bifunctionalpoly (ethylene glycol) spacer for intracellular tracking and delivery. Int J Nanomed 1:51–57. https://doi.org/10.2147/nano.2006.1.1.51

Shinde NC (2012) Nanoparticles: advances in drug delivery systems. Res J Pharmaceut Biol Chem Sci 1:132–137

Shrivastava S, Dash D (2009) Applying nanotechnology to human health: revolution in biomedical sciences. J Nanotechnol. Article ID 184702. https://doi.org/10.1155/2009/184702

Singh P, Kim YJ, Yang DC (2015a) A strategic approach for rapid synthesis of gold and silver nanoparticles by *Panax ginseng* leaves. Artif Cells Nanomed Biotechnol 44:1949–1957. https://doi.org/10.3109/21691401.2015.1115410

Singh R, Shedbalkar UU, Wadhwani SA, Chopade BA (2015b) Bacteriagenic silver nanoparticles: synthesis, mechanism, and applications. Appl Microbiol Biotechnol 99:4579–4593. https://doi.org/10.1007/s00253-015-6622-1

Sonker AS, Pathak J, Kannaujiya V, Sinha R, Pathak J, Kannaujiya V (2017) Characterization and *in vitro* antitumor, antibacterial and antifungal activities of green synthesized silver nanoparticles using cell extract of *Nostoc* sp. strain HKAR-2. Canadian J Biotechnol 1:26–37

Spagnoletti FN, Spedalieri C, Kronberg F, Giacometti R (2019) Extracellular biosynthesis of bactericidal Ag/AgCl nanoparticles for crop protection using the fungus *Macrophomina phaseolina*. J Environ Manag 231:457–466. https://doi.org/10.24870/cjb.2017-000103

Sreelakshmi C, Datta KKR, Yadav JS, Reddy BV (2011) Honey derivatized Au and Ag nanoparticles and evaluation of its antimicrobial activity. J Nanosci Nanotechnol 11:6995–7000. https://doi.org/10.1166/jnn.2011.4240

Sreelakshmy V, Deepa MK, Mridula P (2016) Green synthesis of silver nanoparticles from *Glycyrrhiza glabra* root extract for the treatment of gastric ulcer. J Dev Drugs 5:152

Sun JB, Duan JH, Dai SL (2007) *In vitro* and *in vivo* antitumor effects of doxorubicin loaded with bacterial magnetosomes (DBMs) on H22 cells: the magnetic bionanoparticles as drug carriers. Cancer Lett 258:109–117. https://doi.org/10.1016/j.canlet.2007.08.018

Suresh K, Prabagaran SR, Sengupta S, Shivaji S (2004) *Bacillus indicus* sp. nov., an arsenic-resistant bacterium isolated from an aquifer in West Bengal, India. Int J Syst Evolut Microbiol 54:1369–1375. https://doi.org/10.1099/ijs.0.03047-0

Sveinbjornsdottir S (2016) The clinical symptoms of Parkinson's disease. J Neurochem 139:318–324. https://doi.org/10.1111/jnc.13691

Szajewska H, Skorka A, Dylag M (2007) Meta-analysis: *Saccharomyces boulardii* for treating acute diarrhoea in children. Aliment Pharmacol Therapeut 25:257–264. https://doi.org/10.1111/j.1365-2036.2006.03202.x

Talekar S, Joshi G, Chougle R, Nainegali B, Desai S, Joshi A, Kambale S, Kamat P, Haripurkar R, Jadhav S, Nadar S (2014) Preparation of stable cross-linked enzyme aggregates (CLEAs) of NADH-dependent nitrate reductase and its use for silver nanoparticle synthesis from silver nitrate. Catal Comm 53:62–66. https://doi.org/10.1016/j.catcom.2014.05.003

Torabian P, Ghandehari F, Fatemi M (2018) Biosynthesis of iron oxide nanoparticles by cytoplasmic extracts of bacteria *Lactobacillus casei*. Asian J Green Chem 2:181–188. https://doi.org/10.22034/AJGC.2018.57914

Van der Speeten K, Stuart OA, Sugarbaker PH (2009) Pharmacokinetics and pharmacodynamics of perioperative cancer chemotherapy in peritoneal surface malignancy. Cancer J 15:216–224. https://doi.org/10.1097/PPO.0b013e3181a58d95

Vijayanandan AS, Balakrishnan RM (2018) Biosynthesis of cobalt oxide nanoparticles using endophytic fungus *Aspergillus nidulans*. J Environ Manag 218:442–450. https://doi.org/10.1016/j.jenvman.2018.04.032

Visha P, Nanjappan K, Selvaraj P, Jayachandran S, Elango A, Kumaresan G (2015) Biosynthesis and structural characteristics of selenium nanoparticles using *Lactobacillus acidophilus* bacteria by wet sterilization process. Int J Adv Vet Sci Technol 4:178–183

Wani IA, Ahmad T (2013) Size and shape dependant antifungal activity of gold nanoparticles: a case study of Candida. Colloids Surf B: Biointerf 101:162–170. https://doi.org/10.1016/j.colsurfb.2012.06.005

Wei D, Qian W (2008) Facile synthesis of Ag and Au nanoparticles utilizing chitosan as a mediator agent. Colloids Surf B: Biointerf 62:136–142. https://doi.org/10.1016/j.colsurfb.2007.09.030

West CE (2016) Probiotics for allergy prevention. Benef Microb 7:171–179. https://doi.org/10.3920/BM2015.0073

Wu Z, Wang G, Pan D, Guo Y, Zeng X, Sun Y, Cao J (2016) Inflammation-related pro-apoptotic activity of exopolysaccharides isolated from *Lactococcus lactis* subsp. *lactis*. Benef Microb 7:761–768. https://doi.org/10.3920/BM2015.0192

Xu C, Guo Y, Qiao L, Ma L, Cheng Y, Roman A (2018a) Biogenic synthesis of novel functionalized selenium nanoparticles by *Lactobacillus casei* ATCC 393 and its protective effects on intestinal barrier dysfunction caused by enterotoxigenic *Escherichia coli* K88. Front Microbiol 9:1129. https://doi.org/10.3389/fmicb.2018.01129

Xu C, Qiao L, Guo Y, Ma L, Cheng Y (2018b) Preparation, characteristics and antioxidant activity of polysaccharides and proteins-capped selenium nanoparticles synthesized by *Lactobacillus casei* ATCC 393. Carbohydr Polym 195:576–585. https://doi.org/10.1016/j.carbpol.2018.04.110

Yao M, Li B, Ye H, Huang W, Luo Q, Xiao H, McClements DJ, Li L (2018) Enhanced viability of probiotics (*Pediococcus pentosaceus* Li05) by encapsulation in microgels doped with inorganic nanoparticles. Food Hydrocol 83:246–252. https://doi.org/10.1016/j.foodhyd.2018.05.024

Yazdi MKS, Davoodabadi A, Zarin HRK, Ebrahimi MT, Dallal MMS (2017) Characterisation and probiotic potential of lactic acid bacteria isolated from Iranian traditional Yogurts. Italian J Animal Sci 16:185–188. https://doi.org/10.1080/1828051X.2016.1222888

Yin Y, Yang X, Hu L, Tan Z, Zhao L, Zhang Z, Liu J, Jiang G (2016) Superoxide-mediated extracellular biosynthesis of silver nanoparticles by the fungus *Fusarium oxysporum*. Environ Sci Technol Lett 3:160–165. https://doi.org/10.1021/acs.estlett.6b00066

Yusof HM, Mohamad R, Zaidan UH (2020) Sustainable microbial cell nanofactory for zinc oxide nanoparticles production by zinc-tolerant probiotic *Lactobacillus plantarum* strain TA4. Microb Cell Factor 19:10. https://doi.org/10.1186/s12934-020-1279-6

Zamani B, Golkar HR, Farshbaf S, Emadi-Baygi M, Tajabadi-Ebrahimi M, Jafari P, Akhavan R, Taghizadeh M, Memarzadeh MR, Asemi Z (2016) Clinical and metabolic response to probiotic supplementation in patients with rheumatoid arthritis: a randomized, double-blind, placebo-controlled trial. Int J Rheum Dis 19:869–879. https://doi.org/10.1111/1756-185X.12888

Zamani B, Farshbaf S, Golkar HR, Bahmani F, Asemi Z (2017) Synbiotic supplementation and the effects on clinical and metabolic responses in patients with rheumatoid arthritis: a randomised, double-blind, placebo-controlled trial. Brit J Nutr 117:1095–1102. https://doi.org/10.1017/S000711451700085X

Zhang X, Yan S, Tyagi RD, Surampalli RY (2011) Synthesis of nanoparticles by microorganisms and their application in enhancing microbiological reaction rates. Chemosphere 82:489–494. https://doi.org/10.1016/j.chemosphere.2010.10.023

Zheng B, Qian L, Yuan H, Xiao D, Yang X, Paau MC, Choi MM (2010) Preparation of gold nanoparticles on eggshell membrane and their biosensing application. Talanta 82:177–183. https://doi.org/10.1016/j.talanta.2010.04.014

Applications of Microbe-Based Nanoparticles in Agriculture: Present State and Future Challenges

N. B. Raj, M. K. Swamy, B. Purushotham, and S. K. Sukrutha

1 Introduction

Of late, the merging of nanoparticle knowledge and natural resources has led to the emergence of a new meadow known as nanobiotechnology. It is engrossed in the synthesis, employment and use of nanoparticles for various biological applications (Goodsell 2004). The term "Nano" is derived from Latin language which means dwarf. A nanoparticle consists of all three dimensions within 1–100 nm. In 1857, Michael Faraday published the first scientific report on properties of nanoparticles (NP) entitled "Experimental relations of gold (and other metals) to light". However, in 1981, the concrete progress in nanotechnology was started by Eric Drexler (Drexler 1981; Prathna et al. 2010). Nanoparticles have drawn attention globally, owing to its unique physicochemical, magnetic and optoelectronic properties (Daniel and Astruc 2004; Zharov et al. 2005; Bogunia-Kubik et al. 2002). The production of tailor-made nanoparticles has expanded the dimensions of their applications in myriad of industries such as agriculture, textile, electronics, optics, medical and food (Mandal et al. 2006). Various methods such as physical, chemical and biological are used for NP synthesis. Nanoparticles are synthesized by two methods, namely bottom-up and top-down method. In a bottom-up method, atoms and molecules are combined together to form the final product. This approach results in the synthesis of distinct nanoparticles of different size and shape (Mittal et al. 2013), and the second one is top-down method, wherein a substance is reduced by physical or chemical techniques (Gade et al. 2010). Evidences suggest that, NPs synthesized by bottom-up approach

N. B. Raj · M. K. Swamy (✉) · B. Purushotham
Department of Biotechnology, East West First Grade College, Bengaluru, Karnataka 560091, India
e-mail: swamy.bio@gmail.com

S. K. Sukrutha (✉)
Department of Microbiology, East West First Grade College, Bengaluru, Karnataka 560091, India
e-mail: sukrutha357@gmail.com

A. Lateef et al. (eds.), *Microbial Nanobiotechnology*, Materials Horizons: From Nature to Nanomaterials, https://doi.org/10.1007/978-981-33-4777-9_12

has thermodynamically controlled chemical compositions than top-down method (Thakkar et al. 2010).

Physical and chemical methods result in controlled synthesis of NPs and can yield large amount of NPs in shorter duration (Alkilany et al. 2015; He et al. 2015). Regardless of the pervasive use of physical and chemical methods for NP synthesis, growing body of evidences suggests that, there have been emerging issues such as hazardous procedures, non-biodegradable, cumbersome, capital intensive, energy intensive, toxic by-products, eutrophication, high monodispersity and requirement for toxic chemicals, solvents, reducing agents, stabilizers, high temperature, metal ions, reducing agents, pH and pressure for NP synthesis (Mallick et al. 2004). Hence, the researchers round the globe are in quest for the ancillary route for efficient and nature-friendly NP synthesis owing to their advantages as stated in Table 1.

2 Biological Production of Nanoparticles

Nanoparticle synthesis by biological route is a green chemistry approach that has led to an exciting new and rational interdisciplinary branch of science with considerable prospective due to their enriched diversity and cosmopolitan nature. Harnessing the microorganisms for green NP synthesis is gaining impetus due to ease of synthesis, eco-friendly in nature, unwavering, non-toxic, efficient, economical route and use of ambient conditions for NP synthesis (Sastry et al. 2003). Different types of nanoparticles can be synthesized using biological sources such as plants, arthropods and various microorganisms such as algae, fungi, yeast, bacteria, actinomycetes and viruses (Lateef et al. 2016a, b; Thakkar et al. 2010). Various parameters, such as temperature, pH, time duration, metal ion concentration, culture medium, biomass and inoculum age, influence the size and shape of the nanoparticle (Guilger-Casagrande and Lima 2009). Wide array of amino acids, peptides, enzymes and organic acids synthesized by plants and microorganisms stabilizes the biological formulation of various microbial-mediated nanostructures (Adelere and Lateef 2016; Lateef et al. 2015a). Furthermore, NPs can be synthesized using various substrates, and many factors are known to have effects on the biosynthesis of NPs (Camargo et al. 2009; Vogtle et al. 2009; Swamy et al. 2015, 2018; Rudramurthy et al. 2016; Rudramurthy and Swamy 2018) (Fig. 1).

2.1 Mechanisms of Synthesis of Nanoparticles from Microorganisms

The actual mechanism of NP synthesis still remains skeptical. It is speculated that, metal ions are trapped on the surface or inside the cell of an organism and are

Table 1 Advantages and disadvantages of the use of microorganisms in the synthesis of nanoparticles

Microorganism	Advantages	Disadvantages	References
Bacteria	• Easy to handle • Genetic manipulation is well studied • Rapid growth	• Requires robust equipment • Capping agents are not well recognized • Expensive • Low yield of nanoparticle synthesis	Chen et al. (2009), Kalishwaralal et al. (2010)
Fungi	• High tolerance to metals • Mycelia/biomass and culture filtrate offers large surface area for metal interaction • Downstream methods are much simpler • Secretes wide array of proteins, which is used as capping agents for stabilizing nanomaterials • Production of extracellular enzymes results in efficient bioreduction of metals • Cell wall binding capacity is high • Ability to uptake metal is high • High yield of nanoparticle synthesis • Economic viability	• Standardization methods and maintenance of growth conditions • Capping proteins results in immune response	Mandal et al. (2006), Honary et al. (2013), Yadav et al. (2015), Silva et al. (2016)
Actinomycetes	• Produces higher amounts of proteins	• Less explored	Sastry et al. (2003), Thakkar et al. (2010)
Algae	• Highly efficient in synthesizing silica NPs	• Less explored	Sharma et al. (2016)
Viruses	• Rapid multiplication	• Relatively unexplored for nanoparticle synthesis • Requires a host for virus • Expression • Very expensive process	Koudelka et al. (2015)

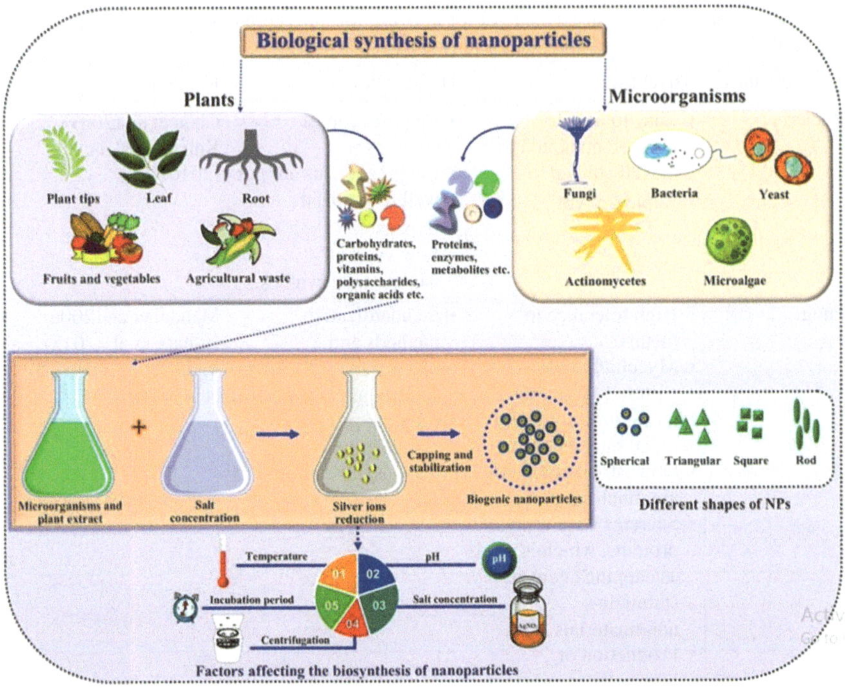

Fig. 1 Synthesis of nanoparticles using different substrates and effect of different factors (adapted from Ali et al. 2020)

reduced by the action of enzymes present on cell wall, in the cytoplasm or in cell filtrate, leading to NPs synthesis (Mukherjee et al. 2002).

2.1.1 Mechanism of Synthesis of AgNPs Using Fungi

In general, when fungi experience various exogenous stress such as exposure to toxic compounds, temperature variations, pressure, nutrient depletion among others, they produce a wide array of biological compounds and enzymes as defense mechanism. After a fungus is exposed to toxic metal ions, it is reduced to non-toxic metallic solid NPs by the action of various extracellular enzymes and metabolites (Vahabi et al. 2011; Durán et al. 2011a, b; Ingle et al. 2008; Mukherjee et al. 2008). First report on synthesis of microbial-based AgNPs was documented in *Fusarium oxysporum* (Ahmad et al. 2003a, b). Growing body of evidences suggests that, mycosynthesis of AgNPs has not yet been completely understood. Various parameters such as nature of fungus, temperature, pH, capping agents and nature of metal ions determine the size and shape of the nanoparticles (Khandel and Shahi 2018; Lee and Jun 2019). The probable steps involved in the mycosynthesis of AgNPs are enumerated as follows:

1. Mechanism of NPs synthesis is linked to cell wall polymers, sugars (responsible for attachment and reduction of metal ions).
2. In the bioreduction step, cell wall surface traps metal ions due to the electrostatic interaction among the metal ions and carboxylate group of the enzymes in the cell wall.
3. This step is followed with metal ions reduction by the enzymes such as NADPH-dependent reductases or nitrate-dependent nitrate reductase, oxidoreductase to form nanoparticles. In addition, quinine derivatives of naphthoquinones and anthraquinones also play roles in the reduction process.

The synthesis of NPs in fungi has been shown to be intracellular, extracellular or take place on the cell surface (Kashyap et al. 2013; Alghuthaymi et al. 2015; Moghaddam et al. 2015; Yadav et al. 2015). Extracellular method is rapid, economical and easy to produce, because during this process, the cell synthesizes enormous amount of enzymes and/or compounds, compared to intracellular synthesis (Nayak et al. 2011; Kashyap et al. 2013). Extracellular AgNPs synthesis in *F. oxysporum* was achieved by reduction of NADPH to $NADP^+$ (Kumar et al. 2007b), enzyme nitrate-dependent reductase and shuttle quinone (Devi and Joshi 2015), in *Aspergillus terreus* using NADH-dependent reductase enzyme (Li et al. 2012). Chan and Mashitah (2012) synthesized AgNPs using different macrofungi and reported a possible role of a diketone compound for silver ion reduction. Sastry et al. (2003) proposed the intracellular AgNPs synthesis in *Verticillium* sp. and involve the following steps:

1. Absorption: Binding of Ag^+ on the surface of the fungal cell wall. This could be possibly due to electrostatic interaction between the Ag^+ and negatively charged carboxylate groups on the enzymes on the cell wall.
2. Reduction of Ag^+: Cell wall enzymes reduce Ag^+ and form a silver nuclei.
3. Synthesis of AgNPs: It involves the growth of the silver nuclei to Ag^0, thus forming AgNPs.
4. They are finally secreted out through the membrane, by a process known as exocytosis (Korbekandia et al. 2013).

The synthesis of AuNPs using fungi comprises two steps. The synthesis of intracellular AuNPs in *Rhizopus oryzae* was reported by Das et al. (2012). *Rhizopus oryzae* mycelia was treated with $HAuCl_4$ solution and incubated for 48 h. This resulted in gradual color change from pale white to purple. The authors proposed the following mechanisms: (1) **Adsorption**: In this step, Au^{3+} binds to the cell wall through electrostatic interaction. Gold ions bind to the fungal mycelia by electrostatic interaction and get reduced to AuNPs through the action of enzymes on the cell wall surface, (2) **Transportation**: In this step, Au^{3+} is transported to the cytoplasm and is reduced by cytoplasmic proteins such as oxidoreductases in a three step process to form Au nanoparticles. Chemical reactions involved in the synthesis of gold nanoparticles are

1. $AuCl_4^- \rightarrow H^+ + AuCl_4^-$
2. $AuCl_4^- \rightarrow AuCl_3 + Cl^-$
3. $AuCl_3 \rightarrow Au^{3+} + 3Cl^-$

The first report of the synthesis of AuNPs was documented by Mukherjee et al. (2002), whereby oxidoreductase enzyme mediated extracellular synthesis of AuNPs in *F. oxysporum*. Nevertheless, the same process mechanism was not observed in *F. moniliforme*. In *Rhodopseudomonas capsulata*, NADH-dependent reductases are involved in AuNPs synthesis. In *Phanerochaete chrysosporium*, extracellular and intracellular AuNPs were synthesized by lignolytic enzymes and proteins (Sanghi et al. 2011).

2.1.2 Factors Affecting the Synthesis of Microbial Nanoparticles

A plethora of known parameters are reported to affect the microbial-mediated synthesis of nanoparticles (Fig. 2). Factors such as pH, temperature, agitation rate, incubation time, enzymes, metal ion concentration, inoculum age and biomass concentration influence the synthesis, monodispersity and application in microbial-based nanoparticles. Some of the important factors that affect nanoparticle biosynthesis are hereby described.

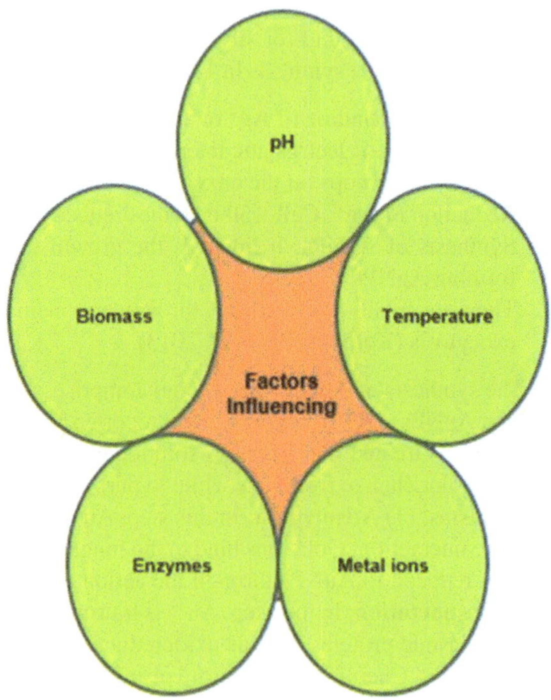

Fig. 2 Factors influencing microbial nanoparticle synthesis

Effect of PH

Variation in pH can regulate the size and shape of the nanoparticle as shown by Gericke and Pinches (2006) in *Verticillium luteoalbum*. At pH 3, spherical-shaped (<10.0 nm) NPs were synthesized. However, the size of the NPs increased with well-defined shapes such as hexagons, triangles, rods and spheres up to pH 5.0. At pH 7 and 9, small spherical as well as bigger particles with irregular and undefined shapes were obtained (Gericke and Pinches 2006; Karbasian et al. 2008). Das et al. (2012) reported that, pH influenced AuNPs synthesis in *Rhizopus oryzae*. At pH 7, the size of the gold nanoparticles was 26 nm, whereas 250 nm were obtained at pH 2.0. Further, they observed that maximum stability was obtained at neutral, basic and slightly acidic conditions, respectively. Similar results were reported in *Penicillium oxalicum* (Du et al. 2015). Concordant results are reported for AgNPs synthesis in several other fungi such as *Epicoccum nigrum*, an endophytic fungus (Qian et al. 2013), *Colleotrichum* sp. ALF2 (Azmath et al. 2016), *Fusarium oxysporum* (Birla et al. 2013), *Aspergillus oryzae* MTCC 1846 (Phanjom and Ahmed 2017), *Aspergillus fumigatus* BTCB10 (Shahzad et al. 2019), *Isaria fumosorosea* (Banu and Balasubramanian 2014), *Arthroderma fulvum* (Xue et al. 2016), and *Sclerotinia sclerotiorum* MTCC 8785 (Saxena et al. 2016).

In another study, modulation of initial pH (5, 8 and 10) in the fungal extract of *Aspergillus terreus* IF0 resulted in elongated-, triangular- and rod-shaped AuNPs. Size range of 10–19 nm AuNPs was obtained at pH 10 (Priyadarshini et al. 2014). Growing body of evidences has shown that microbial-based nanoparticles are also synthesized at neutral or slightly alkaline pH. Balakumaran et al. (2015) obtained well-dispersed and extremely stable 5–30 nm sized, spherical-shaped AgNPs at pH 7 using the endophytic fungus, *Guignardia mangiferae*. In another study, maximum production of AgNPs was achieved at pH 7.2 with a concentration of 136 ppm using *Penicillium oxalicum* GRS-1. The AgNPs obtained was spherical in shape with size ranging from 10 to 40 nm (Rose et al. 2019). Similarly, *Penicillium purpurogenum* synthesized spherical-shaped 8–10 nm size AgNPs at pH 8 (Nayak et al. 2011). Optimum pH for extracellular synthesis of AgNPs synthesized from a marine fungus, *Penicillium fellutanum*, was found to be pH 6 (Kathiresan et al. 2009).

Tidke et al. (2014) studied the influence of different pH on extracellular synthesis of AuNPs in the fungus *Fusarium acuminatum*. Results indicated that, pH from 5 to 7 favored the AuNPs synthesis with a particle size of 8–28 nm with average size of 17 nm. In another study, pH ranging from 7 to 11 was found to be optimum for AuNPs synthesis in *Trichoderma* sp. WL-Go (Qu et al. 2016). Optimum AuNPs synthesis was observed by various workers at neutral and alkaline pH (Mishra et al. 2011a). Synthesis of AuNPs using *Shewanella algae* under anaerobic condition was studied by Konishi et al. (2007). Results suggest that, 10–20 nm of AuNPs was obtained in periplasmic space at pH 7.0, while size of 15–200 nm on the surface at pH 2.8. Mishra et al. (2011b) reported that at pH 2, a highly acidic condition did not favor the synthesis of AuNPs in *Penicillium brevicompactum*. However, maximum yield and a size of 100 nm were observed at pH 6 and 8. Beyond pH 8, slight increase in the size (200 nm) and decrease in AuNPs synthesis was observed. Also, purified lignin peroxidase enzyme from *Acinetobacter* sp. SW 30 was used to synthesize spherical crystalline AuNPs (10 ± 2 nm particle size) and amorphous SeNPs (100 ± 10 nm particle size). The enzyme had maximum activity at pH 2 and temperature of 40 °C (Wadhwani et al. 2017). In addition, Quester et al. (2013) biosynthesized AuNPs using *Neurospora crassa* at 60 °C under various pH conditions. Results indicated that AuNPs synthesized displayed different shapes like triangles, hexagons, pentagons and sizes 10–200, 6–23, 3–12 nm at pH values of 3, 5.5 and 10, respectively. *Shewanella* algae synthesized 10–20 nm size AuNPs at pH 7. However, large size nanoparticles with variable size were observed at lower pH (He et al. 2007; Konishi et al. 2007). Similar results were reported in *Enterobacter* (Sinha and Khare 2012). Conversely, contrasting results were reported in *Escherichia coli* and *Desulfovibrio desulfuricans* (Deplanche and Macaskie 2008). These results attest to the influence of pH on the morphology and sizes of nanoparticles during microbial synthesis, thus necessitating a careful choice of pH for the biofabrication of nanoparticles.

Effect of Temperature

Temperature plays a pivotal role in controlling the activity of the fungus (Dhillon et al. 2012). Ahluwalia et al. (2014) synthesized AgNPs with a size range of 51.10 nm from the culture filtrate of *Trichoderma harzianum* within 4 h at 40 °C. However, there was no increase in rate of reaction from 10 to 30 °C. In *Fusarium oxysporum*, ideal temperature for AgNPs synthesis was 40–60 °C, and a particle size of 21 and 40 nm was produced (Birla et al. 2013). Phanjom and Ahmed (2017) studied the influence of different temperature (10, 30, 50, 70 and 90 °C) on AgNPs synthesis in *Aspergillus oryzae* MTCC 1846. Results indicated that increase in temperature affected the particle size and reduced the rate of the AgNPs synthesis. Spherical-shaped AgNPs with particle size of 4–9 nm diameter were obtained at 90 °C within 20 min, whereas the size and rate of the reaction increased from 5 to 24 nm and up to 6 h with reduction in temperature. According to Xue et al. (2016) in a report on *Arthroderma fulvum*, temperature of 55 °C resulted in the synthesis of spherical AgNPs with a size of 20.56 nm in diameter at the substrate concentration of 1.5 mM, alkaline pH and reaction time of 10 h. Saxena et al. (2016) also observed temperature of 80 °C resulted in enhanced spherical-shaped AgNPs synthesis with a range of 10–15 nm in size in *Sclerotinia sclerotiorum* MTCC 8785. In *Rhizopus stolonifer*, AgNPs with size of 2.86, 25.89 and 48.43 nm were synthesized at various temperatures of 40, 20 and 60 °C (Abdel Rahim et al. 2017). According to Shahzad et al. (2019), *Aspergillus fumigatus* BTCB10 synthesized AgNPs of 322.8 nm at 25 °C and the size increased to 1073.45 nm at 55 °C.

Spheroidal AuNPs with a diameter ranging between 5 and 10 nm at 75 and 100 °C were synthesized using the culture filtrate of phytopathogenic fungus *Botrytis cinerea* (Castro et al. 2013). Spherical- and polycrystalline-shaped, 15–40 nm-sized AuNPs were synthesized in *Gordonia amarae* using 2 mM chloroauric acid ($HAuCl_4$) at pH 10.0 and 90 °C for 20 min (Bennur et al. 2016). In another study, a spherical-, triangular- and hexagonal-shaped, 10–60 nm size extracellular AuNPs were synthesized at 30 °C with different reaction rate ranging from 12 to 72 h using the fungus *Penicillium brevicompactum* (Mishra et al. 2011b). In yet another study by Mishra et al. (2012) on *Penicillium rugulosum,* spherical-, triangular-, hexagonal-shaped, 20–80 nm size AuNPs were synthesized at 30 °C with different reaction rate ranging from 8 to 24 h. In *Epicoccum nigrum,* AuNPs were synthesized both by intra and extracellular mechanisms between the temperature ranging from 27 to 29 °C for 72 h to obtain 5–50 nm size AuNPs (Sheikhloo et al. 2011). Das et al. (2012) synthesized spherical-shaped, 16–25 nm-sized AuNPs from *Rhizopus oryzae* at 30 °C for 24 h. Intracellular synthesis of spherical-shaped, 15–20 nm size AuNPs at 30 °C was reported in *Saccharomyces cerevisiae* (Sen et al. 2011).

Spherical-shaped AuNPs with ~2–50 nm at 30 for 24 h were synthesized using *Shewanella oneidensis* (Suresh et al. 2011a, b). Kumar et al. (2007a, b) reported the synthesis of CdSeNP using *F. oxysporum* and synthesized spherical-shaped 9–15 nm size at room temperature. Narayanan and Sakthivel (2011) synthesized anisotrophic nanostructures such as triangle, hexagonals, decahedral and isotrophic spherical-shaped AuNPs with 25 nm in size at 80 °C for 1 h using different concentration of cell-free extract of *Sclerotium rolfsii.* Sharma et al. (2012) synthesized monodispersed AuNPs with an average of 10 nm size from marine bacterial strain of *Marinobacter pelagius* at room temperature for 24 h. In another study, agglomerated AuNPs with 100 nm size were synthesized by *T. harzianum* at 28 °C for 72 h (Tripathi et al. 2014). Extracellular secretion of *Bacillus niabensis* 45 was used to synthesize face-centered-cubic-shaped AuNPs with a size of 10–20 nm at pH 6 and temperature of 60 °C (Li et al. 2016). Pourali et al. (2017) investigated that *F. oxysporum* and *Bacillus cereus* culture supernatants produced AuNPs which were spherical, hexagonal, and octagonal shape with irregular contours and sizes of around 20–50 nm at 30 °C for 72 h and 37 °C for 24 h, respectively.

Using *Streptomyces* sp. VITDDK3, Gopal et al. (2013) synthesized hexagonal-, cubical-, brick- and irregular-shaped AuNPs of 90 nm size at pH 9 under room temperature in 48 h. Synthesis of crystalline- and spherical-shaped AuNPs with an average size in the range of 14 nm at room temperature within 96 h was synthesized using cell-free extract of *Bacillus marisflavi* (Nadaf et al. Nadaf and Kanase 2019). In another study, *Chlorella pyrenoidusa* extract was used for synthesis of spherical-shaped AuNPs having 25–30 nm in size at pH 8 and 100 °C (Oza et al. 2012). *Spirulina platensis* was used to synthesize spherical-shaped AuNPs, having an average size of ~5 nm at room temperature for 48 h (Suganya et al. 2015). Formation of spherical-, hexagonal- and triangle-shaped AuNPs with size ranged from 18.7 to 93.7 nm was synthesized at room temperature within ten minutes using brown alga, *Stoechospermum marginatum* biomass (Rajathi et al. 2012). Dhas et al. (2012) synthesized well-defined, polydispersed (triangular and spherical) and crystalline AuNPs from *Sargassum myriocystum* with an average size of 15 nm within 15 min at 76 °C. The various investigations have shown the impact of operating temperature on the biofabrication of nanoparticles by microorganisms.

Effect of Metal Concentration

In a study conducted by Prasad and Jha (2010), two bacterial species namely *Lactobacillus* sp. and *Saccharomyces cerevisiae* were used to determine the effect of $CdCl_2$ (0.25 M) on CdSNPs synthesis at 60 °C for 10–20 min. Results indicated that, the particle size was found to be 4.93 ± 0.23 nm, 3.57 ± 0.21 nm in *Lactobacillus* and yeast, respectively. Further, the authors explained that, the particle size in yeast was less because it has a good detoxification machinery mechanisms and organization at the cellular level than *Lactobacillus* sp., a prokaryote. In another study, spherical-shaped iron nanoparticles with particle size ranging between 10 and 24.6 nm were synthesized from *Aspergillus oryzae* TFR9 using salt solution of $FeCl_3$ (0.001 M) at 28 °C for 12 h (Tarafdar and Raliya 2013). *Aspergillus tubingensis* TFR-3 utilized 10^{-3} M of $Mg(NO_3)_2{\cdot}6H_2O$ salt solution at 28° C for 48 h to synthesize magnesium oxide nanoparticle with 5.8 nm (Raliya et al. 2014a, b). Sheikhloo (2011b) employed 1 mM aqueous $HAuCl_4$ solution with pH 2.5 at 27–29 °C for 72 h to synthesize intracellular AuNPs having spherical, triangle and rod shapes with the size range of 5–100 nm using the fungus *Penicillium chrysogenum.* A rapid approach for the biosynthesis of AuNPs by *Trichoderma viride* MTCC 5661 and *Hypocrea lixii* MTCC 5659 using 250 mg/L Chloroauric 5 acid was reported by Mishra et al. (2014b). Gold nanoparticles with 20–30 and 120 nm in size were synthesized within 10 min at 30 °C in *T. viride*, which was similar to *Hypocrea lixii* with 20 nm size AuNPs but at 100 °C, respectively.

Abdel Rahim et al. (2017) synthesized 2.86 nm-sized AgNPs at 10 mM $AgNO_3$, while bigger sizes of 54.67 and 14.23 nm were synthesized at 100 and 1 mM, respectively, from *Rhizopus stolonifer*. Salunkhe et al. (2011) synthesized spherical-shaped AgNPs using *Cochliobolus lunatus* with size ranging from 5 to 100 nm at pH 7 and 28 °C using 635 ppm silver in. In another study, Fayaz et al. (2010) synthesized spherical-shaped, 2–4 nm size AgNPs in *T. viride* using 1 mM Ag^+ in dark condition at 40 °C. Aziz et al. (2016) reported spherical-shaped 5–15 nm size AgNPs using 1 mM silver nitrate in *Mucor hiemalis.* Barabadi et al. (2017) synthesized spherical-shaped AgNPs in *Penicillium aculeatum* with the average particle size of 60 nm at 28 °C for 24 h using 3 mM $HAuCl_4$. Also, *Candida albicans* mediated spherical-shaped AgNPs with 20–40 and 60–80 nm size, and non-spherical AuNPs were obtained using 1 mM aqueous $HAuCl_4$ in 24 h (Chauhan et al. 2011). Various concentrations of silver nitrate (1–10 mM), pH (1–10) and temperature (30–80 °C) conditions were investigated for AgNPs synthesis in *Guignardia mangiferae.* Among the conditions, 1 mM silver nitrate was found to be optimum and produced spherical-shaped AgNPs with 5–30 nm size at pH 7 and 30 °C within 12 h (Balakumaran et al. 2015). Concordant results were obtained with 1 mM concentration of silver nitrate for AgNPs synthesis in *E. nigrum* and *Penicillium* sp (Qian et al. 2013; Singh et al. 2014).

In another study, plant pathogenic fungus *Helminthosporum solani* was treated with 1 mM aqueous solution of chloroaurate ions at 37 ± 1 °C for 72 h to produce a diverse mixture of intracellular AuNPs with the size ranging from 2 to 70 nm. A plurality of polydisperse spheres (44%), but a significant number are homogeneously sized rods (12%), triangle (30%), hexagonal (7%), pyramids (5%) and stars (2%) were observed. Further, polydispersed spherical-shaped particles had size range of 2–70 nm, and the rods averaged 12–15 nm, the hexagons with 30–42 nm, the triangles with 25–40 and 5–20 nm in star-shaped were obtained (Kumar et al. 2008). Algal extract of *Turbinaria conoides* was used to synthesize small spherical-, triangle- and pseudo-spherical-shaped particles AuNPs having 6–10 nm size using 1 mM aqueous solution of gold chloride at room temperature for 24 h (Rajeshkumar et al. 2013). Similarly, in *Calothrix* algae, 0.1 mm $AuCl_4$ solution was used to synthesize truncated-, triangular- and spherical-shaped AuNPs in the size range of 30–120 nm at 22–25 °C for 3 days (Kumar et al. 2016). Thus, concentration of precursors can influence the formation of nanoparticles with defined effects on size and morphology.

Effect of Enzymes

Extracellular enzymes produced by microorganisms are well documented and play an important role as reducing agents in the production of nanoparticles. Increasing body of evidences suggests that, the cofactors such as nicotinamide adenine dinucleotide (NADH) and reduced form of Nicotinamide adenine dinucleotide phosphate (NADPH)-dependent enzymes act as reducing agents and transfer the electrons from NADH by NADH-reliant enzymes, which act electron carriers (Zomorodian et al. 2016; Baymiller et al. 2017). Extracellular synthesis of AuNPs by *R. capsulata* was achieved by the reduction of gold by NADH-reliant reductase enzymes (He et al. 2007). Similar enzyme machinery was used to synthesize AuNPs in *Stenotrophomonas maltophilia* (Nangia et al. 2009). Sanghi et al. (2011) reported that *P. chrysosporium* synthesized extracellular AuNPs using laccase enzyme and intracellularly by ligninase within 90 min at 37 °C. Ahmad et al. (2002) reported that *F. oxysporum* synthesized CdS nanoparticles via extracellular secretion of sulfate reductases enzyme (Table 2).

Likewise, Gholami-Shabani et al. (2015) and Gholami-Shabani et al. (2016) synthesized AuNPs using sulfite oxidoreductase enzyme in *Escherichia coli* and *Fusarium oxysporum*. In yet another study by Gholami-Shabani et al. (2014), AgNPs were synthesized from *Fusarium oxysporum* using the enzyme nitrate reductase. Similar studies on AgNPs synthesis using the enzyme nitrate reductase and quinones were documented in *Sclerotium rolfsii* (Narayanan et al. 2011), *Fusarium oxysporum* (Kumar et al. 2007a, b; Hamedia et al. 2017), *Bacillus clausii* (Mukherjee et al. 2018), *B. subtilis* (Saifuddin et al. 2009), *E. coli* (Gurunathan et al. 2009), *Pseudomonas aeruginosa* (Kumar and Mamidyala Kumar and Mamidyala 2011a, b) and in *Bacillus cereus* strain HMH1 for magnetic iron oxide nanoparticle synthesis (Fatemi et al. 2018). Also, lignin peroxidase enzyme was used in the synthesis of spherical crystalline AuNPs (10 ± 2 nm) and amorphous-shaped SeNPs (100 ± 10 nm) in *Acinetobacter* sp. SW30 (Wadhwani et al. 2017). These studies established roles being played by different enzymes in the biosynthesis of nanoparticles; thus, their modulation would influence the generation of nanoparticles.

Table 2 Some metal nanoparticles synthesized using various enzymes in fungi and bacteria

Microorganisms	Type of NPs	Size (nm)	Morphology	Enzyme	References
Fusarium oxysporum	Silver	20–40	Spherical	NADH-dependent	Mukherjee et al. (2002)
Lentinus edodes	Gold	5–50	Spherical	Laccase Tyrosinase Peroxidase	Vetchinkina et al. (2014)
Pseudomonas aeruginosa	Silver	13	Spherical	Nitrate reductase	Kumar and Mamidyala (2011a, b)
Shewanella oneidensis	Silver	24	Spherical	c-Type cytochromes	Ng et al. (2013)
Stenotrophomonas maltophilia	Silver	93	Cubic	Chromium reductase	Oves et al. (2013)
Escherichia coli	Gold	10	Spherical	Sulfite reductase	Gholami-Shabani et al. (2015)
Lentinus edodes	Silver	50–100	Walnut	Laccase	Lateef et al. (2015a)
Bacillus safensis LAU 13	Silver	5–30	Spherical	Keratinase	Lateef et al. (2015b)
Aspergillus niger L3 (NE) and *Trichoderma longibrachiatum* L2 (TE)	Silver-Gold	6.98–52.51	Spherical, oval, irregular	Xylanases	Elegbede et al. (2019)
A. niger L3 (NEA) and *T. longibrachiatum* L2 (TEA)	Silver-Gold	4.88–123.99	Spherical, flower-shaped	Xylanases	Elegbede et al. (2020)

Effect of Biomass

Amount of biomass influences the production and properties of microbial-mediated nanoparticles. Gold nanoparticles with spherical-shaped, size range of 85.1–210 nm were synthesized from 5 g of biomass in *A. fumigatus* (Bathrinarayanan et al. 2013). Live biomass (5 g) of *Spirulina platensis* produced 11.6 nm size AgNPs (Mahdieh et al. 2012), while intracellular 5–35 nm-sized AuNPs were obtained from the alga *Tetraselmis kochinensis* using 10 g of biomass (Senapati et al. 2012). Zhang et al. (2014) reported the formation of spherical-shaped AgNPs with size of ~6 nm using 10 g of dry biomass of *Lactobacillus fermentum*. Seshadri et al. (2011) synthesized spherical-shaped PbSNPs from active biomass of *Rhodosporidium diobovatum*, with size range of 2–5 nm. Stable, monodisperse 10 nm size AuNPs were produced by *Marinobacter pelagius* (Sharma et al. 2012).

Overall, these evidences suggest that various synthesis parameters can lead to the synthesis of nanoparticles having different properties. However, the influence of various factors still remains understudied, thus warranting further research studies for tailor-made microbial-mediated nanoparticle synthesis, particularly through robust optimization techniques such as Taguchi approach, response surface methodology and artificial intelligence techniques.

2.2 *Nanoparticle Synthesis Using Bacteria*

In the midst of diverse population of microorganisms, bacteria have a notable competence to reduce heavy metal ions and therefore are considered as the promising candidates for microbial-based nanoparticle synthesis. Advantages of bacteria-mediated nanoparticle research include their availability, high growth rates, easy manipulation and their ability to adapt to wide ranges of environmental and culture conditions. Various researchers round the globe have harnessed various growth parameters to produce tailor-made nanoparticles, which could be exploited for various applications (Table 3).

Kumar et al. (2014) synthesized 11.3 nm spherical-shaped AuNPs from *Delftia* sp. strain KCM-006. Honary et al. (2012) studied the synthesis of spherical-shaped, 11.8–130 nm size AuNPs using *E. coli*. Similarly, Du et al. (2007) synthesized 20–30 nm dimension, hexagonal-, triangle-shaped AuNPs using *E. coli*. Exopolysaccharide from *Bacillus megaterium* MSBN04 was used in the synthesis of 5–20 nm size and spherical-shaped AuNPs by Sathiyanarayanan et al. (2014). Baker et al. (2016) synthesized AuNPs using *R capsulata* and obtained 10–20 nm size, spherical-shaped nanoparticles. Similarly, He et al. (2007) reported 10–20 nm dimensions, irregular-shaped AuNPs by *R capsulata*. Alkali-tolerant *Rhodococcus* was used to synthesize 5–15 nm in size AuNPs (Ahmad et al. 2003a). Otari et al. (2017) synthesized spherical-shaped AuNPs with 25 nm size in *Lactococcus lactis*. In another study, spherical- and nanotriangle-shaped, 2–6 nm AuNPs were produced from *Marinobactor pelagius* (Sharma et al. 2012).

Table 3 Synthesis of some nanoparticles using bacteria

Bacterial species	Metal	Size (nm)	Shape	References
Serratia nematodiphila	Ag	10–31	Crystalline and spherical	Malarkodi et al. (2013)
Nocardiopsis sp. MBRC-1	Ag	45 ± 0.15	Spherical	Manivasagan et al. 2013
Novosphingobium sp. HG-C3	Ag	8–25	Spherical and crystalline	Du et al. (2016)
Bacillus safensis LAU 13	Ag–Au	13–80	Spherical	Ojo et al. (2016)
Bacillus safensis LAU 13	Ag	5–95	Spherical	Lateef et al. (2015c)
Kinneretia THG-SQ14	Ag	15–20	Monodisperse, FCC, spherical	Singh et al. (2016a)
Lactobacillus casei subsp. *Casei*	Ag	25–50	Spherical	Korbekandi et al. (2012)
Acinetobacter calcoaceticus	Ag	8–12	Spherical	Singh et al. (2013a, b)
Acinetobacter calcoaceticus	Ag	4–40	Spherical	Gaidhani et al. (2013)
Klebsiella pneumoniae	Ag	15–37	Spherical	Duraisamy and Yang (2013)
Pseudomonas aeruginosa	Ag	8–24	Spherical	Kumar and Mamidyala (2011a, b)
Stenotrophomonas maltophilia	Ag	93	Cuboidal	Oves et al. (2013)
Exiguobacterium sp.	Ag	5–50	Spherical	Tamboli and Lee (2013)
Bacillus licheniformis	Ag	50	Spherical	Sriram et al. (2012)
Escherichia coli VM1	Ag	10–15	Cuboid	Maharani et al. (2016)
Ochrobactrum sp.	Ag	38–85	Spherical	Thomas et al. (2014)
Escherichia coli	Au	11.8–130	Spherical	Honary et al. (2012)
Escherichia coli	Au	20–30	Hexagonal, Triangle	Du et al. (2007)
Escherichia coli	Au	5–20	Spherical	Gopal et al. (2013)
*Pseudomonas fluorescens*417	Au	5–50	Spherical	Husseiny et al. (2007)
Thermomonospora sp.	Au	2–6	Spherical	Khan and Ahmad (2014)
Bacillus safensis LAU 13	Au	10–45	Spherical	Ojo et al. (2016)
Shewanella oneidensis	Au	12	Spherical	Suresh et al. (2011a, b)

The first report on bacterial AgNPs synthesis was studied in *Pseudomonas stutzeri* AG259. Result indicated the presence of 200 nm size with equilateral- and triangular-shaped AgNPs, accumulated in the periplasm (Klaus et al. 1999). In another study, *Streptococcus thermophilus* was used to synthesize spherical-shaped, 28–122 nm size AgNPs (El-Shanshoury et al. 2011). Juibari et al. (2011) synthesized 10–100 nm size, spherical-shaped AgNPs by using *Ureibacillus thermosphaericus*. Otari et al. (2014) exercised the synthesis of spherical-shaped AgNPs having the dimensions of 10–15 nm size in *Rhodococcus* sp. In yet another study, monodispersed, 5–15 nm size AuNPs were reported in *Rhodococcus* sp. (Ahmad et al. 2003a, b). *Rhodopseudomonas palustris* was reported to synthesize 8.0 ± 0.25 nm size CdSNPs in the cytoplasm (Bai et al. 2009). Sinha et al. (2012) synthesized spherical-shaped mercury-based nanoparticle in the size range of 2–5 nm sized by *Enterobacter* species. Formation of 30 nm size zinc sulfide nanoparticle was achieved by using *Desulforibrio caledoiensis* (Qi et al. 2013).

2.3 Nanoparticle Synthesis Using Fungi

Myconanotechnology is a developing, pragmatic division of nanotechnology which deals with the study of fungi for the nanoparticle synthesis (Kashyap et al. 2013). The ability of the fungi to multiply on inexpensive substrates and synthesis of economically important metabolites, proteins and enzymes have attracted considerable interest to harness them as promising candidate for tailor-made nanoparticles (Table 4).

Prasad and Jha (2010) reported rapid method of biosynthesis of spherical-shaped, 3.57 ± 0.21 nm size CdS nanoparticle using *Saccharomyces cerevisiae*. For the first time, biomass of *Aspergillus flavus* was used to synthesize oval- and spherical-shaped, 62–74 nm size TiO_2NPs (Rajakumara et al. 2012). Vijayanandan and Balakrishnan (2018) reported that, *Aspergillus nidulans* synthesized spherical-shaped, 20–29 nm size cobalt oxide nanoparticles using sulfur-bearing proteins acting as a capping agent. Zinc oxide nanoparticles with spherical-shaped 53–69 nm in size were biosynthesized using culture filtrates of *A. niger* (Kalpana et al. 2018). *Azoarcus* sp. CIB synthesized spherical-shaped SeNPs having average size of 123 ± 35 nm (Fernández-Llamosas et al. 2016). In another study, for the first time, Sarkar et al. (2011) reported the synthesis of spherical-shaped, SeNPs of 30–15 nm size using the culture filtrate of *Alternaria alternata*. Moghaddam et al. (2017) synthesized hexagonal-shaped ZnONPs having 10–61 nm size in *Pichia kudriavzevii*. In another study, AgNPs was synthesized from *Saccharomyces boulardii* with 3–10 nm size (Kaler et al. 2013). Also, spherical-shaped, 20–50 nm size AgNPs were fabricated using *Fusarium oxysporum* (Durán et al. 2005).

Mohammadian and Shojaosadati (2007) biosynthesized spherical-shaped 5–60 nm in size AgNPs using *F. oxysporum*, while Maliszewska et al. (2013) utilized *Penicillium nalgiovense* AJ15 for spherical-shaped AgNPs synthesis with the particle size 25.2 ± 2.8 nm. In another study, Singh et al. (2014) exercised endophytic

Table 4 Nanoparticles synthesis by some fungi

Fungi	Metal	Size (nm)	Shape	References
Aspergillus fumigatus BTCB10	Ag	322.8	Spherical	Shahzad et al. (2019)
Aspergillus oryzae MTCC 1846	Ag	7–27	Spherical	Phanjom and Ahmed (2017)
Fusarium oxysporum	Ag	24	Spherical	Hamedia et al. (2017)
Trichoderma harzianum	Ag	51.10	Spherical	Ahluwalia et al. (2014)
Trichoderma viride	Ag	5–40	Spherical	Fayaz et al. (2009)
Isaria fumosorosea	Ag	51.31–111.02	Spherical	Banu and Balasubramanian (2014)
Penicillium oxalicum GRS-1	Ag	10–40	Spherical	Rose et al. (2019)
Penicillium purpurogenum	Ag	8–10	Spherical	Nayak et al. (2011)
Penicillium oxalicum	Ag	36.53	Spherical	Du et al. (2015)
Verticillium sp	Ag	25–12	–	Sastry et al. (2003)
Saccharomyces boulardii	Ag	3–10	–	Kaler et al. (2013)
Phanerochaete chrysogenum	Au	10–100	Spherical	Sheikhloo et al. (2011b)
Neurospora crassa	Au	3–200	Different shapes	Quester et al. (2013)
Alternaria alternata	Au	12 ± 5	Spherical, triangular, hexagonal	Sarkar et al. (2012)
Rhizopus oryzae	Au	5–65	Spherical	Das et al. (2012)
Neurospora crassa	Au	32	Spherical	Castro-Longoria et al. (2011)
Sclerotium rolfsii	Au	25	Spherical	Das et al. (2012)
Phanerochaete chrysosporium	Cd	2.56	Face-centered cubic	Chen et al. (2014)
Rhodotorula mucilaginosa	Cu	10.5	Spherical	Salvadori et al. (2014)
Cryptococcus humicola	Fe	8–9	Regular	Vainshtein et al. (2014)
Saccharomyces cerevisiae	Fe	–	Irregular	Vainshtein et al. (2014)

fungi *Penicillium* sp. to synthesize spherical-shaped, 25–30 nm size silver nanoparticles. Further, endophytic fungi *Aspergillus conicus, Penicillium janthinellum* and *Phomosis* sp. synthesized AgNPs having polydispersed spherical- or hexagonal-shaped and 6–12 mm, 8–14 mm 10–16 mm in size, respectively, (Bharathidasan and Panneerselvam 2012). Verma et al. (2010) synthesized polydispersed spherical or hexagonal, 10–25 nm in size AgNPs from an endophytic fungus *Aspergillus clavatus. Cladosporium cladosporioides* was exercised for the formation of AgNPs having polydispersed and spherical-shaped, 10–100 nm size (Balaji et al. 2009). In another study, spherical-shaped AgNPs ranging from 25 to 50 nm sizes were synthesized using *Aspergillus tamarii* (Raliya et al. 2014a, b).

2.4 Nanoparticle Synthesis Using Algae

Algae have also been explored for the synthesis of metal nanoparticles. Senapati et al. (2012) synthesized intracellular spherical- and triangular-shaped AuNPs with dimensions of 5–35 nm in size using *Tetraselmis kochinensis.* Kannan et al. (2013) synthesized AgNPs having 3–44 nm size using marine macroalga *Chaetomorpha linum.* In another study, rapid synthesis of spherical-, hexagonal- and triangle-shaped AuNPs in brown alga, *Stoechospermum marginatum* with dimensions from 18.7 to 93.7 nm was achieved in ten minutes (Rajathi et al. 2012). Kumar et al. (2012) synthesized spherical-shaped AgNPs with dimensions of 20 nm using *Sargassum tenerrimum.* Similarly, in *Sargassum muticum,* spherical-shaped AgNPs 43–79 nm size was synthesized by Madhiyazhagan et al. (2015). Formation of AgNPs having spherical to oval shape with 31 nm in size was synthesized by *Laminaria japonica* (Kim et al. 2018). Jena et al. (2014) synthesized AgNPs using *Scenedesmus* sp. (IMMTCC-25). Results indicated that, living cells synthesized AgNPs having 3–35 nm in size and spherical shape. However, 5–10 nm size and spherical shape was obtained with raw algal extract, respectively (Table 5).

Namvar et al. (2015) used the extract of *Sargassum muticum* and synthesized spherical and crystalline shape AuNPs with dimensions less than 10 nm. In another study, *Pithophora oedogonia* was used for the rapid synthesis of 32.06 nm size AuNPs within 1 h (Li et al. 2016). Biosynthesis of stable spherical-shaped AuNPs with 25–30 nm in size at pH 8 was achieved using *Chlorella pyrenoidusa* extract (Oza et al. 2012). Lengke et al. (2006) studied intracellular and extracellular synthesis of cubical and octahedral, 10–25 nm size AuNPs in *Plectonema boryanum* UTEX 48. A marine green microalga, *Tetraselmis suecica,* was exercised for spherical-shaped AuNPs of size range 51–120 nm by Shakibaie et al. (2010). Ramakritinan et al. (2013) synthesized 22–30 nm size, Ag/Au bimetallic spherical-shaped nanoparticle from *Gracilaria.*

In another study, *Tetraselmis suecica* was used to synthesize 51–120 nm size, spherical-shaped AuNPs (Shakibaie et al. 2010). *Phormidium tenue* was exercised for tailor-made AuNPs and resulted in the synthesis of 14.84 nm size spherical- and

Table 5 Nanoparticles synthesis from some algae

Algal species	Metal	Size (nm)	Shape	References
Ulva lactuca	Ag	–	Spherical	Murugan et al. (2015)
Leptolyngbya valderianum	Ag	2–20	Spherical	Roychoudhury and Pal (2014)
Spirogyra varians	Ag	35	Squasi-spheres	Salari et al. (2016)
Laminaria japonica	Ag	31	Spherical to oval	Kim et al. (2018)
Sargassum tenerrimum	Ag	20	Spherical	Kumar et al. (2012)
Sargassum muticum	Ag	43–79	Spherical	Madhiyazhagan et al. (2015)
Porphyra vietnamensis	Ag	13 ± 03	Spherical	Venkatpurwar and Pokharkar (2011)
Scenedesmus sp.	Ag	15–20	Spherical, crystalline	Jena et al. (2014)
Pithophora oedogonia	Ag	25–44	Cubical, hexagonal	Sinha et al. (2015)
Padina gymnospora	Au	18.7–93.7	Spherical, hexagonal and triangular	Singh et al. (2013a, b)
Acanthophora spicifera	Au	27–35	Spherical	Swaminathan et al. (2011)
Padina pavonica	Au	30–100	Spherical	Isaac and Renitta (2015)
Chara zeylanica	Au	24	Triangular	Parial et al. (2012)
Turbinaria conoides	Au	6–10	Spherical	Rajeshkumar et al. (2013)
Sargassum incisifolium	Au	12.38	Spherical	Mmola et al. (2016)
Ecklonia cava	Au	30 ± 0.25	Spherical and triangular	Castro et al. (2014)
Padina tetrastromatica	Au	8–10	Spherical	Rajeshkumar et al. (2017)
Pithophora oedogonia	Au	32.06	Spherical	Nangia et al. (2009)
Prasiola crispa	Au	5–25	Spherical	Dhamotharan et al. (2010)
Chlorella pyrenoidusa	Au	25–30	Spherical	Oza et al. (2012)
Sargassum bovinum	Pd	5–10	Octahedral	Momeni and Nabipour (2015)
Chlorococcum sp. MM11	Fe	20–50	Spherical	Vigneshwaran et al. (2006)

irregular-shaped was observed (Parial et al. 2012). MubarakAli et al. (2013) synthesized spherical-shaped, 30 nm-sized AuNPs using *Coelastrella* sp. In a work done by Dahoumane et al. (2014), 5.7 ± 0.9 nm spherical-shaped AuNPs were synthesized using *Cosmarium impressulum*. Formation of 5–25 nm size and spherical-shaped AuNPs was achieved using *Prasiola crispa* (Sharma et al. 2014). Similarly, *Spirogyra submaxima* was used to synthesize 2–50 nm size and spherical-, triangular-, hexagonal-shaped AuNPs by Roychoudhury and Pal (2014). In another study, 3.85–77.13 nm size, spherical-shaped AuNPs were synthesized using *Galaxaura elongata* (Abdel-Raouf et al. 2013). Formation of 60 nm size, triangle-, rectangle- and square-shaped AuNPs was achieved using *Turbinaria conoides* (Rajeshkumar et al. 2013). Abboud et al. (2014) synthesized 5–45 nm size, spherical- and elongated-shaped CuO using *Bifurcaria bifurcata*. In another study, *Scenedesmus*-24 synthesized 120–175 nm size, oval-shaped CdNPs (Jena et al. 2015). Formation of 66–95 nm size, rod-shaped ZnONPs was synthesized using *Gracilaria edulis* (Priyadharshini et al. 2014). Similarly, *Sargassum muticum* was used to synthesize ZnONPs with 3–57 nm size and hexagonal wurtzite shape (Azizi et al. 2014). In yet another study, Nagarajan and Kuppusamy (2013) synthesized 36 nm size, spherical-, triangle-, radial-, hexagonal-, rod-shaped ZnONPs using *Sargassum myriocystum*. Synthesis of Fe_3O_4 nanoparticle using *Sargassum muticum* was investigated by Mahdavi et al. (2013). Results indicated the presence of 18 ± 4 nm size, cubic shape Fe_3O_4 nanoparticles.

2.5 Nanoparticle Synthesis Using Actinomycetes

Alkali-tolerant *Rhodococcus* was used to synthesize 5–15 nm AuNPs on the mycelial surface and cytoplasmic membrane (Ahmad et al. 2003a). In a similar study by the authors on *Thermomonospora* sp., investigation resulted in the synthesis of spherical-shaped AuNPs under alkaline condition (Ahmad et al. 2003b). In another study, formation of 5–50 nm, polygonal-shaped, polygonal AuNPs were synthesized from S*treptomyces* sp. (Karthik et al. 2013). Manikprabhu and Lingappa (2013) exercised the synthesis of AuNPs having 28–50 nm, irregular-shaped AuNPs in *Streptomyces coelicolor*. In another study, 10–20 nm size spherical-shaped AuNPs were produced from *Streptomyces hygroscopicus* (Waghmare et al. 2014). Subashini et al. (2014) synthesized 20–70 nm size, spherical-shaped AgNPs from *Streptomyces* sp. (Table 6). AgNPs having 20–60 nm in size and spherical shape were synthesized in *Streptomyces* sp. (Thenmozhi et al. 2013).

3 Microbial Nano-biological Control Agents in Agriculture

In the agriculture industry, crop protection from the attack of various phytopathogens and efforts at improving the crop growth for a better yield are prime areas of

Table 6 Nanoparticles synthesis by some actinomycetes

Actinomycete	Metal	Size (nm)	Shape	References
Nocardiopsis sp. MBRC-1	Ag	45	Spherical	Manivasagan et al. (2013)
Streptomyces sp. VITPK1	Ag	20–45		Sanjenbam et al. (2014)
Rhodococcus sp.	Ag	10	Spherical	Otari et al. (2012)
Streptomyces sp. JAR1	Ag	68.13	Crystalline	Chauhan et al. (2013)
Streptomyces coelicolor	Ag	28–50	Irregular	Manikprabhu and Lingappa (2013)
Hansenula anomala	Au	18–20	Spherical and rod	Kumar et al. (2011)
Streptomyces sp. VITDDK3	Au	90	Hexagonal, cubical, brick and irregular	Nejad et al. (2015)
Streptomyces sp.	Au	5–50	Polygonal	Karthik et al. (2013)

constant research. In this context, microbial nanotechnology is emerging as an ancillary platform to deliver fertilizers, herbicides and pesticides more efficiently. In the following section, recent applications of microbial-based nanoparticles in agriculture are discussed.

3.1 NPs as Antimicrobials

Mounting evidences suggest the potentials of microbial-based NPs at attenuating phytopathogenic fungi and pests. For instance, silver nanoparticles synthesized from *Trichoderma harzianum* inhibited sclerotia germination and mycelial growth of the white mold *Sclerotinia sclerotiorum* (Guilger et al. 2017). Balakumaran et al. (2015) reported antifungal activity of phytopathogenic fungi such as *Colletotrichum* sp., *Rhizoctonia solani* and *Curvularia lunata* using AgNPs (1 mg/ml) synthesized from endophytic fungus *Guignardia mangiferae*. Moussa et al. (2013) reported that fungal chitosan mediated nano-irradiated silver composite at 125 μg/ml concentration which was coated on strawberries and then infected with *Botrytis cinerea* Pers, a gray mold fungus. Results showed 90% growth inhibition of the mycopathogen after 7 days of storage due to moderate lysis in the fungal hyphae. However, in the control group, the fruits were completely infected with *B. cinerea,* respectively. Antifungal activity against plant pathogenic fungus, *F. oxysporum* at 8 μg/ml AgNPs, exercised from *Bacillus* strain GP-23 has been reported (Gopinath and Velusamy 2013). Silver nanoparticle biosynthesized from *A. alternata* promoted the antifungal activity of fluconazole. Maximum antifungal activity was observed against *Trichoderma* sp. and *Phoma glomerata*. However, no significant activity was seen in *Phoma herbarum* and *Fusarium semitectum* (Gajbhiye et al. 2009). Joshi et al. (2019) reported that, mycogenic SeNPs synthesized from *T. atroviride* showed exceptional in vitro antifungal property against *Pyricularia grisea* and further inhibited the infection of *Colletotrichum capsici* and *Alternaria solani* on chili and tomato leaves, respectively.

Nanofibrous mats were prepared using fungal chitosan and *Trichoderma viride*. At 28 °C, *T. viride* multiplies rapidly and competes for space and nutrients against phytopathogens such as *Fusarium* sp. and *Alternaria* sp. (Spasova et al. 2011). *Penicillium duclauxii*-mediated AgNPs (concentration ranging from 50 to 200 ppm) were used to test antifungal activity against *Bipolaris sorghicola*. Results showed that maximum percentage of inhibition was found to be 70% at 200 ppm AgNPs concentration (Almaary et al. 2020). Ibrahim et al. (2020) biosynthesized AgNPs from endophytic bacterium *Pseudomonas poae* strain CO, isolated from *Allium sativum* which inhibited the mycelium growth, spore germination, the length of the germ tubes and the mycotoxin production of the wheat *Fusarium* head blight pathogen *Fusarium graminearum*. Biosynthesized AgNPs from endophytic bacterium *Bacillus siamensis* strain C1 inhibited *Xanthomonas oryzae* pv. *oryzae* (Xoo) strain LND0005, a rice bacterial leaf blight and *Acidovorax oryzae* (Ao) strain RS-1, bacterial brown stripe pathogen effectively and promoted the plant growth (Ibrahim et al. 2019). In another study, different concentrations of AgNPs (2, 4 and 10 μg/ml) synthesized using *Serratia* sp. BHU-S4 completely inhibited germination of conidia in *Bipolaris sorokiniana* and encouraged plant growth in pathogen challenged plant (Mishra et al. 2014a).

Antibacterial activity of different concentrations of AgNPs (0.1–0.3 mM) obtained from *Spirulina platensis* was effective against various plant pathogens such as *Pseudomonas solanocearum, P. syringae, Xanthomonas malvacearum* and *X. campestris*. Further, it was also reported that addition of 1 μg of ciprofloxacin in conjunction with 0.2 mM AgNPs enhanced the antibacterial activity against the tested bacterial pathogens (Mala et al. 2012). Silver nanoparticles synthesized from *Pseudomonas rhodesiae* have also been reported to inhibit *Dickeya dadantii*, a soft rot pathogen at 50 μg/mL AgNPs concentration (Hossain et al. 2019). It is clearly evident that nanoparticles that were produced by microbes could be exploited for nanocontrol of plant pathogens to forestall disease outbreak, spoilage and loss of agricultural produce.

3.2 *Microbial Nano-biosensor*

Wang et al. (2010) synthesized SeNPs using *Bacillus subtilis*. The results indicated that bacterial proteins served as templates and eventually converted selenium from monoclinic nano-sphere to trigonal nanowires. The biosensor thus developed had high sensitivity and affinity for H_2O_2, and the detection limit for H_2O_2 was found to be 8×10^{-8} M. Similarly, studies on *Bacillus pumilus* sp. BAB-3706 resulted in the synthesis of SeNPs biosensor. The detection limit was found to be 3.00 μM (Prasad et al. 2015). In another study, yeast cells were used to synthesize Au-Ag alloy nanoparticles-based electrochemical sensor, which was highly sensitive toward the electrochemical oxidation of vanillin. Results suggest that, biosensor was able to detect vanillin in vanilla bean and vanilla tea samples at a lower concentration limit of 40 nM (Zheng et al. 2010). Thus, microbial fabricated nanoparticles can be used for biosensing applications in agriculture for quality control and monitoring of deterioration or contamination of agricultural produce.

3.3 Microbial Nanopesticides

Helan et al. (2013) reported that maximum activity against *Spodoptera litura* was observed from chitosan coated metabolite of entomopathogenic fungi *Nomurae rileyi*. In yet another study, chitosan nanoparticles with insecticidal protein beauvericin reduced pupal and adult emergence and displayed pesticidal activity in *Spodoptera litura* in all the life stages. However, maximum death was observed in early larval instar phase (Arvind et al. 2014).

Nanoparticles synthesized using destruxin from *Metarhizium anisopliae* significantly reduced the infestations of the adult female grasshopper, *Hetiracris littoralis* compared to control (Sabbour 2014). In a similar study, nano-extracted destruxin from *M. anisopliae* decreased the number of eggs laid/female of *Plodia interpunctella* (Sabbour 2015). Babu et al. (2014) evaluated different concentrations of AgNPs (ranging from 5 to 25 mg/ml) synthesized from marine bacteria *Shewanella algae bangaramma* against *Lepidiota mansueta*. Results indicated that, on 15th day, maximum LC_{50} and LC_{90} values with 95% confidential limits were observed with 3rd-instar larvae at 25.0 mg/ml AgNPs concentration. In conclusion, microbe-mediated nanoparticles have shown potentials for application as pesticides in agriculture.

3.4 Plant Growth Promoters

Pour et al. (2019) investigated the effect of nano-encapsulated plant growth-promoting bacteria such as *Pseudomonas fluorescens* VUPF5 and *Bacillus subtilis* VRU1 and their metabolites in promoting UCB1 pistachio micropropagation. The results revealed that the use of bacterial-mediated nanocapsules substantially enhanced the root length and proliferation. The nanoformulation of the VUPF5 metabolite led to the highest root (6.26 cm) and the largest shoot length (3.34 cm) compared to the control (0.9 cm and 3.3 cm), respectively. In addition, significant increase in the average shoot length and the fresh weight of plant (0.18 compared to the control) was also observed during encapsulation of bacterial metabolites produced by *P. fluorescens* VUPF5 and *B. subtilis* VRU1. The effect of different concentrations of *Laminaaria japonica* extract-mediated AgNPs on seed germination and plant growth was investigated in *Triticum aestivum* and *Phaseolus mungo*. Results indicated that, AgNPs seed germination was not affected in both plants. However, up to the concentration of 30 ppm AgNPs, the roots of *T. aestivum* and *P. mungo* were found to be longer than the control. At the concentration of 40 ppm, the growth of roots and shoot was decreased when compared to control. Further treatment of seedlings with 60 and 80 ppm AgNPs experienced phytotoxicity, as shoot and root growth was significantly affected.

Ibrahim et al. (2019) reported that AgNPs was synthesized from *Bacillus siamensis* strain C1, an endophytic bacterium from *Coriandrum sativum* for elucidating the effect of plant growth-promoting activity in rice. Results showed the tremendous increase in treated rice seedlings having the root length (7.4 cm), shoot length (11.6 cm), fresh weight (0.24 g) and dry weight (0.04 g), compared to the root length (5.1 cm), shoot length (6.5 cm), fresh weight (0.12 g) and dry weight (0.03 g) of the control, respectively. These results indicate the possibility of using microbe-mediated nanoparticles as plant growth stimulators.

3.5 *Microbial Nanobiofertilizer*

Mycogenic-nanoparticles serve as an indispensable resource and could be used as fertilizer. *Aspergillus fumigatus* TFR-9-mediated ZnO nanoparticles (10 ppm) were foliar sprayed on the leaf of 2-week-old cluster bean plants and studied for its plant growth-promoting property. Results showed enhanced plant biomass (27.1%), shoot length (31.5%), root length (66.3%), root area (73.5%), chlorophyll content (276.2%), total soluble leaf protein (27.1%), rhizospheric microbial population (ranging from 11 to 14%), enzyme activities such as acid phosphatase (73.5%), alkaline phosphatase (48.7%), and phytase (72.4%) in the rhizosphere over control in a week-old plants. After maturity, the gum content and its viscosity in cluster bean seeds increased by 7.5%. Overall, these results suggest that, *Aspergillus fumigatus* TFR-9 mediated ZnONP could serve as a potential candidate for microbial nanofertilizer production (Raliya and Tarafdar 2013). Similarly, 10 mg/l of TiO_2NP synthesized from *Aspergillus flavus* TFR 7 were foliar sprayed on the leaves of two weeks old mung bean plants. Results showed a significant improvement in shoot length (17.02%), root length (49.6%), root area (43%), root nodule (67.5%), chlorophyll content (46.4%) and total soluble leaf protein (94%). Furthermore, microbial population such as bacteria, fungi and actinomycetes increased between 21.4 and 48.1% in the rhizosphere in six week-old plant. Increase in enzyme activity was observed over control in six week-old plant. The authors predicted that, increase in the enzyme activity may be due to increase in the microflora in the rhizosphere (Raliya et al. 2015). Raliya et al. (2014b) investigated the influence of MgO_2NP at 15 mg/l concentration synthesized from *A. flavus* strain TFR-12 on 2-week-old *Cyamopsis tetragonoloba*. They concluded that, chlorophyll content increased by 76.1%. Further, remarkable increase in the shoot–root growth (18.2–49.2%) and chlorophyll photosynthetic pigment (76.1%) was observed. These results have shown that the nanoparticles can aptly act as fertilizers in promoting the plant growth.

4 Current Challenges

The greatest challenge in using microbial-based NPs is the state of reproducibility. Genetic instability and various stress response encountered by a microbial cell may lead to differences in the composition, reproducibility and yield in the nanoparticle synthesis. Further, growing lines of evidences support existence of a vast difference within the microbial species toward nanoparticle synthesis. Optimization of ideal conditions is understudied for microbial growth as well as sample preparation and processing, which are prerequisite to maintain homogeneity. Efficient control of size, shape and dispersity of nanoparticle still remains unanswered. Comprehensive field trials are essential to comprehend the nanotoxicity in plants regarding uptake and translocation of nanoparticles by plants and effect of the NPs in the rhizosphere and on root surfaces. Nanoparticles may also interact with non-targeted plants causing undesirable outcomes. Influence of NPs on beneficial microorganisms, water eco-system, animals, birds, insects, human and environment remains untapped area. Toxicity residues and biotransformed products of NPs in plants, plant-based products and in the soil are not yet well studied. Awareness to the general public and to policy makers regarding the possible outcomes of application of microbial nanotechnology is still lacking. These factors at the moment limit the applications of microbe-mediated nanoparticles for various applications in agriculture.

5 Future Prospects

The convergence of chemistry, material engineering and microorganisms is considered as futuristic approach for eco-friendly nanoparticle synthesis. Globally, tremendous developments have been advanced in the field of microbial nanotechnology, and their applications in myriad areas in general and agriculture sector, which is the backbone of the developing nation like India in particular, are explored incessantly. Nevertheless, there are certain gaps and limitations in designing optimum synthesis of nanoparticles which need to be solved by the scientific community. Maximum productivity can be achieved by optimizing various biological, physical and chemical parameters such as type of microorganism used, enzyme concentration, protein concentration, inoculum age, pH, temperature, concentration of metal ions, culture medium and reaction time among others.

Another relatively unexplored area is mechanism of microbial-mediated nanoparticle synthesis. Thus, more exhaustive investigations are warranted to outline the mechanism, proteins, enzymes and other factors involved in the nanoparticle synthesis. At present, only few types of NPs such as Au, Ag, CdS, Zr, Mg, Pt, Pd, Se, Zn, Cu, Fe and Ti are being synthesized from microbial sources. Hence, other types of NP synthesis through biological route must be encouraged. Many reports suggest that, among the microbial world, fungi are more explored for NPs synthesis. Therefore, unmapped microbial cell factories from various sources such as marine, extremophiles and plant dwelling endophytes ought to be explored systematically. Stability of NPs synthesized from various microorganisms still remains unexplored. Therefore, systematic research has to focus on their stability under laboratory and pilot scale production. It is imperative to gain insight on inexpensive downstream processing technologies for sustainable production of microbial-based nanoparticles. Future studies must focus on industrial scale production of microbial nanofactories. Comprehensive research studies should be carried on microbial-based nano-biosensors, nanoherbicides, nanopesticides, plant growth promoters, precision farming, and the effect of exposure of antimicrobial nanoparticles against the plant pathogens needs to be well documented. Regulations and ethical issues need to be dealt meticulously. Economics of NPs scale up must be addressed, and this will be a roadmap for the feasibility and sustainability of tailor-made microbial-based NPs in the market in the years to come.

6 Conclusion

This chapter has uncovered the current status and a panoramic view of microbial nanotechnology in the scope of agricultural science and engineering. Current mechanisms, sources, advances, potential applications challenges and future perspectives have been discussed. Nanoparticle synthesis using microorganisms is an eco-friendly approach which could serve as an ancillary platform to physical and chemical methods that release harmful chemicals to the environment. With the ongoing process and efforts in improving the characteristics and mode of synthesis of NPs, it could be possible for the implementation of these approaches on a commercial scale in the near future.

References

Abboud Y, Saffaj T, Chagraoui A, Bouari AE, Brouzi K, Tanane O, Ihssane B (2014) Biosynthesis, characterization and antimicrobial activity of copper oxide nanoparticles (CONPs) produced using brown alga extract (*Bifurcaria bifurcata*). Appl Nanosci 4:571–576. https://doi.org/10.1007/s13204-013-0233-x

Abdel Rahim K, Mahmoud SY, Ali AM, Almaary KS, Mustafa AE, Husseiny SM (2017) Extracellular biosynthesis of silver nanoparticles using *Rhizopus stolonifer*. Saudi J Biol Sci 24:208–216. https://doi.org/10.1016/j.sjbs.2016.02.025

Abdel-Raouf N, Al-Enazi NM, Ibraheem IBM (2013) Green biosynthesis of gold nanoparticles using *Galaxaura elongata* and characterization of their antibacterial activity. Arab J Chem. https://doi.org/10.1016/j.arabjc.2013.11.044

Adelere IA, Lateef A (2016) A novel approach to the green synthesis of metallic nanoparticles: the use of agro-wastes, enzymes and pigments. Nanotechnol Rev 5:567–587. https://doi.org/10.1515/ntrev-2016-0024

Ahluwalia V, Kumar J, Sisodia R, Shakil NA, Walia S (2014) Green synthesis of silver nanoparticles by *Trichoderma harzianum* and their bio-efficacy evaluation against *Staphylococcus aureus* and *Klebsiella pneumonia*. Ind Crops Prod 55:202–206. https://doi.org/10.1016/j.indcrop.2014.01.026

Ahmad A, Mukherjee P, Mandal D, Senapati S, Khan MI, Kumar R, Sastry M (2002) Enzyme mediated extracellular synthesis of CdS nanoparticles by the fungus, *Fusarium oxysporum*. J Am Chem Soc 16(124):12108–12109. https://doi.org/10.1021/ja027296o

Ahmad A, Senapati S, Khan MI, Kumar R, Ramani R, Srinivas V, Sastry M (2003a) Intracellular synthesis of gold nanoparticles by a novel alkalotolerant actinomycete, *Rhodococcus* species. Nanotechnology 14:824. https://doi.org/10.1088/0957-4484/14/7/323

Ahmad A, Senapati S, Khan MI, Kumar R, Sastry M (2003b) Extracellular biosynthesis of monodisperse gold nanoparticles by a novel extremophilic actinomycete, *Thermomonospora* sp. Langmuir 19:3550–3553. https://doi.org/10.1021/la026772l

Alghuthaymi MA, Almoammar H, Rai M, Said-Galiev E, Abd-Elsalam KA (2015) Myconanoparticles: synthesis and their role in phytopathogens management. Biotechnol. Biotechnol Equip 29:221–236. https://doi.org/10.1080/13102818.2015.1008194

Ali M, Ahmed T, Wu W, Hossain A, Hafeez R, Islam Masum M, Wang Y, An Q, Sun G, Li B (2020) Advancements in plant and microbe-based synthesis of metallic nanoparticles and their antimicrobial activity against plant pathogens. Nanomaterials 6:1146. https://doi.org/10.3390/nano10061146

Alkilany AM, Bani Yaseen AI, Kailani MH (2015) Synthesis of monodispersed gold nanoparticles with exceptional colloidal stability with grafted polyethylene glycol-g-polyvinyl alcohol. J Nanomat 1:2015. http://doi.org/10.1155/2015/712359

Almaary KS, Sayed SRM, Abd-Elkader OM, Dawoud TM, Orabi NFE, Elgorban AM (2020) Complete green synthesis of silver-nanoparticles applying seed-borne *Penicillium duclauxii*. Saudi J of Biol Sci 27:1333–1339. https://doi.org/10.1016/j.sjbs.2019.12.022

Arvind RS, Karthick R, Namasivayam S, Sai Shankar S (2014) Biocompatible chitosan nanoparticles incorporated pesticidal protein Beauvericin (Csnp-Bv) preparation for the improved pesticidal activity against major groundnut defoliator *Spodoptera litura* (Fab.) (Lepidoptera; Noctuidae). Int J ChemTech Res 6:5007–5012

Aziz N, Pandey R, Barman I, Prasad R (2016) Leveraging the attributes of Mucor hiemalis-derived silver nanoparticles for a synergistic broad-spectrum antimicrobial platform. Front Microbiol 7:1984. https://doi.org/10.3389/fmicb.2016.01984

Azizi S, Ahmad MB, Namvar F, Mohamad R (2014) Green biosynthesis and characterization of zinc oxide nanoparticles using brown marine macroalga *Sargassum muticum* aqueous extract. Mater Lett 116:275–277. https://doi.org/10.1016/j.matlet.2013.11.038

Azmath P, Baker S, Rakshith D, Satish S (2016) Mycosynthesis of silver nanoparticles bearing antibacterial activity. Saudi Pharm J. 24:140–146. https://doi.org/10.1016/j.jsps.2015.01.008

BabuY, Janaki V, Ramakritinan M, Umarani R Nayimabanu T, Arumugam K (2014) Application of biosynthesized silver nanoparticles in agricultural and marine pest control. Curr Nanosci 10:374–381. https://doi.org/10.1007/s13355-015-0349-4

Bai HJ, Zhang ZM, Guo Y, Yang GE (2009) Biosynthesis of cadmium sulfide nanoparticles by photosynthetic bacteria *Rhodopseudomonas palustris*. Colloids Surf B 70:142–146. https://doi.org/10.1016/j.colsurfb.2008.12.025

Baker S, Nagendra MN, Satish Prasad S (2016) Endogenic mediated synthesis of gold nanoparticles bearing bactericidal activity. J Microsc Ultrastruct 4:162–166. http://dx.doi.org/10.1016/j.jmau.2016.01.004

Balaji DS, Basavaraja S, Deshpande R, Mahesh DB, Prabhakar BK, Venkataraman A (2009) Extracellular biosynthesis of functionalized silver nanoparticles by strains of *Cladosporium cladosporioides* fungus. Coll Surf B 68:88–92. https://doi.org/10.1016/j.colsurfb.2008.09.022

Balakumaran MD, Ramachandran R, Kalaichelvan PT (2015) Exploitation of endophytic fungus, *Guignardia mangiferae* for extracellular synthesis of silver nanoparticles and their in vitro biological activities. Microbiol Res 178:9–17. https://doi.org/10.1016/j.micres.2015.05.009

Banu AN, Balasubramanian C (2014) Optimization and synthesis of silver nanoparticles using *Isaria fumosorosea* against human vector mosquitoes. Parasitol Res 113:3843–3851. https://doi.org/10.1007/s00436-014-4052-0

Barabadi H, Honary S, Mohammadi MA, Ahmadpour E, Rahimi MT, Alizadeh A, Naghibi F, Saravanan M (2017) Green chemical synthesis of gold nanoparticles by using *Penicillium aculeatum* and their scolicidal activity against hydatid cyst protoscolices of *Echinococcus granulosus*. Env Sci Pollut Res 24:5800–5810. https://doi.org/10.1007/s11356-016-8291-8

Bathrinarayanan PV, Thangavelu D, Muthukumarasamy VK, Munusamy C, Gurunathan B (2013) Biological synthesis and characterization of intracellular gold nanoparticles using biomass of *Aspergillus fumigatus*. Bull Mat Sci 36:1201–1205. https://doi.org/10.1007/s12034-013-0599-0

Baymiller M, Gobble A, Huang F (2017) Rapid one-step synthesis of gold nanoparticles using the ubiquitous coenzyme NADH. Matters 3:e201705000007. https://doi.org/10.19185/matters.201705000007

Bennur T, Khan Z, Kshirsagar R, Javdekar V, Zinjarde S (2016) Biogenic gold nanoparticles from the Actinomycete *Gordonia amarae*: application in rapid sensing of copper ions. Sens Actuators B: Chem 233:684–690. https://doi.org/10.1016/j.snb.2016.04.022

Bharathidasan R, Panneerselvam A (2012) Biosynthesis and characterization of silver nanoparticles using endophytic fungi Aspergillus concius, *Penicillium janthinellum* and *Phomosis* sp. Int J Pharm Sci Res 3:3163–3169. https://doi=10.1.1.278.5118

Birla SS, Gaikwad SC, Gade AK, Rai MK (2013) Rapid synthesis of silver nanoparticles from *Fusarium oxysporum* by optimizing physicocultural conditions. Sci World J 10:2013. https://doi.org/10.1155/2013/796018

Bogunia-Kubik K, Sugisaka M (2002) From molecular biology to nanotechnology and nanomedicine. Biosystems 65:123–138. https://doi.org/10.1016/s0303-2647(02)00010-2

Camargo PH, Satyanarayana KG, Wypych F (2009) Nanocomposites: synthesis, structure, properties and new application opportunities. Mat Res 12:1–39. https://doi.org/10.1590/S1516-14392009000100002

Castro L, Blázquez ML, Muñoz JA, González F, Ballester A (2013) Biological synthesis of metallic nanoparticles using algae. IET Nanobiotechnol 7:109–116. https://doi.org/10.1049/iet-nbt.2012.0041

Castro ME, Cottet L, Castillo A (2014) Biosynthesis of gold nanoparticles by extracellular molecules produced by the phytopathogenic fungus *Botrytis cinerea*. Mater Lett 115:42–44. http://doi.org/10.1016/j.matlet.2013.10.020

Castro-Longoria E, Vilchis-Nestor AR, Avalos-Borja M (2011) Biosynthesis of silver, gold and bimetallic nanoparticles using the filamentous fungus *Neurospora crassa*. Coll Surf B Biointer 83:42–48. https://doi.org/10.1016/j.colsurfb.2010.10.035

Chan YS, Mashitah MD (2012) Instantaneous biosynthesis of silver nanoparticles by selected macro-fungi. Aust J Basic App Sci 6:222–226

Chauhan A, Zubair S, Sherwani STA, Sajid M, Raman SC, Azam A, Owais M (2011) Fungus-mediated biological synthesis of gold nanoparticles: potential in detection of liver cancer. Int J Nanomed 6:2305–2319. https://doi.org/10.2147/ijn.s23195

Chauhan R, Kumar A, Abraham J (2013) A biological approach to the synthesis of silver nanoparticles with Streptomyces sp JAR1 and its antimicrobial activity. Sci Pharmaceut 81:607–624. https://dx.doi.org/10.3797%2Fscipharm.1302-02

Chen G, Yi B, Zeng G, Niu Q, Yan M, Chen A, Du J, Huang J, Zhang Q (2014) Facile green extracellular biosynthesis of CdS quantum dots by white rot fungus *Phanerochaete chrysosporium*. Colloids Surf B Biointerf 117:199–205. https://doi.org/10.1016/j.colsurfb.2014.02.027

Chen YL, Tuan HY, Tien CW, Lo WH, Liang HC, Hu YC (2009) Augmented biosynthesis of cadmium sulfide nanoparticles by genetically engineered *Escherichia coli*. Biotechnol prog 25:1260–1266. https://doi.org/10.1002/btpr.199

Dahoumane SA, Yéprémian C, Djédiat C, Couté A, Fiévet F, Coradin T, Brayner R (2014) A global approach of the mechanism involved in the biosynthesis of gold colloids using micro-algae. J Nanopart Res 16:2607. https://doi.org/10.1007/s11051-014-2607-8

Daniel MC, Astruc D (2004) Gold nanoparticles: assembly, supramolecular chemistry, quantum-size-related properties, and applications toward biology, catalysis, and nanotechnology. Chem Rev 104:293–346. https://doi.org/10.1021/cr030698+

Das SK, Dickinson C, Lafir F, Brougham DF, Marsili E (2012) Synthesis, characterization and catalytic activity of gold nanoparticles biosynthesized with *Rhizopus oryzae* protein extract. Green Chem 14:1322–1334. https://doi.org/10.1039/C2GC16676C

Deplanche K, Macaskie LE (2008) Biorecovery of gold by Escherichia coli and Desulfovibrio desulfuricans. Biotechnol Bioeng 99:1055–1064. https://doi.org/10.1002/bit.21688

Devi LS, Joshi SR (2015) Ultrastructure of silver nanoparticles biosynthesized using endophytic fungi. J Microsco Ultrastruc 3:29–37

Dhamotharan R, Punitha D, Murugesan S, Subha TS (2010) Brown algal biomass mediated biosynthesis of gold nanoparticles. Int J NanoSci Nanotechnol 1:33–40

Dhas TS, Kumar VG, Abraham LS, Karthick V, Govindaraju K (2012) *Sargassum myriocystum* mediated biosynthesis of gold nanoparticles. Spectrochim Acta Part A: Mol And Biomol Spect 99:97–101. https://doi.org/10.1016/j.saa.2012.09.024

Dhillon GS, Brar SK, Kaur S, Verma M (2012) Green approach for nanoparticle biosynthesis by fungi: current trends and applications. Crit Rev Biotechnol 32:49–73. https://doi.org/10.3109/07388551.2010.550568

Drexler KE (1981). Molecular engineering: an approach to the development of general capabilities for molecular manipulation. PNAS 78:5275–5278. https://doi.org/10.1073/pnas.78.9.5275

Du J, Singh H, Yi TH (2016) Biosynthesis of silver nanoparticles by *Novosphingobium* sp. THG-C3 and their antimicrobial potential. Artif Cells Nanomed Biotechnol 45:211–217. http://doi.org/10.1080/21691401.2016.1178135

Du L, Jiang H, Liu X, Wang E (2007) Biosynthesis of gold nanoparticles assisted by *Escherichia coli* DH5 and its application on direct electrochemistry of hemoglobin. Electrochem Com 9:1165–1170. https://www.x-mol.com/paperRedirect/1197185

Du L, Xu Q, Huang M, Xian L, Feng JX (2015) Synthesis of small silver nanoparticles under light radiation by fungus *Penicillium oxalicum* and its application for the catalytic reduction of methylene blue. Mater Chem Phys 15:40–47. https://doi.org/10.1016/j.matchemphys.2015.04.003

Duraisamy K, Yang SL (2013) Synthesis and characterization of bactericidal silver nanoparticles using cultural filtrate of simulated microgravity grown *Klebsiella pneumoniae*. Enzyme Microb Technol 52:151–156. https://doi.org/10.1016/j.enzmictec.2012.12.006

Durán N, Marcato PD, Alves OL, DeSouza G, Esposito E (2005) Mechanistic aspects of biosynthesis of silver nanoparticles by several *Fusarium oxysporum* strains. J Nanobiotechnol 3:1–8. https://doi.org/10.1186/1477-3155-3-8

Durán N, Marcato PD, Durán M, Yadav A, Gade A, Rai M (2011) biogenic synthesis of extracellular metal nanoparticles by peptides, bacteria, fungi, and plants. Appl Microbiol Biotechnol 90: 1609–1624. https://doi.org/10.1007/s00253-011-3249-8

Durán N, Marcato PD, Durán M, Yadav A, Gade A, Rai M (2011b) Mechanistic aspects in the biogenic synthesis of extracellular metal nanoparticles by peptides, bacteria, fungi, and plants. Appl Microbiol Biotechnol 90:1609–1624. https://doi.org/10.1007/s00253-011-3249-8

Elegbede JA, Lateef A, Azeez MA, Asafa TB, Yekeen TA, Oladipo IC, Aina DA, Beukes LS, Gueguim-Kana EB (2020) Biofabrication of gold nanoparticles using xylanases through valorization of corncob by *Aspergillus niger* and *Trichoderma longibrachiatum*: antimicrobial, antioxidant, anticoagulant and thrombolytic activities. Waste Biomass Valoriz 11:781–791. https://doi.org/10.1007/s12649-018-0540-2

Elegbede JA, Lateef A, Azeez MA, Asafa TB, Yekeen TA, Oladipo IC, Abbas SH, Beukes LS, Gueguim-Kana EB (2019) Silver-gold alloy nanoparticles biofabricated by fungal xylanases exhibited potent biomedical and catalytic activities. Biotechnol Prog 35:e2829. https://doi.org/10.1002/btpr.2829

El-Shanshoury AER, ElSilk SE, Ebeid ME (2011) Extracellular biosynthesis of silver nanoparticles using *Escherichia coli* ATCC 8739, *Bacillus subtilis* ATCC 6633, and *Streptococcus thermophilus*

ESh1 and their antimicrobial activities. ISRN Nanotechnol 2011. https://doi.org/10.5402/2011/385480

Faraday M (1857) Experimental relations of gold (and other metals) to light. Philos Trans R Soc Lond 147:145–181. https://doi.org/10.1080/14786445708642424

Fatemi M, Mollania N, Momeni-Moghaddam M, Sadeghifar F (2018) Extracellular biosynthesis of magnetic iron oxide nanoparticles by *Bacillus cereus* strain HMH1: characterization and *in vitro* cytotoxicity analysis on MCF-7 and 3T3 cell lines. J Biotechnol 270:1–11. https://doi.org/10.1016/j.jbiotec.2018.01.021

Fayaz M, Tiwary CS, Kalaichelvan PT, Venkatesan R (2010) Blue orange light emission from biogenic synthesized silver nanoparticles using *Trichoderma viride*. Colloids Surf B: Biointerf 75:175–178. https://doi.org/10.1016/j.colsurfb.2009.08.028

Fayaz MA, Balaji K, Kalaichelvan PT, Venkatesan R (2009) Fungal based synthesis of silver nanoparticles-an effect of temperature on the size of particles. Colloids Surf B: Biointerf. 74:123–126. https://doi.org/10.1016/j.colsurfb.2009.07.002

Fernández-Llamosas H, Castro L, Blázquez ML, Díaz E, Carmona M (2016) Biosynthesis of selenium nanoparticles by *Azoarcus* sp. CIB. Microb Cell Fact 15:109. https://doi.org/10.1186/s12934-016-0510-y

Gade A, Ingle A, Whiteley C, Rai M (2010) Mycogenic metal nanoparticles: progress and applications. Biotechnol Lett 32:593–600. https://doi.org/10.1007/s10529-009-0197-9

Gaidhani S, Singh R, Singh D, Patel U, Shevade K, Yeshvekar R, Chopade BA (2013) Biofilm disruption activity of silver nanoparticles synthesized by *Acinetobacter calcoaceticus* PUCM 1005. Mater Lett 108:324–327. https://doi.org/10.1016/j.matlet.2013.07.023

Gajbhiye M, Kesharwani J, Ingle A, Gade A, Rai M (2009) Fungus-mediated synthesis of silver nanoparticles and their activity against pathogenic fungi in combination with fluconazole. Nanomedicine 5:382–386. https://doi.org/10.1016/j.nano.2009.06.005

Gericke M, Pinches A (2006) Biological synthesis of metal nanoparticles. Hydrometallurgy 83:132–140. https://doi.org/10.1016/j.hydromet.2006.03.019

Gholami-Shabani M, Akbarzadeh A, Norouzian D, Amini A, Gholami-Shabani Z, Imani A, Chiani M, Riazi G, Shams-Ghahfarokhi M, Razzaghi-Abyaneh M (2014) Antimicrobial activity and physical characterization of silver nanoparticles green synthesized using nitrate reductase from *Fusarium oxysporum*. Appl Biochem Biotechnol 172:4084–4098. https://doi.org/10.1007/s12010-014-0809-2

Gholami-Shabani M, Imani A, Shams-Ghahfarokhi M, Gholami-Shabani Z, Pazooki A, Akbarzadeh A, Riazi G, Razzaghi-Abyaneh M (2016) Bioinspired synthesis, characterization and antifungal activity of enzyme-mediated gold nanoparticles using a fungal oxidoreductase. J Ir Chem Soc 13:2059–2068. https://doi.org/10.1007/s13738-016-0923-x

Gholami-Shabani M, Shams-Ghahfarokhi M, Gholami-Shabani Z, Akbarzadeh A, Riazi G, Ajdari S, Amani A, Razzaghi-Abyaneh M (2015) Enzymatic synthesis of gold nanoparticles using sulfi te reductase purifi ed from *Escherichia coli:* a green eco-friendly approach. Proc Biochem 50:1076–1085. https://doi.org/10.1016/j.procbio.2015.04.004

Goodsell DS (2004) Bionanomedicine in action. In: Bionanotechnology: lessons from nature. Wiley, Hoboken, NJ, USA

Gopal JV, Thenmozhi M, Kannabiran K, Rajakumar G, Velayutham K, Rahuman AA (2013) Actinobacteria mediated synthesis of gold nanoparticles using *Streptomyces* sp. VITDDK3 and its antifungal activity. Mat Lett 93:360–362. https://doi.org/10.1016/j.matlet.2012.11.125

Gopinath V, Velusamy P (2013) Extracellular biosynthesis of silver nanoparticles using *Bacillus* sp. GP-23 and evaluation of their antifungal activity towards *Fusarium oxysporum*. Spectrochim Acta Part A: Mol Biomol Spectr 106:170–174. https://doi.org/10.1016/j.saa.2012.12.087

Guilger M, Pasquoto-Stigliani T, Bilesky-Jose N, Grillo R, Abhilash PC, Fraceto LF, Lima RD (2017) Biogenic silver nanoparticles based on *Trichoderma harzianum*: synthesis, characterization, toxicity evaluation and biological activity. Sci. Rep. 7:44421. https://doi.org/10.1038/srep44421

Guilger-Casagrande M, De Lima R (2009) Synthesis of silver nanoparticles mediated by fungi: a review. Front Bioeng Biotechnol 7:287. https://doi.org/10.3389/fbioe.2019.00287

Gurunathan S, Kalishwaralal K, Vaidyanathan R, Venkataraman D, Pandian SRK, Muniyandi J, Hariharan N, Eom SH (2009) Biosynthesis, purifi cation and characterization of silver nanoparticles using *Escherichia coli*. Colloids Surf B Biointerf 74:328–335. https://doi.org/10.1016/j.colsurfb.2009.07.048

Hamedia S, Ghaseminezhad M, Shokrollahzadeh S, Shojaosadati SB (2017) Controlled biosynthesis of silver nanoparticles using nitrate reductase enzyme induction of filamentous fungus and their antibacterial evaluation. Artif Cells Nanomed Biotechnol 45:1588–1596. https://doi.org/10.1080/21691401.2016.1267011

He M, Protesescu L, Caputo R, Krumeich F, Kovalenko MV (2015) A general synthesis strategy for monodisperse metallic and metalloid nanoparticles (In, Ga, Bi, Sb, Zn, Cu, Sn, and their alloys) via in situ formed metal long-chain amides. Chem Mater 27:635–647. https://doi.org/10.1021/cm5045144

He S, Guo Z, Zhang Y, Zhang S, Wang J, Gu N (2007) Biosynthesis of gold nanoparticles using the bacteria *Rhodopseudomonas capsulata*. Mater Lett 61:3984–3987. https://doi.org/10.1016/j.matlet.2007.01.018

Helan JC, Anand LFA, Karthick SRN, Bharani RSA (2013) Improved pesticidal activity of fungal metabolite from Nomureae rizeyi with chitosan nanoparticles. In: Proceedings of the "International conference on advanced nanomaterials and emerging engineering technologies", India, pp 387–390

Honary S, Barabadi H, Gharaei-Fatahabad E, Naghibi F (2013) Green synthesis of silver nanoparticles induced by the fungus *Penicillium citrinum*. Trop J Pharmaceut Res 12:7–11. https://doi.org/10.4314/tjpr.v12i1.2

Honary S, Gharaei-Fathabad E, Paji ZK, Eslamifar M (2012) A novel biological synthesis of gold nanoparticle by Enterobacteriaceae family. Trop J Pharmaceut Res 11:887–891. https://doi.org/10.4314/tjpr.v11i6.3

Hossain A, Hong X, Ibrahim E, Li B, Sun G, Meng Y, Wang Y, An Q (2019) Green synthesis of silver nanoparticles with culture supernatant of a Bacterium *Pseudomonas rhodesiae* and their antibacterial activity against soft rot pathogen *Dickeya dadantii*. Molecules 24:2303. https://doi.org/10.3390/molecules24122303

Husseiny MI, Abd El-Aziz M, Badr Y, Mahmoud MA (2007) Biosynthesis of gold nanoparticles using *Pseudomonas aeruginosa*. Spectrochim Acta Part A: Mol Biomol Spectr 67:1003–1006. https://doi.org/10.1016/j.saa.2006.09.028

Ibrahim E, Fouad H, Zhang M, Zhang Y, Qiu W, Yan C, Li B, Mo J, Chen J (2019) Biosynthesis of silver nanoparticles using endophytic bacteria and their role in inhibition of rice pathogenic bacteria and plant growth promotion. RSC Adv 9:29293–29299. https://doi.org/10.1039/c9ra04246f

Ibrahim E, Zhang M, Zhang Y, Hossain A, Qiu W, Chen Y, Wang Y, Wu W, Sun G, Li B (2020) Green-synthesization of silver nanoparticles using endophytic bacteria isolated from garlic and its antifungal activity against wheat *Fusarium* head blight pathogen *Fusarium graminearum*. Nanomaterials 10:219. https://doi.org/10.3390/nano10020219

Ingle A, Gade A, Pierrat S ,Sönnichsen C, Rai M (2008) Mycosynthesis of silver nanoparticles using the fungus fusarium acuminatum and its activity against some human pathogenic bacteria. Cur Nanosci 4:141–144

Isaac G, Renitta RE (2015) Brown Algae mediated synthesis, characterization of gold nanoparticles using *Padina pavonica* and their antibacterial activity against human pathogen. Int J Pharm Tech Res 8:31–40

Jena J, Pradhan N, Aishvarya V, Nayak RR, Dash BP, Sukla LB, Panda PK, Mishra BK (2015) Biological sequestration and retention of cadmium as CdS nanoparticles by the microalga *Scenedesmus*-24. J Appl Phycol 27:2251–2260. https://doi.org/10.1007/s10811-014-0499-8

Jena J, Pradhan N, Nayak RR, Dash BP, Sukla LB, Panda PK, Mishra BK (2014) Microalga Scenedesmus sp.: a potential low-cost green machine for silver nanoparticle synthesis. J Microbiol Biotechnol 24:522–533. https://doi.org/10.4014/jmb.1306.06014

Joshi SM, Britto SD, Jogaiah S, Ito S (2019) Mycogenic selenium nanoparticles as potential new generation broad spectrum antifungal Molecules. Biomolecules 9:419. https://doi.org/10.3390/biom9090419

Juibari MM, Abbasalizadeh S, Jouzani GS, Noruzi M (2011) Intensified biosynthesis of silver nanoparticles using a native extremophilic *Ureibacillus thermosphaericus* strain. Mater Lett 65:1014–1017. https://doi.org/10.1016/j.matlet.20. https://doi.org/10.12.056

Kaler A, Jain S, Banerjee UC (2013) Green and rapid synthesis of anticancerous silver nanoparticles by *Saccharomyces boulardii* and insight into mechanism of nanoparticle synthesis. Biomed Res Int 1:2013. https://doi.org/10.1155/2013/872940

Kalishwaralal K, Gopalram S, Vaidyanathan R, Deepak V, Kumar SR, Gurunathan PS (2010) Optimization of α-amylase production for the green synthesis of gold nanoparticles. Colloids Surf B: Biointerf 77:174–180. https://doi.org/10.1016/j.colsurfb.20. https://doi.org/10.01.018

Kalpana VN, Kataru BA, Sravani N, Vigneshwari T, Panneerselvam A, Rajeswari VD (2018) Biosynthesis of zinc oxide nanoparticles using culture filtrates of *Aspergillus niger*: antimicrobial textiles and dye degradation studies. OpenNano 3:48–55. https://doi.org/10.1016/j.onano.2018.06.001

Kannan RRR, Arumugam R, Ramya D, Manivannan K, Anantharaman P (2013) Green synthesis of silver nanoparticles using marine macroalga *Chaetomorpha linum*. Appl Nanosci 3:229–233. https://doi.org/10.1007/s13204-012-0125-5

Kashyap PL, Kumar S, Srivastava AK, Sharma AK (2013) Myconanotechnology in agriculture: a perspective. World J Microbiol Biotechnol 29:191–207. https://doi.org/10.1007/s11274-012-1171-6

Karbasian M, Atyabi SM, Siadat SD, Momem SB, Norouzian D (2008) Optimizing nano-silver formation by Fusarium oxysporum (PTCC 5115) employing response surface methodology. Am J Agri Biol Sci 3:433–437. https://doi.org/10.3844/ajabssp.2008.433.437

Karthik L, Gaurav K, Tarun K, Arindam B, Palakashi R, Bhaskara R (2013) Marine actinobacterial mediated gold nanoparticles synthesis and their antimalarial activity. Nanomedicine 9:951–960. https://doi.org/10.1016/j.nano.2013.02.002

Kathiresan K, Manivannan S, Nabeel MA, Dhivya B (2009) Studies on silver nanoparticles synthesized by a marine fungus, *Penicillium fellutanum* isolated from coastal mangrove sediment. Colloids Surf B: Biointerf 71:133–137. https://doi.org/10.1016/j.colsurfb.2009.01.016

Khan SA, Ahmad A (2014) Enzyme mediated synthesis of water-dispersible, naturally prtein capped, monodispersed gold nanoparticles; their characterization and mechanistic aspects.RSC Adv 4:7729–7734. https://doi.org/10.1039/c3ra43888k

Khandel P, Shahi SK (2018) Mycogenic nanoparticles and their bio-prospective applications: current status and future challenges. J Nanostr Chem 8:369–391. https://doi.org/10.1007/s40097-018-0285-2

Kim DY, Saratale RG, Shinde S, Syed A, Ameen F, Ghodake G (2018) Green synthesis of silver nanoparticles using *Laminaria japonica* extract: characterization and seedling growth assessment. J Cleaner Prod 172:2910–2918. https://doi.org/10.1016/j.jclepro.2017.11.123

Klaus T, Joerger R, Olsson E, Granqvist CG (1999) Silver-based crystalline nanoparticles, microbially fabricated. Proc Natl Acad Sci 96:13611–13614. https://doi.org/10.1073/pnas.96.24.13611

Konishi Y, Tsukiyama T, Tachimi T, Saitoh N, Nomura T, Nagamine S (2007) Microbial deposition of gold nanoparticles by the metal-reducing bacterium *Shewanella* algae. Electrochim Acta 5:186–192. https://doi.org/10.1016/j.electacta.2007.02.073

Korbekandia H, Asharia Z, Iravanib S, Abbasi S (2013) Optimization of biological synthesis of silver nanoparticles using fusarium oxysporum. Iran J Pharm Res 12: 289–298

Korbekandi H, Iravani S, Abbasi S (2012) Optimization of biological synthesis of silver nanoparticles using *Lactobacillus casei* subsp. *casei*. J Chem Technol Biotechnol 87:932–937. https://doi.org/10.1002/jctb.3702

Koudelka KJ, Pitek AS, Manchester M, Steinmetz NF (2015) Virus-based nanoparticles as versatile nanomachines. Ann Rev Virol 2:379–401. https://doi.org/10.1146/annurev-virology-100114-055141

Kumar B, Smita K, Sánchez E, Guerra S, Cumbal L (2016) Ecofriendly ultrasound-assisted rapid synthesis of gold nanoparticles using *Calothrix* algae. Adv Nat Sci: Nanosci Nanotechnol 7:025013. https://doi.org/10.1088/2043-6262/7/2/025013

Kumar CG, Mamidyala SK (2011) Extracellular synthesis of silver nanoparticles using culture supernatant of *Pseudomonas aeruginosa*. Colloids Surf B Biointerf 84:462–426. https://doi.org/10.1016/j.colsurfb.2011.01.042

Kumar CG, Poornachandra Y, Mamidyala SK (2014) Green synthesis of bacterial gold nanoparticles conjugated to resveratrol as delivery vehicles. Colloids Surf B: Biointerf 123:311–317. https://doi.org/10.1016/j.colsurfb.2014.09.032

Kumar CG, Mamidyala SK (2011b) Extracellular synthesis of silver nanoparticles using culture supernatant of *Pseudomonas aeruginosa*. Colloids Surf B Biointerf 84:462–466. https://doi.org/10.1016/j.colsurfb.2011.01.042

Kumar P, Senthamil Selvi S, Lakshmi Prabha A, Prem Kumar K, Ganeshkumar RS, Govindaraju M (2012) Synthesis of silver nanoparticles from *Sargassum tenerrimum* and screening phytochemicals for its antibacterial activity. Nano Biomed Eng 4:12–16. https://doi.org/10.5101/nbe.v4i1.p12-16

Kumar SA, Abyaneh MK, Gosavi SW, Kulkarni SK, Pasricha R, Ahmad A, Khan MI (2007a) Nitrate reductase-mediated synthesis of silver nanoparticles from $AgNO_3$. Biotechnol Lett 29:439–445. https://doi.org/10.1007/s10529-006-9256-7

Kumar SA, Ansary AA, Ahmad A, Khan MI (2007) Extracellular biosynthesis of CdSe quantum dots by the fungus, *Fusarium oxysporum*. J Biomed Nanotechnol 3:190–194. https://doi.org/10.1166/jbn.2007.027

Kumar SA, Peter YA, Nadeau JL (2008) Facile biosynthesis, separation and conjugation of gold nanoparticles to doxorubicin. Nanotechnology 19:495101. https://doi.org/10.1088/0957-4484/19/49/495101

Kumar SK, Arumugam RAP, Berchmans S (2011) Synthesis of gold nanoparticles: an ecofriendly approach using *Hansenula anomala*. ACS Appl Mater Interf 3:1418–1425. https://doi.org/10.1021/am200443j

Lateef A, Adelere IA, Gueguim-Kana EB, Asafa TB, Beukes LS (2015a) Green synthesis of silver nanoparticles using keratinase obtained from a strain of *Bacillus safensis* LAU 13. Int Nano Lett 5:29–35. https://doi.org/10.1007/s40089-014-0133-4

Lateef A, Akande MA, Ojo SA, Folarin BI, Gueguim-Kana EB, Beukes LS (2016b) Paper wasp nest-mediated biosynthesis of silver nanoparticles for antimicrobial, catalytic, anti-coagulant and thrombolytic applications. 3Biotech 6:140. http://dx.doi.org/10.1007/s13205-016-0459-x

Lateef A, Ojo SA, Elegbede JA (2016a) The emerging roles of arthropods and their metabolites in the green synthesis of metallic nanoparticles. Nanotechnol Rev 5:601–622. https://doi.org/10.1515/ntrev-2016-0049

Lateef A, Ojo SA, Akinwale AS, Azeez L, Gueguim-Kana EB, Beukes LS (2015c) Biogenic synthesis of silver nanoparticles using cell-free extract of *Bacillus safensis* LAU 13: antimicrobial, free radical scavenging and larvicidal activities. Biologia 70:1295–1306. http://dx.doi.org/10.1515/biolog-2015-0164

Lee SH, Jun BH (2019) Silver nanoparticles: synthesis and application for nanomedicine. Int J Mol Sci 20:E865. https://doi.org/10.3390/ijms20040865

Lengke MF, Fleet ME, Southam G (2006) Synthesis of platinum nanoparticles by reaction of filamentous cyanobacteria with platinum (IV)-chloride complex. Langmuir 22:7318–7323. https://doi.org/10.1021/la060873s

Li G, He D, Qian Y, Guan B, Gao S, Cui Y, Yokoyama K, Wang L (2012) Fungus-mediated green synthesis of silver nanoparticles using Aspergillus terreus. Int J of Mol Sci 13: 466–476. https://doi.org/10.3390/ijms13010466

Li Y, Li Y, Li Q, Fan X, Gao J, Luo Y (2016) Rapid biosynthesis of gold nanoparticles by the extracellular secretion of *Bacillus niabensis* 45: characterization and antibiofilm activity. J Chem 1:2016. https://doi.org/10.1155/2016/2781347

Madhiyazhagan P, Murugan K, Kumar AN, Nataraj T, Dinesh D, Panneerselvam C, Subramaniam J, Kumar PM, Suresh U, Roni M, Nicoletti M (2015) *Sargassum muticum*-synthesized silver nanoparticles: an effective control tool against mosquito vectors and bacterial pathogens. Parasitol Res 114:4305–4317. https://doi.org/10.1007/s00436-015-4671-0

Maharani V, Sundaramanickam A, Balasubramanian T (2016) *In vitro* anticancer activity of silver nanoparticle synthesized by *Escherichia coli* VM1 isolated from marine sediments of Ennore southeast coast of India. Enzyme Microb Technol 95:146–154. https://doi.org/10.1016/j.enzmictec.2016.09.008

Mahdavi M, Namvar F, Ahmad MB, Mohamad R (2013) Green biosynthesis and characterization of magnetic iron oxide (Fe_3O_4) nanoparticles using seaweed (*Sargassum muticum*) aqueous extract. Molecules 18:5954–5964. https://doi.org/10.3390/molecules18055954

Mahdieh M, Zolanvari A, Azimee AS (2012) Green biosynthesis of silver nanoparticles by *Spirulina platensis*. Scientia Iranica 19:926–929. https://doi.org/10.1016/j.scient.2012.01.010

Mala R, Arunachalam P, Sivasankari M (2012) Synergistic bactericidal activity of silver nanoparticles and ciprofloxacin against phytopathogens. J Cell Tissue Res 12:3249–3254

Malarkodi C, Rajeshkumar S, Paulkumar K, Vanaja M, Jobitha GDG, Annadurai G (2013) Bactericidal activity of bio mediated silver nanoparticles synthesized by *Serratia nematodiphila*. Drug Invent Today 5:119–125. https://doi.org/10.1016/j.dit.2013.05.005

Maliszewska I, Juraszek A, Bielska K (2013) Green synthesis and characterization of silver nanoparticles using Ascomycota fungi *Penicillium nalgiovense* AJ12. J Clust Sci 25:989–1004. https://doi.org/10.1007/s10876-013-0683-z

Mallick K, Witcomb MJ, Scurrell MS (2004) Polymer stabilized silver nanoparticles: a photochemical synthesis route. J Mat Sci 39:4459–4463

Mandal D, Bolander ME, Mukhopadhyay D, Sarkar G, Mukherjee P (2006) The use of microorganisms for the formation of metal nanoparticles and their application. Appl Microbiol Biotechnol 69:485–492. https://doi.org/10.1007/s00253-005-0179-3

Manikprabhu D, Lingappa K (2013) Antibacterial activity of silver nanoparticles against methicillin-resistant *Staphylococcus aureus* synthesized using model *Streptomyces* sp. Pigment by photo-irradiation method. J Pharmaceut Res 6:255–260. https://doi.org/10.1016/j.jopr.2013.01.022

Manivasagan P, Venkatesan J, Senthilkumar K, Sivakumar K, Kim S (2013) Biosynthesis, antimicrobial and cytotoxic effect of silver nanoparticles using a novel *Nocardiopsis* sp. MBRC-1. BioMed Res Int 2013. http://dx.doi.org/10.1155/2013/287638

Mishra A, Kumari M, Pandey S, Chaudhry V, Gupta KC, Nautiyal CS (2014b) Biocatalytic and antimicrobial activities of gold nanoparticles synthesized by *Trichoderma* sp. Bioresour Technol 166:235–242. https://doi.org/10.1016/j.biortech.2014.04.085

Mishra A, Tripathy SK, Wahab R, Jeong SH, Hwang I, Yang YB, Kim YS, Shin HS, Yun SI (2011a) Microbial synthesis of gold nanoparticles using the fungus Penicillium brevicompactum and their cytotoxic effects against mouse mayo blast cancer C 2 C 12 cells. Appl Microbiol Biotechnol 92:617–630. https://doi.org/10.1007/s00253-011-3556-0

Mishra A, Tripathy SK, Yun SI (2011b) Bio-synthesis of gold and silver nanoparticles from *Candida guilliermondii* and their antimicrobial effect against pathogenic bacteria. J Nanosci Nanotechnol 11:243–248. https://doi.org/10.1166/jnn.2011.3265

Mishra A, Tripathy SK, Yun SI (2012) Fungus mediated synthesis of gold nanoparticles and their conjugation with genomic DNA isolated from *Escherichia coli* and *Staphylococcus aureus*. Process Biochem 47:701–711. https://doi.org/10.1016/j.procbio.2012.01.017

Mishra S, Singh BR, Singh A, Keswani C, Naqvi AH, Singh HB (2014a) Biofabricated silver nanoparticles act as a strong fungicide against *Bipolaris sorokiniana* causing spot blotch disease in wheat. PLoS ONE 9:e97881. https://doi.org/10.1371/journal.pone.0097881

Mittal AK, Chisti Y, Banerjee UC (2013) Synthesis of metallic nanoparticles using plant extracts. Biotechnol Adv 31:346–356. https://doi.org/10.1016/j.biotechadv.2013.01.003

Mmola M, Roes-Hill ML, Durrell K, Bolton JJ, Sibuyi N, Meyer ME, Beukes DR, Antunes E (2016) Enhanced antimicrobial and anticamcer activity of silver and gold nanoparticles synthesized using *Sargassum incisifolium* aqueous extracts. Molecules 21:1633. https://doi.org/10.3390/molecules21121633

Moghaddam AB, Moniri M, Azizi S, Rahim RA, Ariff AB, Saad WZ, Namvar F, Navaderi M, Mohamad R (2017) Biosynthesis of ZnO nanoparticles by a new Pichia kudriavzevii yeast strain and evaluation of their antimicrobial and antioxidant activities. Molecules 22:872. https://doi.org/10.3390/molecules22060872

Moghaddam AB, Namvar F, Moniri M1, Tahir PM, Azizi S, Mohamad R (2015) Nanoparticles biosynthesized by fungi and yeast: a review of their preparation, properties, and medical applications. Molecules 20:16540–16565. https://doi.org/10.3390/molecules200916540

Mohammadian A, Shojaosadati Rezaee MH (2007) *Fusarium oxysporum* mediates photogeneration of silver nanoparticles. Scientia Iranica 14:323–326

Momeni S, Nabipour I (2015) A simple green synthesis of palladium nanoparticles with *Sargassum* alga and their electrocatalytic activities towards hydrogen peroxide. Appl Biochem Biotechnol 176:1937–1949. https://doi.org/10.1007/s12010-015-1690-3

Moussa SH, Tayel AA, Alsohim AS, Abdallah RR (2013) Botryticidal activity of nanosized silver-chitosan composite and its application for the control of gray mold in strawberry. J Food Sci 78:M1589–M1594. https://doi.org/10.1111/1750-3841.12247

MubarakAli D, Arunkumar J, Nag KH, SheikSyedIshack KA, Baldev E, Pandiaraj D, Thajuddin N (2013) Gold nanoparticles from Pro and eukaryotic photosynthetic microorganisms—comparative studies on synthesis and its application on biolabelling. Colloids Surf B: Biointerf 103:166–173. https://doi.org/10.1016/j.colsurfb.2012.10.014

Mukherjee K, Gupta R, Kumar G, Kumari S, Biswas S, Padmanabhan P (2018) Synthesis of silver nanoparticles by *Bacillus clausii* and computational profiling of nitrate reductase enzyme involved in production. J Genet Eng Biotechnol 6:527–536. https://doi.org/10.1016/j.jgeb.2018.04.004

Mukherjee P, Roy M, Mandal BP, Dey GK, Mukherjee PK, Ghatak J, Tyagi AK, Kale SP (2008) Green synthesis of highly stabilized nanocrystalline silver particles by a non-pathogenic and agriculturally important fungus T. asperellum. Nanotechnol 19:1–7

Mukherjee P, Senapati S, Mandal D, Ahmad A, Khan MI, Kumar R, Sastry M (2002) Extracellular synthesis of gold nanoparticles by the fungus *Fusarium oxysporum*. ChemBioChem 3:461–463. https://doi.org/10.1002/1439-7633(20020503)3:5%3c461::AID-CBIC461%3e3.0.CO;2-X

Murugan K, Samidoss CM, Panneerselvam C, Higuchi A, Roni M, Suresh U, Chandramohan B, Subramaniam J, Madhiyazhagan P, Dinesh D, Rajaganesh R (2015) Seaweed-synthesized silver nanoparticles: an eco-friendly tool in the fight against Plasmodium falciparum and its vector *Anopheles stephensi*? Parasitol Res 114:4087–4097. http://doi.org/10.1007/s00436-015-4638-1

Nadaf NY, Kanase SS (2019) Biosynthesis of gold nanoparticles by *Bacillus marisflavi* and its potential in catalytic dye degradation. Arab J Chem 12:4806–4814. https://doi.org/10.1016/j.arabjc.2016.09.020

Nagarajan S, Kuppusamy KA (2013) Extracellular synthesis of zinc oxide nanoparticle using seaweeds of Gulf of Mannar, India. J Nanobiotechnol 11:39. https://doi.org/10.1186/1477-3155-11-39

Namvar F, Rahman HS, Mohamad R, Rasedee A, Yeap SK, Chartrand MS, Azizi S, Tahir PM (2015) Apoptosis induction in human leukemia cell lines by gold nanoparticles synthesized using the green biosynthetic approach. J Nanomater 2015:642621. https://doi.org/10.1155/2015/642621

Nangia Y, Wangoo N, Goyal N, Shekhawat G, Suri CR (2009) A novel bacterial isolate Stenotrophomonas maltophilia as living factory for synthesis of gold nanoparticles. Microb Cell Fact 8:39. https://doi.org/10.1186/1475-2859-8-39

Narayanan KB, Sakthivel N (2011) Facile green synthesis of gold nanostructures by NADPH-dependent enzyme from the extract of *Sclerotium rolfsii*. Colloids Surf A: Physicochem Eng Aspects 380:156–161. https://doi.org/10.1016/j.colsurfa.2011.02.042

Nayak RR, Pradhan N, Behera D, Pradhan KM, Mishra S, Sukla LB, Mishra BK (2011) Green synthesis of silver nanoparticle by *Penicillium purpurogenum* NPMF: the process and optimization. J Nanopart Res 13:3129–3137. https://doi.org/10.1007/s11051-010-0208-8

Nejad S, Shahidi Bonjar M, Hosein G, Naimeh K (2015) Biosynthesis of gold nanoparticles by *Strep-tomyces fulvissimus*. Nanomed J 2:153–159. https://doi.org/10.7508/nmj.2015.02.007

Ng CK, Sivakumar K, Liu X, Madhaiyan M, Ji L, Yang L, Tang C, Song H, Kjelleberg S, Cao B (2013) Infl uence of outer membrane c-type cytochromes on particle size and activity of extracellular nanoparticles produced by Shewanella oneidensis. Biotechnol Bioeng 110:1831–1837. https://doi.org/10.1002/bit.24856

Ojo SA, Lateef A, Azeez MA, Oladejo SM, Akinwale AS, Asafa TB, Yekeen TA, Akinboro A, Oladipo IC, Gueguim-Kana EB, Beukes LS (2016) IEEE Trans NanoBiosci 15:433–442. https://doi.org/10.1109/TNB.2016.2559161

Otari SV, Kumar M, Kim I-W, Lee JH, Lee JK (2017) Rapid, thermostable antimicrobial peptide-mediated synthesis gold nanoparticles as highly efficient charge trapping medium for sol-gel-derived thin film. Mater Lett 188:375–378. https://doi.org/10.1016/j.matlet.2016.11.104

Otari SV, Patil RM, Nadaf NH, Ghosh SJ, Pawar SH (2012) Green biosynthesis ofsilver nanoparticles from an actinobacteria *Rhodococcus* sp. Mater Lett 72:92–94. https://doi.org/10.1016/j.matlet.2011.12.109

Otari SV, Patil RM, Nadaf NH, Ghosh SJ, Pawar SH (2014) Green synthesis of silver nanoparticles by microorganism using organic pollutant: its antimicrobial and catalytic application. Environ Sci Pollut Res 21:1503–1513. https://doi.org/10.1007/s11356-013-1764-0

Oves M, Khan MS, Zaidi A, Ahmed AS, Ahmed F, Ahmad E, Sherwani A, Owais M, Azam A (2013) Antibacterial and cytotoxic effi cacy of extracellular silver nanoparticles biofabricated from chromium reducing novel OS4 strain of *Stenotrophomonas maltophilia*. PLoS ONE 8:e59140. https://doi.org/10.1371/journal.pone.0059140

Oza G, Pandey S, Mewada A, Kalita G, Sharon M, Phata J, Ambernath W, Sharon M (2012) Facile biosynthesis of gold nanoparticles exploiting optimum pH and temperature of fresh water algae *Chlorella pyrenoidusa*. Adv Appl Sci Res 3:1405–1412

Parial P, Patra HK, Dasgupta AKP, Pal P (2012) Screening of different algae for synthesis of gold nanoparticles. Eur J Phycol 47:22–29. https://doi.org/10.1080/09670262.2011.653406

Phanjom P, Ahmed G (2017) Effect of different physicochemical conditions on the synthesis of silver nanoparticles using fungal cell filtrate of Aspergillus oryzae (MTCC No. 1846) and their antibacterial effect. Adv Nat Sci: Nanosci Nanotechnol 8:045016. https://doi.org/10.1088/2043-6254/aa92bc

Pour MM, Saberi-Riseh R, Mohammadinejad R, Hosseini A (2019) Investigating the formulation of alginate-gelatin encapsulated *Pseudomonas fluorescens* (VUPF5 and T17-4 strains) for controlling *Fusarium solani* on potato. Int J Biol Macromol 133:603–613. https://doi.org/10.1016/j.ijbiomac.2019.04.071

Pourali P, Badiee SH, Manafi S, Noorani T, Rezaei A, Yahyaei B (2017) Biosynthesis of gold nanoparticles by two bacterial and fungal strains, *Bacillus cereus* and *Fusarium oxysporum*, and assessment and comparison of their nanotoxicity in vitro by direct and indirect assays. Electron J Biotechnol 29:86–93. https://doi.org/10.1016/j.ejbt.2017.07.005

Prasad K, Jha AK (2010) Biosynthesis of CdS nanoparticles: an improved green and rapid procedure. J Colloids Interf Sci 342:68–72. https://doi.org/10.1016/j.jcis.2009.https://doi.org/10.003

Prasad KS, Vaghasiya JV, Soni SS, Patel J, Patel R, Kumari M, Jasmani F, Selvaraj K (2015) Microbial selenium nanoparticles (SeNPs) and their application as a sensitive hydrogen peroxide biosensor. Appl Biochem Biotechnol 177:1386–1393. https://doi.org/10.1007/s12010-015-1814-9

Prathna TC, Mathew L, Chandrasekaran N, Raichur AM, Mukherjee A (2010) Biomimetic synthesis of nanoparticles: science, technology & applicability. In Biomimetics learning from nature 2010 Mar 1. Intechopen. https://doi.org/10.5772/8776

Priyadarshini E, Pradhan N, Sukla LB, Panda PK (2014) Controlled synthesis of gold nanoparticles using *Aspergillus terreus* IF0 and its antibacterial potential against Gram negative pathogenic bacteria. J Nanotechnol 1:2014. https://doi.org/10.1155/2014/653198

Priyadharshini RI, Prasannaraj G, Geetha N, Venkatachalam P (2014) Microwave-mediated extracellular synthesis of metallic silver and zinc oxide nanoparticles using macro-algae (*Gracilaria edulis*) extracts and its anticancer activity against human PC3 cell lines. Appl Biochem Biotechnol 174:2777–2790. https://doi.org/10.1007/s12010-014-1225-3

Qi P, Zhang D, Wan Y (2013) Sulfate-reducing bacteria detection based on the photocatalytic property of microbial synthesized ZnS nanoparticles. Anal Chim Acta 800:65–70. https://doi.org/10.1016/j.aca.2013.09.015

Qian Y, Yu H, He D, Yang H, Wang W, Wan X, Wang L (2013) Biosynthesis of silver nanoparticles by the endophytic fungus *Epicoccum nigrum* and their activity against pathogenic fungi. Bioproc Biosyst Eng 36:1613–1619. https://doi.org/10.1007/s00449-013-0937-z

Qu Y, Shen W, Pei X, Ma F, You S, Li S, Wang J, Zhou J (2016) Biosynthesis of gold nanoparticles by *Trichoderma* sp. WL-Go for azo dyes decolorization. J Environ Sci 56:79–86. http://dx.doi.org/10.1016/j.jes.2016.09.007

Quester K, Avalos-Borja M, Vilchis-Nestor AR, Camacho-López MA, Castro-Longoria E (2013) SERS properties of different sized and shaped gold nanoparticles biosynthesized under different environmental conditions by *Neurospora crassa* extract. PLoS ONE 8:e77486. https://doi.org/10.1371/journal.pone.0077486

Rajakumara G, Rahumana AA, Roopanb SM, Khannac VG, Elangoa G, Kamaraja C, Zahira AA, Velayuthama K (2012) Fungus-mediated biosynthesis and characterization of TiO_2 nanoparticles and their activity against pathogenic bacteria. Spectrochim Acta A Mol Biomol Spectr 91:23–29. https://doi.org/10.1016/j.saa.2012.01.011

Rajathi FAA, Parthiban C, Kumar VG, Anantharaman P (2012) Biosynthesis of antibacterial gold nanoparticles using brown alga, *Stoechospermum marginatum* (Kutzing). Spectrochim Acta Part A 99:166–173. https://doi.org/10.1016/j.saa.2012.08.081

Rajeshkumar S, Kumar SV, Malaakodi C, Vanaja M, Paulkumar K, Annadurai G (2017) Optimized synthesis of gold nanoparticles using green chemical process and its in vitro anticancer activity against HepG2 and A549 cell lines, mechanics. Mater Sci and Eng. ISSN 2412–5954. https://doi.org/10.2412/mmse.95.26.479

Rajeshkumar S, Malarkodi C, Gnanajobitha G, Paulkumar K, Vanaja M, Kannan C, Annadurai G (2013) Seaweed-mediated synthesis of gold nanoparticles using *Turbinaria conoides* and its characterization. J Nanostruct Chem 3:44. https://doi.org/10.1186/2193-8865-3-44

Raliya R, Biswas P, Tarafdar JC (2015) TiO2 nanoparticle biosynthesis and its physiological effect on mung bean (*Vigna radiata* L.). Biotechnol Rep 5:22–26. https://doi.org/10.1016/j.btre.2014.10.009

Raliya R, Tarafdar JC (2013) ZnO nanoparticle biosynthesis and its effect on phosphorous-mobilizing enzyme secretion and gum contents in clusterbean (*Cyamopsis tetragonoloba* L.). Agri Res 2:48–57. https://doi.org/10.1007/s40003-012-0049-z

Raliya R, Tarafdar JC, Choudhary K, Mal P, Raturi A, Gautam R, Singh SK (2014a) Synthesis of MgO nanoparticles using *Aspergillus tubingensis* TFR-3. J Bionanosci 8:34–38. https://doi.org/10.1166/jbns.2014.1195

Raliya R, Tarafdar JC, Singh SK, Gautam R, Choudhary K, Maurino VG, Saharan V (2014b) MgO nanoparticles biosynthesis and its effect on chlorophyll contents in the leaves of Clusterbean (*Cyamopsis tetragonoloba* L.). Adv Sci Eng Med 6:538–545. https://doi.org/10.1166/asem.2014.1540

Ramakritinan CM, Kaarunya E, Shankar S, Kumaraguru AK (2013) Antibacterial effects of Ag, Au and bimetallic (Ag-Au) nanoparticles synthesized from red algae. Trans Tech Publications, pp 211–230. https://doi.org/10.4028/www.scientific.net/SSP.201.211

Rose GK, Soni R, Rishi P, Soni SK (2019) Optimization of the biological synthesis of silver nanoparticles using *Penicillium oxalicum* GRS-1 and their antimicrobial effects against common food-borne pathogens. Green Process Synth 8:144–156. https://doi.org/10.1515/gps-2018-0042
Roychoudhury P, Pal R (2014) Synthesis and characterization of nanosilver—a blue green approach. Indian J Appl Res 4:54–56
Rudramurthy GR, Swamy MK (2018) Potential applications of engineered nanoparticles in medicine and biology: an update. J Biol Inorg Chem 23:1185–1204. https://doi.org/10.1007/s00775-018-1600-6
Rudramurthy GR, Swamy MK, Sinniah UR, Ghasemzadeh A (2016) Nanoparticles: alternatives against drug-resistant pathogenic microbes. Mol J 21:836. https://doi.org/10.3390/molecules21070836
Sabbour MM (2014) Evaluating toxicity of nano-extracted Destruxin from *Metarhizium anisopliae* against the grasshopper *Hetiracris littoralis* in Egypt. J Egypt Acad Soc Env Dev 15:1–7. https://doi.org/10.13140/rg.2.1.2116.3365
Sabbour MM (2015) Laboratory and store efficacy of nano-extracted Destruxin from *Metarhizium anisopliae* against Indian meal moth Plodia interpunctella (Lepidoptlera-Pyralidae). J Nanosci Nanoeng 1:142–147
Saifuddin N, Wong CW, Yasumira AAN (2009) Rapid biosynthesis of silver nanoparticles using culture supernatant of bacteria with microwave irradiation. Eur J Chem 6:61–70. https://doi.org/10.1155/2009/734264
Salari Z, Danafar F, Dabaghi S, Ataei SA (2016) Sustainable synthesis of silver nanoparticles using macroalgae *Spirogyra varians* and analysis of their antibacterial activity. J Saudi Chem Soc 20:459–464. http://doi.org/10.1016/j.jscs.2014.10.004
Salunkhe RB, Patil SV, Salunke BK, Patil CD, Sonawane AM (2011) Studies on silver accumulation and nanoparticle synthesis by *Cochliobolus lunatus*. Appl Biochem Biotechnol 165:221–234. https://doi.org/10.1007/s12010-011-9245-8
Salvadori MR, Ando RA, Oller do Nascimento CA, Corrêa B (2014) Intracellular biosynthesis and removal of copper nanoparticles by dead biomass of yeast isolated from the wastewater of a mine in the Brazilian Amazonia. PLoS One 29–9(1):e87968. https://doi.org/10.1371/journal.pone.0087968
Sanghi R, Verma P, Puri S (2011) Enzymatic formation of gold nanoparticles using *Phanerochaete chrysosporium*. Adv Chem Eng Sci 1:154. https://doi.org/10.4236/aces.2011.13023
Sanjenbam P, Gopal JV, Kannabiran K (2014) Anticandidal activity of silver nanoparticles synthesized using *Streptomyces* sp.VITPK1. J Med Mycol 24:211–219. https://doi.org/10.1016/j.mycmed.2014.03.04
Sarkar J, Dey P, Saha S, Acharya K (2011) Mycosynthesis of selenium nanoparticles. IET Micro Nano Lett 6:599–602. https://doi.org/10.1049/mnl.2011.0227
Sarkar J, Ray S, Chattopadhyay D, Laskar A, Acharya K (2012) Mycogenesis of gold nanoparticles using a phytopathogen Alternaria alternata. Bioprocess Biosyst Eng 35:637–643. https://doi.org/10.1007/s00449-011-0646-4
Sastry M, Ahmad A, Khan MI, Kumar R (2003) Biosynthesis of metal nanoparticles using fungi and actinomycete. Curr Sci 25:162–170
Sathiyanarayanan G, Vignesh V, Saibaba G, Vinothkanna A, Dineshkumar K, Viswanathan MB, Selvin J (2014) Synthesis of carbohydrate polymer encrusted gold nanoparticles using bacterial exopolysaccharide: a novel and greener approach. RSC Adv 4:22817–22827. https://doi.org/10.1039/c4ra01428f
Saxena J, Sharma PK, Sharma MM, Singh A (2016) Process optimization for green synthesis of silver nanoparticles by Sclerotinia sclerotiorum MTCC 8785 and evaluation of its antibacterial properties. SpringerPlus 5:1–10. https://doi.org/10.1186/s40064-016-2558-x
Sen K, Sinha P, Lahiri S (2011) Time dependent formation of gold nanoparticles in yeast cells: a comparative study. Biochem Eng J 55:1–6. https://doi.org/10.1016/j.bej.2011.02.014

Senapati S, Syed A, Moeez S, Kumar A, Ahmad A (2012) Intracellular synthesis of gold nanoparticles using alga *Tetraselmis kochinensi*. Mater Lett 79:116–118. https://doi.org/10.1016/j.matlet.2012.04.009

Seshadri S, Saranya K, Kowshik M (2011) Green synthesis of lead sulfide nanoparticles by the lead resistant marine yeast, *Rhodosporidium diobovatum*. Biotechnol Progr 27:1464–1469. https://doi.org/10.1002/btpr.651

Shahzad A, Saeed H, Iqtedar M, Hussain SZ, Kaleem A, Abdullah R, Sharif S, Naz S, Saleem F, Aihetasham A, Chaudhary A (2019) Size-controlled production of silver nanoparticles by *Aspergillus fumigatus* BTCB10: likely antibacterial and cytotoxic effects. J Nanomat 1:2019. https://doi.org/10.1155/2019/5168698

Shakibaie M, Forootanfar H, Mollazadeh-Moghaddam K, Bagherzadeh Z, Nafissi-Varcheh N, Shahverdi AR, Faramarzi MA (2010) Green synthesis of gold nanoparticles by the marine microalga *Tetraselmis suecica*. Biotechnol Appl Biochem 57:71–75. https://doi.org/10.1042/BA20100196

Sharma A, Sharma S, Sharma K, Chetri SPK, Vashishtha A, Singh P, Kumar R, Rathi B, Agrawal V (2016) Algae as crucial organisms in advancing nanotechnology: a systematic review. J Appl Phycol 28:1759–1774. https://doi.org/10.1007/s10811-015-0715-1

Sharma B, Purkayastha DD, Hazra S, Gogoi L, Bhattacharjee CR, Ghosh NN, Rout J (2014) Biosynthesis of gold nanoparticles using a freshwater green alga, *Prasiola crispa*. Mater Lett 116:94–97. http://doi.org/10.1016/j.matlet.2013.10.107

Sharma N, Pinnaka AK, Raje M, Ashish FN, Bhattacharyya MS, Choudhury AR (2012) Exploitation of marine bacteria for production of gold nanoparticles. Microbial Cell Fact 11:86. https://doi.org/10.1186/1475-2859-11-86

Sheikhloo Z, Salouti M, Katiraee F (2011) Biological synthesis of gold nanoparticles by fungus Epicoccumnigrum. J Clust Sci 22:661–665. https://doi.org/10.1007/s10876-011-0412-4

Silva LP, Bonatto CC Polez VLP (2016) Green synthesis of metal nanoparticles by fungi: current trends and challenges, ch.4, pp 71–89

Singh D, Rathod V, Ninganagouda S, Hiremath J, Singh AK, Mathew J (2014) Optimization and characterization of silver nanoparticle by endophytic fungi *Penicillium* sp. isolated from Curcuma longa (turmeric) and application studies against MDR *E. coli* and *S. aureus*. Bioinorg Chem Appl 1:2014. https://doi.org/10.1155/2014/408021

Singh H, Du J, Yi TH (2016a) Kinneretia THG-SQI4 mediated biosynthesis of silver nanoparticles and its antimicrobial efficacy. Artif. Cells Nanomed Biotechnol 45:602–608. http://doi.org/10.3109/21691401.2016.1163718

Singh M, Kalaivani R, Manikandan S, Sangeetha N, Kumaraguru AK (2013a) Facile green synthesis of variable metallic gold nanoparticle using *Padina gymnospora*, a brown marine macroalga. Appl Nanosci 3:145–151. https://doi.org/10.1007/s13204-012-0115-7

Singh R, Wagh P, Wadhwani S, Gaidhani S, Kumbhar A, Bellare J, Chopade BA (2013b) Synthesis, optimization, and characterization of silver nanoparticles from *Acinetobacter calcoaceticus* and their enhanced antibacterial activity when combined with antibiotics. Int J Nanomed 8:4277. https://doi.org/10.2147/ijn.s48913277-4290

Singhal U, Khanuja M, Prasad R and Varma A (2017) Impact of Synergistic Association of ZnO-Nanorods and Symbiotic Fungus *Piriformospora indica* DSM 11827 on *Brassica oleracea* var. *botrytis* (Broccoli). Front Microbiol 8:1909. https://doi.org/10.3389/fmicb.2017.01909

Sinha A, Khare SK (2012) Mercury bioremediation by mercury accumulating *Enterobacter* sp. cells and its alginate immobilized application. Biodegradation 23:25–34. https://doi.org/10.1007/s10532-011-9483-z

Sinha SN, Paul D, Halder N, Sengupta D, Patra SK (2015) Green synthesis of silver nanoparticles using fresh water green alga *Pithophora oedogonia* (Mont.) Wittrock and evaluation of their antibacterial activity. Appl Nanosci 5:703–709. https://doi.org/10.1007/s13204-014-0366-6

Spasova M, Manolova N, Naydenov M, Kuzmanova J (2011) Electrospun biohybrid materials for plant biocontrol containing chitosan and *Trichoderma viride* spores. J Bioact Compat Polym 26:48–55. https://doi.org/10.1177/0883911510391446

Sriram MI, Kalishwaralal K, Gurunathan S (2012) Biosynthesis of silver and gold nanoparticles using *Bacillus licheniformis*. In: Soloviev M (ed) Nanoparticles in biology and medicine: methods and protocols. Springer, Dordrecht, pp 33–43. https://doi.org/10.1007/978-1-61779-953-2_3

Subashini J, Khanna VG, Kannabiran K (2014) Anti-ESBL activity of silver nanoparticles biosynthesized using soil *Streptomyces* species. Bioproc Biosyst Eng 37:999–1006. https://doi.org/10.1007/s00449-013-1070-8

Suganya KU, Govindaraju K, Kumar VG, Dhas TS, Karthick V, Singaravelu G, Elanchezhiyan M (2015) Blue green alga mediated synthesis of gold nanoparticles and its antibacterial efficacy against Gram positive organisms. Mater Sci Eng 47:351–356. https://doi.org/10.1016/j.msec.2014.11.043

Suresh AK, Pelletier DA, Wang W, Broich ML, Moon JW, Gu B, Allison DP, Joy DC, Phelps TJ, Doktycz MJ (2011a) Biofabrication of discrete spherical gold nanoparticles using the metal-reducing bacterium *Shewanella oneidensis*. Acta Biomat 7:2148–2152. https://doi.org/10.1016/j.actbio.2011.01.023

Suresh AK, Pelletier DA, Wang W, Broich ML, Moon JW, Gu B, Allison DP, Joy DC, Phelps TJ, Doktycz MJ (2011) Biofabrication of discrete spherical gold nanoparticles using the metal-reducing bacterium *Shewanella oneidensis*. Acta Biomater 7:2148–2152. https://doi.org/10.1016/j.actbio.2011.01.023

Swaminathan S, Murugesan S, Damodarkumar S, Dhamotharan R, Bhuvaneshwari S (2011) Synthesis and characterization of gold nanoparticles from alga *Acanthophora spicifera* (VAHL) Boergesen. Int J Nanosci Nanotechnol 2:85–94

Swamy MK, Akhtar MS, Mohanty SK, Sinniah UR (2015) Synthesis and characterization of silver nanoparticles using fruit extract of Momordica cymbalaria and assessment of their in vitro antimicrobial, antioxidant and cytotoxicity activities. Spectrochim Acta A: Mol Biomol Spectr 151:939–944. https://doi.org/10.1016/j.saa.2015.07.009

Swamy MK, Rudramurthy GR, Patra JK, Sinniah UR (2018). Microbe-based metallic nanoparticles synthesis and biomedical applications: an update. In: Jayanta Kumar P, Gitishree D, Shin H-S (eds) Microbial biotechnology: application in food and pharmacology, vol 2. Springer International, Singapore, pp 395–434. https://doi.org/10.1007/978-981-10-7140-9_19

Tamboli DP, Lee DS (2013) Mechanistic antimicrobial approach of extracellularly synthesized silver nanoparticles against gram positive and gram negative bacteria. J Hazard Mater 260:878–884. https://doi.org/10.1016/j.jhazmat.2013.06.003

Tarafdar JC, Raliya R (2013) Rapid, low-cost, and ecofriendly approach for iron nanoparticle synthesis using *Aspergillus oryzae* TFR9. J. Nanoparticles Article ID:141274. https://doi.org/10.1155/2013/141274

Thakkar KN, Mhatre SS, Parikh RY (2010) Biological synthesis of metallic nanoparticles. Nanomed: nanotechnol Biol Med 6:257–262. http://doi.org/10.1016/j.nano.2009.07.002

Thenmozhi M, Kannabiran K, Kumar R, Khanna VG (2013) Antifungal activity of Streptomyces sp. VITSTK7 and its synthesized Ag2O/Ag nanoparticles against medically important Aspergillus pathogens. J Med Mycol 23:97–103. https://doi.org/10.1016/j.mycmed.2013.04.005

Thomas R, Janardhanan A, Varghese RT, Soniya E, Mathew J, Radhakrishnan E (2014) Antibacterial properties of silver nanoparticles synthesized by marine *Ochrobactrum* sp. Braz J Microbiol 45:1221–1227. https://doi.org/10.1590/s1517-83822014000400012

Tidke PR, Gupta I, Gade AK, Rai M (2014) Fungus-mediated synthesis of gold nanoparticles and standardization of parameters for its biosynthesis. IEEE Trans Nanobiosci13:397–402. http://dx.doi.org/10.1109/TNB.2014.2347803

Tripathi RM, Gupta RK, Singh P, Bhadwal AS, Shrivastav A, Kumar N, Shrivastav BR (2014) Ultrasensitive detection of mercury (II) ions in water sample using gold nanoparticles synthesized by Trichoderma harzianum and their mechanistic approach. Sens Actuators B: Chem 204:637–646. https://doi.org/10.1016/j.snb.2014.08.015

Vahabi K, Mansoori GA, Karim S (2011) Biosynthesis of silver nanoparticles by fungus Trichoderma reesei. InSci J 1:65–79. https://doi.org/10.5640/insc.010165

Vainshtein M, Belova N, Kulakovskaya T, Suzina N, Sorokin V (2014) Synthesis of magnetosensitive iron-containing nanoparticles by yeasts. J Ind Microbiol Biotechnol 41:657–663. https://doi.org/10.1007/s10295-014-1417-4

Venkatpurwar V, Pokharkar V (2011) Green synthesis of silver nanoparticles using marine polysaccharide: study of in-vitro antibacterial activity. Mater Lett 65:999–1002. https://doi.org/10.1016/j.matlet.2010.12.057

Verma VC, Kharwar RN, Gange AC (2010) Biosynthesis of antimicrobial silver nanoparticles by the endophytic fungus *Aspergillus clavatus*. Nanomedicine 5:33–40. https://doi.org/10.2217/nnm.09.77

Vetchinkina EP, Loshchinina EA, Burov AM, Dykman LA, Nikitina VE (2014) Enzymatic formation of gold nanoparticles by submerged culture of the basidiomycete *Lentinus edodes*.J Biotechnol 82:37–45. https://doi.org/10.1016/j.jbiotec.2014.04.018

Vigneshwaran N, Kathe AA, Varadarajan PV, Nachane RP, Balasubramanya RH (2006) Biomimetics of silver nanoparticles by white rot fungus, Phaenerochaete chrysosporium. Colloids Surf B: Biointerf 53(1):55–59. https://doi.org/10.1016/j.colsurfb.2006.07.014

Vijayanandan AS, Balakrishnan RM (2018) Biosynthesis of cobalt oxide nanoparticles using endophytic fungus *Aspergillus nidulans*. J. Env. Mgt. 218:442–450. https://doi.org/10.1016/j.jenvman.2018.04.032

Vogtle F, Richardt G, Werner N (2009) Types of Dendrimers and their Syntheses. W. Wiley-VCH Verlag GmbH & Co. KGaA, Germany (Ed.) Dendrimer chemistry: concepts, syntheses, properties, applications, pp 81–167. https://doi.org/10.1002/9783527626953.ch4

Wadhwani SA, Shedbalkar UU, Singh R, Chopade BA (2017) Biosynthesis of gold and selenium nanoparticles by purified protein from *Acinetobacter* sp. SW 30. Enzyme Microbial Technol 111:81–86. https://doi.org/10.1016/j.enzmictec.2017.10.007

Waghmare SS, Deshmukh AM, Sadowski Z (2014) Biosynthesis, optimization, purification and characterization of gold nanoparticles. Afr J Microbiol Res 8:138–146. https://doi.org/10.5897/AJMR10.143

Wang T, Yang L, Zhang B, Liu J (2010) Extracellular biosynthesis and transformation of selenium nanoparticles and application in H_2O_2 biosensor. Colloids Surf B: Biointerf 80:94–102. https://doi.org/10.1007/s12010-015-1814-9

Xue B, He D, Gao S, Wang D, Yokoyama K, Wang L (2016) Biosynthesis of silver nanoparticles by the fungus Arthroderma fulvum and its antifungal activity against genera of *Candida, Aspergillus* and *Fusarium*. Int J Nanomed 11:1899. https://doi.org/10.2147/IJN.S98339

Yadav A, Kon K, Kratosova G, Duran N, Ingle AP, Rai M (2015) Fungi as an efficient mycosystem for the synthesis of metal nanoparticles: progress and key aspects of research. Biotechnol Lett 37:2099–3120. https://doi.org/10.1007/s10529-015-1901-6

Zhang M, Zhang K, De Gusseme B, Verstraete W, Field R (2014) The antibacterial and anti-biofouling performance of biogenic silver nanoparticles by *Lactobacillus fermentum*. Biofouling 30:347–357. https://doi.org/10.1080/08927014.2013.873419

Zharov VP, Kim JW, Curiel DT, Everts M (2005) Self-assembling nanoclusters in living systems: application for integrated photothermal nanodiagnostics and nanotherapy. Nanomed: Nanotechnol Biol Med 1:326–345. https://doi.org/10.1016/j.nano.2005.10.006

Zheng D, Hu C, Tian G, Dang X, Hu S (2010) Preparation and application of a novel vanillin sensor based on biosynthesis of Au–Ag alloy nanoparticles. Sens Actuators B-Chem 148:247–252. https://doi.org/10.1016/j.snb.2010.04.031

Zomorodian K, Pourshahid S, Sadatsharifi A, Mehryar P, Pakshir K, Rahimi MJ, Arabi Monfared A (2016) Biosynthesis and characterization of silver nanoparticles by *Aspergillus* species. Biomed Res Int 29:5435397. https://doi.org/10.1155/2016/5435397

Microbial Nanobiotechnology in Nanocatalysis: Degradation of Pollutants and Sensing Applications

A. Donia, N. Malik, and H. Bokhari

1 Introduction

Due to rapid increase in environmental pollution, scientists have now begun to opt for green approaches for industrial processes. Green chemistry in association with biotechnology can be employed to develop efficient methods for manufacturing nanomaterials (Ribeiro et al. 2016). Nanomaterials manufactured using microbial nanotechnology could result in a favorable environment-friendly substitute to not only chemical but also physical methods of synthesis of nanoparticles (Li et al. 2011; Khan et al. 2018). In this review, various applications of nanomaterials in sensing as well as degradation of pollutants have been discussed.

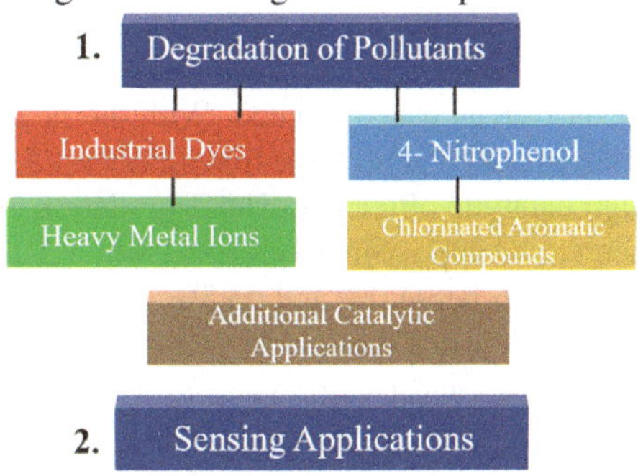

A. Donia · N. Malik · H. Bokhari (✉)
Biosciences Department, Faculty of Science, COMSATS University, Islamabad, Pakistan
e-mail: habib@comsats.edu.pk

A. Lateef et al. (eds.), *Microbial Nanobiotechnology*, Materials Horizons: From Nature to Nanomaterials, https://doi.org/10.1007/978-981-33-4777-9_13

2 Degradation of Pollutants

The microbial biosynthesis of various nanoparticles such as metal oxides, metallic, and non-metallic nanoparticles by multiple bacterial strains, molds, microalgae, and yeast has been studied previously (Hulkoti and Taranath 2014). Large production of pesticides, pharmaceuticals, chemicals, and dyes used in various industries has led to pollution of water and soil (Doshi et al. 2018; Hu et al. 2018). The waste produced by these big industries is rich in metalloids and heavy metals that have been proven hazardous for the environment (Schröfelet al. 2014; Choi et al. 2018). By exploiting the waste as a means to manufacture nanoparticles through a biological mechanism, degradation of these pollutants can be made possible (Seifan et al. 2018; Huang et al. 2018; Zhang et al. 2018).

2.1 Degradation of Industrial Dyes

Dyes, very frequently and in very large quantities, are used in numerous industries such as textile, pharmaceutical, and food processing. Although many techniques have been designed and employed to eliminate residual dye present in industrial waste, none has been able to yield efficient results. This poses a severe threat to a large number of organisms. Dyes have proven to be dangerous owed to their toxicity and ability of eutrophication.

Biosynthesized nanoparticles in comparison with polycrystalline materials have not only a bigger specific surface area but also a larger number of active sites (Fang et al. 2019). The photocatalytic effectiveness of biosynthesized SnO_2 nanoparticles in degradation of three commonly used dyes was evaluated. With the help of absorption spectra for the breakdown of dyes, taken at multiple time intervals, SnO_2 nanoparticles were observed to have high catalytic efficiency. The degradation efficiency of the Erichrome Black T was seen to be approximately 97.8%, while that of methyl orange and methylene blue was seen to be 94.0% and 93.3%, respectively. Another valuable feature of SnO_2 nanoparticles is that they can be removed from the reaction mixture effortlessly through centrifugation (Srivastava and Kowshik 2017). Dyes such as rhodamine B are difficult to be degraded by reason of their resistance against UV light irradiation and bio-mediated degradation. In a study performed to evaluate the photo-degradation of RhB, photocatalytic efficiency of biosynthesized ZnS nanoparticles was evaluated, where tert-butanol was used as hydroxyl radicals, whereas ammonium oxalate was used as scavengers of holes. These photo-generated holes were seen to have played a central role in decolorizing. It was also observed that ZnS nanoparticles carry high degrading ability for RhB (Fang et al. 2019).

In another study investigating the photocatalytic capability of Se nanoparticles for degradation of RhB, Se nanorods and nanospheres which were reduced using *Lysinibacillus* sp. ZYM-1 were used. While the nanorods did not show substantial catalytic effect on Rhb degradation, the nanospheres-H2O2 system exhibited greater visible light-mediated photocatalytic efficiency producing phthalic acid and three intermediates of *N*-deethylation as the degradation products. (Che et al. 2017). Also, metal tolerance potential of soil fungus *Cladosporium oxysporum* AJP03 for effective production of gold nanoparticles with excellent catalytic activity for rhodamine B degradation within 7 min of reaction time under experimental conditions has been reported (Bhargava et al. 2016). A newly isolated strain *Trichoderma* sp. WL-Go was similarly utilized in the green synthesis of gold nanoparticles (AuNPs) which greatly improved the decolorization rate of several azo dyes (Qu et al. 2017).

For the methylene blue degradation, biosynthesized PbS nanocrystals in the shape of nanoparticles, nanocuboids, and nanosheets have been studied. According to the study, in the presence of H2O2, PbS acts as an oxidizer and an electron capture agent. The results suggested that nanocuboids had the highest dye degradation efficiency, although they had the lowest specific surface area. The researchers revealed the possibility that the high levels of catalytic activity of PbS nanocuboids could have been influenced by the crystal plane (Yue et al. 2016). A recent study found that the *Cordyceps militaris* zinc oxide nanoparticles, in the presence of UV radiation, has the ability to degrade hazardous dye methylene blue approximately 97%. This study underscores the potential of using *C. militaris* zinc oxide nanoparticles as ideal photo catalyst for the safeguard of the environment from hazardous effects of water pollution (Li et al. 2019).

In a study aimed to synthesize safe, cost-effective, and novel silver nanoparticles without using any synthetic capping and reducing agents, silver nanoparticles synthesized from endophytic fungal extract of *Pestalotiopsis versicolor* (Speg.) Steyaert showed strong antibacterial activity against both gram-positive and gram-negative bacteria and also exhibited good azo dye degrading potential against Orange G, rhodamine B, and Congo red (Rajput et al. 2017). A recent study found that the *Bacillus pumillis* silver nanoparticles showed great potential to improve bioremediation for Congo red dye removal up to 13% (Modi et al. 2015).

In a study aimed to investigate the green biosynthesis of silver–gold and gold alloy nanoparticles via cell-free extract of *Bacillus safensis* LAU 13 strain, researchers found that gold and silver–gold alloy nanoparticles exhibited significant degradation (more than 90%) of malachite green after 48 h (Ojo et al. 2016). Silver nanoparticles synthesized from xylanases of *Trichoderma longibrachiatum* L2 (TEA) and *Aspergillus niger* L3 (NEA) were also able to degrade malachite green (78.97%) and methylene blue (25.30%) (Elegbede et al. 2018). Silver–gold alloy nanoparticles, which were synthesized from xylanases of *Aspergillus niger* L3 (NE) and *Trichoderma longibrachiatum* L2 (TE), were found to degrade malachite green (91.39%) and methylene blue (47.10%) (Elegbede et al. 2019).

2.2 Degradation of 4-Nitrophenol

4-nitrophenol is a frequently used nitro-aromatic pollutant found in synthetic pesticides, dyes, herbicides, and other chemical products. 4-nitrophenol functions to stimulate and inhibit various nerve endings and can produce undesirable consequences affecting the human central nervous system. The degradation of 4-nitrophenol could be effortlessly measured through UV–Vis spectroscopy; hence, it has been used to gage the catalytic efficiency of nanoparticles (Zhang and Hu 2017; Hosseinpour et al. 2018). In a recent study, marine bacterium *Bacillus* sp. GP was used to create Pd/Au nanoparticles. Thereafter, the catalytic efficiency of the Pd/Au nanoparticles to reduce 4-nitrophenol was evaluated. The results suggested that biosynthesized Pd/Au nanoparticles were weak compared to the chemically synthesized kind. Hence, AlO_3, SiO_2, Fe_3O_4 and various other metal oxides were utilized in order to improve the catalytic activity. Resultantly, it was seen that the enhanced activity was dependent on alloy formation and amount of metal ions because of a number of reasons. Firstly, ions of metal oxides could be considered as the Lewis acids, promoting the reaction and secondly, due to alloy formation that disconcerted the action of noble metal sites (Zhang et al. 2018).

In another study, using *Shewanella oneidensis* MR-1 on rGO, Au (gold) nanoparticles were biosynthesized in bother culture as well as ambient conditions. The biosynthesized Au-NPs/rGO showed promising impacts on the 4-nitrophenol microbial-mediated reduction compared with its chemical parallel. They also displayed higher levels of catalytic activity and could be reused to degrade other pollutants, e.g., nitrobenzene (Cumbal et al. 2003). After ten reduction cycles, catalytic efficiency was consistent at 72% as of initial value. The rise in performance and reusability was credited to the synergic effects of Au nanoparticles and rGo. Firstly, because rGO elevated the concentration level of 4-nitrophenol surrounding the Au nanoparticles while providing a guide to stop the Au nanoparticles from clustering and secondly, because of the improvement in electron transport because of conductivity of rGo (Dong et al. 2015).

In the presence of bio-rGO or in the absence of catalyst, the peak corresponding to nitrophenolate anion at 400 nm was constant over time. With the addition of bio-AuNPs, bioAuNPs/rGO, or chem-AuNPs/rGO, researchers found that there was a decrease in the absorbance at 400 nm and an increase in the new peak at 300 nm related to 4-aminophenol, suggesting that there is a catalytic reduction of 4-nitrophenol. They also found linear correlations between $\ln(A_t/A_0)$ (A_t and $A_0$are the absorbance at 400 nm at time t and zero, respectively) and the reaction time in the presence of various nanomaterials, suggesting that the reactions followed pseudo-first-order kinetics. The first-order rate constant normalized by Au molar concentration (k) of bio-AuNPs/rGO was found to be 138.45 $s^{-1}\ M^{-1}$, which was approximately ten times higher than that of bio-AuNPs (13.47 $s^{-1}\ M^{-1}$). Moreover, the normalized k values of other nanohybrids of AuNPs and carbon-nanomaterials that were previously formed using chemical methods ranged from 1.96 to 33.54 $s^{-1}\ M^{-1}$, which were also much lower than that of bio-AuNPs/rGO (Dong et al. 2015).

2.3 Degradation of Chlorinated Aromatic Compounds

The usage of chlorinated aromatic compounds at different levels in industries has increased vastly owing to their resistance to oxidation, flame resistance, and low water solubility. This exaggerated use has a principal role in water, air, and soil pollution. These compounds are quite difficult to remove, and while various redox approaches have been employed to degrade their firm structure, it is imperative to devise an effortless yet effective treatment for the dehalogenation of these xenobiotics. Biosynthesized Pd-based nanoparticles have shown incredible results in this regard. In a study aimed to synthesize Pd nanoparticles, cell surfaces of *Desulfovibrio vulgaris*, *Desulfovibrio desulfuricans* and *Desulfovibrio* sp. 'Oz-7' were used, where hydrogen acted as electron donor. The bacteria species was proved to affect both the dehalogenation activities of Pd nanoparticles along with bioreductivity of Pd^{2+}. In comparison with chemical Pd nanoparticles, biogenic Pd nanoparticles have been found to have a specific rate of dechlorination 30 times higher. This dissimilarity in catalytic efficiency was ascribed to chemical composition, 3-D array, and functional group structures (Baxter-Plant et al. 2003).

Pd-NPs were biosynthesized, by another group, using *Desulfovibrio desulfuricans* and hydrogen as an electron donor. In this study, the resulting chloride release was used to indicate dehalogenation. The breakdown of 2-chlorophenol biphenyls and polychlorinated biphenyls was used for the approximation of dehalogenation. 2,3,4,5-tetrachloro biphenyl was dehalogenized, and biosynthesized Pd-NPs were used to gage the dehalogenation impact in comparison with the chemically reduced control. The chemically reduced Pd nanoparticles had a chloride release rate of 5% in comparison with the biosynthesized Pd nanoparticles (Baxter-Plant et al. 2004).

Shewanella oneidensis MR-1 has been used to synthesize Pd nanoparticles. In the study, nanoparticles were produced both intracellularly and extracellularly. The electron donors used in this experimentation were H2, lactate, ethanol, pyruvate, and formate. In both solid and liquid conditions, the resultant Pd nanoparticles demonstrated exceptional dehalogenation of polychlorinated biphenyl congeners. Catalytically, the results produced by just 50 mg/L of biosynthesized Pd nanoparticles gave results as good as 500 mg/L of chemical Pd nanoparticle powder (Windt et al. 2005).

Microbial reduction of soluble Pd(II) via cells of *Shewanella oneidensis* MR-1 and of an auto aggregating mutant led to precipitation of palladium Pd(0) nanoparticles inside the periplasmic space (bioPd) and on the cell wall. There were recoveries with a percentage higher than 90% of Pd associated to biomass due to biosorption and following bioreduction of Pd(II) with H2, lactate, formate, pyruvate, or ethanol as electron donors. Therefore, the bioPd(0) nanoparticles produced can reductively dehalogenate polychlorinated biphenyl congeners in sediment and aqueous matrices. In assays with concentrations up to 1000 mg Pd(II) l^{-1} with depletion of soluble Pd(II) of 77.4% and higher, bioreduction was observed. Researchers found that there was more than 90% reduction of PCB 21 (2,3,4-chloro biphenyl) coupled to production of its dechlorination products PCB 1 (2-chloro biphenyl) and PCB 5

(2,3-chloro biphenyl) at a concentration of 1 mg l^{-1} within 5 h at 28 °C. Bioreductive precipitation of bipod using *Shewanella oneidensis* cells mixed with sediment samples contaminated with a mixture of polychlorinated biphenyl congeners led to dechlorination of both lightly and highly chlorinated polychlorinated biphenyl congeners adsorbed to the contaminated sediment matrix within 48 h at 28 °C. Fifty milligrams per liter of bioPd led to a catalytic effect that was comparable to what seen with 500 mg l^{-1} commercial Pd(0) powder. The high reactivity of 50 mg l^{-1} bioPd observed in the soil suspension was reflected in the reduction of the sum of seven most toxic polychlorinated biphenyls to 27% of their initial concentration (Windt et al. 2005).

3 Removal of Heavy Metals Ions

The huge quantity of wastewater produced by industries such as hydrometallurgy and mining contains toxic heavy metal ions (Rüdelet al. 2015). Experiments for the elimination of these pollutants have continued to be carried out for a long time as these heavy metal ions are a serious threat to human health. However, while most of these methods are expensive, others have harsh environmental side effects. There is a rise in the number of researchers invigorated to find the ideal solution to remove heavy metal ions from the environment by employing biological techniques (Ojoawo et al. 2017; Vilardi et al. 2018; Vilela et al. 2018).

The toxicity of both chromium (VI) and vanadium (V) was reduced simultaneously using *Shewanella loihica* PV-4 as, due to vanadium ore smelting, they are frequently present in wastewater together. Vanadium (V) and chromium (IV) bio-reduction has been observed to be indirectly proportional as the bio-reduction of vanadium (V), when elevated, decreases the bio-reduction of chromium (VI). Twenty-seven days after the experiment, the efficiency of removal was observed to be 71.3% for vanadium (V) while 91.2% for chromium (VI) (Wang et al. 2017). Biosynthesized Pd nanoparticles were utilized to remove Cr^{6+} from wastewater. Biologically manufactured nanoscale Pd nanoparticles are expected to have better catalytic efficiency in comparison with chemically reduced Pd nanoparticles because of their smaller sizes and better surface-to-volume ratios. However, unlike their chemical counterparts, biosynthesized Pd nanoparticles had lower catalytic ability for Cr (VI) reduction without ultrasonic treatment. It has been clarified by numerous studies that achieving a high catalytic productivity and mild extract environment cannot be achieved by the reductive method of Cr^{6+}viabacterially produced Pd nanoparticles (Ha et al. 2016).

In a study performed to assess the potential of nanoparticles from microalgae to absorb zinc and lead from pharmaceutical effluent, the silver nanoparticles of *Nannochloropsis* sp. and *Chlorella vulgaris* exhibited percentage reduction of 70.35% and 74.62%, respectively, for zinc. Researchers also found that gold nanoparticles of *Nannochloropsis* sp. and *Chlorella vulgaris* had percentage reduction of 60.32% and 66.83%, respectively, for zinc. The concentration of lead was lowered by *Chlorella vulgaris* silver nanoparticles (66.10%), *Nannochloropsis* sp. silver nanoparticles (68.86%), *Chlorella vulgaris* gold nanoparticles (57.41%), and *Nannochloropsis* sp. gold nanoparticles (66.53%) (Adenigba et al. 2020). The improved performance of both gold and silver nanoparticles compared to their biomass could be attributed to the increased reaction surface area, increasing binding sites that are available for the metals to adhere to the cells (Taman et al. 2015).

Iron sulfide material, which is formed via sulfate-reducing bacteria, is considered to be best adsorbent for many heavy metals. It has the ability to adsorb between 100 and 400 mg g^{-1}, and residual levels in solutions can be of the order of pg per liter. Iron sulfide material has strong magnetic forms which could be produced and removed from the suspension with the adsorbate through magnetic separation. Before drying, the surface area of both the strongly magnetic and weakly magnetic iron sulfides is of the order of 400–500 m^2 g^{-1} as reported by the neutron scattering, magnetic characters, and the adsorption of several heavy metals. The strongly magnetic and weakly magnetic iron sulfides are considered to be effective adsorbents for heavy metals (Watson et al. 2000).

4 Additional Catalytic Applications

Multiple other applications of biosynthesized nanoparticles have been reported. Hydrogen was produced from hypophosphite using biosynthesized Pd nanoparticles as a catalyst. The resulting release of hydrogen was taken as a measure to estimate the catalytic effect. The Pd nanoparticles in this project were procured by bacteria biomass and cell-free methods. Immediately after adding hypophosphite to bioPd nanoparticles, hydrogen was liberated with a high initial release rate. The percentage of release of hydrogen however leveled off after roughly 10 min. It was seen that the cell-free Pd nanoparticles exhibited a higher hydrogen release rate and performed better than the bacteria biomass control. Researchers that carried out this study believe that the difference was supposedly because a fraction of bioPd nanoparticles fixed in the cell envelope was unreachable (Bunge et al. 2010).

Cu nanoparticles were biosynthesized intracellularly using *Shewanella oneidensis* MR-1 by a group. For the assessment of the catalytic azide-alkyne cycloadditions performance of Cu nanoparticles, the production of multiple triazole derivatives was done via Cu^{+}-catalyzed cycloaddition of benzyl azide (Kimber et al. 2018). It was suggested by the authors that when Cu nanoparticles came in direct contact with air, a partially oxidized Cu_2O thin shell will be developed. The Cu nanoparticles showed good catalytic effect toward copper catalyzed azide-alkyne cycloaddition suggesting it as a green and simple synthesis approach for production of reactive Cu nanocatalysts. Also, stabilizing compounds or inorganic substances could be another contributing factor toward this significant catalytic activity (Velmurugan et al. 2017).

It was proven in another study that the electro-catalytic oxidation of hydrazine was notably upheld by bio-Ag nanoparticles. Through $AgNO_3$ precursor reduction coupled with the existence of lead-resistant MKH1 bacteria, the required Ag nanoparticles were synthesized. Electrochemical hydrazine oxidation led to the development of association between current density peak and the Ag catalyst content. The authors revealed that as a result of the accumulation of Ag nanoparticles, no increase was observed in the current density even when Ag nanoparticle content surpassed the threshold value as showed by the forward scan. Ag nanoparticles displayed high-level stability as well as a significant catalytic efficiency. The impedance results and chronoamperometry tests of the biosynthesized Ag nanoparticles were an indicator of their capability in hydrazine fuel cells (Rostami et al. 2018).

5 Sensing Applications

Sensing pollutants are pivotal to protect the environment from their disastrous effects. Many protection agencies have determined and set the maximum tolerable levels of known pollutants in the world. Determining the concentration and type of the pollutant requires chemical, biological, and physical assays. Sampling and the subsequent laboratory analysis of the samples are time-consuming, expensive, and laborious, resulting to be the main limitations of these assays. It is challenging to ascertain the precise nature and composition of the pollutant under field circumstances with the traditional methods. A sensor could help in monitoring the infield environment but at the cost of specificity and detection limit. Ongoing research mainly focuses on ways to increase the specificity, selectivity, and sensitivity of the sensors for monitoring of the environment. This can be achieved in two possible ways: The first one is to target the binding between the recognition element and the contaminant, whereas the second one is to improve the electronic interface and transduction to the sensing layer (Mehndiratta et al. 2013).

Developing the detection methods will provide fast and accurate detection of pollutants. The sensors should be remote and portable to facilitate pollutant detection over large field areas. A sensor is a tool that can detect compounds (chemical or biological) by generating a digital electronic signal via interaction with the compound it has been designed to detect. Conventional sensors can identify the pollutants in biological samples, air, water, soil, and chemical compounds. They are also capable of detecting industrial products at low levels up to ppm and ppb (Kumar et al. 2017).

Using nanoparticles, these detection levels could be increased. Nanoparticle sensors can be designed to have high specificity and accuracy. Even at significantly low concentrations, nanoparticle sensors can uncover microbial pathogens, heavy metals, and organic compounds (Formoso et al. 2016). The reactivity increases with the increase in surface area to volume ratio, which also enhances sensitivity. Multiplex sensors have the ability to detect multiple compounds, based on the idea that the small size improves the quantity of reactive sites on the sensor. The development of nanosensors in the shape of nanotubes and nanowires helps in monitoring of the environment. Detection of NH_3 and NO_2 using single-walled carbon nanotubes has already been proven successfully, showing faster response in comparison with solid-state sensors (Kong et al. 2000). Solid-state sensors have the ability to be operated at 200–600 °C, whereas single-walled carbon nanotubes can be performed at room temperature. The enhanced reactivity shown by single-walled carbon nanotubes can be explained by the fact that gaseous atoms can interact directly to these nanotubes leading to a decrease or increase of electrically generated signal. Detection of glucose in water has successfully been done using boron-doped silicon nanowires (Shao et al. 2005). Detection of calcium, proteins, and antibodies using these boron-doped silicon nanowires was also achieved (Cui et al. 2001; Patolsky and Lieber 2005). Nanowires can be employed in the real-time sensing to detect pathogens, chemical, and biological agents in water, edibles, and air.

Nanotechnology has an essential role in targeting the previously stated problems via several approaches. Nanoparticles can be treated with numerable kinds of biological as well as chemical ligands, which assist in improving their specificity. Many ligands such as proteins, DNA, and enzymes have the ability to bind to nanoparticles with high specificity. Scientists can control the surface-to-volume ratio through altering the shape and size of nanoparticles, increasing control over the interaction quality with the analytes. The capability to fabricate nanoparticles of different metals aids in increasing both conductivity and sensitivity. Nanoparticle sensors may be synthesized to sense organic or inorganic contaminants and biological organisms. Using porous silicon, semiconductor nanostructures can be used to detect organic contaminants. Photoluminescence, which is a feature of porous silicon, could be quenched in the existence of inorganic or organic molecules.

Using the previous approach, pesticide concentration as low as 1 ppm can be easily detected (Stefano et al. 2005). Organic contaminants can quench photoluminescence due to the transfer of energy to the triple substrate of the organic substrate and the transfer of electron from the conduction silicon nanocrystal band to the empty quencher orbital. Carbon nanotube-based sensors can be employed to sense several gases such as NO_2, NH_3, or O_3. Electrical resistance of the nanotubes significantly changes upon contact with the previous gases. Therefore, it can be easily detected. Several heavy metals such as Cd, Pb, or Hg could be detected using nanoparticles via calorimetric or fluorescence-based method. The previous studies have functionalized gold nanoparticles with 11-mercaptoundecanoic acid or chitosan to detect heavy metal ions such as Cd, Pb, or Hg (Kim et al. 2001; Sugunan et al. 2005). The binding of heavy metal ions to the metal chelators such as chitosan and mercaptoundecanoic acid leads to aggregation of the nanoparticles, making wavelength absorption shift and change of color from red to blue. Although this is not only for a specific metal ion, it is very sensitive toward the recognition of heavy metal ions in general. Specific sensors were developed based on a similar principle. A sensor for lead detection was developed using gold nanoparticles, which were coated with lead-dependent DNA enzymes (Liu and Lu 2004). In the absence of lead, enzymes triggered aggregation of nanoparticles. Alternatively, in the presence of lead, activated enzyme cleaved the substrate and there is inhibition of the aggregation, resulting in a wavelength absorption shift.

A gold nanoparticle (AuNP)-based sensor aimed to detect Hg(II) ions in aqueous solution has been developed (Huang and Chang 2006). rhodamine B (RhB) molecules, which are extremely fluorescent in bulk solution, can fluoresce poorly when adsorbed onto AuNPs surfaces due to fluorescence resonance energy transfer as well as collision with AuNPs. The existence of Hg in the solution leads to rhodamine release and hence restores the fluorescence. It is noteworthy that the intensity of restoring the fluorescence is linked with the amount of Hg in the solution. Modifying the surface of nanoparticles with thiol ligands and addition of a chelating agent to the solution provide specificity toward mercury. This nanosensor could detect an exceptionally low levels of mercury concentration (2 ppb) with less than 10 min, making it highly sensitive and selective.

Sensor for *Escherichia coli* detection using mannose encapsulated gold nanoparticles was developed because affinity of mannose toward type I pili of *E. coli* is very high (Lin et al. 2002). In comparison with free mannose, mannose encapsulated in nanoparticles showed greater affinity for *E. coli*. Instead of using a sensor that is specific for a specific microorganism, silver nanoparticles have been proposed based on a universal system to detect bacteria (Yonzon et al. 2005). Silver nanoparticles were functionalized with maltose, which exhibited affinity toward lectin or concanavalin A (carbohydrate-binding protein). Magnetic nanoparticles functionalized with amine can be used to remove bacteria from the media (Huang et al. 2010). The positive charge present on nanoparticle surface exhibited electrostatic interaction with negative charge present on the surface of the bacteria. Bacterial removal efficacy from water samples was 88.5%. Gold nanoparticles were coated with IgG, and the binding of IgG with protein was explored to sense pathogenic microorganisms (Ho et al. 2004). This method may be applied for sensing *Staphylococcus aureus* and *Staphylococcus saprophyticus*. IgG antibody was coated on magnetic nanoparticle surface to purify water. Applying an external magnet helps in removing the trapped microorganisms. To detect microorganisms, quantum dots may possibly be used as a fluorescent labeling system. Conjugation of an antibody with quantum dots could be applied to detect pathogenic microorganisms' viz. *Giardia* and *cryptosporidium* (Zhu et al. 2004). A study was successful in the concurrent sensing of four different toxins, viz. cholera toxin, ricin, staphylococcal enterotoxin B, and shiga-like toxin 1, using antibodies and quantum dot shellsquantum dot shells as probes (Goldman et al. 2004).

The virulence of *Bacillus anthracis* is encoded by a tripartite exotoxin and capsule. The sensor for *Bacillus* spores uses calcium dipicolinic acid as a biomarker. Calcium dipicolinic acid is especially found in the spores without interfering molecules. Deposition of spores on the silver film surface over nanosphere resulted in a surface-enhanced Raman scattering spectrum unique to calcium dipicolinic acid (Yonzon et al. 2005). Method of anthrax spore detection was developed using a surface-assisted fluorescent sensing system (Yilmaz et al. 2010).Exotoxin synthesized by *Bacillus* was detected using a nanosensor (Zhang et al. 2011).

6 Challenges/future Prospects of Microbial-Synthesized NPs in Environmental Applications

In spite of all the advantages, there are still very limited applications of microbial nanobiotechnology (Luo et al. 2015). Synthesis of nanomaterials by bacteria can be either through intracellular or extracellular mechanisms which generally have opposite advantages and disadvantages in terms of purification and dispersity of metal nanoparticles. In general, nanoparticles produced via extracellular mechanisms are more polydispersed (with a great variability in size) than nanoparticles produced via intracellular mechanisms. Yet, less downstream extraction/purification procedures (e.g., detergent uses and ultrasound treatment) are required in extracellular nanomaterial productions. Therefore, synthesis via extracellular mechanisms, which are reported for molds and yeast, can significantly simplify the steps of purification. This is also considered as an advantage for a possible reuse of microorganisms for further biosynthesis cycles. But, identification and characterization of the enzymes responsible for nanobiosynthesis in molds are still missing. The photoautotrophic metabolism of cyanobacteria and microalgae is based on light (as energy source), carbon dioxide (as carbon source), water, and inorganic nutrients. This condition lowers the culture media costs, if compared to culture media used for the growth of bacteria, yeasts, and molds. This may have the potential for future increasing the scale from the laboratory to the industrial scale, also by the design and the advancement of solar photobioreactors for the reduction and fixation of carbon dioxide in the atmospheric (Grasso et al. 2020).

7 Conclusion

In the light of recent literature herein reported, microbial nanobiotechnology holds great potential in nanocatalysis including pollutant degradation and sensing applications. Pollutants such as industrial dyes, nitrophenol, chlorinated aromatic compounds, and heavy metals ions could be successfully degraded via nanoparticles produced by different microorganisms. Microbial nanobiotechnology could be deployed to detect organic compounds, microbial pathogens, and heavy metals even at very low concentrations. It is a field that could really spur innovation in nanomanufacturing holding a great potential in different areas.

References

Adenigba VO, Omomowo IO, Oloke JK, Fatukasi BA, Odeniyi MA, Adedayo AA (2020). Evaluation of microalgal-based nanoparticles in the adsorption of heavy metals from wastewater. IOP Conf Ser Mater Sci Eng 805(1):012030. https://doi.org/10.1088/1757-899X/805/1/012030

Baxter-Plant VS, Mikheenko IP, Macaskie LE (2003) Sulphate-reducing bacteria, palladium and the reductive dehalogenation of chlorinated aromatic compounds. Biodegradation 14(2):83–90. https://doi.org/10.1023/A:1024084611555

Baxter-Plant VS, Mikheenko IP, Robson M, Harrad SJ, Macaskie LE (2004) Dehalogenation of chlorinated aromatic compounds using a hybrid bioinorganic catalyst on cells of *Desulfovibrio desulfuricans*. Biotechnol Lett 26(24):1885–1890. https://doi.org/10.1007/s10529-004-6039-x

Bhargava A, Jain N, Khan MA, Pareek V, Dilip RV, Panwar J (2016) Utilizing metal tolerance potential of soil fungus for efficient synthesis of gold nanoparticles with superior catalytic activity for degradation of rhodamine B. J Environ Manag 183:22–32. https://doi.org/10.1016/j.jenvman.2016.08.021

Bunge M, Søbjerg, LS, Rotaru AE, Gauthier D, Lindhardt AT, Hause G, Finster K, Kingshott P, Skrydstrup T, Meyer RL (2010) Formation of palladium (0) nanoparticles at microbial surfaces. Biotechnol Bioeng 107(2):206–215. https://doi.org/10.1002/bit.22801

Che L, Dong Y, Wu M, Zhao Y, Liu L, Zhou H (2017) Characterization of selenite reduction by *Lysinibacillus* sp. ZYM-1 and photocatalytic performance of biogenic selenium nanospheres. ACS Sustain Chem Eng 5(3):2535–2543. https://doi.org/10.1021/acssuschemeng.6b02889

Choi S, Johnston M, Wang GS, Huang CP (2018) A seasonal observation on the distribution of engineered nanoparticles in municipal wastewater treatment systems exemplified by TiO_2 and ZnO. Sci Total Environ 625:1321–1329. https://doi.org/10.1016/j.scitotenv.2017.12.326

Cui Y, Wei Q, Park H, Lieber CM (2001) Nanowire nanosensors for highly sensitive and selective detection of biological and chemical species. Science 293(5533):1289–1292. https://doi.org/10.1126/science.1062711

Cumbal L, Greenleaf J, Leun D, SenGupta, AK (2003) Polymer supported inorganic nanoparticles: characterization and environmental applications. React Funct Polym 54(1–3):167–180. https://doi.org/10.1016/S1381-5148(02)00192-X

De Stefano L, Moretti L, Rendina I, Rotiroti L (2005) Pesticides detection in water and humic solutions using porous silicon technology. Sens Actuators B Chem 111:522–525. https://doi.org/10.1016/j.snb.2005.03.047

Dong B, Liu G, Zhou J, Wang A, Wang J, Jin R, Lv H (2015) Biogenic gold nanoparticles-reduced graphene oxide nanohybrid: synthesis, characterization and application in chemical and biological reduction of nitroaromatics. RSC Adv 5(118):97798–97806. https://doi.org/10.1039/C5RA19806B

Doshi B, Sillanpää M, Kalliola S (2018) A review of bio-based materials for oil spill treatment. Water Res 135:262–277. https://doi.org/10.1016/j.watres.2018.02.034

Elegbede JA, Lateef A, Azeez MA, Asafa TB, Yekeen TA, Oladipo IC, Abbas SH, Beukes LS, Gueguim-Kana EB (2019) Silver-gold alloy nanoparticles biofabricated by fungal xylanases exhibited potent biomedical and catalytic activities. Biotechnol Progr 35:e2829. https://doi.org/10.1002/btpr.2829.

Elegbede JA, Lateef A., Azeez MA, Asafa TB, Yekeen TA, Oladipo IC, Adebayo EA, Beukes LS, Gueguim-Kana EB (2018) Fungal xylanases-mediated synthesis of silver nanoparticles for catalytic and biomedical applications. IET Nanobiotechnol 12(6):857–863. https://doi.org/10.1049/iet-nbt.2017.0299

Fang X, Wang Y, Wang Z, Jiang Z, Dong M (2019) Microorganism assisted synthesized nanoparticles for catalytic applications. Energies 12(1):190. https://doi.org/10.3390/en12010190

Formoso P, Muzzalupo R, Tavano L, De Filpo G, Pasquale Nicoletta F (2016) Nanotechnology for the environment and medicine. Mini Rev Med Chem 16(8):668–675. https://doi.org/10.2174/1389557515666150709105129

Goldman ER, Clapp AR, Anderson GP, Uyeda HT, Mauro JM, Medintz IL, Mattoussi H (2004) Multiplexed toxin analysis using four colors of quantum dot fluororeagents. Anal Chem 76(3):684–688. https://doi.org/10.1021/ac035083r

Grasso G, Zane D, Dragone R (2020) Microbial nanotechnology: challenges and prospects for green biocatalytic synthesis of nanoscale materials for sensoristic and biomedical applications. Nanomaterials 10(1):11. https://doi.org/10.3390/nano10010011

Ha C, Zhu NW, Shang R, Shi CH, Cui JY, Sohoo I, Wu PX, Cao YL (2016) Biorecovery of palladium as nanoparticles by *Enterococcus faecalis* and its catalysis for chromate reduction. Chem Eng J 288:246–254. https://doi.org/10.1016/j.cej.2015.12.015

Ho KC, Tsai PJ, Lin YS, Chen YC (2004) Using biofunctionalized nanoparticles to probe pathogenic bacteria. Anal Chem 76(24):7162–7168. https://doi.org/10.1021/ac048688b

Hosseinpour M, Fatemi S, Ahmadi SJ, Morimoto M, Akizuki M, OshimaY, Fumoto E (2018). The synergistic effect between supercritical water and redox properties of iron oxide nanoparticles during in-situ catalytic upgrading of heavy oil with formic acid. Isotopic study. Appl Catal B 230:91–101. https://doi.org/10.1016/j.apcatb.2018.02.030

Hu X, Ren C, Kang W, Mu L, Liu X, Li X, Wang T, Zhou Q (2018) Characterization and toxicity of nanoscale fragments in wastewater treatment plant effluent. Sci Total Environ 626:1332–1341. https://doi.org/10.1016/j.scitotenv.2018.01.180

Huang CC, Chang HT (2006) Selective gold-nanoparticle-based "turn-on" fluorescent sensors for detection of mercury (II) in aqueous solution. Anal Chem 78(24):8332–8338. https://doi.org/10.1021/ac061487i

Huang YF, Wang YF, Yan XP (2010) Amine-functionalized magnetic nanoparticles for rapid capture and removal of bacterial pathogens. Environ Sci Technol 44(20):7908–7913. https://doi.org/10.1021/es102285n

Huang Z, Zeng Z, Chen A, Zeng G, Xiao R, Xu P, He K, Song Z, Hu L, Peng M, Huang T, Chen G (2018) Differential behaviors of silver nanoparticles and silver ions towards cysteine: bioremediation and toxicity to *Phanerochaete chrysosporium*. Chemosphere 203:199–208. https://doi.org/10.1016/j.chemosphere.2018.03.144

Hulkoti NI, Taranath TC (2014) Biosynthesis of nanoparticles using microbes—a review. Colloids Surf B 121:474–483. https://doi.org/10.1016/j.colsurfb.2014.05.027

Khan T, Abbas S, Fariq A, Yasmin A (2018) Microbes: Nature's cell factories of nanoparticles synthesis. In: Exploring the realms of nature for nanosynthesis. Springer, Cham, pp 25–50. https://doi.org/10.1007/978-3-319-99570-0_2

Kim Y, Johnson RC, Hupp JT (2001) Gold nanoparticle-based sensing of "spectroscopically silent" heavy metal ions. Nano Lett 1(4):165–167. https://doi.org/10.1021/nl0100116

Kimber RL, Lewis EA, Parmeggiani F, SmithK, Bagshaw H, Starborg T, Joshi N, Figueroa AI, van der Laan G, Cibin G, Diego Gianolio, Haigh SJ, Pattrick RAD, Turner NJ, Lloyd JR (2018) Biosynthesis and characterization of copper nanoparticles using *Shewanella oneidensis*: application for click chemistry. Small 201814:1–8. https://doi.org/10.1002/smll.201703145

Kong J, Franklin NR, Zhou C, Chapline MG, Peng S, Cho K, Dai H (2000) Nanotube molecular wires as chemical sensors. Sci. 287(5453):622–625.https://doi.org/10.1126/science.287.5453.622

Kumar R, Pathak A, Chauhan M, Negi S (2017) Measurement of environmental pollution: types and techniques. In: Advances in environmental biotechnology. Springer, Singapore, pp 13–37. https://doi.org/10.1007/978-981-10-4041-2_2

Li JF, Rupa EJ, Hurh J, Huo Y, Chen L, Han Y, Ahn JC, Park JK, Lee HA, Mathiyalagan R, Yang DC (2019) *Cordyceps militaris* fungus mediated Zinc Oxide nanoparticles for the photocatalytic degradation of Methylene blue dye. Optik 183:691–697. https://doi.org/10.1016/j.ijleo.2019.02.081

Li X, Xu H, Chen ZS, Chen G (2011) Biosynthesis of nanoparticles by microorganisms and their applications. J Nanomater. https://doi.org/10.1155/2011/270974

Lin CC, Yeh YC, Yang CY, Chen CL, Chen GF, Chen CC, Wu YC (2002) Selective binding of mannose-encapsulated gold nanoparticles to type 1 pili in *Escherichia coli*. J Am Chem Soc 124(14):3508–3509. https://doi.org/10.1021/ja0200903

Liu J, Lu Y (2004) Accelerated color change of gold nanoparticles assembled by DNAzymes for simple and fast colorimetric Pb2+ detection. J Am Chem Soc 126(39):12298–12305. https://doi.org/10.1021/ja046628h

Luo CH, Shanmugam V, Yeh CS (2015) Nanoparticle biosynthesis using unicellular and subcellular supports. Asia Mater 7(8):e209–e209. https://doi.org/10.1038/am.2015.90

Mehndiratta P, Jain A, Srivastava S, Gupta N (2013) Environmental pollution and nanotechnology. Environ Pollut 2(2):49. https://doi.org/10.5539/ep.v2n2p49

Modi S, Pathak B, Fulekar MH (2015) Microbial synthesized silver nanoparticles for decolorization and biodegradation of azo dye compound. J Environ Nanotechnol 4(2):37–46. https://doi.org/10.13074/jent.2015.06.152149

Ojo SA, Lateef A, Azeez MA, Oladejo SM, Akinwale AS, Asafa TB, Yakeen TA, Akinboro A, Oladipo IC, Gueguim-Kana EB, Beukes LS (2016) Biomedical and catalytic applications of gold and silver-gold alloy nanoparticles biosynthesized using cell-free extract of *Bacillus safensis* LAU 13: antifungal, dye degradation, anti-coagulant and thrombolytic activities. IEEE Trans Nanobiosci 15(5):433–442. https://doi.org/10.1109/tnb.2016.2559161

Ojoawo SO, Lateef A, Oyeniran FA, Kupoluyi OT, Opatola OS, Daramola JO (2017) Bioaccumulation of heavy metals in steel processing industrial effluents using *Bacillus safensis* LAU 13. J Environ Biotechnol Res 6(1):58–63

Patolsky F, Lieber CM (2005) Nanowire nanosensors. Mater Today 8(4):20–28. https://doi.org/10.1016/S1369-7021(05)00791-1

Qu Y, Shen W, Pei X, Ma F, You S, Li S, Wang J, Zhou, J (2017) Biosynthesis of gold nanoparticles by *Trichoderma* sp. WL-Go for azo dyes decolorization. J Environ Sci (China) 56:79–86. https://doi.org/10.1016/j.jes.2016.09.007

Rajput K, Agrawal S, Sharma J, Agrawal PK (2017). Mycosynthesis of silver nanoparticles using endophytic fungus and investigation of its antibacterial and azo dye degradation efficacy. Kavaka 49:65–71. https://doi.org/10.36293/sfj.2019.0006

Ribeiro BD, Coelho MAZ, de Castro AM (2016) Principles of green chemistry and white biotechnology. In: White biotechnology for sustainable chemistry. The Royal Society of Chemistry, London, pp 1–8. https://doi.org/10.1039/9781782624080-00001

Rostami H, Khosravi F, Mohseni M, Rostami AA (2018) Biosynthesis of Ag nanoparticles using isolated bacteria from contaminated sites and its application as an efficient catalyst for hydrazine electrooxidation. Int J Biol Macromol 107:343–348. https://doi.org/10.1016/j.ijbiomac.2017.08.179

Rüdel H, DíazMuñiz C, Garelick H, Kandile NG, Miller BW, Pantoja Munoz L, Peijnenburg WJGM, Purchase D, Shevah Y, van Sprang P, Vijver M, Vink JPM (2015) Consideration of the bioavailability of metal/metalloid species in freshwaters: experiences regarding the implementation of biotic ligand model-based approaches in risk assessment frameworks. Environ Sci Pollut Res 22:7405–7421. https://doi.org/10.1007/s11356-015-4257-5

Schröfel A, Kratošová G, Šafaˇrík I, Šafaˇríková M, Raška I, Shor LM (2014) Applications of biosynthesized metallic nanoparticles—a review. Acta Biomater 10:4023–4042. https://doi.org/10.1016/j.actbio.2014.05.022

Seifan M, Ebrahiminezhad A, Ghasemi Y, Samani AK, Berenjian A (2018) The role of magnetic iron oxide nanoparticles in the bacterially induced calcium carbonate precipitation. Appl Microbiol Biotechnol 102:3595–3606. https://doi.org/10.1007/s00253-018-8860-5

Shao MW, Shan YY, Wong NB, Lee ST (2005) Silicon nanowire sensors for bioanalytical applications: glucose and hydrogen peroxide detection. Adv Funct Mater 15(9):1478–1482. https://doi.org/10.1002/adfm.200500080

Srivastava P, Kowshik M (2017) Fluorescent lead(IV) sulfide nanoparticles synthesized by *Idiomarina* sp. strain PR58-8 for bioimaging applications. Appl Environ Microbiol 83:e03091-e3116. https://doi.org/10.1128/AEM.03091-16

Sugunan A, Thanachayanont C, Dutta J, Hilborn JG (2005) Heavy-metal ion sensors using chitosan-capped gold nanoparticles. Sci Technol Adv Mater 6(3–4):335. https://doi.org/10.1016/j.stam.2005.03.007

Taman R, Ossman ME, Mansour MS, Farag HA (2015). Metal oxide nano-particles as an adsorbent for removal of heavy metals. J Adv Chem Eng 5(3):1–8. https://doi.org/10.4172/2090-4568.1000125

Velmurugan P, Hong SC, Aravinthan A, Jang SH, Yi PI, Song YC, Jung ES, Park JS, Sivakumar S (2017) Comparison of the physical characteristics of green-synthesized and commercial silver nanoparticles: evaluation of antimicrobial and cytotoxic effects. Arab J Sci Eng 42:201–208. https://doi.org/10.1007/s13369-016-2254-8

Vilardi G, Mpouras T, Dermatas D, Verdone N, Polydera A, Di Palma L (2018) Nanomaterials application for heavy metals recovery from polluted water: The combination of nano zero-valent iron and carbon nanotubes. Competitive adsorption non-linear modeling. Chemosphere 201:716–729. https://doi.org/10.1016/j.chemosphere.2018.03.032

Vilela P, Liu H, Lee SC, Hwangbo S, Nam KJ, Yoo CK (2018) A systematic approach of removal mechanisms, control and optimization of silver nanoparticle in wastewater treatment plants. Sci Total Environ 633:989–998. https://doi.org/10.1016/j.scitotenv.2018.03.247

Wang G, Zhang B, LiS, Yang M, Yin C (2017) Simultaneous microbial reduction of vanadium (V) and chromium (VI) by *Shewanella loihica* PV-4. Bioresour Technol 227: 353–358. https://doi.org/10.1016/j.biortech.2016.12.070

Watson JHP, Cressey BA, Roberts AP, Ellwood DC, Charnock JM, Soper AK (2000). Structural and magnetic studies on heavy-metal-adsorbing iron sulphide nanoparticles produced by sulphate-reducing bacteria. J Magn Magn Mater 214(1–2):13–30. https://doi.org/10.1016/S0304-8853(00)00025-1

Windt WD, Aelterman, P, Verstraete W (2005) Bioreductive deposition of palladium (0) nanoparticles on *Shewanella oneidensis* with catalytic activity towards reductive dechlorination of polychlorinated biphenyls. Environ Microbiol 7(3):314–325. https://doi.org/10.1111/j.1462-2920.2005.00696.x

Yonzon CR, Stuart DA, Zhang X, McFarland AD, Haynes CL, Van Duyne RP (2005) Towards advanced chemical and biological nanosensors—an overview. Talanta 67(3):438–448. https://doi.org/10.1016/j.talanta.2005.06.039

Yue L, Wang J, Zhang Y, Qi S, Xin B (2016) Controllable biosynthesis of high-purity lead-sulfide (PbS) nanocrystals by regulating the concentration of polyethylene glycol in microbial system. Bioprocess Biosyst Eng. 39(12):1839–1846. https://doi.org/10.1007/s00449-016-1658-x

Zhang D, Anderson MJ, Huarng MC, Alocilja EC (2011) Nanoparticle-based biobarcoded DNA sensor for the rapid detection of pagA gene of *Bacillus anthracis*. IEEE Trans Nanotechnol 10(6):1433–1438. https://doi.org/10.1109/TNANO.2011.2165965

Zhang H, Hu X (2017) Rapid production of Pd nanoparticle by a marine electrochemically active bacterium: *Shewanella* sp. CNZ-1 and its catalytic performance on 4-nitrophenol reduction. RSC Adv 7:41182–41189. https://doi.org/10.1039/C7RA07438G

Zhang Y, Qiang L, Yuan Y, Wu W, Sun B, Zhu L (2018) Impacts of titanium dioxide nanoparticles on transformation of silver nanoparticles in aquatic environments. Environ Sci Nano 5:1191–1199. https://doi.org/10.1039/C8EN00044A

Zhu L, Ang S, Liu WT (2004) Quantum dots as a novel immunofluorescent detection system for *Cryptosporidium parvum* and *Giardia lamblia*. Appl Environ Microbiol 70(1):597–598. https://doi.org/10.1128/AEM.70.1.597-598.2004

Application of Microbial-Synthesized Nanoparticles in Food Industries

A. P. Ingle, R. Philippini, S. E. Martiniano, F. A. F. Antunes, T. M. Rocha, and S. S. da Silva

1 Introduction

Current population of the entire world is about 7.8 billion, and it is expected reach more than 8 billion in next couple of years (Worldometers 2020). On one hand, providing plenty and good quality food for everyone is very challenging task, and on the other hand, food wastage leads to major losses in the food industry. According a report of Food and Agriculture Organization (FAO) of the United Nations, more than 1.3 billion metric tons of consumable food is lost or get wasted every year at various stages because of inefficient post-harvest techniques, lack of proper storage and transport facilities, and market and consumer wastage of food (Nile et al. 2020). Therefore, significant increase in food crop production is very necessary; simultaneously, management of food wastage is also essentially required to overcome the food crisis caused by the emerging population and environmental issues (Nile et al. 2020). There are various biotic (microbial infections) and abiotic (environmental conditions) factors which are responsible for food wastage. Among these, food spoilage by microbial contamination is most predominant which usually results in reduction of the food quality and also affects food security, decreases shelf life of food products, and increases the risks of food-borne diseases (Sperber and Doyle 2009).

Considering these concerns, there is urgent need to search for novel and effective strategies to increase food production and minimize the food wastage. Currently, nanotechnology is emerging as promising science having widespread applications in

A. P. Ingle (✉) · R. Philippini · S. E. Martiniano · F. A. F. Antunes · T. M. Rocha ·
S. S. da Silva (✉)
Department of Biotechnology, Engineering School of Lorena, University of Sao Paulo, Lorena, SP, Brazil
e-mail: inghleavinash14@gmail.com

S. S. da Silva
e-mail: silviosilverio@gmail.com

A. Lateef et al. (eds.), *Microbial Nanobiotechnology*, Materials Horizons: From Nature to Nanomaterials, https://doi.org/10.1007/978-981-33-4777-9_14

food industries and has sustainable impact on various sectors including food industries (Santos et al. 2020). Nanomaterials which are the building blocks of nanotechnology reported to have revolutionary uses for different purposes in food industry like food processing, food packaging and many more (Rai et al. 2019).

Nanomaterials have attracted a considerable attention due to their unique physicochemical properties which makes them exceptionally suitable for food processing and improved food packaging. The nanomaterials are usually synthesized using physical, chemical, and biological methods, among which, biologically synthesized nanoparticles have potential applications in food industries because of their high stability and surface properties. To date, different biological systems like plants, bacteria, fungi, algae, cyanobacteria, arthropods among others have been successfully exploited for the synthesis of a variety of nanomaterials (Adelere and Lateef 2016; Lateef et al. 2016a). Among these, microbially synthesized nanomaterials have attracted a great of attention over plant-based synthesis because microorganisms can be grown in laboratory using simple media, so there is no need to hack the plants.

Nano-based food packaging offers many benefits over conventional packaging that uses plastic barriers, and at the same time, its biological properties such as antimicrobial activities provide an increased shelf life without altering the quality of food products (Sekhon 2010). Moreover, nanomaterials can also be used for the development of smart and intelligent packaging systems, i.e., sensory labels which facilitates localization, sensing, reporting about food items leading to improved efficiency and security. In addition, nano-based delivery systems also play pivotal role in the improvement of nutraceutical values of the food components (Ezhilarasi et al. 2013; Singh et al. 2017).In the present chapter, we have summarized microbial synthesis of different nanomaterials using different microorganisms; moreover, recent developments in the field of nanotechnology and its role in food industries, particularly in food processing and food packaging, are discussed.

2 Microbial Synthesis of Nanoparticles

As mentioned earlier, among biological agents, synthesis of a variety of nanoparticles from microbial systems such as bacteria, fungi, algae, and cyanobacteria is preferred to other biological systems such as plants. To date, different microorganisms have been successfully exploited for the synthesis of nanoparticles which are briefly discussed.

2.1 Bacterial Synthesis

Because of adverse effects of chemically synthesized nanoparticles, extensive studies have been carried out to develop more sustainable and eco-friendly approaches using biological systems. Among microorganisms, bacteria are known to synthesize nanoparticles both intracellularly and extracellularly. To date, several types of nanoparticles such as gold, silver, silver–gold alloy, zirconia, cadmium, and many more have been synthesized using different bacteria (Mubarak et al. 2012; Asmathunisha and Kathiresan 2013; Ojo et al. 2016; Oladipo et al. 2017; Debnath et al. 2020). Recent advances demonstrate several fields of application for these nanoparticles, e.g., pharmaceutical (Yadav et al. 2011), bioremediation (Debnath et al. 2020), anticancer activity (Ebrahimzadeh et al. 2020), tumor targeting (Park et al. 2008) as well as anticoagulant and thrombolytic activities (Lateef et al. 2018a, b; Elegbede and Lateef 2019). For instance, Debnath et al. (2020) has shown the applicability of zirconia nanoparticles (ZrO_2NPs) produced by *Pseudomonas aeruginosa* in the removal of tetracycline by adsorption from contaminated waste water. From that, they conducted the assays varying different process parameter such as contact time, pH and ZrO_2NPs concentration in order to establish the most suitable conditions to achieve a higher efficiency. The aforementioned variables presented optimum conditions at pH 6.0, ZrO_2 at 1.2 g/L and contact time of 30 min; these combined removed up to 98% of total target.

In alignment to the strategy of controlling microbial availability to prevent infectious diseases, Jayaseelan et al. (2012) reported the use of non-toxic ZnO nanoparticles synthesized by *Aeromonas hydrophila* against pathogenic bacteria and fungi. A series of various microorganisms were tested to examine the antimicrobial and antifungal activity through the method of well diffusion and minimum inhibitory concentration, and controls were subjected to antibiotics as a comparison. The significant zone of inhibition was found in the case of *P. aeruginosa* and *A. flavus* along with maximum measured halo of 22 and 19 mm, respectively, with minimum inhibitory concentration of 2.9 μg/ml was achieved. In brief, advantageous findings of *A. hydrophila* nanoparticle synthesis pathway enabled scientists to perform rapid ZnONPs biosynthesis at cost-effective and feasible conditions. Still, the phenomenon which elucidates such antimicrobial activity is associated to the surface–volume ratio, size, and shape of nanoparticles, in such manner that the microbial membrane permeability is modified by introducing gaps or pits therein hindering respiratory metabolism that leads to apoptosis of the cells (Baker and Satish 2015). In the case of gram-positive bacteria as it naturally has a thicker layer of peptidoglycan cell wall which confers overwhelming resistance toward many biochemicals, it has been shown that biosynthesized gold nanoparticles (AuNPs) also present a capability of acting as an inhibitor to its cellular survival, thus offering an alternative over the use of antibiotics (Menon et al. 2017).

In a combination approach, silver and gold nanoparticles were biosynthesized by *Sporosarcina koreensis* DC4. Particularly, with respect to enhancement of antimicrobial growth, activities of several commercial antibiotics improved by approximately five-fold in the presence of silver nanoparticles (AgNPs) against significantly resistant microorganisms such as *E. coli*, *V. parahaemolyticus* and *S. enterica* (Singh et al. 2016). Similarly, Lateef et al. (2015a) reported improvement of 7.4–142.9% for some antibiotics against bacteria due to synergistic effect with AgNPs biosynthesized by *Bacillus safensis*. The AgNPs also showed anti-candida, anticoagulant, and thrombolytic activities in another study (Lateef et al. 2016b), while its keratinase facilitated the synthesis of AgNPs with antibacterial property (Lateef et al. 2015b). Furthermore, Singh et al. (2016) carried out investigation to evaluate the catalytic activity of gold nanoparticles. Remarkably, it was found to reduce 4-nitrophenol to 4-aminophenol with AuNPs acting as a catalyst.

A similar co-production of AuNPs and AgNPs was carried out in an investigation by Sowani and colleagues (Sowani et al. 2016a, b). Interestingly, in this case, an actinomycete characterized as *Gordonia amicalis* HS-11 was cultivated in n-hexadecane in order to produce glycolipid which was attributed and confirmed to be the reducing agent for the conversion of $HAuCl_4$ and $AgNO_3$ into Au^0 and Ag^0, respectively. Thereafter, these synthesized AuNPs and AgNPs were individually evaluated by their potential antioxidant scavenging activity, with approximately 90 and 95% inhibitory effect toward hydroxyl radicals by using, respectively, AuNPs and AgNPs. In this context, it is of great importance to emphasize the fact that structural properties such as size, shape, and mono-dispersity of nanoparticles strongly influence their antioxidant activity (Ramamurthy et al. 2013).

Another strategy was employed by Otari et al. (2012) to synthesize AgNPs by subjecting a culture of *Rhodococcus* sp. to a batch flask fermentation along with $AgNO_3$ solution at final concentration of 3×10^{-3} M. Product generation was monitored by UV-vis spectrophotometer at 420 nm, and the change in color from colorless to brownish indicated the synthesis of AgNPs. It was reported that nitrate reductase and set of enzymes found in *Rhodococcus* sp. were responsible for the synthesis under anaerobic condition rather than respiratory type. Additionally, free-cell broth did not show any reduction potential in $AgNO_3$ solution, whereby no change in color was visualized. This fact supports the assumption that synthesis of nanoparticles may selectively occur intracellularly.

Moreover, Fayaz et al. (2011) investigated the use of extremophilic bacterium, namely *Geobacillus stearothermophilus* to mediate the biosynthesis of AuNPs and AgNPs, to elucidate the set and types of proteins secreted extracellularly through a polyacrylamide gel electrophoresis. The results comprised a set of not less than seven different enzymes of molecular weight ranging between 12 and 98 kDa which suggestively participate in the reduction reaction of metallic ions and play a crucial role in providing electrons. Attractively, it was found that amide residual linkages bound and coated the produced nanoparticles, to enhance their stability and also serve possibly as a support for activating the nanoparticles for several profitable applications, e.g., immobilization of enzymes (Ashtari et al. 2012; Ingle et al. 2017), by substituting the usual chemically activation methods in a sustainable and cost-effective strategy.

2.2 Fungal Synthesis

Like bacteria, fungi have emerged as an efficient system for the microbial synthesis of nanoparticles due to distinctive characteristics such as high wall binding capacity, easy to culture, simpler biomass handling, and most importantly extracellular synthesis. To date, several fungi have been used for the synthesis of a variety of nanoparticles (Yadav et al. 2015). Among metal nanoparticles, AgNPs have been extensively studied due to their promising biological activities, particularly antimicrobial activity. AgNPs reported to have antimicrobial activity against a variety of pathogens including food spoiling organisms. Devi and Joshi (2015) evaluated the potential of three different endophytic fungi, viz. *Aspergillus tamarii* PFL2, *Aspergillus niger* PFR6, and *Penicillium ochrochloron* PFR8 towards the synthesis of AgNPs. The results obtained showed that *A. tamarii* PFL2 synthesized smallest size nanoparticles (i.e., average particle size of 3.5 ± 3 nm) compared to the nanoparticles synthesized by the other two fungi *A. niger* PFR6 (8.7 ± 6 nm) and *P. ochrochloron* PFR8 (7.7 ± 4.3 nm). Similarly, Guilger-Casagrande and de Lima (2019) reviewed the role of various fungi in the synthesis of AgNPs.

Apart from AgNPs, other nanoparticles such as gold, copper, silver–gold alloy, etc. have been produced using different fungi. Dhanasekar et al. (2015) demonstrated extracellular synthesis of AuNPs using *Alternaria* sp. Further, they also studied the effect of different concentration of gold chloride solution on the size of nanoparticles. The TEM analysis performed revealed that nanoparticles with varied shape such as spherical, rod, square, pentagonal, and hexagonal were synthesized for 1 mM gold chloride. However, formation of quasi-spherical and spherical nanoparticles/heart shape nanoparticles having size in the range of 7–13 and 15–18 nm were observed for 0.3 and 0.5 mM gold chloride solution, respectively. Molnár et al. (2018) demonstrated extracellular synthesis of AuNPs having size range of 6–40 nm using thermophilic filamentous fungi. It has been shown that xylanases produced by *A. niger* and *Trichoderma longibrachiatum* fabricated AgNPs, AuNPs, and Ag-AuNPs with potent antimicrobial and antioxidant activities among others (Elegbede et al. 2018, 2019, 2020).

Zielonka and Klimek-Ochab (2017) reviewed the role of various species of *Fusarium* in the synthesis of different nanoparticles. The authors concluded that among fungi, *F. oxysporum* can be used as efficient microbial system for the simple, rapid and large-scale synthesis of nanoparticles. Recently, Noor et al. (2020) proposed an eco-friendly, cost-effective, and easily scalable methods for the biological synthesis of copper nanoparticles using *A. niger* strain STA9. Further, authors evaluated their anticancer, antidiabetic, and antibacterial activities. It was reported that these nanoparticles showed significant anticancer and antidiabetic activities. Moreover, they were also found to be effective against different bacteria such as *Escherichia coli*, *Staphylococcus aureus*, *Klebsiella pneumoniae*, *Micrococcus luteus*, and *Bacillus subtilis*.

In addition, there are many reports on the synthesis of some other nanoparticles from fungi which mainly include synthesis of zinc oxide nanoparticles from *A. niger* (Shamim et al. 2019), and selenium nanoparticles using *Trichoderma atroviride* (Joshi et al. 2019). All the above studies clearly indicate that fungi can be efficiently used for the synthesis of different nanoparticles which can be used for various purposes including applications in food industries.

2.3 Algal Synthesis

Similar to various above-mentioned microbial systems, algae also have some advantages over other biological processes because it is easy to produce nanoparticles on large scale using algae as bulk biomass of algae can be produced easily. To date, more than a hundred different micro- and macro-algae have been used for both intracellular and extracellular synthesis of nanoparticles (Dhavale et al. 2020). The synthesis of AgNPs was demonstrated using cell free extract of green alga *Chlorella vulgaris* (Xie et al. 2007). Similarly, Vivek et al. (2011) found that a red alga *Gelidiella acerosa* has ability to reduce aqueous silver ions into spherical AgNPs with an average size of 22 nm. In another study, Tsibakhashvili et al. (2011) studied the extracellular synthesis of AgNPs using extract of *Spirulina platensis*. Moreover, Abdel-Raouf et al. (2019) reported synthesis of AgNPs by the reduction of aqueous solutions of silver nitrate using powder and solvent extracts of *Padina pavonica*. The results revealed that the used algae has promising potential in the synthesis of AgNPs producing 49.58–86.37 nm sized particles of variable shapes i.e. spherical, triangular, rectangle, polyhedral and hexagonal. More recently, *Nannochloropsis* sp, *Chlorella vulgaris,* and *Neodesmus pupukensis* have been employed to synthesize AgNPs for antimicrobial, antioxidant, and bioremediation applications (Adenigba et al. 2020; Omomowo et al. 2020).

Algae are well known for their ability to hyperaccumulate heavy metal ions and possess an excellent capability to modify them into more malleable forms (Fawcett et al. 2017). Due to these abilities, algae have attracted a considerable attention from scientific community and being used as model microbial system for the synthesis of bio-nanomaterials. The extract of algae usually consists of carbohydrates, proteins, minerals, oil, fats, polyunsaturated fatty acids, and other bioactive compounds such as antioxidants (polyphenols, tocopherols), and pigments such ascarotenoids (carotene, xanthophyll), chlorophylls, and phycobilins (phycocyanin, phycoerythrin) (Michalak and Chojnacka 2015; Khanna et al. 2019). These active compounds act as reducing and stabilizing agents during the synthesis of nanoparticles.

2.4 Synthesis from Cyanobacteria

Cyanobacteria have recently been a wise choice for mediating nanoparticles synthesis as they are capable of bio-remediating toxic metals and further convert them into attractive and amenable products (Ebrahimzadeh et al. 2020). One advantageous aspect of this utilization is the simple nutrient media required by those types of organisms and can easily be cultivated in large scale (Yu et al. 2013). In alignment with the emerging trend of anticancer activity provided specifically by AgNPs, Ebrahhimzadeh and co-workers guided a study of green synthesis of AgNPs through *Anabaena flos-aquae* which is a simple, rapid, inexpensive, and ecofriendly manner to achieve the target. In their work, cytotoxicity tests against human breast cancer cells (T47D) were performed using biosynthesized AgNPs and cell extract from *A. flos-aquae*. Among the trials, upon application of concentration of 100 μg/mL of both AgNPs and cell extract, cancer cells inhibition was significantly pronounced and a final cellular viability of 8.64 and 29% was encountered, for the respective cases.

The anticancer property is commonly related to the release of silver ions inside the living cells by that reactive oxygen species are generated causing oxidative damage to biomolecules such as proteins, DNA, and lipids, thus resulting in cell death associated with physiological oxidative stress (AbdalDayem et al. 2017). Therefore, the usage of several types of nanoparticles may play a guideline role in scientific advances. The application varies in accordance to the metal of their composition, whereas the efficacy is extensively correlated to the morphological properties of each of the nanoparticles. Overall, along with ceaseless efforts and bacteria mediation ability as a key factor for providing reductant agents, nanoparticles represent a valuable tool and a promising strategy for realistic society problem resolution by the means of an ecological and economically attractive production.

3 Applications of Nanoparticles in Food Industries

Nanomaterials can be effectively used for different applications in food industries, but among them they have been found to play important roles in food processing and food packaging.

3.1 Food Processing

Food additives are substances that are added to foods to maintain or modify their taste, improve their appearance, aroma, or prolong their shelf life. These substances act as antioxidants, coloring materials, and emulsifiers, as well as flavoring, foaming, gelling, and humectant agents. They also act as food preservatives and sweeteners (Kumar et al. 2019). These substances must conform to guidelines set by regulatory agencies like FDA in the US and EUPHA in European countries, as some of them have considerable side effects in both animal and human health.

The employment of nanoparticles in the food industry is relatively new when compared to other areas, such as pharmaceutical, medical, and technological industries. Even though, the utilization of nanoparticles is presented in several food types, like beverages, cereals, milk powder, and sausages, promoting the general food appearance, while they also improve the texture, taste, and other sensorial attributes, such as absorption and bioavailability of nutrients (Huang et al. 2010; Carrillo-Inungaray et al. 2018). Some metal oxides chemicals like silicon dioxide (SiO_2), magnesium oxide (MgO), and titanium dioxide (TiO_2) are permitted by the US FDA as an anti-caking agent, food flavor carrier, and food color additives (He et al.

2018). Other characteristics such as sensorial attributes (texture, taste, and flavoring) as well as process ability and stability may increase the shelf life of a determined product (Huang et al. 2010). The development of nanoparticles for the utilization in food processing can be elaborated by the utilization of synthetic nanoparticles or biosynthesized materials. The biosynthesized nanomaterials (or green synthesized) can be elaborated from plants, presenting some advantages, such as (1) decrease use of hazardous materials, as employed biological systems act like capping, reducing, and stabilizing agents; (2) reduction of energy resources and chemicals, as biosynthesis reactions usually occur in neutral pH, room temperature and pressure; and (3) biocompatibility and low toxicity of most synthesized materials, due to surface functionalization (He et al. 2018). Figure 1 depicts some properties and applications of traditional and biosynthesized nanoparticles.

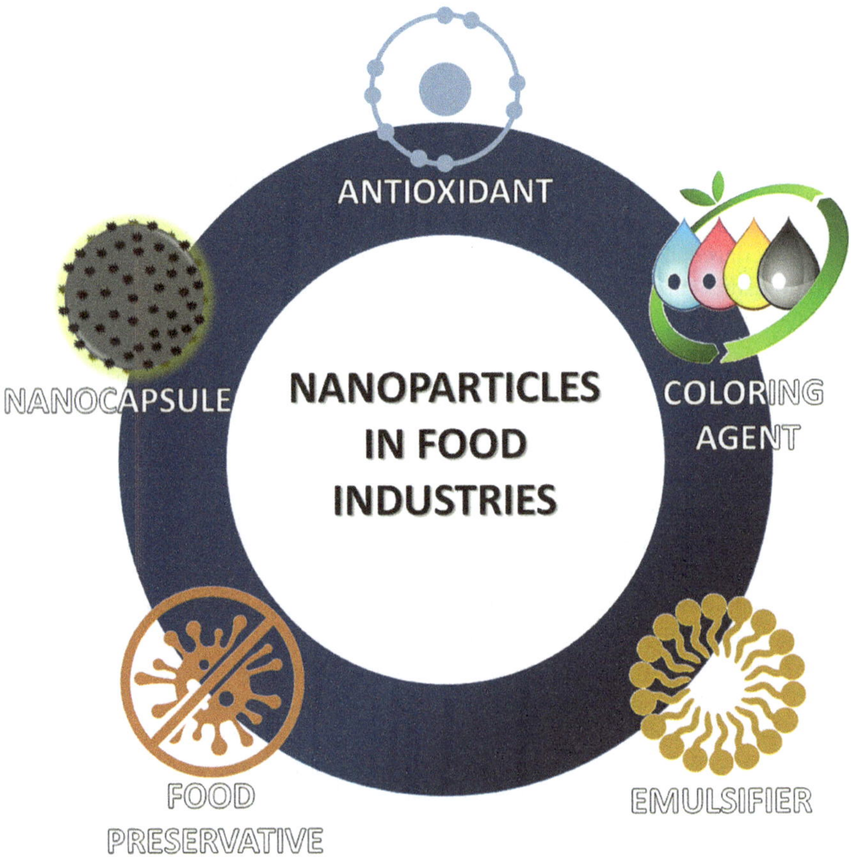

Fig. 1 Properties and applications of traditional and biosynthesized nanoparticles

3.1.1 Antioxidants

An antioxidant is a substance described as a component that is added in small amounts and slows/inhibits the autoxidation of oxidizable molecules (Cui et al. 2018). The utilization of synthetic antioxidants in processed foods is wide, preventing the formation of free radicals. However, it is suggested that they can cause several diseases like asthma, joint pain, dermatitis, and stomach and eye problems (Kumar et al. 2019). Nanoparticles can be utilized as nanoantioxidants becuase they have ability to slow down the autoxidation overall rate, by trapping the carrying radicals, or by decreasing the initiation of oxidative events (Valgimigli et al. 2018), presenting strong and persistent interactions with biomolecules that are more effective against free radical-induced damage (Sandhir et al. 2015).

Several metals present the nanoantioxidant properties (palladium, silver, platinum, and gold), as well as some metal oxides, acting like catalase mimics (CAT-mimics), by decomposing hydrogen peroxide into water and oxygen when in neutral or basic pH values (Valgimigli et al. 2018). In acidic pH values, the prooxidant effect is observed in a behavior similar to peroxidase enzymes, where HO· is released from hydrogen peroxide in a Fenton-like reaction (Nosaka and Nosaka 2017). AgNPs biosynthesized from plants like *Cassia tora* and *Artocarpus altilis* leaves, as well as from brown algae *Cystoseira trinodis* displayed antioxidant activities (Saravanakumar et al. 2015; Ravichandran et al. 2016; Gu et al. 2018). Similarly, AgNPs biosynthesized using extracts of *Cola nitida, Theobroma cacao, Synsepalum dulcificum, Hyptis suaveolens, Petiveria alliacea,* and *Chasmanthera dependens* (Lateef et al. 2016c, d, e, 2018a, b,2020) have displayed potent antioxidant activities.

Silica is considered a biochemically inert compound, allowing its utilization for the elaboration of food-grade nanoantioxidants, such as caffeic acid and gallic acid (Valgimigli et al. 2018), as well as *Citrus* and *Syzygium* essential oils (Cui et al. 2017; Himed et al. 2019). Magnesium oxide nanoparticles are highly ionic particulates that presents high surface areas, unique optical, electronic, magnetic, thermal, mechanical, and chemical properties due to its characteristic structures (Ravishankar and Jamuna 2011; Ramanujam and Sundrarajan 2014). Dobrucka (2016) studied the biosynthesis of nanoantioxidant particles using aqueous extract of *Artemisia abrotanum* in the neutralization of DPPH (2,2-diphenyl-1-picrylhydrazyl) radical, in different solution ranges, while Sharmila et al. (2019) evaluated the nanoantioxidant activity of MgO using *Pisonia alba* leaf extract, assessed for its free radical scavenging activity and reducing power activity.

Some polyphenols have also been utilized as strong antioxidants, neutralizing the free radicals by direct scavenging or quenching of oxygen free radicals and inhibition of oxidative enzymes that generate reactive oxygen species (Jana et al. 2017). Palacio et al. (2020) studied the encapsulation of three polyphenols, epigallocatechin-3-gallate (EGCG), propyl gallate (PG), and gallic acid (GA), using an ionic cross-linking method on succinyl-chitosan nanoparticles for the evaluation of the drug encapsulation efficiency and loading capacity.

3.1.2 Coloring Materials

Food coloring agent is any substance that is added to some product aiming to modify or enhance its color. Artificial colorants like Brilliant Blue FCF, Allura Red AC, Tartrazine, and some others are used in dairy products, snacks, sweets, and some drinks. Some of them have been banned in many countries around the world due to carcinogenic potential, as well as asthma and chromosomal aberrations (Kumar et al. 2019). The utilization of metal oxides as a color additive nanomaterial such as TiO_2 is allowed since it corresponds to less than 1% by weight of food (https://www.ecfr.gov/).

Natural pigments can be incorporated with nanoparticles and applied as colorants in foods. Carotenoids (carotenes and xanthophylls) are a basic source of red, orange, and yellow pigments, which present several benefits to human health, like anti-inflammatory, antioxidant, anticancer agents, and skin protection (Jana et al. 2017). Sowani et al. (2016a, b) developed gold and silver nanoparticles from the orange pigments produced by the actinomycete *Gordonia amicalis*, as De Campo et al. (2019) also incorporated zeaxanthin nanoparticles (Zea-NP) in yogurt for the evaluation of color, textural parameters, viscosity, and syneresis via in vitro digestion.

3.1.3 Emulsifiers

In the food processing industry, the utilization of emulsifier products is largely used in many products, such as baked goods, beverages, candies, creamy sauces, ice cream, and mayonnaise. Some of the traditionally employed emulsifiers like carboxymethylcellulose and polysorbate-80 present pharmacological toxicity effects, promoting some undesirable effects like irritable bowel syndrome (IBS), disruption of gut bacteria, delayed immune responses, and obesity (Kumar et al. 2019).

A nanoemulsifier is composed of oil, water, although it is necessary for the addition of surfactant as a stabilizer (McClements 2012). Lipophilic functional foods can be delivered using oil in water (O/W) nanoemulsion systems, where the oil droplet is surrounded by surfactant droplets (Xiong et al. 2020). The study of rice-bran protein-based nanoemulsion was developed by Chen et al. (2020), which showed high surface activity, and it was used as a carrier agent of quercetin, improving bioavailability and cell permeability of this molecule in in vitro digestion with stimulation of gastric secretion. Duarte et al. (2020) reported the nano-emulsification of chamomile (*Matricaria chamomilla*) essential oil evaluating the droplet size and droplet growth. In another study, Chen et al. (2019) evaluated the effects of self-nano-emulsions using *Sonchus oleraceus* Linn extract, using surfactants, and co-surfactants in four oils (castor oil, soybean oil, safflower oil, and sunflower oil) for the evaluation of in vitro simulated digestion.

3.1.4 Food Preservatives

The utilization of some traditional food preservatives for long durations may result in headache, gradual loss of mental concentration, and low immune response (Kumar et al. 2019). Some nanoparticles can be used for the production of nano-preservatives, as an alternative strategy to inhibit or reduce the microbial growth control, as this increases the products shelf life (He and Hwang 2016). Several biosynthetic nano-preservatives have been produced and tested for the control of bacteria and fungi, using metals such as silver and gold, as well as some metal oxides like Fe_3O_4, MgO, and TiO_2.

The employment of biocompatible AgNPs for the control of gram-negative and gram-positive bacteria, like *Escherichia coli* and *Staphylococcus aureus* was elaborated by García-Ruiz et al. (2015), presenting a growth-inhibitory effect of two nanoparticles (PEG-Ag NPs and GSH-Ag NPs) on wine. The utilization of *Candida lusitaniae* in the production of AgNPs for the bacterial control of gram-negative *Klebsiella pneumoniae* and gram-positive *Staphylococcus aureus* was elaborated by Eugenio et al. (2016).

The utilization of supernatant media from cellular biomass culture for the preparation of nano-preservatives has been described in the literature, as some yeasts present the nitrate reductase activity. For example, Fernández et al. (2016) evaluated the antifungal activity of AgNPs produced from supernatant culture media of *Cryptococcus laurentii* and *Rhodotorula glutinis* against the phytopathogenic fungus *Botrytis cinerea, Penicillium expansum, Aspergillus niger, Alternaria* sp., and *Rhizopus* sp.

The utilization of metal oxides like MgO was studied by Das et al. (2018), where leaf extracts of *Bauhinia purpurea* were used for the production of nanoflakes with antibacterial activity on *Staphylococcus aureus.* The elaboration of supermagnetic Fe_3O_4-MgO nanoparticles using the essential oil from nutmeg has been proven to present antibacterial and antifungal activities (Mousavi et al. 2019). The reactive oxygen species (ROS) in TiO_2 photocatalysis (TPC) is capable of causing the inactivation of microorganisms, by damaging irreversibly the cell wall and bacterial membrane, causing ions and small molecules leakage, degradation of cells component and finally death (Zhu et al. 2018). TiO_2 is usually used in sweets and presents great potential of application in bottles, milk boxes, and packing foils (Ghosh and Das 2015). TiO_2 can be also applied in drinking water (Donovan et al. 2016), fresh vegetables (Pignata et al. 2017), and drinks (Chai et al. 2014).

3.1.5 Nanocapsules

The nanoencapsulation is defined as the technology of packaging nanoparticles of solid, liquid, or gas (core or active), within a secondary material (matrix or shell), to form the nanocapsules (Augustin and Hemar 2009). Nanocapsules are structures that vary from 0.1 nm to 1 μm size, and due to its increased range of applications, this core may contain active ingredients, which may have several benefits, such as antioxidant agent, volatile ingredients retainer, stability and taste enhancer, as well as ensuring ingredient continuous delivery (Hamad et al. 2017). As nanocapsules are capable of carrying components through digestive system, they may resist the stomach acid attacks and are being employed in the delivery of lipophilic health supplements, vitamins, and minerals (Maftoonazad and Ramaswamy, 2018; Xiong et al., 2020). In fact, nanoencapsulation packs substances in miniature making use of techniques such as nanocomposite, nano-emulsification, and nano-structuration, and provides final product functionality that includes controlled release of the core (Sekhon 2010).

Several food-grade ingredients can be nano-encapsulated, such as essential oils (Cui et al. 2017; Hussein et al. 2020), chitosan (Palacio et al. 2020), polyphenols (Chen et al. 2020), some organic acids (Sandhir et al. 2015), vitamins (Khayata et al. 2012) for increasing bioavailability of foods. Although the encapsulation of these substances improves not only bioavailability, stability as it controls the release of substance, but also reduces odor and unwanted taste of the compound (Maftoonazad and Ramaswamy 2018). Nanocapsules can be also employed for the development of biodegradable sensors, controlling the temperature and humidity on food, as well as for the development of nano-films which prevents the contact with oxygen that causes food spoilage (He and Hwang 2016; Carrillo-Inungaray et al. 2018).

3.2 Food Packaging

Packaging systems are very important in the food supply chain as they serve as food containers during the transport, preventing damages and improving shelf life, safety, and quality of the food (Vilela et al. 2018; Yildirim et al. 2018). However, since conventional packaging methods cannot effectively prevent damages caused by biological and physical–chemical factors (e.g., microbial growth, oxidation, light, and moisture), active packaging materials are used for improving quality, safety, and shelf life of foods (Khaneghah et al. 2018a).

3.2.1 Active Food Packaging Systems

Nanotechnology is a growing tendency in the food industry and plays an important role in the development of active packaging (AP) systems. AP nanomaterials are usually composed of metals, essential oils, biopolymers, organic acids, plant extracts, sugars, and enzymes, but their compositions vary according to their active functions. Moreover, nanomaterials can act as nanocarriers for delivering bioactive compounds (Bahrami et al. 2019). Furthermore, biodegradable AP materials can replace plastic packaging and reduce environmental impacts, since micro- and nano-plastics enter the food chain and cause serious disturbance in the ecosystem (Pandey et al. 2020). AP materials can be divided according to their active functions in emitters (e.g., antimicrobial, antioxidants, and CO_2) and scavengers/absorbers (e.g., O_2, moisture,

UV light) (Vilela et al. 2018; Yildirim et al. 2018), but some active food packaging systems are developed to act in both forms (Table 1). The following sections comprise the main applications of nanoparticles in AP systems.

3.2.2 Active Emitters

The antimicrobial properties of AP materials are widely studied due to the negative impacts of uncontrolled microbial growth in food. Antimicrobial packaging systems are generally used in meat, seafood, fresh and processed vegetables and fruits, cereals, grains, bakery, and dairy products to inhibit cell growth and extend shelf life (Yildirim et al. 2018). In antimicrobial AP system, antimicrobial agents, such as essential oils, plant extracts, organic acids, metals, enzymes, biopolymers, bacteriocins (antimicrobial proteins), proteins, and other antimicrobial agents, are added into the package system (Khaneghah et al. 2018b). Likewise, antioxidants emitters packaging systems are used for improving the oxidative stability of meat, seafood, nuts, high-oil-content food, and oils, while CO_2 emitters prevent microbial growth and extend shelf life of meat (Yildirim et al. 2018).

Natural compounds can be associated with metal-based nanoparticles either for their biosynthesis or as a matrix, improving their antimicrobial and antioxidant activities. Essential oils are extracted from aromatic plants and present several medicinal properties. These compounds are widely explored in the food industry due to antimicrobial properties; in addition, terpenoids, phenols, and aldehydes are their main components and their hydrophobicity contributes to the damage of microbial cells (Khaneghah et al. 2018a). Mohammadi et al. (2020) produced an AP film based on whey protein isolate associated with cinnamon essential oil and chitosan nanofiber. In this study, cinnamon oil (a natural antimicrobial and antioxidant) and chitosan nanofiber were both structured as lipid nanocarriers, and the use of chitosan nanofiber improved the film mechanical properties, while cinnamon oil increased antibacterial activity against the common food pathogens *Escherichia coli*, *Staphylococcus aureus*, and *Pseudomonas aeruginosa*.

Biopolymers and amino acids are among the most studied compounds in AP systems due to their antimicrobial properties. Ibrahim et al. (2019) evaluated the use of lysine amino acid conjugated with a nano-biodegradable polymer (polystyrene sulfate) as an edible coat in paper sheets and observed an enhancement of the antimicrobial activity against several bacteria and yeast pathogens, as well as the improvement of its mechanical and physical properties. Pandey et al. (2020) studied chitosan, a natural polymer, in the form of nano-layers in AP systems for meat packaging and demonstrated an effective antimicrobial activity against *Listeria monocytogenes* and *Escherichia coli* bacteria. In another study, Liu et al. (2020) produced nanocomposite films using a soybean soluble polysaccharide, zinc nanoparticles, and micro-fibrillated cellulose, exhibiting an efficient antibacterial activity.

Among nanomaterials, AgNPs are widely studied for their antimicrobial properties, and their production using plant extracts is an alternative green method of

Table 1 Some examples of nanomaterials used in active packaging system and their functional properties

Active packaging system	Nano-based material	Active function	References
Emitters	Cinnamon oil nanoemulsion and chitosan nanofiber	Antioxidant and activity against bacteria (*Escherichia coli*, *Staphylococcus aureus*, and *Pseudomonas aeruginosa*)	Mohammadi et al. (2020)
	Conjugated lysine with nano-polymer polystyrene sulfate matrix	Activity against bacteria (*Bacillus subtilis*, *S. aureus*, *P. aeruginosa*, and *E. coli*) and yeast (*Candida albicans*)	Ibrahim et al. (2019)
	Silver nanoparticles in chitosan matrix	Activity against bacteria (*E. coli*, and *Listeria monocytogenes*)	Pandey et al. (2020)
	ZnO nanoparticles in soluble soybean polysaccharide and micro fibrillated cellulose matrix	Activity against bacteria (*E. coli*, and *S. aureus*)	Liu et al. (2020)
	Silver nanoparticles and thymol in gelatin-modified montmorillonite and cellulose-acetate matrix	Antioxidant and activity against bacteria (*E. coli*, *S. aureus*, *Pseudomonas aeruginosa*, and *Salmonella enterica*) and fungi (*Aspergillus niger* and *Aspergillus flavus*)	Dairi et al. (2019)
Scavengers/absorbers	SiO_2 nanoparticles in low-density polyethylene	Oxygen scavenger; prevented the browning process; enhanced the phenolic and sugar content of loquat fruit	Wang et al. (2020)
	Iron nanoparticles in microemulsion	Oxygen scavenger; improved the shelf life of roasted sunflower seed and walnuts	Mu et al. (2013)
	Iron nanoparticles blended in a silicone matrix	Oxygen scavenger and moisture absorber	Foltynowicz et al. (2017)
	Silver nanoparticles and thymol in gelatin-modified montmorillonite and cellulose-acetate matrix	UV light barrier	Dairi et al. (2019)
	ZnO nanoparticles in starch-kefiran matrix	UV light barrier and moisture absorber	Shahabi-Ghahfarrokhi and Babaei-Ghazvini (2018)

synthesis, as demonstrated by Dairi et al. (2019). In this study, the authors synthesized AgNPs using *Curcuma longa* extract and produced nanocomposite films with gelatin-modified montmorillonite and cellulose acetate, demonstrating a synergistic antimicrobial effect against pathogenic bacteria and fungi, as well as high antioxidant properties.

3.2.3 Active Scavengers and Absorbers

Several food products are sensitive to oxygen damages since its presence can decrease the quality and shelf life of the products by enabling the microbial growth and the oxidation process, leading to sensorial, physical, and mechanical changes in packaged foods (Vilela et al. 2018; Yildirim et al. 2018). Iron nanocomposites are widely studied in active food packaging due to their potential to act as oxygen scavengers. Foltynowicz et al. (2017) demonstrated the potential of nanoscale zero-valent iron particles as oxygen scavengers and humidity absorbers dispersed in silicon matrix to act as multilayer films for AP systems in food. Mu et al. (2013) developed iron nanoparticles by liquid-phase reduction method in microemulsions systems to evaluate their potential as oxygen scavenger in high-oil-content food (roast sunflower seed and walnuts), obtaining a high oxygen absorption and increasing the storage time. Nano-SiO_2 packing films increased loquat fruits antioxidant activity, retarding internal browning, and extending the storage life (Wang et al. 2020).

Renewable sources are alternatives to green synthesis of nanocompounds. Plant extracts have high antioxidant, anti-inflammatory, and antimicrobial properties and are often used in nanoparticles biosynthesis. Dairi et al. (2019) synthesized AgNPs using *Curcuma longa* extract in an organoclay matrix with thymol; the nanoparticles presented spherical shape and about 7–40 nm size; the produced films demonstrated an improvement in UV barrier and good antioxidant activity. Biopolymers also demonstrated great potential as a matrix for nanoparticle association for AP absorber/scavenger and emitter systems. A polymer matrix of starch-kefiran containing ZnO nanoparticles decreased the moisture content of the biofilms and improved the UV light absorption and might be used as an alternative UV-protective material in food (Shahabi-Ghahfarrokhi and Babaei-Ghazvini 2018).

Nanotechnology helps to solve challenges and improve the efficacy of AP for food. Moreover, when their production occurs in association with eco-friendly methods, besides improving food quality and safety, the use of nanotechnology also presents environmental benefits, such as non-toxicity and biodegradability (Pandey et al. 2020).

3.3 *Smart Food Packaging Systems*

The smart intelligent packing is a system encompassed by the nanotechnology (nanosensors), developed for biochemical detecting changes, e.g., by identifying some grown pathogens or gases released by the food in conditions of spoiling, in order to warn the consumer of its environments (Kuswandi 2016). The smart food packaging system comprises monitoring, being remarkable in industry food due to possibility to overhaul itself and/or inform the final user of contaminations and some undesirable microorganisms presence, being able to reveal microbial pathogens in food products in 2–7 days as well as for detection of toxin, pesticide, and spoilage (Pal 2017). While active package protects the foods, the aim of smart food packaging system is the communication among manufacturer, retailer, and final consumer, showing the feasibility to track and inform the inside and outside conditions of the package, by indicators, e.g., time-temperature, microorganism spoilage sensor, physical shock, leakage sensor, allergen sensor, oxygen, humidity, and others (Kuswandi et al. 2011). Figure 2 presents an outline of main characteristics of smart packaging systems.

Indeed, improved packaging purposes the incorporation of nanomaterials in order to increase physical characteristics performance, while the active packaging points out the intentional release of the nanocompounds aimed at the taste and self-life improvement. Additionally, with these features, the smart packaging stands out with use of nanosensors to monitor adequately the product, being able to verify some undesirable concerns during all its self-life. Besides, smart packaging also offers the possibility of active tags and traceability system by biosensors indicators, where it uses radiofrequency connection to transfer or transmit data through barcodes (Abad et al. 2009; Bagchi 2012).

Food expiration date is usually estimated based on mild conditions such as temperature storage. However, unpredictable factors must be taken into account, e.g., if the food was submitted to high temperatures and humidity for long times that could reduce this date. Within this context, a nanosensor could indicate with more precision the real condition of this food along its expiration date. Thus, nanosensors can track

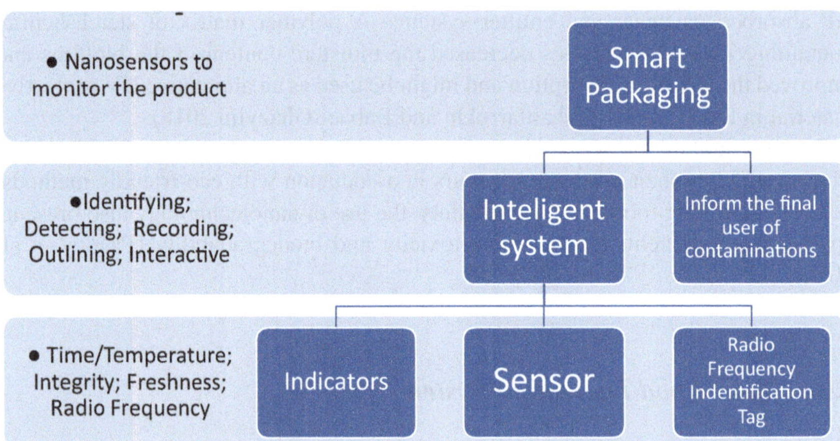

Fig. 2 Outline of main characteristics of smart packaging systems

the internal and external plastic packaging, serving as electronic tongue or noses, e.g., showing in situ information generated by food temperature and humidity, as well as degradation products, spoilage, or microbial contamination by color changes (Bouwmeester et al. 2009). Among main commercially biosensor, the company Toxin Alert (Ontario, Canada) developed the Toxin Guard™, a system based on antibody–antigen insertion in the packaging that reacts with some microorganisms (parasites such as *Cyclospora*, bacteria as *Salmonella, Escherichia coli* O157, and many other), changing its color or shape, indicating the concern. Additionally, the SIRA Technologies (Pasadena, CA-USA) established the Food Sentinel System®, a technology focused on immunochemical method with defined antibody added to a membrane coupled to a specific commercial universal product code bar system, such that when in contact with some microorganisms like *Escherichia coli* 0157:57H7, *Salmonella* ssp., or *Listeria monocytogenes*, it forms a specific dark smudge in the bar code, enabling to verify the presence of the pathogen (Lee and Mijanur Rahman 2013).

Additionally, the full temperature monitoring of the product since its manufacturing and transport until storage is pivotal for the total quality control of the sold product. Within this context, the time-temperature indicators (TTIS) are introduced as an interested key to obtain information of the real situation of the material. TTIS are classified in three types: (i) abuse indicators (specific temperature register), (ii) partial temperature history indicators (this point is registered just if the temperature overloads a defined limit), and (ii) full temperature history indicators (register of temperature along all product time) (Sharma et al. 2017).

A variety of materials can be applied in smart packaging, by using different indicators. For example, TiO_2 can be used to photosensitize the reduction of methylene blue (MB) under UV exposure that changes color when in contact with O_2 (Lee et al. 2002). Moreover, due to the volatile amine released by the spoiled food, the polyaniline film presents visible color changes (Kuswandi et al. 2012). Additionally, a system by using humidity as indicator was developed by applying a color nanocrystalline cellulose film able to interact with the electromagnetic field that changes color according to the moisture (Zhou 2013). However, there are a vast amount of material in literature that can be applied such: Xanthine and hypoxanthine chemical indicator; TiO_2-coated oriented-polypropylene; carbon nanotubes; absorbent pads containing Ag nanoparticles (NPs); xanthine amperometric sensor; Ag montmorillonite NPs; polyvinyl chloride (PVC) with ZnO NPs, low-density polyethylene (LDPE) films loaded with Ag and ZnO NPs or with Ag and ZnO; cellulose Ag nanoparticles (AgNPs); carbon nanotubes with allyl isothiocyanate and cellulose; low-density polyethylene with AgNPs; ethylene vinyl alcohol (EVOH) with AgNPs; pullulan with AgNPs polyethylene with Ag, TiO_2; nanoclays with matrix of polyamide; sodium alginate with $CaCl_2$ and AgNPs; isotactic polypropylene (iPP) with $CaCO_3$ nanofiller; polyethylene with Ag and TiO_2 NPs and others (Sharma et al. 2017). Regarding the consumer safety by possible packaging compounds mitigations to the food, all materials need to be studied and take into account according to their maximum limit absorption by humans (Silvestre et al. 2011). According to Fuertes et al. (2016), there are many companies with commercial technology for (i) time and temperature indicators such as Pymah Corp, Timestrip Plc, Colour-Therm,

Monitor Mark, 3 M™, Minnesota, Ciba Specialty Chemical and FreshPoint, Temptime Corp., Thermographic Measurements Ltd., and Vitsab., (ii) Integrity Indicators such as Insignia Technologies Ltd., Timestrip Ltd., FreshPoint Lab and Mitsubishi Gas Chemical Inc., (iii) Freshness Indicators such as COX Technologies DSM NV and Food Quality Sensor RipSense™ and ort Research, and (iv) Radio Frequency Indicators such as CAEN RFID Srl, Mondi Pic, Convergence Systems Ltd., and Temptrip LLC. Indeed, the world smart packaging market was appraised at around USD 6.65 billion in 2015, with estimated growth at 8.91% (CAGR) from 2016 to 2024.

4 Conclusion

The applications of nanoparticles for the development of food quality in food processing have significantly increased in the last few years. With the help of nanoparticles, it will be possible to produce a different range of foods and supplements, with specific dietary necessities. The increased antioxidant and coloring properties, as well as the resistance to microorganisms promoted by the use of nanoparticles will have a direct effect in future foods, directing new trends and proportioning new tastes, aromas and increased shelf life. Through this way, the employment of nanoparticles in food scenario presents a great influence in both human and animal health, as it contributes to the economy and environment with the development of inexpensive and cleaner technologies.

References

Abad E, Palacio F, Nuin M, De Z'arate AG, Juarros A, G'omez J, Marco S (2009) RFID smart tag for traceability and cold chain monitoring of foods: demonstration in an intercontinental fresh fish logistic chain. J Food Eng 93(4):394–399. https://doi.org/https://doi.org/10.1016/j.jfoodeng.2009.02.004

AbdalDayem A, Hossain MK, Lee SB, Kim K, Saha SK, Yang GM, Cho HY, Cho SG (2017) The role of reactive oxygen species (ROS) in the biological activities of metallic nanoparticles. Int J Mol Sci 18(1):120. https://doi.org/10.3390/ijms18010120

Abdel-Raouf N, Al-Enazi NM, Ibraheem IBM, Alharbi RM, Alkhulaifi MM (2019) Biosynthesis of silver nanoparticles by using of the marine brown alga *Padina pavonia* and their characterization. Saudi J Biol Sci 26:1207–1215. https://doi.org/10.1016/j.sjbs.2018.01.007

Adelere IA, Lateef A (2016) A novel approach to the green synthesis of metallic nanoparticles: the use of agro-wastes, enzymes, and pigments. Nanotechnol Rev 5(6):567–587. https://doi.org/10.1515/ntrev-2016-0024

Adenigba VO, Omomowo IO, Oloke JK, Fatukasi BA, Odeniyi MA, Adedayo AA (2020) Evaluation of microalgal-based nanoparticles in the adsorption of heavy metals from Wastewater. IOP Conf Series: Mater Sci Eng C 805: 012030. https://.doi.org/https://doi.org/10.1088/1757-899X/805/1/012030.

Ashtari K, Khajeh K, Fasihi J, Ashtari P, Ramazani A, Vali H (2012) Silica-encapsulated magnetic nanoparticles: enzyme immobilization and cytotoxic study. Int J Biol Macromol 50(4):1063–1069. https://doi.org/10.1016/j.ijbiomac.2011.12.025

Asmathunisha N, Kathiresan K (2013) A review on biosynthesis of nanoparticles by marine organisms. Colloids Surf B: Biointerf 103:283–287. https://doi.org/10.1016/j.colsurfb.2012.10.030

Augustin MA, Hemar Y (2009) Nano- and micro-structured assemblies for encapsulation of food ingredients. Chem Soc Rev 38(4):902–912. https://doi.org/10.1039/b801739p

Bagchi A (2012) Intelligent sensing and packaging of foods for enhancement of shelf life: concepts and applications. Int J Sci Eng Res 3:10

Bahrami A, Delshadi R, Jafari SM, Williams L (2019) Nano-encapsulated nisin: an engineered natural antimicrobial system for the food industry. Trends Food Sci Technol 94:20–31. https://doi.org/10.1016/j.tifs.2019.10.002

Baker S, Satish S (2015) Biosynthesis of gold nanoparticles by *Pseudomonas veronii* AS41G inhabiting *Annona squamosa* L. Spectrochim Acta A: Mol Biomol Spectr 150:691–695. https://doi.org/10.1016/j.saa.2015.05.080

Bouwmeester H, Dekkers S, Noordam MY, Hagens WI, Bulder AS, de Heer C, ten Voorde SECG, Wijnhoven SWP, Marvin HJP, Sips AJAM (2009) Review of health safety aspects of nanotechnologies in food production. Reg Toxicol Pharm 53(1):52–62. https://doi.org/10.1016/j.yrtph.2008.10.008

Carrillo-Inungaray ML, Trejo-Ramirez JA, Reyes-Munguia A, Carranza-Alvarez C (2018) Use of nanoparticles in the food industry: advances and perspectives. In: Grumezescu AM, Holban AM (eds) Impact of nanoscience in the food industry, pp 419–444. doi:https://doi.org/10.1016/b978-0-12-811441-4.00015-7

Chai C, Lee J, Lee Y, Na S, Park J (2014) A combination of TiO_2–UV photocatalysis and high hydrostatic pressure to inactivate *Bacillus cereus* in freshly squeezed *Angelica keiskei* juice. LWT Food Sci Technol 55(1):104–109. https://doi.org/10.1016/j.lwt.2013.08.015

Chen L, Lin X, Xu X, Chen Y, Li K, Fan X Pang J, Teng H (2019) Self-nano-emulsifying formulation of *Sonchus oleraceus*Linn for improved stability: implications for phenolics degradation under in vitro gastro-intestinal digestion: food grade drug delivery system for crude extract but not single compound. J Funct Food 53:28–35. https://doi.org/https://doi.org/10.1016/j.jff.2018.12.009

Cui H, Yuan L, Li W, Lin L (2017) Antioxidant property of SiO_2-eugenol liposome loaded nanofibrous membranes on beef. Food Packag Shelf Life 11:49–57. https://doi.org/10.1016/j.fpsl.2017.01.001

Dairi N, Ferfera-Harrar H, Ramos M, Garrigós MC (2019) Cellulose acetate/AgNPs-organoclay and/or thymol nanobiocomposite films with combined antimicrobial/antioxidant properties for active food packaging use. Int J Biol Macromol 121:508–523. https://doi.org/10.1016/j.ijbiomac.2018.10.042

Das B, Moumita S, Ghosh S, Khan MI, Indira D, Jayabalan R, Tripathy SK, Mishra A, Balasubramanian P (2018) Biosynthesis of magnesium oxide (MgO) nanoflakes by using leaf extract of *Bauhinia purpurea* and evaluation of its antibacterial property against *Staphylococcus aureus*. Mater Sci Eng: C 91:436–444. doi:https://doi.org/10.1016/j.msec.2018.05.059.

De Campo C, Assis RQ, da Silva MM, Costa TMH, Paese K, Guterres SS, Rios AO, Flôres SH (2019) Incorporation of zeaxanthin nanoparticles in yogurt: influence on physicochemical properties, carotenoid stability and sensory analysis. Food Chem 125230. doi:https://doi.org/10.1016/j.foodchem.2019.125230

Debnath B, Majumdar M, Bhowmik M, Bhowmik KL, Debnath A, Roy DN (2020) The effective adsorption of tetracycline onto zirconia nanoparticles synthesized by novel microbial green technology. J Environ Manag 261:110235. https://doi.org/10.1016/j.jenvman.2020.110235

Devi LS, Joshi SR (2015) Ultrastructures of silver nanoparticlesbiosynthesized using endophytic fungi. J Microsco Ultrastruc 3:29–37. https://doi.org/10.1016/j.jmau.2014.10.004

Dhanasekar NN, Rahul G, Narayanan KB et al (2015) Green chemistry approach for the synthesis of gold nanoparticles using the fungus *Alternaria* sp. J Microbiol Biotechnol 25(7):1129–1135. https://doi.org/10.4014/jmb.1410.10036

Dhavale R, Jadhav S, Sibi G (2020) Microalgae mediated silver nanoparticles (Ag-NPs) synthesis and their biological activities. J Crit Rev 7(2):15–20

Dobrucka R (2016) Synthesis of MgO nanoparticles using Artemisia abrotanum herba extract and their antioxidant and photocatalytic properties. Iran J Sci Technol Trans A: Sci 42(2):547–555. https://doi.org/10.1007/s40995-016-0076-x

Donovan AR, Adams CD, Ma Y, Stephan C, Eichholz T, Shi H (2016) Single particle ICP-MS characterization of titanium dioxide, silver, and gold nanoparticles during drinking water treatment. Chemosphere 144:148–153. https://doi.org/10.1016/j.chemosphere.2015.07.081

Duarte JL, Bezerra DC, da ConceiçãoEC MRHV, Fernandes CP (2020) Self-nano-emulsification of chamomile essential oil: a novel approach for a high value phytochemical. Colloid Interface Sci Commun 34:100225. https://doi.org/10.1016/j.colcom.2019.100225

Ebrahimzadeh Z, Salehzadeh A, Naeemi AS, Jalali A (2020) Silver nanoparticles biosynthesized by *Anabaena flos-aquae* enhance the apoptosis in breast cancer cell line. Bull Mater Sci 43(1):1–7. https://doi.org/10.1007/s12034-020-2064-1

Elegbede JA, Lateef A (2019) Green synthesis of silver (Ag), gold (Au) and silver-gold (Ag-Au) alloy nanoparticles: a review on recent advances, trends and biomedical applications. In: Verma DK, Goyal MR, and Suleria HAR (eds) Nanotechnology and nanomaterial applications in food, health and biomedical sciences. Apple Academic Press Inc. /CRC Press, Taylor and Francis Group, Oakville, Ontario, Canada, pp 3–89. https://doi.org/https://doi.org/10.1201/9780429425660-1. ISBN 978–1–77188–764–9

Elegbede JA, Lateef A, Azeez MA, Asafa TB, Yekeen TA, Oladipo IC, Aina DA, Beukes LS, Gueguim-Kana EB (2020) Biofabrication of gold nanoparticles using xylanases through valorization of corncob by *Aspergillus niger* and *Trichoderma longibrachiatum*: antimicrobial, antioxidant, anticoagulant and thrombolytic activities. Waste Biomass Valor 11(3):781–791. https://doi.org/10.1007/s12649-018-0540-2

Elegbede JA, Lateef A, Azeez MA, Asafa TB, Yekeen TA, Oladipo IC, Hakeem AS, Beukes LS, Gueguim-Kana EB (2019) Silver-gold alloy nanoparticles biofabricated by fungal xylanases exhibited potent biomedical and catalytic activities. Biotechnol Prog 35(5):e2829. https://doi.org/10.1002/btpr.2829

Elegbede JA, Lateef A, Azeez MA, Asafa TB, Yekeen TA, Oladipo IC, Adebayo EA, Beukes LS, Gueguim-Kana EB (2018) Fungal xylanases-mediated synthesis of silver nanoparticles for catalytic and biomedical applications. IET Nanobiotechnol 12(6):857–863. https://doi.org/10.1049/iet-nbt.2017.0299

Eugenio M, Müller N, Frasés S, Almeida-Paes R, Lima LMTR, Lemgruber L, Farina M, Souza W, Sant'Anna C (2016) Yeast-derived biosynthesis of silver/silver chloride nanoparticles and their antiproliferative activity against bacteria. RSC Adv 6(12):9893–9904. doi:https://doi.org/10.1039/c5ra22727e

Ezhilarasi PN, Karthik P, Chhanwal N, Anandharamakrishnan C (2013) Nanoencapsulation techniques for food bioactive components: a review. Food Bioprocess Technol 6:628–647. doi:https://doi.org/10.1007/s11947-012-0944-0.

Fayaz AM, Girilal M, Rahman M, Venkatesan R, Kalaichelvan PT (2011) Biosynthesis of silver and gold nanoparticles using thermophilic bacterium *Geobacillus stearothermophilus*. Process Biochem 46(10):1958–1962. https://doi.org/10.1016/j.procbio.2011.07.003

Fernández JG, Fernández-Baldo MA, Berni E, Camí G, Durán N, Raba J, Sanz MI (2016) Production of silver nanoparticles using yeasts and evaluation of their antifungal activity against phytopathogenic fungi. Process Biochem 51(9):1306–1313. https://doi.org/10.1016/j.procbio.2016.05.021

Foltynowicz Z, Bardenshtein A, Sängerlaub S, Antvorskov H, Kozak W (2017) Nanoscale, zero valent iron particles for application as oxygen scavenger in food packaging. Food Packag Shelf Life 11:74–83. https://doi.org/10.1016/j.fpsl.2017.01.003

García-Ruiz A, Crespo J, López-de-Luzuriaga JM, Olmos ME, Monge M, Rodríguez-Álfaro MP, Martín-Álvarez PJ, Bartolome B, Moreno-Arribas MV (2015) Novel biocompatible silver nanoparticles for controlling the growth of lactic acid bacteria and acetic acid bacteria in wines. Food Control 50:613–619. https://doi.org/10.1016/j.foodcont.2014.09.035

Ghosh S, Das AP (2015) Modified titanium oxide (TiO_2) nanocomposites and its array of applications: a review. Toxicol Environ Chem 97(5):491–514. https://doi.org/10.1080/02772248.2015.1052204

Gu H, Chen X, Chen F, Zhou X, Parsaee Z (2018) Ultrasound-assisted biosynthesis of CuO-NPs using brown alga *Cystoseira trinodis*: Characterization, photocatalytic AOP, DPPH scavenging and antibacterial investigations. Ultrasonic Sonochem 41:109–119. https://doi.org/10.1016/j.ultsonch.2017.09.006

Guilger-Casagrande M, de Lima R (2019) Synthesis of silver nanoparticles mediated by fungi: a review. Front Bioeng Biotechnol 7:287. https://doi.org/10.3389/fbioe.2019.00287

He X, Hwang HM (2016) Nanotechnology in food science: functionality, applicability, and safety assessment. J Food Drug Anal 24(4):671–681. https://doi.org/10.1016/j.jfda.2016.06.001

He X, Deng H, Hwang H (2018) The current application of nanotechnology in food and agriculture. J Food Drug Anal. https://doi.org/10.1016/j.jfda.2018.12.002

Himed L, Merniz S, Monteagudo-Olivan R, Barkat M, Coronas J (2019) Antioxidant activity of the essential oil of *Citrus limon* before and after its encapsulation in amorphous SiO_2. Sci Afri. e00181. doi:https://doi.org/10.1016/j.sciaf.2019.e00181

Huang Q, Yu H, Ru Q (2010) Bioavailability and delivery of nutraceuticals using nanotechnology. J Food Sci 75(1):R50–R57. https://doi.org/10.1111/j.1750-3841.2009.01457.x

Hussein A, Abdel-Mottaleb MA, El-assal M, Sammour O (2020) Novel biocompatible essential oil-based lipid nanocapsules with antifungal properties. J Drug Delivery Sci Technol 101605. doi:https://doi.org/10.1016/j.jddst.2020.101605

Ibrahim S, Saied HE, Hasanin M (2019) Active paper packaging material based on antimicrobial conjugated nano-polymer/amino acid as edible coating. J King Saud Univ Sci. 31:1095–1102. doi: https://doi.org/https://doi.org/10.1016/j.jksus.2018.10.007.

Ingle AP, Rathod J, Pandit R, da Silva SS, Rai M (2017) Comparative evaluation of free and immobilized cellulase for enzymatic hydrolysis of lignocellulosic biomass for sustainable bioethanol production. Cellulose 24(12):5529–5540. https://doi.org/10.1007/s10570-017-1517-1

Jana S, Gandhi A, Jana S (2017) Nanotechnology in bioactive food ingredients. Nanotechnol Appl Food 21–41. doi:https://doi.org/10.1016/b978-0-12-811942-6.00002-9

Jayaseelan C, Rahuman AA, Kirthi AV, Marimuthu S, Santhoshkumar T, Bagavan A, Gaurav K, Karthik L, Rao BKV (2012) Novel microbial route to synthesize ZnO nanoparticles using *Aeromonas hydrophila* and their activity against pathogenic bacteria and fungi. Spectrochim. Acta Part A: Mol Biomol Spectr 90:78–84. https://doi.org/10.1016/j.saa.2012.01.006

Joshi SM, Britto SD, Jogaiah S, Ito S (2019) Mycogenic selenium nanoparticles as potential new generation broad spectrum antifungal molecules. Biomolecules 9:419. doi:https://doi.org/10.3390/biom9090419.

Khaneghah AM, Hashemi SMB, Limbo S (2018) Antimicrobial agents and packaging systems inantimicrobial active food packaging: an overview of approaches and interactions. Food Bioprod Process 111:1–19. https://doi.org/10.1016/j.fbp.2018.05.001

Khaneghah AM, Hashemi SMB, Eş I, Fracassetti D, Limbo S (2018) Efficacy of antimicrobial agents for food contact applications: biological activity, incorporation into packaging, and assessment methods: a review. J Food Protect 81:1142–1156. https://doi.org/10.4315/0362-028X.JFP-17-509

Khanna P, Kaur AA, Goyal D (2019) Algae-based metallic nanoparticles: Synthesis, characterization and applications. J Microbiol Methods 163:105656. https://doi.org/https://doi.org/10.1016/j.mimet.2019.105656

Khayata N, Abdelwahed W, Chehna MF, Charcosset C, Fessi H (2012) Preparation of vitamin E loaded nanocapsules by the nanoprecipitation method: from laboratory scale to large scale using a membrane contactor. Int J Pharm 423(2):419–427. https://doi.org/10.1016/j.ijpharm.2011.12.016

Kumar N, Singh A, Sharma DK, Kishore K (2019) Toxicity of food additives. Food Safety Human Health 67–98. doi:https://doi.org/10.1016/b978-0-12-816333-7.00003-5

Kuswandi B, Wicaksono Y, Jayus AA, Heng LY, Ahmad M (2011) Smart packaging: sensors for monitoring of food quality and safety. Sens Instrum Food Qual 5:137–146. https://doi.org/10.1007/s11694-011-9120-x

Kuswandi B (2016) Nanotechnology in food packaging. In: Ranjan S, Dasgupta, N, Lichtfouse E (eds) Nanoscience in food and agriculture. Springer International Publishing Switzerland. doi: https://doi.org/10.1007/978-3-319-39303-2

Kuswandi B, Jayus RA, Abdullah A, Heng LY, Ahmad M (2012) A novel colorimetric food package label for fish spoilage based on polyaniline film. Food Cont 25:184. https://doi.org/10.1016/j.foodcont.2011.10.008

Lateef A, Ojo SA, Akinwale AS, Azeez L, Gueguim-Kana EB, Beukes LS (2015) Biogenic synthesis of silver nanoparticles using cell-free extract of *Bacillus safensis* LAU 13: antimicrobial, free radical scavenging and larvicidal activities. Biologia 70(10):1295–1306. https://doi.org/10.1515/biolog-2015-0164

Lateef A, Adelere IA, Gueguim-Kana EB, Asafa TB, Beukes LS (2015) Green synthesis of silver nanoparticles using keratinase obtained from a strain of *Bacillus safensis* LAU 13. Int Nano Lett 5(1):29–35. https://doi.org/10.1007/s40089-014-0133-4

Lateef A, Akande MA, Azeez MA, Ojo SA, Folarin BI, Gueguim-Kana EB, Beukes LS (2016) Phytosynthesis of silver nanoparticles (AgNPs) using miracle fruit plant (*Synsepalum dulcificum*) for antimicrobial, catalytic, anticoagulant, and thrombolytic applications. Nanotechnol Rev 5(6):507–520. https://doi.org/10.1515/ntrev-2016-0039

Lateef A, Azeez MA, Asafa TB, Yekeen TA, Akinboro A, Oladipo IC, Azeez L, Ojo SA, Gueguim-Kana EB, Beukes LS (2016) Cocoa pod husk extract-mediated biosynthesis of silver nanoparticles: its antimicrobial, antioxidant and larvicidal activities. J Nanostruct Chem 6(2):159–169. https://doi.org/10.1007/s40097-016-0191-4

Lateef A, Azeez MA, Asafa TB, Yekeen TA, Akinboro A, Oladipo IC, Azeez L, Ajibade SE, Ojo SA, Gueguim-Kana EB, Beukes LS (2016) Biogenic synthesis of silver nanoparticles using a pod extract of *Cola nitida*: Antibacterial and antioxidant activities and application as a paint additive. J Taibah Univ Sci 10(4):551–562. https://doi.org/10.1016/j.jtusci.2015.10.010

Lateef A, Folarin BI, Oladejo SM, Akinola PO, Beukes LS, Gueguim-Kana EB (2018) Characterization, antimicrobial, antioxidant and anticoagulant activities of silver nanoparticles synthesized from *Petiveria alliacea* L. leaf extract. Preparative Biochem Biotechnol 48(7):646–652. https://doi.org/https://doi.org/10.1080/10826068.2018.1479864.

Lateef A, Ojo SA, Elegbede JA (2016) The emerging roles of arthropods and their metabolites in the green synthesis of metallic nanoparticles. Nanotechnol Rev 5(6):601–622. https://doi.org/10.1515/ntrev-2016-0049

Lateef A, Ojo SA, Elegbede JA, Akinola PO, Akanni EO (2018) Nanomedical applications of nanoparticles for blood coagulation disorders. In: Dasgupta N, Ranjan S, and Lichtfouse E (eds) Environmental nanotechnology. Environmental chemistry for a sustainable world, vol 14. Springer International Publishing AG, Cham, Switzerland, pp 243–277. https://doi.org/https://doi.org/10.1007/978-3-319-76090-2_8. ISBN 978–3–319–76089–6

Lateef A, Ojo SA, Oladejo SM (2016) Anti-candida, anti-coagulant and thrombolytic activities of biosynthesized silver nanoparticles using cell-free extract of *Bacillus safensis* LAU 13. Process Biochem 51(10):1406–1412. https://doi.org/10.1016/j.procbio.2016.06.027

Lateef A, Oladejo SM, Akinola PO, Aina DA, Beukes LS, Folarin BI, Gueguim-Kana EB (2020) Facile synthesis of silver nanoparticles using leaf extract of *Hyptis suaveolens* (L.) Poit for environmental and biomedical applications. IOP Conf Ser: Mater Sci Eng 805:012042. https://doi.org/https://doi.org/10.1088/1757-899X/805/1/012042

Lee SJ, Mijanur Rahman ATM (2013) Intelligent packaging for food products. In: Han JH (ed) Innovations in food packaging. Academic Press, Cambridge, Massachusetts, USA, 624 p

Lee SW, Mao C, Flynn CE, Belcher AM (2002) Ordering of quantum dots using genetically engineered viruses. Science 296:892–895

Liu J, Liu C, Zheng X, Chen M, Tang K (2020) Soluble soybean polysaccharide/nano zinc oxide antimicrobial nanocomposite films reinforced with microfibrillated cellulose. Int J Biol Macromol 159:793–803. https://doi.org/10.1016/j.ijbiomac.2020.05.084

Maftoonazad N, Ramaswamy H (2018) Novel techniques in food processing: bionanocomposites. Curr Opin Food Sci 23:49–56. https://doi.org/10.1016/j.cofs.2018.05.010

McClements DJ (2012) Nanoemulsions versus microemulsions: terminology, differences, and similarities. Soft Matter 8(6):1719–1729. https://doi.org/10.1039/c2sm06903b

Menon S, Rajeshkumar S, Kumar V (2017) A review on biogenic synthesis of gold nanoparticles, characterization, and its applications. Resour Effi Technol 3(4):516–527. https://doi.org/10.1016/j.reffit.2017.08.002

Mohammadi M, Mirabzadeh S, Shahvalizadeh R, Hamishehkar H (2020) Development of novel active packaging films based on whey protein isolate incorporated with chitosan nanofiber and nano-formulated cinnamon oil. Int J Biol Macromol 149:11–20. https://doi.org/10.1016/j.ijbiomac.2020.01.083

Molnár Z, Bódai V, Szakacs G, Erdélyi B, Fogarassy Z, Sáfrán G, Varga T, Kónya Z, Tóth-Szeles E, Szűcs R, Lagzi I (2018) Green synthesis of gold nanoparticles by thermophilic filamentous fungi. Sci Reports 8:3943. https://doi.o.rg/https://doi.org/10.1038/s41598-018-22112-3

Mousavi SM, Hashemi SA, Ramakrishna S, Esmaeili H, Bahrani S, Koosha M, Babapoor A (2019) Green synthesis of supermagnetic Fe_3O_4–MgO nanoparticles via Nutmeg essential oil toward superior anti-bacterial and anti-fungal performance. J Drug Deliv Sci Technol 101352. doi:https://doi.org/10.1016/j.jddst.2019.101352

Mu H, Gao H, Chen H, Tao F, Fang X, Ge L (2013) A nanosized oxygen scavenger: preparation and antioxidant application to roasted sunflower seeds and walnuts. Food Chem 136:245–250. https://doi.org/10.1016/j.foodchem.2012.07.121

Nile SH, Baskar V, Selvaraj D, Nile A, Xiao J, Kai G (2020) Nanotechnologies in food science: applications, recent trends, and future perspectives. Nano-Micro Lett 12:45. https://doi.org/https://doi.org/10.1007/s40820-020-0383-9

Noor S, Shah Z, Javed A, Ali A, Hussain SB, Zafar S, Ali H, Muhammad SA (2020) A fungal based synthesis method for copper nanoparticles with the determination of anticancer, antidiabetic and antibacterial activities. J Microbiol Methods 174:105966. https://doi.org/https://doi.org/10.1016/j.mimet.2020.105966

Nosaka Y, Nosaka AY (2017) Generation and detection of reactive oxygen species in photocatalysis. Chem Rev 117(17):11302–11336. https://doi.org/10.1021/acs.chemrev.7b00161

Ojo SA, Lateef A, Azeez MA, Oladejo SM, Akinwale AS, Asafa TB, Yekeen TA, Akinboro A, Oladipo IC, Gueguim-Kana EB, Beukes LS (2016) Biomedical and catalytic applications of gold and silver-gold alloy nanoparticles biosynthesized using cell-free extract of *Bacillus safensis* LAU 13: Antifungal, dye degradation, anti-coagulant and thrombolytic activities. IEEE Trans Nanobiosci 15(5):433–442. https://doi.org/10.1109/TNB.2016.2559161

Oladipo IC, Lateef A, Elegbede JA, Azeez MA, Asafa TB, Yekeen TA, Akinboro A, Gueguim-Kana EB, Beukes LS, Oluyide TO, Atanda OR (2017) *Enterococcus* species for the one-pot biofabrication of gold nanoparticles: characterization and nanobiotechnological applications. J Photochem Photobiol B 173:250–257. https://doi.org/10.1016/j.jphotobiol.2017.06.003

Omomowo IO, Adenigba VO, Ogunsola SB, Adeyinka GC, Oluyide OO, Adedayo AA, Fatukasi BA (2020) Antimicrobial and antioxidant activities of algal-mediated silver and gold nanoparticles. IOP Conf Ser: Mater Sci Eng C805:012010. https://doi.org/https://doi.org/10.1088/1757-899X/805/1/012010

Otari SV, Patil RM, Nadaf NH, Ghosh SJ, Pawar SH (2012) Green biosynthesis of silver nanoparticles from an actinobacteria *Rhodococcus* sp. Mater Lett 72:92–94. https://doi.org/10.1016/j.matlet.2011.12.109

Pal J (2017) A New approach in food packaging. J Food: Microbiol Safe Hyg 2:2. https://doi.org/10.4172/2476-2059.1000121

Palacio J, Monsalve Y, Ramírez-Rodríguez F, López B (2020) Study of encapsulation of polyphenols on succinyl-chitosan nanoparticles. J Drug Deliv Sci Technol 101610. doi:https://doi.org/10.1016/j.jddst.2020.101610

Pandey VK, Upadhyay SN, Niranjan K, Mishra PK (2020) Antimicrobial biodegradable chitosan-based composite nano-layers for food packaging. Int J Biol Macromol 157:212–219. https://doi.org/10.1016/j.ijbiomac.2020.04.149

Park JH, von Maltzahn G, Zhang L, Schwartz MP, Ruoslahti E, Bhatia SN, Sailor MJ (2008) Magnetic iron oxide nanoworms for tumor targeting and imaging. Adv Mater 20(9):1630–1635. https://doi.org/10.1002/adma.200800004

Pignata C, D'Angelo D, Fea E, Gilli G (2017) A review on microbiological decontamination of fresh produce with non-thermal plasma. J Appl Microbiol 122(6):1438–1455. https://doi.org/10.1111/jam.13412

Rai M, Ingle AP, Gupta I, Pandit R, Paralikar P, Gade A, Chaud MV, Santos CAD (2019) Smart nanopackaging for the enhancement of food shelf life. Environ Chem Lett 17:277–290. https://doi.org/10.1007/s10311-018-0794-8

Ramamurthy CH, Padma M, Mareeswaran R, Suyavaran A, Kumar MS, Premkumar K, Thirunavukkarasu C (2013) The extra cellular synthesis of gold and silver nanoparticles and their free radical scavenging and antibacterial properties. Colloids Surf B: Biointerfaces 102:808–815. https://doi.org/10.1016/j.colsurfb.2012.09.025

Ravichandran V, Vasanthi S, Shalini S, Shah SAA, Harish R (2016) Green synthesis of silver nanoparticles using *Atrocarpusaltilis* leaf extract and the study of their antimicrobial and antioxidant activity. Mater Lett 180:264–267. https://doi.org/10.1016/j.matlet.2016.05.172

Sandhir R, Yadav A, Sunkaria A, Singhal N (2015) Nano-antioxidants: an emerging strategy for intervention against neurodegenerative conditions. Neurochem Int 89:209–226. https://doi.org/10.1016/j.neuint.2015.08.011

Santos CAD, Ingle AP, Rai M (2020) The emerging role of metallic nanoparticles in food. Appl Microbiol Biotechnol 104:2373–2383. https://doi.org/10.1007/s00253-020-10372-x

Saravanakumar A, Ganesh M, Jayaprakash J, Jang HT (2015) Biosynthesis of silver nanoparticles using *Cassia tora* leaf extract and its antioxidant and antibacterial activities. J Ind Eng Chem 28:277–281. https://doi.org/10.1016/j.jiec.2015.03.003

Sekhon BS (2010) Food nanotechnology: an overview. Nanotechnol Sci Appl 3:1–15. https://doi.org/10.2147/NSA.S8677

Shahabi-Ghahfarrokhi I, Babaei-Ghazvini A (2018) Using photo-modification to compatibilize nano-ZnO in development of starch-kefiran-ZnO green nanocomposite as food packaging material. Int J Biol Macromol 124:922–930. https://doi.org/10.1016/j.ijbiomac.2018.11.241

Shamim A, Abid MB, Mahmood T (2019) Biogenic synthesis of zinc oxide (ZnO) Nanoparticles using a fungus (*Aspergillusniger*) and their characterization. Int J Chem 11(2):119–126. https://doi.org/10.5539/ijc.v11n2p119

Sharma C, Dhiman R, Rokana N, Panwar H (2017) Nanotechnology: an untapped resource for food packaging. Front Microbiol 8:1735. https://doi.org/10.3389/fmicb.2017.01735

Sharmila G, Muthukumaran C, Sangeetha E, Saraswathi H, Soundarya S, Kumar NM (2019) Green fabrication, characterization of *Pisonia alba* leaf extract derived MgO nanoparticles and its biological applications. Nano-Struct Nano-Object 20:100380. https://doi.org/10.1016/j.nanoso.2019.100380

Silvestre C, Duraccio D, Sossio C (2011) Food packaging based on polymer nanomaterials. Prog Polymer Sci 36:1766–1782. https://doi.org/10.1016/j.progpolymsci.2011.02.003

Singh T, Shukla S, Kumar P, Wahla V, Bajpai VK, Rather IA (2017) Application of nanotechnology in food science: perception and overview. Front Microbiol 8:1501. doi: https://doi.org/10.3389/fmicb.2017.01501

Singh P, Singh H, Kim YJ, Mathiyalagan R, Wang C, Yang DC (2016) Extracellular synthesis of silver and gold nanoparticles by *Sporosarcina koreensis* DC4 and their biological applications. Enzyme Microb Technol 86:75–83. https://doi.org/10.1016/j.enzmictec.2016.02.005

Sowani H, Mohite P, Damale S, Kulkarni M, Zinjarde S (2016a) Carotenoid stabilized gold and silver nanoparticles derived from the actinomycete *Gordonia amicalis* HS-11 as effective free radical scavengers. Enzyme Microb Technol 95:164–173. https://doi.org/10.1016/j.enzmictec.2016.09.016

Sowani H, Mohite P, Munot H, Shouche Y, Bapat T, Kumar AR et al (2016b) Green synthesis of gold and silver nanoparticles by an actinomycete *Gordonia amicalis* HS-11: mechanistic aspects and biological application. Process Biochem 51(3):374–383

Sperber WH, Doyle MP (2009) Introduction to the microbiological spoilage of foods and beverages. Compendium of the microbiological spoilage of foods and beverages. Springer International Publishing, New York, pp 1–40

Tsibakhashvili NY, Kirkesali EI, Pataraya DT, Gurielidze MA, Kalabegishvili TL, Gvarjaladze DN, Tsertsvadze GI, Frontasyeva MV, Zinicovscaia II, Wakstein MS, Khakhanov SN (2011) Microbial synthesis of silver nanoparticles by *Streptomyces glaucus* and *Spirulina platensis*. Adv Sci Lett 4:1–10

Valgimigli L, Baschieri A, Amorati R (2018) Antioxidant activity of nanomaterials. J Mater Chem B 6(14):2036–2051. https://doi.org/10.1039/c8tb00107c

Vilela C, Kurek M, Hayouka Z, Röcker B, Yildirim S, Antunes MC, Nilsen-Nygaard J, Pettersen MK, Freire CSR (2018) A concise guide to active agents for active food packaging. Trends Food Sci Technol 80:212–222. https://doi.org/10.1016/j.tifs.2018.08.006

Vivek M, Kumar PS, Steffi S, Sudha S (2011) Biogenic silver nanoparticles by *Gelidiella acerosa* extract and their antifungal effects. Avicenna J Med Biotechnol 3:143–148

Wang L, Shao S, Madebo MP, Hou Y, Zheng Y, Jin P (2020) Effect of nano-SiO_2 packing on postharvest quality and antioxidant capacity of loquat fruit under ambient temperature storage. Food Chem 315:126295. https://doi.org/10.1016/j.foodchem.2020.126295

Worldometers (2020) available at: https://www.worldometers.info/world-population/#:~:text=The%20current%20world%20population%20is,currently%20living)%20of%20the%20world. Accessed on 26 July 2020

Xie J, Lee JY, Wang D, Ting YP (2007) Identification of active biomolecules in the high-yield synthesis of single crystalline gold nanoplates in algal solutions. Small 3:672–682. https://doi.org/10.1002/smll.200600612

Xiong K, Zhou L, Wang J, Ma A, Fang D, Xiong L, Sun Q (2020) Construction of food-grade pH-sensitive nanoparticles for delivering functional food ingredients. Trends Food Sci Technol 96:102–113. https://doi.org/10.1016/j.tifs.2019.12.019

Yadav A, Kon K, Kratosova G, Duran N, Ingle AP, Rai M (2015) Fungi as an efficient mycosystem for the synthesis of metal nanoparticles: progress and key aspects of research. Biotechnol Lett 37:2099–2120. https://doi.org/10.1007/s10529-015-1901-6

Yadav A, Ghune M, Jain DK (2011) Nano-medicine based drug delivery system. J Adv Pharm Edu Res 1(4):201–213

Yildirim S, Röcker B, Pettersen MK, Nilsen-Nygaard J, Ayhan Z, Rutkaite R, Radusin T, Suminska P, Marcos B, Coma V (2018) Active packaging applications for food. Compr Rev Food Sci Food Safety 17:165–199. https://doi.org/10.1111/1541-4337.12322

Yu Y, You L, Liu D, Hollinshead W, Tang YJ, Zhang F (2013) Development of *Synechocystis* sp. PCC 6803 as a phototrophic cell factory. Marine drugs 11(8):2894–2916. doi: https://doi.org/10.3390/md11082894

Zhou C (2013) Theoretical analysis of double-microfluidic-channels photonic crystal fiber sensor based on silver nanowires. Opt Commun 288:42–46. https://doi.org/10.1016/j.optcom.2012.09.060

Zhu Z, Cai H, Sun DW (2018) Titanium dioxide (TiO_2) photocatalysis technology for non-thermal inactivation of microorganisms in foods. Trends Food Sci Technol 75:23–35. https://doi.org/10.1016/j.tifs.2018.02.018

Zielonka A, Klimek-Ochab M (2017) Fungal synthesis of size-defined nanoparticles. Adv Nat Sci: Nanosci Nanotechnol 8:043001. https://doi.org/10.1088/2043-6254/aa84d4

The manufacturer's authorised representative in the EU is Springer Nature Customer Service Centre GmbH, Europaplatz 3, 69115 Heidelberg, Germany. If you have any concerns regarding our products, please contact ProductSafety@springernature.com

Printed and bound by CPI Group (UK) Ltd, Croydon, CR0 4YY
30/09/2026
02225288-0012